Dahlem Workshop Reports
Life Sciences Research Report 29
The Biology of Learning

The goal of this Dahlem Workshop is:
to reconcile learning theory and
natural behavior

Life Sciences Research Reports
Editor: Silke Bernhard

Held and published on behalf of the
Stifterverband für die Deutsche Wissenschaft

Sponsored by:
Senat der Stadt Berlin
Stifterverband für die Deutsche Wissenschaft

The Biology of Learning

P. Marler and H. S. Terrace,
Editors

Report of the Dahlem Workshop on
The Biology of Learning
Berlin 1983, October 23-28

Rapporteurs:
P. C. Holland · D. E. Kroodsma · J. C. Marshall
R. Menzel · J. Morton

Program Advisory Committee:
P. Marler and H. S. Terrace, Chairpersons
T. Bever · L. L. Cavalli-Sforza · J. L. Gould
K. Immelmann · R. F. Thompson · A. R. Wagner

Springer-Verlag
Berlin Heidelberg New York Tokyo 1984

Copy Editors: M. A. Cervantes-Waldmann, K. Geue
Photographs: E. P. Thonke

With 4 photographs, 17 figures and 1 table

ISBN-13:978-3-642-70096-5 e-ISBN-13:978-3-642-70094-1
DOI: 10.1007/978-3-642-70094-1

CIP-Kurztitelaufnahme der Deutschen Bibliothek
The biology of learning :
report of the Dahlem Workshop on the Biology of Learning,
Berlin, October 23 – 28, 1983 /
P. Marler and H. S. Terrace, ed. Rapporteurs: P. C. Holland ...
[Held and publ. on behalf of the Stifterverb. für d. Dt. Wiss.
Sponsored by: Senat d. Stadt Berlin ; Stifterverb. für d. Dt. Wiss.].
– Berlin ; Heidelberg ; New York ; Tokyo : Springer, 1984. –
(Life sciences research report ; 29)
(Dahlem Workshop reports)

2131/3020-5 4 3 2 1 0

Table of Contents

The Dahlem Konferenzen

Founders

Recognizing the need for more effective communication between scientists, especially in the natural sciences, the Stifterverband für die Deutsche Wissenschaft*, in cooperation with the Deutsche Forschungsgemeinschaft**, founded Dahlem Konferenzen in 1974. The project is financed by the founders and the Senate of the City of Berlin.

Name

Dahlem Konferenzen was named after the district of Berlin called "Dahlem", which has a long-standing tradition and reputation in the arts and sciences.

Aim

The task of Dahlem Konferenzen is to promote international, interdisciplinary exchange of scientific information and ideas, to stimulate international cooperation in research, and to develop and test new models conducive to more effective communication between scientists.

Dahlem Workshop Model

Dahlem Konferenzen organizes four workshops per year, each with a limited number of participants. Since no type of scientific meeting proved effective enough, Dahlem Konferenzen had to create its own concept. This concept has been tested and varied over the years, and has evolved into its present form which is known as the *Dahlem Workshop Model.* This model provides the framework for the utmost possible interdisciplinary communication and cooperation between scientists in a given time period.

**The Donors Association for the Promotion of Sciences and Humanities*

***German Science Foundation*

The main work of the Dahlem Workshops is done in four interdisciplinary discussion groups. Lectures are not given. Instead, selected participants write background papers providing a review of the field rather than a report on individual work. These are circulated to all participants before the meeting to provide a basis for discussion. During the workshop, the members of the four groups prepare reports reflecting their discussions and providing suggestions for future research needs.

Topics
The topics are chosen from the fields of the Life Sciences and the Physical, Chemical, and Earth Sciences. They are of contemporary international interest, interdisciplinary in nature, and problem-oriented. Once a year, topic suggestions are submitted to a scientific board for approval.

Participants
For each workshop participants are selected exclusively by special Program Advisory Committees. Selection is based on international scientific reputation alone, although a balance between European and American scientists is attempted. Exception is made for younger German scientists.

Publication
The results of the workshops are the Dahlem Workshop Reports, reviewed by selected participants and carefully edited by the editor of each volume. The reports are multidisciplinary surveys by the most internationally distinguished scientists and are based on discussions of new data, experiments, advanced new concepts, techniques, and models. Each report also reviews areas of priority interest and indicates directions for future research on a given topic.

The Dahlem Workshop Reports are published in two series:
1) Life Sciences Research Reports (LS), and
2) Physical, Chemical, and Earth Sciences Research Reports (PC).

Director
Silke Bernhard, M.D.

Address
Dahlem Konferenzen
Wallotstrasse 19
1000 Berlin (West) 33

1984. Berlin, Heidelberg, New York, Tokyo: Springer-Verlag.

Introduction

P. Marler* and H.S. Terrace**
*The Rockefeller University Field Research Center
Millbrook, NY 12545
**Dept. of Psychology, Columbia University
New York, NY 10027, USA

For the first half of this century, theories of animal conditioning were regarded as the most promising approach to the study of learning - both animal and human. For a variety of reasons, disillusionment with this point of view has become widespread during recent years.

One prominent source of disenchantment with conditioning theory is a large body of ethological observations of both learned and unlearned natural behavior. These challenge the generality of principles of animal learning as derived from the intensive study of a few species in specialized laboratory situations. From another direction, the complexities of human language acquisition, surely the most impressive of learned achievements, have prompted developmental psychologists to doubt the relevance of principles of animal learning.

Even within the realm of traditional studies of animal learning, it has become apparent that no single set of currently available principles can cope with the myriad of new empirical findings. These are emerging at an accelerating rate from studies of such phenomena as selective attention and learning, conditioned food aversion, complex problem-solving behavior, and the nature of reinforcement. Not very surprisingly, as a reaction against the long-held but essentially unrealized promise of general theories of learning, many psychologists have asked an obvious question: does learning theory have a future?

THE FUNDAMENTAL IMPORTANCE OF LEARNING

We believe that the proper answer to the above question is affirmative. It is an inescapable fact that organisms learn about their environments and that the remarkable abilities involved cry out for scientific explanation. The current disarray of learning theory should not detract from the fundamental goal of understanding the general principles of learning, even though the form of a fully viable theory of learning remains to be determined.

Numerous discoveries have given added impetus to efforts to reformulate and supplement traditional theories of learning in order to encompass findings they cannot presently explain. Learning theory has also retrenched from its original, perhaps overambitious, aims. In so doing, it has developed new and sophisticated models of a variety of compelling phenomena.

Contemporary studies of Pavlovian conditioning provide numerous examples of the kind of approach currently followed by students of animal learning (see Hearst, Jenkins, and Terrace, all this volume). Empirical findings concerning contingency learning, overshadowing and blocking, and sophisticated theoretical analyses of these phenomena (cf. (2, 6, 11)) have made clear the scope and complexity of Pavlovian conditioning. This research has also revealed the inadequacy of traditional approaches when they have focused exclusively on preparations in which mere temporal contiguity between conditioned and unconditioned stimuli was regarded as sufficient for learning to occur and in which the conditioned response was regarded as a form of the unconditioned response.

For the first time, students of animal learning have undertaken a systematic study of cognition in animals. This line of inquiry, which would have been unthinkable a decade ago, promises to provide a basis for comparing cognitive processes in animals and man (Lea and Menzel, both this volume; also see (12, 13)). New insights are also flowing into learning theory from both ethology and physiology. These areas in turn derive inspiration from burgeoning psychological research on such topics as autoshaping and the conditioning of food aversions.

Recent advances in the study of animal learning would undoubtedly be seen in a more positive light were it not for a strong and widespread disappointment over the failure of general learning theory to make good on its all-embracing promissory notes. While it is probably premature to anticipate a new generation of general theories of learning in the near future, students of animal learning can derive some comfort from the opportunities currently available for investigating associative principles of learning with unprecedented rigor. At the same time it is important to recognize that principles of association, which provide the basis for the only model of animal learning that currently enjoys a significant following, are by themselves an insufficient basis for a truly general theory.

BIOLOGICAL "CONSTRAINTS"

Some of the findings not readily accommodated by classical principles of association are referred to widely under the rubric of "biological constraints on learning." Several examples are reviewed by Bolles, Revusky, and Shettleworth (all this volume). They fall into two general categories. One group comes from demonstrations of how readily specific stimuli can be associated with one another, how selective attention determines the effectiveness of a stimulus in particular situations, and how an animal behaves as a result of forming a particular association.

One of the most robust and widely cited examples of this type of biological constraint is food aversion learning. Indeed, thanks to the ease of establishing conditioned food aversion, a variety of empirical findings have stimulated learning theorists to conceptualize the relevant variables as "parameters" of associative learning. It should, however, be noted that this point of view begs the question of why, on a priori grounds, particular organisms attend to and associate specific stimuli and why they are then disposed to take particular patterns of action. While it is heartening to discover that several features of food aversion learning are consistent with other instances of associative learning (Revusky, this volume), one must not lose sight of the fact that we still lack any general means of predicting, even in this instance, whether a particular stimulus will be an effective signal.

A second category of so-called "biological constraints" is exemplified by instances of "specialized" learning. Primary illustrations of such learning are imprinting and song learning (Bateson ("The Neural Basis of Imprinting"), Immelmann, Konishi, and Marler, all this volume). Examples of specialized learning pose different and perhaps more fundamental problems for association theory than does food aversion learning. As yet there is virtually no evidence that these phenomena are instances of associative learning. If they are not, new principles of learning must be formulated to account for them. Indeed, instances of specialized learning suggest that the very term biological constraints is a misnomer. There are many indications of brain mechanisms that have positively and explicitly evolved for particular learning tasks. Such findings are hardly consistent with the view that specialized learning is simply a constrained version of a more generalized, and thus by implication, more significant form of learning.

PSYCHOLOGICAL VS. BIOLOGICAL APPROACHES TO LEARNING

One aim of this Dahlem Workshop on "The Biology of Learning" was to reach beyond the traditional boundaries of learning theory and search for common ground with those engaged in formulating novel biological approaches to behavioral plasticity. Thus, aside from recognizing specialized instances of learning as phenomena that warrant study without preconceptions as to their nature, we also sought to bring to bear on the study of learning a wide range of issues, for example, the neurophysiological and cellular bases of learning, behavioral ecology, and the general natural history of learning.

A major dichotomy is evident between psychologists concerned with learning theory and biologists studying behavioral plasticity. Learning theorists tend to favor preparations in which learning can be studied at a particular plateau of growth and development. In their view, the essence of learning can be found in the ability of organisms to adapt to any problem that might be thrust on them by changing and unpredictable environments, whether natural or synthetic ones of the experimenter's choosing. Typically, the learning theorist's choice of problem is guided more by theory-based logic than by a knowledge of the organism's natural history.

Biologists, on the other hand, have tended to think of behavioral plasticity in the larger context of the organism's daily round of activities for survival and reproduction. These activities are conducted in environments that are unique for each species and that often vary drastically from one phase of the life cycle to another. In satisfying its requirements, an animal must actively seek out microenvironments, the optima for which change with the time of day and the season, and even more radically as the organism proceeds from birth to maturity.

The life cycle approach of the ethologist focusses especially on the plasticity of natural behavior and the relationship between learning and the dynamics of natural patterns of development. A parallel can be drawn in human infant studies. Learning theorists may study conditioning simply to verify that certain effects can be obtained at a particular stage. Developmental psychologists and psycholinguists, on the other hand, are likely to relate an infant's behavior and learning abilities to particular phases of emotional or cognitive development. Such concerns have led those who favor a naturalistic approach to learning to be especially sensitive to the functional necessity for behavioral plasticity to vary with age and with stages of neuropsychological development. It is no accident that research on imprinting, song learning, and infant development in particular, has illuminated the underlying mechanisms and the functional significance of sensitive phases in behavioral development.

ORGANISMAL STATE AND SPECIES

Observations of many kinds, originating in the field and subsequently transposed to the laboratory for experimentation, leave little doubt that comparative studies of natural learning processes call for a stronger emphasis upon species differences than appears relevant when organisms are learning to cope with more arbitrary paradigms.

The mechanisms that underlie imprinting and song learning cannot be understood without first acknowledging the pervasive role of unlearned, functionally adaptive predispositions to associate particular classes of stimuli. These are often species-specific and state-dependent. Such predispositions influence motor aspects of behavior and bias the formation of certain stimulus-response associations over others. As the state of the organism changes, so rates of associative learning vary accordingly, in harmony with the organism's adaptive needs at the time. It should also be recognized that state dependence extends beyond passive receptivity. It influences recall processes in functionally relevant ways and drives organisms to seek out situations that, on the basis of previously learned expectations, have the potential to satisfy current needs.

Changing states also prime the organism for responsiveness to stimuli specific to certain requirements. Thus, ducklings typically imprint while following an object that they have sought out. Following is initiated most effectively by an object that, while it may or may not satisfy specific visual requirements, also emits the maternal call of the species. Such learning has later impact on the choice of partners and not, for example, on the choice of food objects. Similarly, song-learning birds are innately predisposed to favor sounds that incorporate species-specific elements. In turn, such elements trigger memorization of associated stimuli. These stimuli are remembered in a particular context and linked by state-dependent recall with specific and predetermined behaviors in future life. Thus, memorized songs are applied to the task of song development and not, as memorized predator sounds might be, to the recognition of danger.

The intrusion of genetic influences into behavioral development and learning became one of the workshop's most controversial themes. The inevitable question arose, to what extent do imprinting and song learning represent phenomena that are fundamentally different from the more typical paradigms of the animal learning theorist? To biologists at least, it seemed reasonable to ask whether such special learning has less in common with phenomena such as maze running or keypecking than with the learning that occurs in the course of human development, especially during language acquisition. One must also consider the prospect that some aspects of behavioral plasticity in human infancy may be allied more closely with processes of growth and differentiation than with learning in the traditional sense.

SPECIALIZED LEARNING, VISUAL DEVELOPMENT, AND CORTICAL PLASTICITY

Striking commonalities were uncovered in the course of the workshop between song learning and imprinting in birds, development of the mammalian visual system, and the emergence of human speech in infancy. All display clearly defined critical periods or sensitive phases in which early experience has its maximal impact on subsequent behavior. The sensitive phases have a self-terminating quality, as though organisms wait for appropriate stimulation before closing one phase of development and progressing to another. Depriving a cat of certain kinds of visual stimulation postpones termination of the sensitive phase of visual plasticity. In both imprinting and song learning, the sensitive phase can be extended beyond its normal limits by witholding fully optimal stimuli.

Especially intriguing are apparent parallels between the stimulus requirements of experimentally-sensitive single units in the developing visual cortex and the stimulus requirements of auditory mechanisms underlying song learning. Evidence was presented to suggest that songbird brains come equipped with circuitry specifically responsive to song components but held back from guiding vocal development by auditory feedback until activated by specific stimuli. Lacking appropriate stimulation, such mechanisms, although innate, remain latent and the bird must develop its song abnormally, without their help. Similarly, in visual development, preferred orientations of cortical neurons can be changed by visual experience only within certain limits. Neurons that are already highly selective for orientation cannot change their orientation preference as a result of visual experience but will maintain their original response properties if exposed to corresponding visual stimuli. If not, they become unresponsive ((10); see also Singer, this volume).

The possible extent of cortical plasticity during visual development is thus quite limited. It is surely inappropriate to view these limitations as reflecting a fundamental deficiency in cortical plasticity. Rather than invoking constraints, it seems more productive to think in terms of a functionally optimal balance between neural plasticity and genetically preordained neurological connectivity. It is presumably to this end that the response properties of neurons can only be changed within the limits set by genetic programs (9, 10).

THE NEUROPHYSIOLOGY OF LEARNING

For the first time in the history of research on learning, there exist some compelling opportunities for making a choice between psychological and physiological approaches on the basis of empirical developments rather than the bias of individual investigators. As attested to by many contributors to this volume, discoveries in a variety of rapidly growing subdisciplines provide a concrete basis for such now familiar neurophysiological metaphors as "analyzers," "detectors," "cell assemblies," and "centers of excitation" (see Bateson ("The Neural Basis of Imprinting"), Brown, Changeux, Konishi, Quinn, Sahley, Singer, Squire, and Thompson, all this volume).

Most dramatic are the pioneering investigations of Kandel and his associates into the cellular basis of learning in the invertebrate, Aplysia. Thanks to such efforts, Pavlov's goal of delineating the neurophysiological basis of classical conditioning seems within reach (at least of the type exemplified by alpha-conditioning). In Aplysia we now know that in a monosynaptic sensory-motor system, habituation and sensitization result from presynaptic changes in the transmission of sensory stimulation. Through the involvement of additional circuitry, classical conditioning and sensitization have both been shown to involve enhanced transmission of a sensory signal. It will be a major challenge to extend these findings to other preparations, in particular to the variety of phenomena now regarded as instances of classical conditioning (for examples see Hearst, Jenkins, and Terrace, all this volume).

The impressive progress we have witnessed in understanding the neurophysiology of basic behavioral phenomena in invertebrates provides a firm basis for anticipating the discovery of equivalent mechanisms underlying more complex phenomena in vertebrates. It is nevertheless important to strike a balance between recognizing the importance of behavioral, neurological, and physiological specializations for certain types of learning, while at the same time acknowledging that fundamentally similar processes may be involved at the cellular level.

In principle, one can begin to see how to assemble networks from simple "Aplysia-like" circuits that must sustain higher-order learning phenomena such as classical conditioning (4). It is now reasonable to extend such modelling to encompass special cases such as song learning (Konishi, this volume). As this exercise proceeds, there will be increasing demands for specialized "hard-wired" circuits.

One of the more lively and productive controversies in neurobiology in the next few years will focus on whether it is more appropriate to think of learning in terms of "instruction" or "selection." Changeux (this volume) clearly favors the latter. His concept of selective stabilization of innate "pre-representations" as a consequence of experience found a receptive audience, especially among students of learning in songbirds and honey bees (Marler and Gould, both this volume). The analogy between learning and immunological theory is a compelling one (1) and promises to have a far-reaching impact on future research on the mechanisms of plasticity. Such mechanisms are especially relevant to such phenomena as the development of perceptual categories and concepts that fall outside the domain of classical learning theory. Again there are close affinities with developmental neurobiology, especially in the field of visual development where neuroselectional processes appear to play an important role (Singer, this volume).

LOCALIZATION OF BRAIN FUNCTION

One factor contributing to renewed interest in special learning is the growing evidence of localization of brain function. Since Broca and Wernicke, we have been accustomed to the existence of centers for the control of speech. The symptoms of different kinds of aphasia are very much a function of the precise location of brain damage (Brown, this volume).

The issue of localization of function is arising more and more frequently with other kinds of learned behaviors. Both song learning and imprinting involve highly localized brain structures (Bateson ("The Neural Basis of Imprinting") and Konishi, both this volume; see also (8)). Knowledge of the neuroanatomical circuitry underlying a given learning paradigm is becoming increasingly important for its interpretation. Garcia interprets the selective association of experience of taste and of nausea as a direct consequence of the convergence of gustatory and visceral afferents in the brain stem, indicating an intimate relationship among taste, ingestion, and emesis (Revusky, this volume). As Thompson notes (this volume), lesioning studies appear to implicate the cerebellum in very specific ways in the control of eyelid conditioning.

Such research is important, not only for its obvious bearing on how the brain sustains learning, but also as a source of new distinctions between fundamentally different learning processes. Clinical studies of amnesic brain-damaged humans appear to indicate two distinct, anatomically separable memory systems, one "procedural," the other "declarative" (Squire, this volume). The latter appears to correspond roughly to perceptual and cognitive functions and the former to learned sequences of stimulus-response interactions. The distinction is especially notable because it is drawn along lines that differ from the traditional dimensions of learning theory (but see Holland et al., this volume).

Mishkin's discovery (7) of what appears to be an equivalent dichotomy in memory mechanisms in the monkey brain lends additional weight to a distinction between two systems of storing information, one a "habit" system, the other a system for storing stimulus representations. Lesion studies indicate that these two neuroanatomically distinct systems normally work in parallel and in such close harmony that it is only as a consequence of the artificial disruption of one that the other suddenly becomes apparent. Cognitively oriented theorists may find it easier to relate to these new neuroanatomical findings than theorists using learning taxonomies based on traditional classical and instrumental training paradigms.

COGNITION AND LEARNING

Given the uniqueness of human language, it is hardly surprising that students of animal and human learning approach their subject matter with different points of view. Yet only during the past few decades has it been shown convincingly that no elaboration of simple S-S or S-R units can explain the complexities of language. A similar conclusion has emerged from recent studies of animal behavior. Thus, for quite different reasons, the behaviorist doctrine, that "higher-order" phenomena such as language and complex problem-solving were reducible to simpler conditioning processes, has been generally abandoned in favor of what has been referred to as the cognitive approach to behavior.

Essential to cognitive theorizing is the concept of the internalized representation of objects and events, and the kinds of transformations such representations undergo in learning and memory. While the modes of representations postulated in the case of human language and memory differ from those that have been proposed in the case of animal behavior, it is important to note the similarity of the logic that, in each instance, has been used to argue for the existence of representations. So long as an organism can make a learned response without guidance from any physically present stimulus as, for example, when drawing something from memory, or when generating "missing" elements of a complex proximal stimulus that has been experimentally impoverished, it is reasonable to appeal to a representation as an internal stimulus that controls the response in question (cf. (3, 5, 13)).

Modern cognitive psychology can be said to have emerged from compelling demonstrations of the importance of representations in explaining behavior. Earlier attempts to study cognitive phenomena met with the objection that they were "mentalistic" and, therefore, not even potentially observable. However, such concepts as information processing, branching, executive control, and so on, which derive from computer programming and artifical intelligence and which are firmly anchored to reliable independent and dependent variables, have effectively exorcised the ghost in the machine.

How representations are to be incorporated into theories of learning remains a controversial question. However, there is widespread agreement in the study of both human and animal behavior that an important first step is to distinguish, on the one hand, between the content of learning and how that content can be changed and, on the other hand, how and when an organism acts upon the contents of learning (cf. Terrace, this volume). By insisting that an organism's knowledge is fully revealed by the organism's performance and by failing to recognize that performance typically underestimates knowledge, behaviorists have failed to come to grips with the essence of cognitive psychology.

Two features of human cognition deserve special attention in the development of general theories of learning. Human learning is best characterized as a process of acquiring information, often for its own sake, about various features and relationships of the environment. The role of any immediate or concrete reward for such information-gathering is often impossible to discern. Thus, the temporal relationships between the acquisition of information, the consequences of that information-gathering and its subsequent utilization appear to be quite different in animal and human learning (Estes, this volume).

Since much human learning is mediated by language, its contents can often be described by reference to linguistically framed propositions. Another basic question is, how much does language learning have in common with other kinds of learning? One important theme of this workshop was the irrelevance of traditional theories of learning to language acquisition (see Gleitman and Osherson and Weinstein, both this volume). Future theorizing on language learning will undoubtedly seek to determine whether, as Chomsky and others have argued, human language acquisition presumes an innate and specific correspondence between a child's learning capacities and the features of language that the child acquires.

By contrast, one of the most compelling aspects of cognitive processes in animals is their nonlinguistic nature. Just how do animals learn to "think" without language? Answers to that question should emerge from contemporary research on animal cognition, e.g., studies of concept formation, cognitive maps (see Lea and Menzel, both this volume), complex feats of memory (13), and communication.

CONCLUSIONS

Despite the breadth of subject matter represented at this Dahlem Workshop on "The Biology of Learning," there was widespread agreement as to the directions that future studies of learning should take. There was also general agreement that, while heuristically useful, the traditional view that all learning can be reduced to one or a combination of two basic processes of conditioning needs to be modified in a number of ways.

At the behavioral level, provision must be made for distinguishing between learning and performance. Theories of learning will have to make greater use of cognitive factors than was previously the case. Also, in specifying what an organism can learn and what the organism does about the knowledge it acquires, it is essential to understand the role of learning in the organism's natural history (see Bolles, Heinrich, Gould, and Shettleworth, all this volume). The study of an organism's natural history inevitably yields insights into the functional significance of what is learned (see Hollis, this volume) and the degree to which an organism learns and how it does so are optimal solutions to the challenges posed by the organism's environment (see "Genes, Evolution, and Learning" by Bateson, this volume). Future studies of learning will also have to specify how traditional theories of association can be modified to accommodate such phenomena as song learning and imprinting, or if they cannot, what new forms of learning need to be postulated.

Unprecedented opportunities exist for studying learning at the physiological level. Just as our knowledge of how the structural properties of the genotype give rise to particular phenotypes has had an influence on the nature of theories of genetics and evolution, progress in understanding the neurological and chemical machinery that makes learning possible will have an inevitable influence on the form of future theories of learning.

Acknowledgements. Preparation of this paper was supported by NIMH grant PHS 114651 to P. Marler and by NSF grant BNS 82-02423, fellowships from All Souls College, Oxford, England, and from the Fulbright Commission to H.S. Terrace.

REFERENCES

(1) Edelman, G. 1978. Group selection and phasic reentrant signaling: a theory of higher brain function. In The Mindful Brain. Cortical Organization and the Group Selective Theory of Higher Brain Function, eds. G. Edelman and V. Mountcastle. Cambridge, MA: MIT Press.

(2) Gibbon, J. 1982. The contingency problem in autoshaping. In Autoshaping and Conditioning Theory, eds. C.M. Locurto, H.S. Terrace, and J. Gibbon. New York, NY: Academic Press.

(3) Gregory, R.L. 1975. Do we need cognitive concepts? In Handbook of Psychobiology, eds. M.S. Gazzaniga and C. Blakemore. New York, NY: Academic Press.

(4) Hawkins, R.D., and Kandel, E.R. 1984. Is there a cell biological alphabet for learning. Psychol. Rev., in press.

(5) Hunter, W.W. 1913. The delayed reaction in animals and children. Behav. Monogr. 2(6).

(6) Jenkins, H.M.; Barnes, R.A.; and Barrera, J. 1982. Why autoshaping depends on trial spacing. In Autoshaping and Conditioning Theory, eds. C.M. Locurto, H.S. Terrace, and J. Gibbon. New York, NY: Academic Press.

(7) Mishkin, M. 1982. A memory system in the monkey. Phil. Trans. R. Soc. Lond. B. 298: 85-95.

(8) Nottebohm, F. 1981. Brain pathways for vocal learning in birds: a review of the first ten years. In Progress in Psychobiology and Physiological Psychology, eds. J.M.S. Sprague and A.N.E. Epstein,ch.9. New York, NY: Academic Press.

(9) Rauschecker, J.P. 1979. The formation of orientation selectivity of the cat's visual cortex. Kybernetika 80: 76-80.

(10) Rauschecker, J.P. 1980. Instructive changes in the kitten's visual cortex and their limitation. Exp. Brain Res. 48: 301-305.

(11) Rescorla, R.A. 1972. Information variables in Pavlovian conditioning. In The Psychology of Learning and Motivation, ed. G.H. Bower. New York, NY: Academic Press.

(12) Roitblat, H.L.; Bever, T.G.; and Terrace, H.S., eds. 1984. Animal Cognition. Hillsdale, NJ: Lawrence Erlbaum.

(13) Terrace, H.S. 1984. Animal cognition. In Animal Cognition, eds. H.L. Roitblat, T.G. Bever, and H.S. Terrace. Hillsdale, NJ: Lawrence Erlbaum.

The Biology of Learning, eds. P. Marler and H.S. Terrace, pp. 15-45. Dahlem Konferenzen 1984. Berlin, Heidelberg, New York, Tokyo: Springer-Verlag.

Animal Learning, Ethology, and Biological Constraints

H.S. Terrace
Dept. of Psychology, Columbia University
New York, NY 10027, USA

Abstract. The subject matter of learning theory and ethology derives from the comparative psychology of the late nineteenth century. Since that time, both disciplines have been transformed decisively. The study of the mental life in animals, which was supposed to provide insight into the evolution of human intelligence, gave way to the study of conditioned behavior. The original approach to the study of instinctive behavior, which amounted to little more than an exercise in taxonomy of instincts, has been replaced by analytic developmental studies which seek to reveal how a species' genotype and particular experiences interact to produce various instances of species-specific behavior. During most of the twentieth century, conditioned behavior was explained by appeal to one or another variety of general learning theory which sought to explain all learned behavior, in all species, by appeal to a simple set of principles. During the past two decades, the goals of learning theory have become less comprehensive. Interest has focused primarily on associations between a cue and a biologically important event (S-S* associations) and between a response and a biologically important event (R-S* associations). Students of animal learning have also investigated various "biological constraints" that violate assumptions about the arbitrariness of the abstract stimuli and responses referred to by learning theory.

Like learning theory, ethology went through an era of general theory construction which attempted to specify instinctive behavior on a broad level - notably the Lorenz-Tinbergen model of the innate releasing mechanism. In response to criticisms of such theories, ethologists have concerned themselves less with theory and more with research on the development of species-specific behavior. Such research has provided students of animal learning with many specialized examples of learning that, along with the literature on biological constraints, need to be accommodated by learning theory. It is suggested, however, that the main obstacles for the advancement of learning theory are problems from within. Specifically, until learning theorists are able to specify how the contents of learning are translated into performance, it seems doubtful that learning theory will be able to specify whether a particular biological constraint should be attributed to a) properties of learning a particular association, b) properties of performing some response, or both a) and b). It is also necessary for ethologists and learning theorists to develop a conceptual system that relates specialized instances of learning such as imprinting and song learning to more flexible varieties of learning.

INTRODUCTION

Despite their common intellectual ancestry (the theory of evolution) and their common subject matter (animal behavior), the disciplines of ethology and animal learning have developed autonomously. While some modest attempts at integration have occurred during the last two decades, it is still too early to assess the viability of such undertakings.

In order to anticipate the outcome of future efforts by ethologists and students of animal learning to coordinate research and theory construction, it is helpful to ask, why has so little integration been attempted previously and what kinds of obstacles stand in the way of future progress? These questions define the major issues that this paper will address.

HISTORICAL BACKGROUND

The Origins and Initial Goals of Comparative Psychology

Comparative psychology emerged as a coherent area of inquiry soon after experimental psychology established itself as a distinct science. There is little question, however, that comparative psychology owes its existence more to Darwin's theory of evolution by natural selection than it does to studies of human sensation, perception, or memory (see (25) and (26) for brief histories of comparative psychology).

In two of his last works, Darwin defined the basic issues that were to become the initial subject matter of comparative psychology: precursors of the human mind in animals and the nature and the function of instincts in animal (and human) behavior (13, 14).

The rationale for studying the continuity of mental functions was stated clearly by Darwin's contemporary, Herbert Spencer. "If the doctrine of evolution is true, the inevitable implication is the Mind can be understood by observing how Mind is evolved" (96)[1]. Romanes, a great admirer of Darwin, not only speculated extensively about the nature of the animal mind (75, 76), but also appealed to Darwin's theory of evolution as the rationale for studying instinctive behavior in a comparative framework. Instinctive behavior, both animal and human, is adaptive. Furthermore, particular forms of instinctive behavior appear to be species-specific. One could therefore argue that instincts were prone to natural selection and that they were as heritable as particular morphological features.

For a variety of (by now) familiar reasons, the seemingly straightforward subject matter of comparative psychology - animal minds and instincts - proved highly refractory to experimental study. Accordingly, this infant discipline was transformed radically - so much so that, in hindsight, its original goals would prove difficult to discern.

Except for a few insightful, but uninfluential, exceptions, comparative psychology's original approach to the study of instincts was little more than an exercise in taxonomy. The question posed by comparative psychologists, as well as by other schools of psychology, was not what variables determined the development and the occurrence of purported instances of instinctive behavior, but how many different kinds of instinctive behavior could one identify?

Instincts were regarded as inherited species-specific behaviors which showed no benefit from practice - either in its initial or in subsequent occurrences. This ancient view of instinct led inexorably to the widespread practice of the nominalist fallacy. In its extreme form, everything from pugnaciousness to altruism was termed instinctive. McDougall, regarded by many as the founder of modern social psychology, argued that all human behavior was instinctive. To substantiate that claim, he postulated more than a thousand human instincts (56).

---

[1] While Spencer advocated a theory of evolution based upon the inheritance of acquired characteristics, that error of judgment is irrelevant to his important insight that the logic of evolutionary theory applies with equal force to such adaptive psychological entities as associations and habits as it does to adaptive morphological characteristics.

Thanks, in large part, to sterile debates as to how many instincts existed in various species, interest in that subject matter waned. Indeed, systematic analyses of instinctive behavior did not begin until the thirties when the seminal publications of Heinroth and Kuo, who independently defined many important lines of inquiry, were translated into fruitful programs of research (30, 43). While the study of instinctive behavior remained in limbo, comparative psychologists directed their attention to a new subject matter: the study of learned behavior.

Interest in learned behavior developed as much in reaction to the problems encountered in studying the animal mind as it did in reaction to the apparent immutability of instinctive behavior. Among the first to recognize the former problem was Morgan who rejected as anecdotal, and therefore unsubstantiated, Romanes' extensive accounts of the intellectual feats of various animals.

Because of his well-known application of Occam's razor to descriptions of animal behavior, it is commonly believed that Morgan denied the existence of mental processes in animals. However, Morgan's intent was simply to seek the simplest psychological level that proved adequate to explain a particular bit of animal behavior (58, 59). While this attitude was a healthy reaction to the excesses of the anecdotal method, it acknowledged unambiguously the existence of mental activity in animals. Indeed, Morgan argued that, since animals cannot themselves introspect, the job of comparative psychologists was to intuit just what kinds of mental operations gave rise to a particular type of behavior, being sure, of course, to select the simplest kinds of operation that could do the job (60). It was in this spirit that Morgan (and many of his contemporaries) advocated the so-called "doubly inductive" method of studying human mental functions - a method which attempted to reconcile the subjective data, obtained from introspective descriptions of mental processes, with objective descriptions of behavior and physiological functions. What Morgan and most of his contemporaries did not anticipate, however, were the insurmountable problems inherent in the application of the introspective method. Foremost were the unreliability of introspective reports and the inability of methods of introspection to capture the contents of unconscious thought processes.

Like their contemporaries in human experimental psychology, comparative psychologists slowly abandoned the introspective method. In so doing, comparative psychology abandoned its only method for studying animal thinking. Accordingly, the vision of Darwin, Spencer, and the first generation of comparative psychologists, of a discipline that would discover the evolution of the human mind, gave way to an outlook that avoided mental functions and instead focused exclusively on the kinds of conditioned behavior that were demonstrated by the classic experiments of Thorndike and Pavlov.[2] As we shall see later, modern studies of animal behavior, again borrowing from methods used to study human cognitive processes, have, at long last, begun to reveal the nature of the animal mind (102).

The Objective Study of Learned Behavior

With few exceptions, the character and the outlook of the various theories that have dominated the study of learned behavior during the 20th century were determined by the writings of three psychologists: Thorndike, Pavlov, and Watson. The publication of Thorndike's monograph and book, both containing the title Animal Intelligence, marked a turning point in comparative psychology and also marked the beginning of a systematic attempt to construct theories of animal learning (103, 104).

From his observations of the problem-solving abilities of a broad variety of vertebrate species, Thorndike formulated a general theory of learned behavior whose main principle has come to be known as the "Law of Effect." Learning amounted to the gradual strengthening (or weakening) of associations between a particular act and the prevailing stimulus conditions. Increases (or decreases) in associative strength were determined by the "satisfying" (or "annoying") effects produced by the

---

[2]Despite the heavy emphasis placed by the first generation of comparative psychologists on identifying mental processes in animals and developing taxonomies of instincts, there were isolated efforts to study intelligent behavior objectively and to analyze the development of instinctive behavior. For example, Small designed a variety of mazes in which he studied the ability of different species to learn arbitrary sequences (91, 92). Spalding performed what appear to be the first systematic "isolation" experiments (93). Loeb's studies of animal activity provided such useful concepts as "forced movements" (50). However, these and related approaches were unable to weigh very heavily against a program of experiments carried out by researchers who were not so much guided by the theory of evolution as by a desire to avoid any mention of mental events and to measure an animal's intelligence in standardized apparatuses that were administered in well controlled laboratory settings.

act. Aside from establishing a methodology in which learning was observed in standardized laboratory settings, Thorndike contributed an important conceptual modification of the traditional view of association by changing its direction from the study of ideas to the study of action[3].

From Thorndike's point of view, the Law of Effect was a general law that applied to all learning in all organisms. While animals differed with respect to "the delicacy, number, complexity, and permanence of associations," objectively observable S-R (stimulus-response) associations sufficed as a common feature of all learning. Thorndike, of course, recognized that, unlike animals, homo sapiens was capable of ideas. He nevertheless believed that the complexity of those ideas would be derived from the principles of association stipulated by the Law of Effect.

Starting from a different tradition and concerned with a different set of problems, Pavlov, the Russian physiologist, independently conceived of an approach to the study of learned behavior which contained many of the features that Thorndike introduced (62). Like Thorndike, Pavlov developed a standardized procedure for the study of animal learning and, on the basis of studying even fewer organisms than did Thorndike, Pavlov proposed his own general principle of learning.

Working in the tradition of reflexology, Pavlov encountered, rather late in his career, a phenomenon which he first referred to as "psychic secretion." His canine subjects salivated before food or other saliva-inducing substances were placed in their mouths. In his search for a mechanistic explanation of psychic secretion, Pavlov controlled the occurrence of salivation by signalling the occurrence of food to such well controlled and presumably arbitrary stimuli as a bell or a light. The outcome of these experiments served as the basis for Pavlov's fundamental contribution to animal learning: the model of the conditioned reflex. Variations of this conditioning paradigm enabled Pavlov to conceptualize many related phenomena that have since defined much of the subject matter of animal learning: conditioned excitation, conditioned inhibition, discrimination learning, extinction, generalization,

[3]Though anticipated by Bain's discussions of sensory-motor associations, it is Thorndike who deserves credit for the view that emphasizes that animals associate movements and stimuli rather than ideas.

and higher-order conditioning, among others (see (5, 53, 68) for further discussions of these concepts).

Despite obvious differences in the mechanisms underlying Thorndike's and Pavlov's conceptualizations of learning, it is important to recognize some important similarities. Neither Thorndike nor Pavlov qualified their laws of learning to take into account the nature of the organism they conditioned, the kind of stimuli that they used, the kind of responses they sought to condition, or the kind of reinforcing stimulus that followed the conditioned response. Thorndike and Pavlov also both regarded the outcome of a learning experiment as the strengthening of an association between an objectively defined stimulus and an objectively defined response. For Pavlov, the response was the same response that was elicited by the unconditioned stimulus (US); for Thorndike it was whatever response solved the problem on hand.

Thorndike's and Pavlov's formulations of learning had two important consequences. The most immediate was the formal founding of the school of animal learning that, through the writings of Watson, has come to be known as "behaviorism." The other was the assumption that the properties of an S-R association were revealed fully by the performance of the response; that is, there was no reason to distinguish between learning and performance. I will argue later that this assumption poses one of the major difficulties learning theory currently faces and that in the absence of a clear basis for distinguishing between learning and performance, the role of "biological constraints" will prove difficult to define.

Watson exerted a powerful influence upon animal learning by articulating more clearly than any other psychologist the essential importance of behavior as the sole subject matter of psychology - both animal and human. While Thorndike and Pavlov were content to say that all mental processes could be reduced to S-R associations, Watson rejected the existence of any central mental processes. The existence of human thought was acknowledged, but only in the form of subvocal speech - a peripheral response that, given adequate amplification, could be observed objectively.

Watson, an outspoken critic of the method of introspection, also rejected consciousness as the subject matter of psychology. All that was left was behavior - conditioned and unconditioned. Though Watson does not appear to have objected to Thorndike's Law of Effect, he clearly would have had difficulty with the subjectively defined states of "satisfaction" and "annoyance"[4]. That Watson implicitly favored Pavlov's approach to animal learning to Thorndike's (perhaps because of the biological orientation of his graduate training) is less important than his founding behaviorism. For Watson, the goals of that movement were to replace mental functions with behavior as the subject matter of psychology, to reduce all animal and human behavior to learned and unlearned responses, and to account for all learned behavior, in all organisms, by appeal to a universal set of laws of conditioning[5].

THE EMERGENCE OF GENERAL LEARNING THEORY

Watson's grandiose claims about the power of behaviorism and the then popular application of the philosophical movements of operationism and logical empiricism provided the ideal climate for the age of general learning theory (6). While the major figures of that age (Guthrie, Tolman, Hull, and Skinner) differed with respect to many fundamental assumptions about the nature and the conditions of learning, they reinforced the views of Thorndike, Pavlov, and Watson that it was possible to establish general laws of learning that did not have to make special provision for particular characteristics of particular species. That attitude was maintained, if not strengthened, by the next generation of learning theorists (9, 17,

---

[4]Thorndike provided an "operational" definition of satisfiers and annoyers, though he seems to have done so as an aside. Most likely, Watson would have been sympathetic to Skinner's version of the Law of Effect and the many applications of that law that Skinner suggested. It should be noted, however, that unlike Watson, Skinner acknowledged the existence of consciousness and mental events and sought to account for them by appeal to the laws of learning (88). Skinner simply rejected mental events as <u>explanations</u> of behavior.

[5]Despite the commonly shared belief in the importance of behavior as the subject matter of psychology, there exist many basic differences between the varieties of behaviorism advocated by Thorndike and Watson and by the next generations of so-called "neo-behaviorists" (Guthrie, Tolman, Hull, Skinner, Spence, and Miller). For lack of a better term, the inexact label "behaviorism" will be retained in this paper. The reader should, however, realize that the assumptions and the philosophies of the different varieties of behaviorism often clash and that it is foolhardy to make sweeping generalizations (cf. Chomsky's broad-brush critique of Skinner's behaviorism (12)).

57, 95), despite the fact that their assumptions about the nature of learning were at least as disparate as those of their predecessors[6].

The conviction that it was possible to formulate species-free laws of learning was also shared by physiological psychologists who paid little heed to the phylogenetic niche of the species they studied in their search for neurophysiological and neurochemical mechanisms of learning. Thus, while it was not surprising that a learning theorist would argue that "... a learned response (R) ... can be made specific to any arbitrarily selected stimulus (S) within the animal's repertoire ... stated abstractly this means that our learning procedures should be able to cause S_1 to elicit R_1, S_2 to elicit R_2, S_3 to elicit R_3, and S_3 to elicit R_1, etc." ((57), p. 315), it does give one pause to read a similar observation by an eminent physiological psychologist: "In effect, in any operant situation, the stimulus, the response, and the reinforcer are completely arbitrary and interchangeable. No one of them bears any biologically built-in fixed connection to the others" ((99), p. 567).

THE EMERGENCE OF ETHOLOGY

The behavioristic view of animal learning flourished mainly in America. Though occasional dissents were heard (e.g., by Kuo (43), Schneirla (79), and Beach (2)), they never coalesced into an alternative school. Such a school did develop in Europe under the leadership of Lorenz and Tinbergen. During the thirties, Lorenz, Tinbergen, and their students, who referred to themselves as ethologists (a term that appears to have been coined by Lorenz's teacher Heinroth), began a research program whose goals were to describe the natural behavior of a species in its natural environment (in the form of so-called "ethograms") and to use those observations to construct an objective theory of instinct. This highly successful enterprise, along with Von Frisch's independently conducted research on bee communication, was recognized by the Nobel Awards Committee who, in 1973, conferred on Lorenz, Tinbergen, and Von Frisch the Nobel Prize for Medicine - the first Nobel award for behavioral research.

---

[6]Two other psychologists who have exerted a strong influence on learning theory are Hebb (29) and Lashley (44-46). However, the contributions of both of these men had more to do with physiological models of learning than with specific proposals for the form learning theory should take.

The Lorenz-Tinbergen theory of instinct centered around "innate releasing mechanisms" in the central nervous system which mediated species-specific responsiveness (to particular "sign stimuli" or "releasers") and innate motor movements (so-called "fixed action patterns"). Of particular importance was Lorenz's and Tinbergen's view of fixed-action patterns. These were considered to be species-specific and highly stereotyped (hence the term "fixed"), unmodifiable, and insensitive to the effect of learning or practice. A fixed-action pattern would occur only in the presence of a particular sign stimulus or, spontaneously, if they had not been released for a long time (52, 106).

Especially because the Lorenz-Tinbergen theory was concerned with "innate" behaviors, it seemed to pose no problem for learning theory. Indeed, the harshest criticism Lorenz advanced concerning the American approach to animal behavior was that it misrepresented itself as a comparative approach to learning (52)[7].

So long as learned and innate behaviors were regarded as two mutually exclusive types of behavior, there was no reason to expect much overlap between the disciplines of animal learning and ethology. For some time, however, it has been clear that this distinction is oversimplistic, if not downright untenable. The heart of the problem was articulated clearly by Lehrman, a student of Schneirla, in his classic critique of the Lorenz-Tinbergen theory of instinct (47). Lehrman, a rare individual who had a sophisticated knowledge of both ethology and learning theory, observed that by designating a particular behavior as "innate," Lorenz and Tinbergen discouraged any further inquiry into the origin or the underlying mechanism of that behavior[8]. What Lehrman suggested as an alternative was an

---

[7]Lorenz was not so much concerned about the species-free approach that characterized the American school of learning theory as he was about claims that certain types of behavior observed in different species were homologous. Hodos and Campbell presented a similar critique of the failure of some comparative psychologists to take into account the role of common ancestry in their discussion of comparative psychology (32).

[8]It is interesting to note that the Lorenz-Tinbergen theory of instinctive behavior bears a curious resemblance with the behaviorist approach to learning theory in that the concept of the innate releasing mechanism is species-free. As Lehrman ((47), p. 35) notes, this point of view does not consider differences in the level of organization of instinctive behavior that may result from evolutionary pressures.

"epigenetic" approach to the study of instinctive behavior. Instead of regarding instinctive behavior as a unitary fixed entity that is independent of external influences, Lehrman proposed a model in which an animal's internal state and the consequences of its behavior interact so as to guide the behavior in question to its next stage of development. In this connection it is important to note the difference between Lehrman's and Lorenz's interpretations of an isolation experiment. For Lorenz, if an experiment in which an organism is isolated at birth yields positive results (the occurrence of a certain behavior), those results provide evidence that a particular behavior is innate. Lehrman argued that an isolation experiment could only provide information regarding the influence of other organisms. Since it is impossible to isolate an organism from itself, an isolation experiment cannot assess the contribution of an animal's feedback from its own behavior.

One important consequence of Lehrman's critique of the Lorenz-Tinbergen theory was the emphasis of developmental studies in ethological research and a shift in the conceptual language of ethology from "innate-" or "instinctive-" to "species-specific"-behavior. As Gottlieb has noted, much of this research has been done in an essentially atheoretical framework (25). We shall later see that the absence of a general theory of species-specific behavior has not prevented ethologists from making a variety of significant discoveries, many of which pose novel problems for learning theory. Such discoveries, in combination with examples of "biological constraints" on learning that, during the mid 60's, were reported with alarming regularity, appeared to threaten the very foundations of learning theory (31, 82, 83). For example, learning theory seemed unable to explain phenomena such as "prepared" associations between particular stimuli (as demonstrated in research on taste aversion (23, 71)), associations between a CS and a US that occur even when the two stimuli are separated by several hours (22), or the classical conditioning of such traditional operants as the pigeon's keypeck (8). Another frequently heard question is the validity of studying behavior in an unnatural "arbitrary" laboratory environment without recognition of the natural ecological pressures that the subject normally encounters.

BIOLOGICAL CONSTRAINTS

"Cue-to-Consequence" Associations

The first systematic studies of the special qualitative properties of particular combinations of CS's and US's were performed by Garcia and his colleagues. Particular CS's were shown to be more effective than other CS's as signals for particular US's (23). In one experiment, pairings of a taste (CS_1) and a nausea-inducing substance (US_1) or of a light-sound compound (CS_2) and electric shock (US_2) were learned more readily than the converse pairings (CS_1 and US_2 or CS_2 and US_1) - hence the term, "cue-to-consequence" associations. Garcia interpreted these results as evidence of an adaptive predisposition to learn that nausea is to be associated with certain tastes and that externally elicited pain is to be associated with external signals.

Initially, these results were interpreted as evidence that certain superthreshold CS's and US's could not become associated by virtue of what would otherwise prove to be an effective contiguous pairing. That claim was later qualified by demonstrations of associations between "unnatural" pairings of particular CS's (e.g., a taste) and particular US's (e.g., an electric shock), albeit with more training than is required for establishing associations between "natural" CS's and US's (72, 112). Thus, in principle, an arbitrary CS could be associated with an arbitrary US so long as provision was made for what might be called a parameter of "biological belongingness"[9]. To date, this approach to cue-to-consequence association has yet to be integrated into a formal model of learning (see (51) for suggestions as to how this might be done).

Associations Formed Under Long CS-US Intervals

Aside from demonstrating particular affinities between particular CS's and US's, it has also been shown that, in the case of certain CS's and US's (especially those used in cue-to-consequence studies), an association between the CS and the US can be established even when the CS precedes the US by as much as several hours (22, 71, 78). Advocates of traditional learning theory have offered two types of rationalizations of the formation of CS-US associations under long delays (cf. Revusky, this volume). One assumes a representation of the CS which persists through the CS-US interval (or which is activated by the US) (71, 94). The other notes that, even in instances where long intervals result in learning, short intervals are more effective than long intervals (70).

At present, the only theory that seems able to accommodate the facts on hand, both in the case of cue-to-consequence and long-delay

---

[9]Thorndike enunciated a principle of belongingness, but that principle was motivated by Gestalt phenomena and by the degree to which the stimulus and response components of a paired-associate item in a verbal learning task matched one another (105).

associations, is some variety of association theory. Admittedly, that theory will prove more complex than those envisioned by currently available models of association, but, unless a simpler theory is available, it is not clear that anything is gained by rejecting association theory on the grounds that one must allow for additional parameters.

Autoshaping

The discovery that a pigeon's keypeck could be conditioned by classical conditioning paradigms and that, in many instances, the keypeck seemed to be more sensitive to stimulus-stimulus (S-S*) than to response-reinforcement (R-S*) contingencies was a rude shock to many students of operant conditioning (28, 49). This is not to say, however, that the keypeck is insensitive to R-S* influences. Even though the keypeck is not the plastic or arbitrary type of response that Skinner and other students of operant behavior once thought it to be, the literature on schedules of reinforcement, ratio schedules in particular (20), and analytical experiments on the role of S-S* versus R-S* contingencies suggest that it is nevertheless influenced by R-S* contingencies (36).

Parenthetically, it is interesting to observe that the classical conditioned keypeck has served as an extremely popular preparation for studying principles of association as involved in defining the necessary conditions for the formation of an S-S* association (21), higher-order conditioning (67), and sensory preconditioning (73). Indeed, just as investigators of operant conditioning have regarded keypecks conditioned by the method of successive approximations as arbitrary responses, investigators who have used autoshaping paradigms have also regarded the keypecks they have conditioned as arbitrary. This state of affairs is consistent with Williams' distinction between the "biology" and the "psychophysics" of association (111). Unexpected discoveries concerning variables that pertain to the former factor do not invalidate the significance of the latter.

Laboratory Versus Field Studies

The methodological criticism that learning theory is based on observations that do not duplicate properly the animal's natural environment suffers from an "either-or" mentality about the proper way in which to carry out research on animal learning. Recent developments suggest that, aside from serving different purposes and answering different questions, knowledge gained from laboratory and field studies complement each other in many helpful ways.

The study of foraging, which was inspired by field observations of the temporal and spatial distributions of food-obtaining habits of many species, has recently been advanced by many carefully controlled experiments that would be difficult, if not impossible, to carry out in the field (41, 42).

Another type of laboratory study, which provided pigeons with periodic deliveries of food, irrespective of their behavior, distinguished between the important categories of "interim" (variable) and "terminal" (stereotyped) behavior (97). This experiment, which investigated the temporal organization of these two types of behavior, clarified the nature of so-called "superstitious" responding. Skinner coined this term to refer to various kinds of stereotyped responses that resulted from the periodic delivery of non-contingent food (89). Had Skinner continued this procedure long enough, he would have observed the transition from interim to terminal behavior reported by Staddon and Simmelhag.

Our understanding of the organization of behavior has also benefited from laboratory studies in which an organism can obtain a variety of reinforcers with any one of a variety of responses (84, 85). These studies showed that particular combinations of responses and reinforcers were more effective than others. The data obtained from such experiments would have been exceedingly difficult to obtain in field studies.

The successful outcome of experiments by ethologists and students of animal learning based upon techniques and concepts borrowed from each others' disciplines bears on the question of the "arbitrariness" of the stimuli and responses used in studies of animal learning. The argument concerning arbitrariness is that the artificiality of laboratory work results in data that lack generality in any natural setting. This criticism is certainly valid in some instances. Bolles, for example, has shown the importance of taking into account an organism's natural reaction to electric shock when attempting to condition avoidance responding (3). At the same time, it is important to recognize the virtue of studying conditioned behavior in situations in which biological factors exert a minimum of influence. How else can one evaluate such an influence? Schwartz, for example, has observed that "The mere fact that a situation is not arbitrary does not imply that the principles it yields will not generalize to other situations ... The central problem lies in understanding just what features of the phenomena are attributable to general principles and which are attributable to situation-specific ones" ((80), p. 195). In a similar vein, Skinner has observed that "... Behavior in a natural habitat would have no special claim to genuineness. What an organism does is a fact about that organism regardless of the conditions under which it does it. A behavioral process is nonetheless real for being exhibited in an arbitrary setting" ((90), p. 1211).

Discussions of arbitrariness have, unfortunately, confused two issues. One is the clearly erroneous "equipotentiality" proposition (82), which holds that any reinforcer can be associated with any response in the context of any stimulus. The other is that, in order to do any kind of analytic research, it is necessary to restrict one's independent and dependent variables. It is, of course, true that the success of analytic research depends upon the care with which one chooses one's variables. Those choices are nevertheless biased.

Significance of Biological Constraints for Theories of Learning

How should biological constraints be interpreted and what issues do they raise for learning theory? In answering these questions, it seems rash to conclude, as some have, that there are no general laws of learning (81) or that one should seek specialized laws of learning that indicate how each species adapts to its ecological niche (78). To accept such views would be tantamount to abandoning, with no alternative in sight, theories that have accounted for the results of thousands of learning experiments in which the behavior of a broad variety of species was reliably predicted and controlled by the independent variables manipulated by the experimenter. To borrow an example from neurophysiology, rejecting the possibility that general laws of learning exist would be like arguing that there are no general laws of neural transmission, in view of the different functions of the electrical activity obtained from whole nerves (logistic) and from single nerve fibers (all-or-none), or the differences in the transmission of impulses by myelinated vs. nonmyelinated fibers, etc.

The significance of biological constraints for learning theory was stated clearly by Marler who observed that, "No organism can be properly thought of as tabula rasa, approaching a learning task as a totally free agent, without constraints. As a first consideration, learning cannot develop without genetic guidance" ((54), p. 254). But just as it is simple-minded to assume tabula rasa, it is also simple-minded to assume that one can predict an organism's behavior solely on the basis of its genotype. What needs to be determined is not whether learning or, by implication, laws of learning exist, but how biological constraints impose themselves on such laws (cf. Bateson, this volume).

A good place to start is with Mayr's suggestion of a continuum of flexibility in the "programs" an organism is equipped with to respond to various kinds of environmental stimulation (55). Mayr's distinction between "open" and "closed" programs of development capture much of the role of biological constraints on learning. In closed programs, an organism relies on a relatively small degree of environmental input in the development of a relatively small number of patterns of adaptive behavior. As a result, its behavior is highly sterotyped and resistant to modification. Examples include imprinting, specific or imitative song learning by birds (cf. (15), and Bateson, Marler, and Immelmann, all this volume) and the digger wasp's ability to remember the degree to which larvae need to be fed, as determined by its first visit to the nest (107).

In open programs, the organism is responsive to a much broader range of stimulation and can do so in a relatively large number of ways. In its traditional sense, learning refers mainly to open-ended programs[10]. Even though open-ended programs are constrained by the genetically guided mechanisms that Marler cites, how learning occurs in the case of such programs remains a special subject matter that requires laws above and beyond whatever principles give rise to those mechanisms in the first place. Thus, discussions of programs for "trial-and-error" learning, "cognitive" learning, and so on (cf. Gould and Marler, this volume) that do not specify how those programs work, beg the basic question, what are the principles of learning?

CURRENT STATUS OF ANIMAL LEARNING THEORY

The conventional wisdom regarding the demise of general learning theory is that one of its fatal flaws was its inability to cope with biological constraints and various types of learning reported by ethologists. There is much reason to believe, however, that the most difficult issues for learning theory come from within and that little progress will be made in understanding the significance of biological constraints until learning

---

[10]In this connection, it is important to recognize that, from a biological point of view, the kinds of open-ended programs that Mayr refers to provide the potential for a degree of biological flexibility not encountered in other biological systems.

theory can put its own house in order. Specifically, learning theory must define the contents of learning, in the form of expectancies and representations, and specify how a given expectancy or representation gives rise to some observable variable.

In view of the many varieties of learning theory, it is difficult to provide a coherent picture of unresolved problems in so short a space (for systematic critiques of the major theorists, consult (4, 5, 19, 61, 68)). For the purposes of the present discussion, I will review briefly four venerable problems and try to indicate briefly their relevance to biological issues (see (68) for a related discussion).

What is Learned?

One unfortunate influence of the philosophical movements of operationism and logical empiricism is the acceptance by many psychologists of the view that the question, "what is learned?", is a pseudo-question. Learning was equated with performance in the sense that what an animal does in a conditioning experiment is presumed to provide an exhaustive account of what is learned (39). This point of view, which is inherent in Thorndike's and Pavlov's S-R models of learning, seems reasonable since performance of some kind is the only known way in which learned behavior manifests itself. The problem, however, is that even when such adjunct principles as stimulus and response generalization are invoked to account for performance in novel situations, performance typically underestimates what an animal has learned during training[11].

Recent research using a variety of paradigms has shown that the question, "what is learned?", can admit to interesting and meaningful answers. For example, Rizley and Rescorla (73), using the sensory preconditioning paradigm introduced by Brogden (7), have shown that pairing S_2 and S_1

---

[11]This problem was by no means unique to studies of animal learning. To a large extent, recent advances in human cognitive psychology and in psycholinguistics can be traced to the explicit recognition of the principle that performance underestimates knowledge. In order to remedy that problem, various types of cognitive structures have been suggested as descriptions of different kinds of knowledge. Because of the verbal nature of such representations, they are not directly applicable to animal learning. That difference, however, does not diminish the importance of asking about the nature of representations that an animal learns by virtue of a particular conditioning paradigm (cf. (102)).

before the US is paired with S_1 results in an association between S_2 and S_1, even though S_2 was never paired with the US and no observable response was detected (or expected) when S_2 was first paired with S_1.

The second-order conditioning paradigm, first used by Pavlov, provides another means of determining what an organism encodes about the US during first-order conditioning. Under this paradigm (as discussed in detail by Rozeboom (77)), S_1 is paired with the US, after which S_2 is paired with S_1. The associative strength of S_1 is then changed, for example, by the operation of extinction. Rashotte et al. and Rescorla have presented evidence that the S_2-S_1 association can be maintained even after the S_1-US association has been extinguished (64, 66)[12]. These studies illustrate that, in a variety of instances of Pavlovian conditioning, the contents of learning can be meaningfully specified as an S-S association which is not directly observable. The validity of including such representations in theories of learned behavior (as well as the necessity of so doing) has been amply demonstrated by a wide variety of experiments on representation in animals (74, 102).

The most unambiguous evidence comes from studies which show that no immediately present stimulus can account for the occurrence of some learned response - an assay for the existence of animal representations first suggested by Hunter (35). Once it is demonstrated that representations of absent stimuli exist, it is but a small step to consider the role of representations of all types of stimuli (cf. (18), p. 186).

Research on the nature of representations in animals has only recently begun. Accordingly, little is known about their properties and how they interact with other psychological processes. For the moment, however, it is important to recognize that representations of critical features

---

[12]These results differ from earlier attempts to demonstrate autonomy between the S_1 and S_2 association and the S_1-US association during second-order conditioning (e.g., (1, 11)). In those instances, it appears as if the response evoked by S_2 was the basis of the association between S_1 and S_2. Rescorla (67) has attributed the contradictory results obtained in experiments using these paradigms to the similarity between S_2 and S_1 (which serves as a basis of encoding the relationship between S_2 and S_1). Specifically, Rescorla hypothesized that "the organism might be encouraged to learn about the stimulus properties of S_1 to the degree that they are similar to S_2" (68).

of a conditioning paradigm provide an important answer to the question, "what is learned?", and that, as distinct from the state of affairs that existed when this question was first raised, it does admit to important answers.

The Conditions for Learning

Aside from specifying what is learned, a theory of learning must also specify the necessary and sufficient conditions for learning to occur. In the case of classical conditioning, experiments by Kamin on blocking (38) and by Rescorla (65) on the distinction between contiguity and contingency with respect to CS-US pairings demonstrated that CS-US contiguity is not sufficient to establish a CR to the CS. The Rescorla-Wagner model of associative learning provides an elegant integration of these and other aspects of classical conditioning (69, 109). This and other more recent models of classical conditioning (e.g., (24, 37)) have sought to characterize the formation of associations as a process in which the CS creates an expectancy that is confirmed by the delivery of the US. On this view, signals which do not create differential expectancies do not become associated with the US.

In the case of instrumental conditioning, the necessity of R-S* contiguity has been questioned by experiments such as those performed by Lett (48) in which an operant discrimination was trained despite the fact that several minutes elapsed between the correct response and the delivery of the reward. The delivery of the reward occurred in a start box at the beginning of the trial that followed the correct choice. This rules out the role of secondary reinforcement as a means of filling the gap between the response and the reinforcing stimulus. The effectiveness of such a procedure can, however, be explained if one postulates that placement in the start box evokes a representation of the correct response that occurred at the end of the preceding trial.

While the role of expectancies and their confirmation have received relatively little attention in the case of instrumental conditioning, the available evidence suggests that these factors may play an important role. For example, in a conditional discrimination, a procedure which provides different reinforcers for each response results in higher acquisition scores than one which provides the same reinforcer for each response (63). In sum, as Tolman speculated some fifty years ago (108), a case can be made that a necessary condition for learning is the confirmation of an S-S* or an R-S* expectancy. A major challenge for learning theory is to characterize the properties of such expectancies.

Learning and Performance

Earlier, we saw that the need for distinguishing between learning and performance was motivated by the fact that, even when allowing for stimulus- and response-generalization, an animal's behavior on $task_1$ offers only a limited view of what it might do on $task_2$. For example, pairing S_1 and S_2 in a sensory preconditioning procedure produces an association that does not manifest itself in a measure of performance until a subsequent phase of the experiment. Consider a more complicated example in which pigeons were trained to respond to four simultaneously presented colors (A,B,C,D) in the sequence, A->B->C->D, irrespective of the physical position of these colors (98). One can ask whether the pigeon learned anything about the relationships between nonadjacent elements (e.g., B and D) or about the ordinal position of a particular element (e.g., B). Recent studies have provided affirmative answers to both questions (102).

In both sensory preconditioning and serial learning experiments, the subject's performance at one stage of training revealed nothing about important knowledge that it had acquired at an earlier stage of training. Yet, that knowledge undoubtedly influenced performance prior to the point at which it was demonstrated. It is that consideration which makes it especially important to distinguish between learning and performance. At present there are no formal theories which show how the contents of learning are translated into performance. Instead, most investigators observe changes in some aspect of a particular response and use those observations as a measure of what the animal has learned.

One important question that will be difficult to answer until a theory of performance and its relation to the contents of learning is developed, is whether the various types of biological constraints that we have discussed have to do with the questions, "what can an animal learn?", "under what conditions will an animal show what it has learned?", or both. For example, it is still not clear in cue-to-consequence studies whether some associations are easier to establish than others, whether there are special conditions that cause an animal to act on certain associations, or whether both factors are operative (cf. (51)).

Learning theorists must also contend with the problem of relating various measures of performance and specifying why different measures of performance fail to agree. This problem is especially apparent in the case of classical conditioning. There is much consensus that, contrary to Pavlov's view, the topography of the CR cannot be predicted from the topography of the UR (e.g., (10, 27, 33, 86, 101, 110, 113)). It is also clear that the CR that is measured is typically but one aspect of a much more comprehensive CR. Thus the CR tells us that conditioning has taken place, but it explains neither the nature of the association that has been formed nor the performance that is observed. Hollis' "prefiguring" hypothesis, which holds that the CS evokes behavior that "... optimize[s] interaction with the forthcoming biologically important event (US)" ((34), p. 3; also see Hollis, this volume) provides an interesting functionalist account of the relationship between the CR and the UR.

How Many Kinds of Learning Are There?

I have restricted discussion thus far to the two most widely discussed forms of learning: classical and instrumental conditioning. In both instances, the contents of learning have been used to define the type of learning. But, even in the case of these well studied categories of learning, it is not clear that reference to S-S* and R-S* associations provides an adequate account of the contents of learning. In instrumental conditioning, for example, there is much evidence that discriminative stimuli play an important role in defining occasions on which the R-S* contingency is in effect (87, 100). It is not clear, however, just how the role of the discriminative stimulus should be represented in the contents of learning and whether situations in which the status of an R-S* contingency is determined by the presence (or absence) of a discriminative stimulus calls for designating a different kind of learning.

The problem of how to specify the contents of learning in the familiar examples of classical and instrumental conditioning is much better defined than it is in the case of examples of learning that come from the ethological literature. Consider, for example, such well documented phenomena as imprinting in ducklings and song learning in the white-crowned sparrow. Both phenomena qualify as examples of learning in the sense that permanent changes in behavior result from particular experiences and that those changes cannot be explained on the basis of maturation per se (see Bateson, Marler, and Immelmann, all this volume). Despite our detailed knowledge of the conditions under which these phenomena occur and the fact that, in both instances, learning must occur during a critical period, it is unclear whether the familiar terms of learning theory apply, e.g., the CS, the CR, S-S* and R-S* contingencies, and so on.

Superficially, there is reason to believe that these kinds of learning differ qualitatively from the more traditional types. It seems doubtful, for example, whether concepts of "expectancy" apply in the case of song learning or imprinting[13]. While there is much reason to regard imprinting and song learning as instances of "closed" programs of learning, referring to them as such simply adds to the urgency of defining the characteristics of such learning and relating them to what is known about the more well understood examples of learning that result from "open" programs. Thus, far from negating the importance of learning theory, examples of learning from the ethological literature add to the importance of broadening learning theory so that it will be able to encompass these and other nontraditional instances of learning[14].

CONCLUSION

A century or so after comparative psychologists began to investigate the nature of instinctive behavior and of animal cognition, we finally have available concepts and techniques for studying both phenomena. Further progress awaits a clearer specification than has been available to date of the contents of learning and its relationship to performance, both in the case of "closed" and "open" types of learning. While biological factors will undoubtedly play a role in a comprehensive theory of the relationship between learning and performance, it is important to keep in mind whether a particular biological factor influences learning, performance, or both.

---

[13]It is interesting to speculate, however, that a more primitive kind of expectancy is relevant to such phenomena. For example, in the case of song learning, the concept of a "template" (cf. (40)) suggests that there may be a neural mechanism that "expects" a particular song. Aside from reference to a physiological level of explanation, the difference between this kind of expectancy and that implied by an S-S* or R-S* association is that, in the latter case, the expectancy is based upon two events. That is to say, song learning may involve a form of nonassociative expectancy.

[14]See Domjan and Galef (16) for a thoughtful discussion of the impact (and lack thereof) of biological constraints on association theory. This review, which appeared after the present paper was completed, notes in particular the failure of individuals who have cited biological constraints as a problem for general learning theory to provide an alternate theoretical point of view.

Acknowledgements. I thank P. Balsam, J. Gibbon, P. Marler, and R. Silver for their helpful comments on an earlier draft. The preparation of this manuscript was supported in part by Grant BNS-82-02423 from the National Science Foundation and by fellowships from All Souls College, Oxford, England, and the Fulbright Commission.

REFERENCES

(1) Amiro, T.W., and Bitterman, M.E. 1980. Second-order appetitive conditioning in goldfish. J. Exp. Psychol.: Anim. Behav. Proc. 6: 41-48.

(2) Beach, F.A. 1950. The snark was a boojum. Am. Psychol. 5: 115-124.

(3) Bolles, R.C. 1972. Reinforcement, expectancy, and learning. Psychol. Rev. 79: 394-409

(4) Bolles, R.C. 1979. Learning Theory. New York: Holt, Rinehart and Winston.

(5) Bower, G.H., and Hilgard, E.R. 1981. Theories of Learning. Englewood Cliffs, NJ: Prentice-Hall.

(6) Bridgman, P.W. 1927. The Logic of Modern Physics. New York: Macmillan.

(7) Brogden, W.J. 1939. Sensory pre-conditioning. J. Exp. Psychol. 25: 323-332.

(8) Brown, P.L., and Jenkins, H.M. 1968. Auto-shaping of the pigeon's key peck. J. Exp. Anal. Behav. 11: 1-8.

(9) Bush, R.R., and Mosteller, F. 1955. Stochastic Models for Learning. New York: Wiley.

(10) Bykov, K.M. 1959. The Cerebral Cortex and the Internal Organs. Moscow: Foreign Languages Publishing House.

(11) Cheatle, M.D., and Rudy, J.W. 1978. Anaylsis of second-order odor-aversion conditioning in neonatal rats: Implications for Kamin's blocking effect. J. Exp. Psychol.: Anim. Behav. Proc. 4: 237-249.

(12) Chomsky, N. 1957. Syntactic Structures. The Hague: Mouton.

(13) Darwin, C. 1871. The Descent of Man. London: John Murray.

(14) Darwin, C. 1872. The Expression of the Emotions in Man and Animals. London: John Murray.

(15) Delius, J.D. 1974. The ontogeny of behavior. In The Neurosciences: Third Study Program, eds. F.O. Schmitt and F.G. Worden. Cambridge, MA: MIT Press.

(16) Domjan, M., and Galef, B.G. 1983. Biological constraints on instrumental and classical conditioning: Retrospect and perspective. Anim. Learn. Behav. 11: 151-161.

(17) Estes, W.K. 1950. Toward a statistical theory of learning. Psychol. Rev. 57: 94-107.

(18) Estes, W.K. 1969. New perspectives on some old issues in associative theory. In Fundamental Issues in Associative Learning, eds. N.J. Mackintosh and W.K. Honig. Halifax: Dalhousie University Press.

(19) Estes, W.K.; Koch, S.; MacCorquodale, K.; Meehl, P.E.; Mueller, C.G., Jr.; Schoenfeld, W.N.; and Verplanck, W.S., eds. 1945. Modern Learning Theory. New York: Appleton-Century-Crofts.

(20) Ferster, C.B., and Skinner, B.F. 1957. Schedules of Reinforcement. New York: Appleton.

(21) Gamzu, E., and Schwartz, B. 1973. The maintenance of keypecking by stimulus contingent and response-independent food presentation. J. Exp. Anal. Behav. 19: 65-72.

(22) Garcia, J.; Ervin, F.R.; and Koelling, R.A. 1966. Learning with prolonged delay of reinforcement. Psychonom. Sci. 5: 121-122.

(23) Garcia, J., and Koelling, R.A. 1966. Relation of cue to consequence in avoidance learning. Psychonom. Sci. 4: 123-124.

(24) Gibbon, J., and Balsam, P. 1981. Spreading association in time. In Autoshaping and Conditioning Theory, eds. C.M. Locurto, H.S. Terrace, and J. Gibbon. New York: Academic Press.

(25) Gottlieb, G. 1979. Comparative psychology and ethology. In The First Century of Experimental Psychology, ed. E. Hearst. Hillsdale, NJ: Erlbaum.

(26) Gould, J.L. 1982. Ethology. New York: Norton.

(27) Hearst, E. 1975. The classical-instrumental distinction: Reflexes, voluntary behavior, and categories of associative learning. In Handbook of Learning and Cognitive Processes: Conditioning and Behavior Theory, ed. W.K. Estes, vol. 2. Hillsdale, NJ: Erlbaum.

(28) Hearst, E., and Jenkins, H.M. 1974. Sign-Tracking: The Stimulus-Reinforcer Relation and Directed Action. Austin, TX: The Psychonomic Society.

(29) Hebb, D.O. 1949. The Organization of Behavior. New York: Wiley.

(30) Heinroth, O. 1911. Beiträge zur Biologie, namentlich Ethologie und Psychologie der Anatiden. Proceedings of the IV. International Ornithologic Congress, Berlin 1910, vol. 5, pp. 589-702.

(31) Hinde, D.E., and Stevenson-Hinde, J., eds. 1973. Constraints on Learning. London: Academic Press.

(32) Hodos, W., and Campbell, C.B.G. 1969. Scala naturae: Why there is no theory in comparative psychology. Psychol. Rev. 76: 337-350.

(33) Holland, P.C. 1979. Conditioned stimulus as a determinant of the form of the Pavlovian conditioned response. J. Exp. Psychol.: Anim. Behav. Proc. 3: 77-104.

(34) Hollis, K.L. 1982. Pavlovian conditioning of signal-centered action patterns and autonomic behavior: A biological analysis of function. Adv. Stud. Behav. 12: 1-64.

(35) Hunter, W.S. 1913. The delayed reaction in animals. Behav. Monog. 2: 6.

(36) Jenkins, H.M. 1977. Sensitivity of different response systems to stimulus-reinforcer and response-reinforcer relations. In Operant-Pavlovian Interactions, eds. H. Davis and H.M.B. Hurwitz. Hillsdale, NJ: Erlbaum.

(37) Jenkins, H.M.; Barnes, R.A.; and Barrera, F.J. 1981. Why autoshaping depends on trial spacing. In Autoshaping and Conditioning Theory, eds. C.M. Locurto, H.S. Terrace, and J. Gibbon. New York: Academic Press.

(38) Kamin, L.J. 1969. Predictability, surprise, attention and conditioning. In Punishment and Aversive Behavior, eds. B.A. Campbell and R.M. Church. New York: Appleton-Century-Crofts.

(39) Kendler, H.H. 1952. "What is learned?" - A theoretical blind alley. Psychol. Rev. 59: 269-277.

(40) Konishi, M. 1965. The role of auditory feedback in the control of vocalization in the white-crowned sparrow. Z. Tierpsych. 22: 770-783.

(41) Krebs, J.R. 1978. Optimal foraging: Decision rules for predators. In Behavioral Ecology: An Evolutionary Approach, eds. J.R. Krebs and N.B. Davies. Sunderland, MA: Sinauer.

(42) Krebs, J.R.; Erichsen, T.J.; Webber, M.I.; and Charnov, E.L. 1977. Optimal prey selection in the great tit (Parus major). Anim. Behav. 25: 30-38.

(43) Kuo, Z.Y. 1921. Giving up instincts in psychology. J. Philos. 18: 645-664.

(44) Lashley, K.S. 1942. An examination of the "continuity theory" as applied to discriminative learning. J. Gen. Psych. 26: 241-265.

(45) Lashley, K.S. 1951. The problem of serial order in behavior. In Cerebral Mechanisms in Behavior, ed. L.A. Jeffries. New York: Wiley.

(46) Lashley, K.S., and Wade, M. 1946. The Pavlovian theory of generalization. Psychol. Rev. 53: 72-87.

(47) Lehrman, D.S. 1953. A critique of Konrad Lorenz's theory of instinctive behavior. Q. Rev. Biol. 28: 337-363.

(48) Lett, B.T. 1975. Long delay learning in the T-maze. Learn. Motiv. 6: 80-90.

(49) Locurto, C.M.; Terrace, H.S.; and Gibbon, J., eds. 1981. Autoshaping and Conditioning Theory. New York: Academic Press.

(50) Loeb, J. 1908. Forced Movements. Philadelphia: Lippincott.

(51) LoLordo, V.M. 1979. Selective associations. In Mechanisms of Learning and Motivation, eds. A. Dickinson and R.A. Boakes. Hillsdale, NJ: Erlbaum.

(52) Lorenz, K.Z. 1950. The comparative method in studying innate behavior patterns. Sympos. Soc. Exp. Biol. 4: 221-268.

(53) Mackintosh, N.J. 1974. The Psychology of Animal Learning. New York: Academic Press.

(54) Marler, P. 1975. On strategies of behavioural development. In Function and Evolution in Behaviour, eds. G. Baerends, C. Beer, and A. Manning. Oxford: Clarendon Press.

(55) Mayr, E. 1974. Behavior programs and evolutionary strategies. Am. Sci. 62: 650-659.

(56) McDougall, W. 1908. An Introduction to Social Psychology. London: Methuen.

(57) Miller, N.E. 1967. Laws of learning relevant to its biological basis. Proc. Am. Philos. Soc. 111: 315-325.

(58) Morgan, C.L. 1896. Habit and Instinct. London: Edward Arnold.

(59) Morgan, C.L. 1899. Introduction to Comparative Psychology. London: Walter Scott.

(60) Morgan, C.L. 1930. The Animal Mind. London: Edward Arnold.

(61) Osgood, C.E. 1953. Method and Theory in Experimental Psychology. New York: Oxford University Press.

(62) Pavlov, I.P. 1927. Conditioned Reflexes. Oxford: Oxford University Press.

(63) Peterson, G.B., and Trapold, M.A. 1980. Effects of altering outcome expectancies on pigeons' delayed conditional discrimination performance. Learn. Motiv. 11: 267-288.

(64) Rashotte, M.E.; Griffin, R.W.; and Sisk, C.L. 1977. Second-order conditioning of the pigeon's key-peck. Anim. Learn. Behav. 5: 25-38.

(65) Rescorla, R.A. 1968. Probability of shock in the presence and absence of CS in fear conditioning. J. Comp. Physiol. Psychol. 66: 1-5.

(66) Rescorla, R.A. 1979. Aspects of the reinforcer learned in second-order Pavlovian conditioning. J. Exp. Psychol.: Anim. Behav. Proc. 5: 79-95.

(67) Rescorla, R.A. 1980. Second-Order Conditioning. Hillsdale, NJ: Erlbaum.

(68) Rescorla, R.A., and Holland, P.C. 1982. Behavior studies of associative learning in animals. In Annual Review of Psychology, eds. M.R. Rosenzweig and L.W. Porter. Palo Alto, CA: Annual Reviews.

(69) Rescorla, R.A., and Wagner, A.R. 1972. A theory of Pavlovian conditioning: Variations in the effectiveness of reinforcement and nonreinforcement. In Classical Conditioning: II. Current Research and Theory, eds. A. Black and W.F. Prokasy. New York: Appleton-Century-Crofts.

(70) Revusky, S.H. 1968. Aversion to sucrose produced by contingent X-irradiation-temporal and dosage parameters. J. Comp. Physiol. Psychol. 65: 17-22.

(71) Revusky, S.H. 1977. Learning as a general process with an emphasis on data from feeding experiments. In Food Aversion Learning, eds. N.W. Milgram, L. Kramer, and T.M. Alloway. New York: Plenum.

(72) Revusky, S.H., and Parker, L.A. 1976. Aversions to unflavored water and cup drinking produced by delayed sickness. J. Exp. Psychol.: Anim. Behav. Proc. 2: 342-353.

(73) Rizley, R., and Rescorla, R.A. 1972. Associations in second-order conditioning and sensory preconditioning. J. Comp. Physiol. Psychol. 81: 1-11.

(74) Roitblat, H.L. 1982. The meaning of representation in animal memory. Behav. Brain Sci. 5: 353-406.

(75) Romanes, G.J. 1884. Mental Evolution in Animals. New York: Appleton.

(76) Romanes, G.J. 1889. Mental Evolution in Man. New York: Appleton.

(77) Rozeboom, W.W. 1958. "What is learned?" An empirical enigma. Psychol. Rev. 65: 22-33.

(78) Rozin, P., and Kalat, J.W. 1972. Learning as a situation-specific adaptation. In Biological Boundaries of Learning, eds. M.E.P. Seligman and J.L. Hager. New York: Appleton-Century-Crofts.

(79) Schneirla, T.C. 1946. Contemporary american animal psychology in perspective. In Twentieth Century Psychology, eds. P.L. Harriman, G.L. Freeman, G.W. Hartmann, K. Lewis, A.H. Maslow, and C.E. Skinner. New York: Philosophical Library.

(80) Schwartz, B. 1974. On going back to nature: A review of Seligman and Hager's Biological Boundaries of Learning. J. Exp. Anal. Behav. 21: 183-198.

(81) Seligman, M.E.P. 1970. On the generality of laws of learning. Psychol. Rev. 77: 406-418.

(82) Seligman, M.E.P., and Hager, J.L., eds. 1972. The Biological Boundaries of Learning. New York: Appleton-Century-Crofts.

(83) Shettleworth, S.J. 1972. Constraints on learning. In Advances in the Study of Behavior, eds. D.S. Lehrman, R.A. Hinde, and E. Shaw. New York: Academic Press.

(84) Shettleworth, S.J. 1975. Reinforcement and the organization of behavior in golden hamsters: Hunger, environment, and food reinforcement. J. Exp. Psychol.: Anim. Behav. Proc. 1: 56-87.

(85) Shettleworth, S.J. 1978. Reinforcement and the organization of behavior in golden hamsters: Punishment of three action patterns. Learn. Motiv. 9: 99-123.

(86) Siegel, R.K. 1979. Natural animal addictions: An ethological perspective. In Psychopathology in Animals: Research and Clinical Implications, ed. J.D. Keene. New York: Academic Press.

(87) Skinner, B.F. 1938. The Behavior of Organisms: An Experimental Analysis. New York: Appleton-Century-Crofts.

(88) Skinner, B.F. 1945. The operational analysis of psychological terms. Psychol. Rev. 52: 270-277.

(89) Skinner, B.F. 1948. Superstition in the pigeon. J. Exp. Psychol. 38: 168-172.

(90) Skinner, B.F. 1966. The phylogeny and ontogeny of behavior. Science 153: 1205-1213.

(91) Small, W.S. 1899. An experimental study of the mental processes of the rat: I. Am. J. Psychol. 11: 133-164.

(92) Small, W.S. 1900. An experimental study of the mental processes of the rat: II. Am. J. Psychol. 12: 206-239.

(93) Spalding, D.A. 1873. Instinct: With original observations on young animals. MacMillan's Magazine 27: 282-293. (Reprinted in Br. J. Anim. Behav. 2: 1-11, 1954).

(94) Spear, N.E. 1978. The Processing of Memories: Forgetting and Retention. Hillsdale, NJ: Erlbaum.

(95) Spence, K.W. 1956. Behavior Theory and Conditioning. New Haven: Yale University Press.

(96) Spencer, H. 1886. The Principles of Psychology, 3rd ed. New York: Appleton.

(97) Staddon, J.E.R., and Simmelhag, V.L. 1971. The "superstition" experiment: A reexamination of its implications for the principles of adaptive behavior. Psychol. Rev. 7: 3-43.

(98) Straub, R.O.; Seidenberg, M.S.; Bever, T.G.; and Terrace, H.S. 1979. Serial learning in the pigeon. J. Exp. Anal. Behav. 32: 137-148.

(99) Teitelbaum, P. 1966. The use of operant methods in the assessment and control of motivational states. In Operant Behavior: Areas of Research and Application, ed. W. Honig. New York: Appleton-Century-Crofts.

(100) Terrace, H.S. 1966. Stimulus control. In Operant Behavior Areas of Research and Application, ed. W.K. Honig. Englewood Cliffs, NJ: Prentice-Hall.

(101) Terrace, H.S. 1973. Classical conditioning. In The Study of Behavior, ed. J.A. Nevin. New York: Scott, Foreman.

(102) Terrace, H.S. 1984. Animal Cognition. In Animal Cognition, eds. H.L. Roitblat, T.G. Bever, and H.S. Terrace. Hillsdale, NJ: Erlbaum.

(103) Thorndike, E.L. 1898. Animal intelligence: An experimental study of the associative processes in animals. Psychol. Monog. 2: (whole No. 8).

(104) Thorndike, E.L. 1911. Animal Intelligence. New York: Macmillan.

(105) Thorndike, E.L. 1932. The Fundamentals of Learning. New York: Teachers College.

(106) Tinbergen, N. 1942. An objectivistic study of the innate behavior of animals. Bibl. Biotheo. 1: 39-98.

(107) Tinbergen, N. 1951. The Study of Instinct. Oxford: Clarendon Press.

(108) Tolman, E.C. 1932. Purposive Behavior in Animals and Men. New York: Appleton-Century.

(109) Wagner, A.R., and Rescorla, R.A. 1972. Inhibition in Pavlovian conditioning: Application of a theory. In Inhibition and Learning, eds. R.A. Boakes and M.S. Holliday. New York: Academic Press.

(110) Wasserman, E.A. 1973. Pavlovian conditioning with heat reinforcement produces stimulus-directed pecking in chicks. Science 181: 875-877.

(111) Williams, D.R. 1981. Biconditional behavior: Conditioning without constraint. In Autoshaping and Conditioning Theory, eds. C.M. Locurto, H.S. Terrace, and J. Gibbon. New York: Academic Press.

(112) Willner, J.A. 1978. Blocking of a taste aversion by prior pairings of exteroceptive stimuli with illness. Learn. Motiv. 9: 125-140.

(113) Zener, K. 1937. The significance of behavior accompanying salivary secretion for theories of the conditioned response. Am. J. Psychol. 50: 384-403.

The Biology of Learning, eds. P. Marler and H.S. Terrace, pp. 47-74. Dahlem Konferenzen 1984. Berlin, Heidelberg, New York, Tokyo: Springer-Verlag.

Ethology and the Natural History of Learning

J.L. Gould* and P. Marler**
*Dept. of Biology, Princeton University
Princeton, NJ 08544, USA
**Rockefeller University Field Research Center
Millbrook, NY 12545, USA

Abstract. In the past there has been a tendency for many ethologists to dismiss laboratory studies of learning as unnatural and irrelevant, while many students of animal learning have seen little relevance in the ethological work on the innate bases of behavior. We argue and, in a preliminary way, attempt to demonstrate that a selective synthesis of these two disciplines offers a potentially powerful perspective on learning and suggests comprehensive and testable hypotheses about the mechanisms, organization, and evolution of learning in animals under natural conditions.

INTRODUCTION

The purpose of this paper is fourfold. First, we wish to introduce and define some essential bits of ethological terminology. Second, we want to consider learning from an ethological perspective - that is, the phenomenon of learning in the natural world and its role in the lives of animals. Third, we will try to interpret field behavior in terms of laboratory work on learning. Finally, we hope to suggest some possible avenues toward an interdisciplinary synthesis.

Perhaps the best place to begin is by defining ethology as the study of animal behavior in relation to the animal's natural surroundings - the animal in its world, as Tinbergen put it. Traditionally this has directed attention toward behavioral and physiological specializations peculiar to each species, specializations which are a consequence of the unique challenges and requirements of each species' niche. The title of Tinbergen's classic book, The Study of Instinct, emphasizes his concern with species-specific behavioral specializations.

EARLY ETHOLOGICAL CONCEPTS

Cue Recognition

From the early work of Lorenz and Tinbergen came the four general concepts (and associated jargon) which still provide a conceptual framework for many ethologists. The first is the idea of innate recognition by animals of important objects or individuals in the environment. Ethologists have tirelessly catalogued an almost endless list of such "releasers." The egg-retrieval behavior of ground-nesting birds such as the Greylag goose is a cogent example (36). The sight of an egg outside the nest captures the attention of the incubating parent and results in the goose's standing, extending its neck toward and over the egg, and then gently rolling it back with the bottom of the bill. The recognition of the egg is innate and highly schematic: geese will readily retrieve batteries, beer bottles, and baseballs. The basic, schematic feature or features of the stimulus object which effect recognition are usually called "sign stimuli."

Although there was a time when many ethologists thought complex patterns could be genetically encoded in minute detail and act as sign stimuli, our growing understanding of the mechanics of sensory processing makes it more reasonable to believe that most sign stimuli are the relatively simple sorts of cues for which ensembles of visual, acoustic, and olfactory feature-detector cells in the CNS are known to code (23). This is particularly obvious in the case of behavioral specialists such as toads and bats (20), where even the hypercomplex feature detectors serve relatively clear-cut functions. The existence of similar detectors in generalists such as cats as opposed to a simple dot-by-dot representation of the retina on the visual cortex argues that here, too, they serve an essential function, the logic of which provides much food for thought. Perhaps they could be used in the manner of olfactory coding such that the pattern of output of a variety of high-order feature detectors with differing "personalities" can uniquely define any object, independent of its apparent size or even (with a little clever wiring) its angle, perhaps even preserving lateral relationships. Thus we could form an enormous library of subconscious, automatic recognition cells. (But this cannot easily account for our introspective sense of the world as a high-resolution television picture.) It is frequently the case that several relatively simple cues are effective simultaneously (thereby making the response more specific), but the various relevant feature detectors, whether simple or complex, most often appear to act independently and additively, a phenomenon known as "heterogeneous summation." When a baby herring gull, for instance, pecks at its parent's bill, it is reacting independently to a vertical bar moving horizontally (the parent's bill) and a moving high-contrast spot (the red spot on the parent's yellow bill) (26, 54). Since both stimuli correspond to well-known classes of feature detectors, even to the extent of displaying optimum responses to particular rates of angular motion, it seems reasonable to entertain the idea that the circuitry may be as simple as it looks.

The consequence of the sign-stimulus strategy is that animals can be prewired to respond selectively to important and predictable objects, individuals, and events with a reasonable degree of certainty, at least so long as they are not transported into environments (or laboratory situations) which are "unnatural" to the species.

Effector Organization

The second of the concepts crucial to ethology's exploration of behavior, and a natural outgrowth of the idea of sign stimuli, was the fixed-action pattern. The goose's egg-retrieval behavior is again a good example. Most of this behavior is highly stereotyped and, once triggered, proceeds to completion even if the egg is removed. Other parts depend on specific proprioceptive feedback to tune the behavior. The sight of a goose delicately rolling the ghost of an egg which had been removed as the bird was extending its neck makes it clear that this highly coordinated piece of motor activity is a prewired behavioral unit. These units are now known as "motor programs," a name which emphasizes their neural basis while removing the implication that the units need to be innate or immune to sensory feedback (23). Indeed, the ability to wire up novel behavioral units with all the characteristics of innate motor programs plays a very large role in the behavior of many animals. Many songbirds, for example, must have auditory feedback to learn a new song, but once their song "crystallizes," they can be deafened without effect (37): although they can no longer hear themselves sing, they produce normal songs (Marler, this volume). Similarly, since the song is produced by two bilaterally separate sets of muscles in the syrinx, when the neural connections to one set are cut, the other side continues to sing its part (and only its part) of the duet just as before (43).

Not only can novel behavior be made virtually automatic in this way, but learning can serve to put several innate motor programs together into the proper order. For instance, a good case has been made that though male rhesus monkeys reared in isolation are equipped innately with all of the critical components for copulation behavior - the sign stimulus recognition circuits and the various motor programs - and "know" that the various components belong to this behavioral situation, they assemble them in an inappropriate way (38). The idea that there can be an innately directed trial-and-error ordering of motor program subroutines may be applied to a wide variety of behaviors, such as the maturation of food-burying behavior in squirrels (19) and song learning in birds (Marler, this volume).

The consequence of the motor-program strategy is that animals can be prewired to perform certain sorts of predictable behavior correctly, or, after learning to perform a motor task, make the task a more or less automatic behavioral unit.

The concept of behavioral units has helped explain many kinds of complex but largely innate feats. Research on animals as diverse as orb-weaving spiders, digger wasps, and nest-building birds (e.g., (12)) appears to point to a hierarchical strategy in which one unit or series of units is cycled through repeatedly until some criterion is reached, whereupon another unit is triggered. An Australian digger wasp, for example, builds a funnel to protect its nest from parasites in just this way ((50); also discussed by Bolles, this volume). The wasp begins by constructing a mud cylinder perpendicular to the ground, whether the substrate it is building on is horizontal, oblique, or vertical. Once she has begun, however, the angle of the substrate or tunnel can be changed but the wasp will continue to build a straight tube. The wasp works on the cylinder until it is 3 cm long and then begins the curving neck. If an experimenter repeatedly buries the bottom of the cylinder so the 3 cm criterion is never met, the wasp can be made to build indefinitely, but once the 3 cm criterion is reached and work on the curved neck begins, burying has no effect. The wasp continues the neck until the plane of its opening is roughly 20° from the horizontal. Experimental manipulation of the angle of the tube can determine how much neck must be constructed before the 20° criterion is reached, but once the wasp begins the next stage of the task, further changes have no effect. And so it goes for the construction of the flange and then the bell of the funnel, each unit an independent step with its own set of releasers and motor programs which allow the animal to perform this very complex task.

A similar hierarchical arrangement appears to underlie the processing of information in at least certain types of complex navigation behavior. In these cases, the various "units" are arranged hierarchically as alternatives, with a primary system backed up by one or more secondary systems. Foraging honey bees, for instance, choose the sun over other directional cues if it is available, but automatically resort to polarized light when the sun is not available (16, 22). Many birds use the sun first, but the earth's magnetic field otherwise (reviewed in (24)). (It is only fair to note, however, that some ethologists (e.g., Bateson, this volume) reject the subunit-interaction approach, and instead liken behavior to a "cake" whose individual constituents are inseparable.)

Terminology

We should point out at this stage that our use of the terms sign stimulus and motor program is purposely broad. On the motor side, we think that to divide behavioral gestures into motor programs and reflexes is artificial - to the extent that they can be reliably distinguished at all. The distinction is merely quantitative. Moreover, the endogenously coordinated performances of simple motor programs doubtless evolved from simple reflexive gestures. On the sensory side, we use sign stimuli when the stimulus, however crude, differentially triggers a specific response while other perceptible stimuli do not. From this crude but organized relationship between stimulus and response, more precise sensory filtering has evolved to optimize the specificity of sign stimuli, just as it has added associative learning at the sensory end and operant motor learning at the effector end. This thorny but important issue is reviewed elsewhere (25). We hope that the reader will bear with our attempt to stress evolutionary links at the occasional expense of everyday usage.

Endogenous Control

Endogenously generated motivation, often referred to as "drive," is of course not the exclusive province of ethology. The term is, in fact, used in many different senses. Motivation, as we intend it here, refers only to endogenous, innately prearranged behavioral switchings.

If we take for granted for the moment that much of behavior, particularly in lower animals, is triggered and directed by sign stimuli and accomplished by motor programs, it becomes clear that some behaviors are performed only during certain periods of an animal's life. Egg rolling, for instance, is a behavioral unit which appears about a week before incubation begins and lasts until about a week after hatching. The drive or "motivation" to respond to the sign stimulus for eggs in this way appears to be absent at other times. It is easy to think of behavioral units of this sort as subroutines whose availability is controlled endogenously by day length, hormone level, social signals, or what have you. Hence, the long-term responsiveness to particular cues is shifted in a predictable and adaptive way as an animal's needs change on the basis of innately recognized "priming stimuli" (e.g., (8, 27, 31)).

Attempts to understand the mechanisms underlying most shorter-term changes in motivation, even at this qualitative level, have not been very successful, though the work that has been done on hunger and thirst is an obvious exception. Unfortunately, motivation is extremely difficult to keep sorted out: recent performance of a behavior, deprivation, displacement activity, the "quality" of the sign stimuli, and the motivation associated with other "needs," all affect it in some tangential but significant way. One of the more promising approaches to the study of motivation envisions currently available behavioral alternatives as competing for (or "time sharing") an animal's "attention" on the basis of the levels of drive associated with each (28, 39). There is a preordained system of priorities, so that painful and/or injurious stimuli, for example, always tend to take precedence; but in addition, behavioral priorities can be adjusted to balance the urgency of an animal's needs against the quality of the opportunities available. When honey bees, for instance, evaluate the "quality" of food sources (34), recruit bees take into account the needs of the colony for pollen (protein for brood rearing) vs. nectar (carbohydrates to fuel adults and to store for the winter). Their readiness to follow dances to these two sources varies with the internal condition of the hive. So, too, recruits weight the sugar concentration of the nectar being advertised against the distance being indicated: more dilute nectar must be closer, to allow the calories gathered per unit time to balance out. Finally, the bee's preference for more concentrated nectar can reverse itself as the hive temperature increases. When coolant is needed, nectar with a low sugar content, which can be spread out and fanned to provide evaporative cooling for the hive, becomes more desirable.

Selective Learning

"Imprinting" is another concept preserved from early ethology. We propose to consider it here as a subset of a more general class of "selective learning" processes. The general characteristics of most selective learning include its frequent temporal specificity (the "sensitive phase"), the innate triggering of learning by specific sign stimuli (innate learning triggers, cf. Marler, this volume), its context-dependent cue specificity such that sometimes only particular, frequently multi-modal features of a stimulus will be easily remembered, its frequent resistance to reversal (especially in imprinting), and again particularly in imprinting its usual lack of obvious external reward.

The adaptive role of selective learning seems clear enough. It tells a naive animal surrounded by a world full of potentially learnable stimuli what it should remember, and what to do with the information subsequently. Baby herring gulls, as we saw, are born knowing to peck at a horizontally moving vertical bar and at a moving red spot - both sign stimuli (ILTs) which add independently. They also display innate context-dependent cue biases in memorizing the details of other particular features of this stimulus complex such as the head, beak shape, and so on, if the stimulus feeds them. The chick will not learn the characteristics of a wholly inappropriate stimulus even when fed by it (Margolis and Gould, in preparation). This particular example of parental imprinting is unusual, since food reinforcement is rarely involved in selective learning, but it demonstrates that different sign stimuli can be involved at different stages in the learning sequence.

The result for gull chicks in the wild is that the reliable but schematic, innately recognized stimuli are replaced eventually by a detailed, visual, and/or acoustic picture of the parents which is then used for individual as well as species recognition. In short, information too detailed, or insufficiently generalized across the environments encountered by the species, or otherwise too unpredictable to be prewired can be specifically acquired and used.

This is not to say that animals necessarily memorize and use complex pictures in all instances. Indeed, Lashley (32), studying form discrimination by rats on his jumping stand apparatus, pointed out that "with very complex, irregular figures, the basis of discrimination is a part figure; the response is to some limited cue and the remainder of the figure is ignored." Different rats focussed on different parts of the figures, choosing some simple but sufficient dichotomy between the choices. Lea (this volume) reports a similar phenomenon in pigeons. The "part figures" are very like sign stimuli, and it is tempting to wonder if, when possible, learning employs such basic units as simple or complex feature detectors to make essential discriminations. Certainly this inductive abstraction of a reliable distinction or correlation makes sense in the context of associative learning, to be discussed below. We suppose that when the distinctions to be made are more subtle, more complex configurational details must be remembered than can be accounted for by a simple collection of independently active feature detectors, and learning performance is often improved. As a result, a multi-modal set or even something like a true mental picture must be involved. Examples of photographic-like selective learning are virtually endless and include some aspects of bee learning (Gould, this volume), song learning (Marler, this volume), parental, sibling, and sexual imprinting (Immelmann, this volume), enemy learning by birds (15), and food-avoidance learning (Sahley and Revusky, both this volume). The probability of a relationship between selective learning and the predictability and elaborateness of what must be known by an animal to survive and reproduce provides a conceptual basis for examining selective learning in general.

"Genetic Constraints"

Before looking at the likely need for learning in the wild, a point should be made about the frequently heard phrase "genetic constraints on learning," which may be misleading. It derives from the overwhelming commitment of most psychologists to general process learning theory, a concept that is now outdated (Jenkins, this volume). The implication of the term, it seems to us, is that animals would be smart if their genes did not constrain their general ability to learn and thereby make them selectively stupid. If one is inclined (as many ethologists are) to think of selective learning as involving specialized behavioral units analogous to motor programs and navigational information-processing subroutines, then evolution would be expected to act to create specific learning abilities where needed rather than specifically editing out the general ability to learn everything (or even anything) from specific contexts. Consider the pattern of egg and offspring learning in birds (review in (29); see also (5, 10, 45, 46)). Most birds do not learn to recognize their eggs, even though they see them many times a day, frequently turn them, and otherwise care for them. Species which do memorize the number and/or appearance of their eggs (and, given the opportunity, can do so in remarkable detail) are either subject to brood parasites or nest so close to other birds that the possibility of a mixup is realistic. We could suppose that originally birds learned everything about the world around them, including what their eggs looked like, but that evolution acted to "constrain" the species for whom this learning was not essential. But for evolution to weed out a behavior with such thoroughness, the behavior would probably have to be actively maladaptive. It is not easy to see how the battery-rolling goose would be a less fit parent if it were able to remember what its eggs looked like. This point is underscored by the subset of parasitized host species that do not imprint on their eggs. It seems more reasonable to suppose that they have, for a variety of possible reasons, not evolved this learning program, than that having it, they lost the ability through some sort of "constraining" process.

Nor does it seem to be the case that the species which memorize egg appearance are just somehow generally smarter than others. Herring gulls, for instance, who seem unable to learn even the color of their eggs, do learn to recognize their young individually within three days. This is just the time at which the chicks begin to wander about and the possibility of confusing one nest's chicks with another's becomes a potentially serious problem. Birds which nest in trees generally do not learn what their young look like until the offspring begin to fledge - again, just the time when the potential for confusion begins. Species such as the cliff-nesting kittiwakes who face neither brood parasites, the possibility of foreign chicks wandering into the nest, nor the need to feed their fledged young, never learn to recognize their offspring at all (14). Again, it is not easy to see that there has been any adaptive edge to insouciance which could easily explain these "constraints."

The counter-intuitive logic of this "constraint" position is perhaps more tellingly obvious when we look at our own species. Have our genes constrained us from memorizing the locations of hundreds of buried nuts (Shettleworth, this volume), or from readily learning to compensate for the sun's movement, or from easily distinguishing all the eggs in an oriole's nest, and so on? The idea that evolution has built selective learning routines where necessary seems more consistent with both the field data and our present understanding of how natural selection operates.

LEARNING IN THE FIELD

Food Acquisition

From this perspective on selective learning, we can make some predictions about the likely need for learning (as opposed to preordained "hardwiring") as a function of the behavioral context and an animal's niche or life history. (Shettleworth (this volume) used the same logic and generates a very similar list.) Consider foraging, for example. Animals may make their livings by being specialists perfectly adapted for harvesting one kind of food, or by being generalists, able to gather or capture a wide range of food, or by adopting a strategy falling anywhere along the continuum in between. Some basic prewiring of behavior might seem adaptive for a specialist, with its constant, narrow food preferences. The very predictability from generation to generation of its food's characteristic appearance, defense, preferred habitat, and other behaviors (if it is animate), should make it a prime candidate for some sort of encoding. We would expect specialists, then, as a rule to be less willing to learn and/or able to learn less about food sources than generalists (1, 4). Digger wasps, for instance, are born able to recognize, capture, and paralyze particular prey, with one species specialized for honey bees, another for crickets, and so on. Bees, on the other hand, though innately able to recognize the sign stimulus for flowers, come programmed to memorize the odor, color, and shape (among other things) of each specific species they forage, as well as learning the most efficient way in which to harvest each blossom.

Although the generalist strategy provides more flexibility for an animal to adapt to a new environment or switch between food types as conditions change, a generalist must apparently sacrifice to some extent the benefit of highly efficient morphological and behavioral specializations. Even a generalist must be able to recognize the general class of objects and individuals in its cluttered environment which might qualify as suitable food (as bees recognize flowers as a class), but then it must experiment with each. Experimentation is, of necessity, time consuming and sometimes dangerous.

Specialized learning sequences which help solve these problems are well-known. Some animals, for instance, imprint on salient aspects of the food they are fed pre- or postnatally (9). Others learn from social experience with their parents or other members of their group (7). The broad predispositions of animals of other species are focussed and refined by purely personal experiences, based on some innate (presumably chemical) recognition (before and/or after tasting) that a particular food item is acceptable. The well-known phenomenon of food-avoidance learning (Revusky, this volume) probably represents a specific safeguard for such essential experimentation.

Reproduction

In addition to finding food, an animal must be able to locate, identify, court, induce reproductive readiness, and mate with the best available member of the opposite sex of its species. In this context all species can be considered as specialists, and examples of prewired recognition and motor behavior abound (23). But in many cases the specialization is so extreme (and crucial) that complete reliance on innate cues seems to have been insufficient. To the extent that sign stimuli really are functionally equivalent to the cues abstracted by ensembles of feature detectors (rather than elaborate innate "photographs"), they may be inadequate to distinguish reliably between morphologically similar species or intraspecific variants. The role of sexual imprinting (including song learning) is probably to use the sign stimuli available from parents and/or siblings early in life, when social experience is adequately restricted, to direct the formation of a more precise image (Immelmann and Marler, both this volume). Obviously this is only possible if the species practices some amount of parental care or other suitable sort of social organization. (Of course, one could also argue that the phenomenon of sexual imprinting may create the problem of morphological similarity which it appears to solve.) The recent discovery of the role of sexual imprinting in the "tuning" of inbreeding in some species (Bateson, this volume) illustrates another way in which this learning process can be exploited (2, 3).

Parental Care

Caring for an animal's units of fitness would be a dangerous context for learning and experimentation - except, of course, for learning to recognize offspring when that knowledge is useful. Critical aspects of the parenting process can include choosing of a nest site, recognizing suitable construction materials, building the nest, recognizing the young, feeding, thermoregulating, providing toilet care, training them in food and predator recognition, and otherwise caring for them, and so on. Although there are several established cases of nest-site imprinting, and although in some cases building and parental care behavior (particularly in primates) improves with practice, most of the rest of parenting appears to be relatively immune to learning in most species. Even in primates, it is difficult to assess the role of learning accurately: the behavior of parent monkeys reared in isolation is grossly atypical, and it is difficult to separate this from the pharmacopoeia of abnormal behavior traits they display. Nevertheless, many primatologists believe that the parenting play that young monkeys of many species engage in with the infants of others does contribute to the proper development of skillful parenting when they mature.

Defense

Escape and defensive behaviors are another important facet of an animal's repertoire. Animals need to recognize danger and react appropriately to the particular threat. There is evidence that many animals react to the paired, forward-looking eyes typical of predators (21), recognize species-specific dangers such as coral snakes or starfish (29, 51), or possess a general suspicion of unfamiliar animals, but enemy recognition in general seems surprisingly dependent on experience. Perhaps danger, like a generalist's food supply, comes in too many unpredictable or morphologically complex forms to be adequately catalogued genetically in most cases.

Curio's discovery that the innately recognized mobbing call triggers what one might call species-avoidance learning may provide a general model for the selective learning about dangerous situations (15). Curio and his colleagues used a rotating, 4-compartment box arranged so that birds in cages on either side of the box could see only the compartment facing them, and not the compartment visible to birds on the other side. When one cage of birds was shown a stuffed owl in the rotating box, its occupants began directing the characteristic mobbing call at the boxes, and the birds in the other cage attempted to mob the object which they could see. In most cases what the second group of birds was shown was something as innocuous as a stuffed honey guide (which they had previously ignored), yet they produced the mobbing call on each subsequent exposure. This aversion could then be passed to other naive birds, and from them to others, all without any bird ever having been attacked by a honey guide. The automatic nature of this kind of adaptive learning subroutine is most tellingly illustrated by the success with which Curio transferred the mobbing behavior to an empty milk bottle. It appears to conform to the type of innately triggered learning that occurs in song acquisition and probably in imprinting (Marler, this volume).

It is particularly interesting that animals can learn to modulate and modify their normal escape or attack responses to suit the particular stimulus. Birds and prairie dogs (33), for instance, appear to differentiate (both by their alarm calls and their subsequent behavior) at least two general classes of predators, while vervet monkeys differentiate at least three (and probably four) (49). Different species within any one danger category may be further discriminated, as seems clear in the way herds of antelope judge how close to permit various carnivores to approach. Young vervets appear to recognize the classes of potential threat by the age of one month, perhaps innately, but only in a very general way; they learn from social experience - that is, the behavior of older vervets - which members of the class really do constitute threats. A young vervet will deliver the aerial predator alarm call for storks and even falling leaves, but slowly learns to restrict it to a particular species of eagle, perhaps tutored by adult alarm calling as are Curio's mobbing birds. Obviously, animals must strike a balance between forever escaping and hiding to be safe, and pursuing their many other essential, risk-incurring activities such as foraging, courting, rearing young, and so on. Similarly, many animals clearly learn where safety is to be found as well as how to modify their escape behavior to match the threat. It may be that learned predator recognition and escape is the optimal answer in the face of relatively unpredictable, context-specific risks.

Navigation

The point that animals can learn where to find safety should remind us that members of many species from bees to primates learn the location of their home and the geography of their range. The nature and organization of this learning is not well understood, although it is clear that pigeons normally imprint on their loft location shortly after fledging (22, 23), that honey bees map their flight range while foraging (Gould, this volume), and that many birds can remember the location of vast numbers of food hoards (13). And in navigating to a goal, birds and bees, at least, must learn about the location and movement of the sun or stars as a function of the time of day, in a sort of navigational "calibration" that appears to be a variant of imprinting (18, 35).

Social Behavior, Social Communication, and "Culture"

All animals must understand the social signals of their species and, in many cases, recognize other individuals and their status. The production and decoding of most social signals appear to be at least initiated innately in ontogeny, but the facial, postural, and vocal signals seen in many birds and mammals are so strikingly intricate that either the recognition wiring must have become unusually complex, or only the dynamic components of the gesture must be attended to, or a kind of social imprinting must be taking place. This ambiguity arises because most sign-stimulus recognition systems are demonstrably crude (e.g., recognition of conspecific adults and infants), and at least the primordia for the recognition of facial expression and body posture appear to be innate (47). However, it seems likely that adult animals are sensitive to the complexities of facial expressions as well as to their simpler features. Innate recognition of such complexities is probably not feasible, but there may well be some highly correlated visual components or associated auditory cue which would innately trigger detailed memorization of these gestures and expressions. In short, there may be more learning in this context than meets the eye, and in any case this is a subject ripe for experimentation.

An indisputable function of learning in many social animals is individual or group recognition, often accompanied by memorization of various relationships such as dominance orders, alliances, pairings, and so on. Many social animals provide extended care for their young, giving them an opportunity for play and social learning. Play allows animals a safe opportunity to practice behavior which will later be crucial, and in many species play marks the beginning of the future dominance hierarchy. Social learning includes most obviously cultural practices in foraging, such as fishing for termites in chimpanzees and harvesting mussels by oyster catchers, although there is considerable controversy over the degree to which observational learning plays a major role outside of the higher primates. Toothpick (40) and rock use (6) are fairly convincing examples of social learning in chimpanzees, and other, less well documented cases such as operating drinking fountains are well-known (7). We must be careful here, however, since most if not all of these behaviors involve innate motor elements. Captive-reared chimpanzees, for instance, spontaneously poke sticks in holes as though searching for termites. Some supposedly learned social behaviors may even depend on sign stimuli: the famous blue tits who learned to open milk bottles to get at the cream harvest food in the wild by peeling bark from trees, and the stimulus and motor program for the bottle-opening behavior are probably very similar. Indeed, even hand-reared tits spontaneously peel wallpaper from the walls of their foster homes. Bolles (this volume) goes so far as to wonder if any motor learning is truly novel.

In short, though it is tempting to assume that social learning is entirely free from innate aid and guidance, we suspect that a strict application of such a distinction might ultimately leave this useful category altogether empty of examples. Human language acquisition, for instance, would certainly no longer qualify as a totally learned behavior ((22), and Gleitman, this volume).

ETHOLOGY OF LABORATORY LEARNING STUDIES

Several general functional categories of learning have traditionally been distinguished, such as habituation sensitization, associative learning (classical conditioning), trial-and-error learning (operant conditioning), and higher-order learning which we have called cognitive trial and error.

Although these distinctions are frequently attacked, they provide a useful organization for comparing ethological and behavioristic observations. As biologists we are interested in underlying mechanisms rather than in superficial "laws" which beg both physiological and evolutionary questions. In an attempt to move from words to wiring, our discussion will focus on quite a different level of analysis than that of, for example, Terrace (this volume).

Nonassociative Learning

Habituation, of course, is the stimulus-specific waning of an individual behavioral response as a result of repeated stimulation, as in the well studied case of Aplysia, whose gill-withdrawal response (motor program) becomes evermore difficult to elicit after it has been repeatedly triggered by a jet of water (a crude sign stimulus) ((28), and Quinn, this volume). The decline of responsiveness is not for the most part a result of sensory adaptation (which we can think of as sensory fatigue, and which is brief and usually confined to the sensory cell itself). Although the Aplysia circuit is probably atypical - the sensory neurons send processes directly to the motor neurons so that habituation is not entirely mediated centrally by interneurons - there seems little doubt that the phenomenon is essentially the same as that seen in other animals (see Quinn, this volume). The role of this very simple sort of learning in the lives of animals seems clear: the threshold for triggering individual units of behavior adjusts to the time-averaged "background level" of the relevant stimuli. Hence, the gill of an Aplysia in calm water is very sensitive to mechanical stimulation, but that of an Aplysia in rough water is sufficiently less sensitive so that it is not continually "withdrawn" - only something really out of the ordinary will trigger this defensive behavior. This behavior-specific threshold-tuning function is crucial and helps animals avoid (what we have labeled in the context of wasp hunting and building) behavioral do-loops (23, 30) by becoming adaptively "bored."

Sensitization, however, serves to reduce or eliminate habituation when a novel stimulus is encountered, and even to heighten an unhabituated animal's responsiveness. Hence, an Aplysia with a habituated gill-withdrawal response will, after being poked in the tail or startled with a light, begin to respond to a previously ignored water jet. This sensitization, though pathway-specific in many cases, can serve as a relatively generalized alerting mechanism. The various curiosities of its time course and so on are probably inevitable (but not necessarily optimal) consequences of its wiring and biochemistry (see Quinn, this volume). Mechanistically, habituation and sensitization appear to be combined to create associative learning in at least Aplysia (again, see Quinn and Menzel et. al., both this volume).

Associative Learning

Associative learning, as first described by Pavlov, involves taking a stimulus-response "reflex" and presenting a novel training stimulus (the CS) immediately before the normal stimulus (the US). In time the response (UR) can be elicited by the novel stimulus alone. And as has been shown repeatedly, the CS does not need to be perfectly correlated with the US: as long as it has substantial predictive value, it will be learned (see Jenkins and Terrace, both this volume). Almost the first person to put Pavlov's discovery to use was von Frisch, who used associative learning to demonstrate that fish can hear and honey bees have color vision.

Because of the work of von Frisch, Tinbergen, and other early researchers, ethologists have little trouble in restating associative learning in ethological terms. The UR "reflex" is a motor program; the innately recognized US is a sign stimulus; and the learned CS is, as Hollis (this volume) agrees, a search image. The adaptive value of this prewired form of inductive reasoning seems clear: relatively crude and schematic but diagnostic sign stimuli are replaced through associative learning with what is potentially a far more detailed picture of the appropriate stimulus (although, as we pointed out earlier, something very like a sign stimulus may still dominate some learned reponses in which a minimum of discrimination is needed). The sign stimulus defines the context, while the inductive programming of the learning routine abstracts from a world full of stimuli the ones which best correlate with and predict the context. As Hollis (this volume) points out, the CS in some cases also allows animals to anticipate the arrival of an individual of interest, as would be the case when an animal heard or smelled a predator. (There are, inevitably, some difficulties in this relatively simple interpretation, but most seem to arise from the frequent intrusion of trial-and-error learning which we discuss below.)

Viewed from this perspective, most of the selective learning studied by ethologists is probably simple or slightly elaborated (e.g., two-step) associative learning. Enemy learning in birds, for instance, depends on the mobbing call (the US or triggering sign stimulus) to define the context; virtually all forms of imprinting are directed by sign stimuli (such as the species-typical exodus call in ducks which leads the ducklings from the nest and initiates imprinting); flower learning in bees is triggered by the taste of sugar water; and so on. The frequent restriction of learning to sensitive phases, and the facilitated recall of learned information to particular contexts or times of day (the Kamin effect) probably represent adaptive, context-specific programming. The same is clearly the case for food-avoidance learning (Revusky, this volume).

In this light, the widespread existence of contextual triggering and cue biases in associative learning seems of a piece with the general pattern of selective learning: the danger-avoidance system (which in pigeons, for example, is prewired to attend more to sounds than to visual cues) can be independent of the food-learning system (which in pigeons focusses instead on visual information to the near exclusion of auditory cues) (22). In most cases of associative learning we should probably expect to find an innate recognition of behavioral contextual triggers so that the animal "knows" when learning is appropriate. Moreover, we probably ought to expect to find (where it would be useful) some degree of cue-specific biases in the learning which are correlated with the situation and the species' associated behavioral repertoire, although the generality of many of the basic features of learning argues for the optimistic view that we are looking at many variations on a fairly small set of themes.

Trial-and-error Learning

Trial-and-error learning is generally distinguished from associative learning in that the animal must perform behavioral experiments by which it modifies (i.e., shapes) its behavior into the most adaptive (i.e., rewarding) form. This kind of self-imposed modification obviously depends on environmental feedback <u>and</u> an animal's ability to recognize when a new behavioral variant, whether intentional or not, is more or less successful at obtaining the particular goal. (It is important to keep in mind that the goal need not be food or even an overt positive reward: avoiding danger or punishment, or obtaining internalized rewards can be an adaptive goal in many contexts.) Trial-and-error learning, then, is in many ways analogous to deductive reasoning and is a very common strategy among animals. Bees, for instance, must solve the problem posed by the distinctive morphology of each new species of flower, improving slowly with every visit to a particular type of blossom (Gould, this volume). Indeed, it is almost certainly the dramatic increase in foraging efficiency which results from trial-and-error learning which accounts for the strong tendency of individual bees to specialize on one or a few species of flower (Heinrich, this volume).

In the case of bees (and, we suspect, most other animals), the behavioral solution frequently becomes a highly efficient, relatively stereotyped, and almost automatic motor program (Bolles, this volume). Even rats running mazes can become so used to a particular route that they will run into newly erected barriers or stop after having run the usual distance down an alley to a food dish even though the dish has been plainly moved a few steps further along (48). As mentioned earlier, the adaptive value of this automation of learned but routine behavior is most likely that it frees the attention of the animal for other concerns - a finch, having solved the problem of how to handle and open sunflower seeds, can devote its attention to watching for predators rather than to the intricacies of cracking open seeds and separating the heart from the husk. Introspective evidence suggests that human illustrations of automatization of learned motor patterns are commonplace, and both Squire (this volume) and Mishkin (41) note that in monkeys and humans procedural/habitual tasks such as shoe tying (which, once learned, becomes unconscious) and declarative/associative tasks are anatomically segregated.

In both the bee and the bird example, and many others besides, we note that associative and trial-and-error learning can work in concert to provide first recognition, and then response. The response, obviously, is specific to the particular contingency - the kind of seed - as well as to the general context - feeding vs. fleeing or courting. Hence, particular associative and trial-and-error "subroutines" will often be tightly linked, as we will mention again shortly. And just as in the case of associative learning, any anomalies may represent species-specific, context-dependent programming.

Anomalies and Biases

Anomalies in both associative and trial-and-error learning seem to us to fall into several well studied categories (review in (48, 53), and see Jenkins, Hearst, Revusky, and Bolles, all in this volume) such as the problems with omission schedules and the like. One can reasonably suppose that animals have associative and response biases, and the economically fairly logical behavior on variable schedules as opposed to that on fixed schedules suggests that the behavior may be pre-tuned to the more natural "probabilistic" structure of the real world. Indeed, we would predict that the response pattern of relatively unselective herbivores like rabbits to a food reward might be less probabilistically adapted.

Other examples of response biases are to be seen in the familiar phenomena of autoshaping in which animals perform spontaneous, unrewarded but contextually logical behavior (Hearst, this volume), the reward-specific associative biases seen in pigeon pecking (Jenkins, this volume), learned helplessness, adjunctive behavior, and in differential conditionability (48). The latter category includes the observation that pigeons, for instance, learn more readily to peck than to barpress for food but can be trained to barpress to avoid danger more easily than to peck. Rats, on the other hand, seem able to learn barpressing for food rather quickly, but are difficult to instruct when the goal is to avoid danger. Running away, on the other hand, is easily taught. It is difficult to avoid the suspicion that there are context-specific motor biases - that rats, who manipulate food with the forepaws, and pigeons, who handle food with the beak, in some sense "expect" to learn with these tools in the context of foraging (i.e., food-rewarded tasks). Similarly, the apparently preferred modes of escape behavior seem at least roughly related to the realities of each species' niche.

Each of these anomalies hints at an innate motor bias, a "default routine" (to use programming jargon) which may serve to guide an animal's initial motor experimentation. An example from word processing is apropos: most programs take for granted that you want, say, a 12-space margin, double spacing, page numbering at the top right, and so on. These are the default values, parameters which are assumed in the absence of information to the contrary. Indeed, it is tempting to place a default value interpretation on the associative biases of animals. Although bees, for instance, can learn that a flower is any color from yellow to ultraviolet, it learns the color of purple flowers far more quickly than any other color of flowers (Gould, this volume). At the same time, bees prefer purple silhouettes to all other colors on a spontaneous preference test. It is as though purple is the default parameter - a probabilistic bias which helps guide bees when they experiment with various flowers while searching for food. A similar pattern is seen in shape and odor learning as well.

We suspect, then, that innate, context-specific behavioral predispositions may play a role in certain stages of trial-and-error learning by providing animals with a degree of adaptive (but by no means absolute) guidance. Songbirds all seem to know to experiment with their syrinxes rather than their wings or feet in order to shape their singing behavior to match the song on which they have imprinted (although even here there is a default program - a crude "innate song" motor program - available; see Marler, this volume). It is also interesting to note that many cases of quite complex motor learning appear to involve the liberal use of innate behavioral elements. The elaborate shell-harvesting behavior of oyster catchers, learned from the parents over the course of several months (and of which there are two very different forms), is composed almost entirely of innate behavioral gestures which are reordered, coordinated, and directed on the basis of learning (42). Such a strategy, which involves building up complex motor behavior out of a "library" of innate elements, has obvious advantages for certain tasks. The possibility that at least some other cases of trial-and-error learning exploit this sort of organization seems worthy of serious consideration. For example, certain songbirds "invent" new song themes by recombining innate song elements (Marler, this volume). It may also be the case that the subset of easily reorganized elements is context-specific (as we suggested could be the case with response biases), so that here too there could be the sort of specific prewired links between associative and trial-and-error learning to which we alluded earlier. The apparently learned ordering of burying behavior by squirrels (19) and copulation by rhesus macaques (38) mentioned earlier appears to fall into this pattern, since the motor elements are for the most part unique to the behavioral situation.

Another interesting example was described by Thorpe (53). He gave birds the problem of retrieving food hung on a string inside a glass tube. It is easy for birds with innate beak/foot coordinations to solve this problem by pulling up hanks of string with the beak and catching them in a foot, and so reaching the food. Great tits, for instance, solve the problem quickly and all in the same way. Canaries, on the other hand, lack useful innate motor program elements and so slowly solve the problem by creating novel behavior. Each individual solves the problem in a slightly different way.

MEMORY AND NEUROSELECTION

Before considering higher-order learning (cognitive trial and error), we wish to speculate briefly on the possible neural organization of selective learning. The widespread occurrence of cue biases in associative learning combined with its frequent innate recognition/response organization suggests to us that in at least some cases both the learning and the storage could be prewired. (This possibility is explored in a preliminary way for flower learning in bees by Gould, this volume, and for song learning by Marler, this volume.) For example, if the feature detectors for the relevant sign stimulus were wired to activate memory cells or, when spatial or temporal relationships need to be preserved, to an array of memory cells, and if the cues to which the particular associative learning routine attended were similarly wired to the cells or array, the repeated temporal pairing of the predictive cues with the sign stimulus could be designed to result in the memorization of the new stimuli. This would be consistent with the wiring arrangement inferred for associative learning in Aplysia (Quinn, this volume).

The idea that learning and storage in at least certain predictable associative contexts might be prewired has been suggested in somewhat different forms by Changeux (11) and Edelman (17) (who calls it "neuronal group selection" by analogy with work on clonal selection in immunology). All the possible "right answers" (the range of possible conditioning stimuli) would need to be (more or less weakly) wired in advance to some sort of executive cell population (equivalent to the "innate releasing mechanism" of classical ethology) which, by its strong input from particular feature detectors, would be responsible for triggering the behavioral response. If such were the case, it is tempting to imagine that such cases of associative learning could be thought of as functionally equivalent to alpha-conditioning. (Alpha-conditioning, simply defined, is associative learning in which there is a preexisting but weak overt response to a conditioning cue. On the neural level it must mean that a potentially learnable cue is prewired more or less strongly into the learning/response circuit. Neuronal selection would require here that such prewiring could exist without the overt behavioral response characteristics of alpha-conditioning.) Such prewiring would be the basis of the conditioning biases and (perhaps) the spontaneous preferences animals display, such that the relative strengths of the various biases might correspond to the relative strength of the preexisting connections. The same general organization can readily be imagined for trial-and-error learning in the cases for which motor biases exist or innate motor elements are assembled to create complex behavior, and moreover, for the likely connections between the associative and trial-and-error routines specific to particular contexts. This perspective leads to any number of comparative predictions.

The greatest challenge for the neuroselectional viewpoint is interpretation of the formation, storage, and recall of search images, whether visual, olfactory, or auditory. Even bees seem to use learned associative search images (Gould, this volume), and it seems likely that a neuronal selection strategy would mean preordaining the resolution and usable cues and setting aside a finite number of image arrays for each task. There is fairly good evidence that the eidetic images formed and used by the Hymenoptera are restricted in these ways, and a cross-correlation search strategy might be used to find the best matches between incoming sensory experience and stored search images (Gould, this volume); bats almost certainly use cross correlation to sort through their noisy acoustic world for the return time of their echolocation sounds. Whether these memory storage and search strategies are used in other animals or in other contexts is an important question.

COGNITIVE TRIAL AND ERROR

Although habituation, sensitization, associative learning, and trial-and-error learning seem to account for most of what is seen in the laboratory and field, some behavior calls for other interpretations. We find it both convenient and useful to distinguish these cases of what is often called higher-order learning by the term "cognitive trial and error." The most familiar manifestations of this phenomenon, exemplified by the rats in Olton's radial mazes, are usually interpreted to imply that animals have the ability to use "cognitive maps" - neural constructs by which animals in some sense "think" about the problem, evaluating behavioral alternatives, or formulating a "plan" of sorts (44). The rat may be thought of as solving the maze problem by mapping its previous behavior and then referring to (and subsequently modifying) this map in the process of making later behavioral choices so as not to visit the same area of the maze twice. Cognitive maps may underlie the abilities of animals even down to the level of honey bees (Gould, this volume) which learn their home range and navigate in it as though they had an internalized map. Other sorts of problem-solving behavior in which overt motor experimentation plays a relatively minor role also seem to fall into the same general category. Animals sometimes appear to evaluate behavioral possibilities internally, as though neural "pictures" of objects could be manipulated mentally, and potential consequences imagined.

One difficulty with cognitive trial-and-error is that it resists easy experimental manipulation: the mind is a relatively private organ, and much high-level processing goes on subconsciously anyway. Another problem is that even conscious, cognitive trial-and-error is the major mental process of which each of us is introspectively aware. As Griffin (25) has pointed out, this may raise the thorny question of animal consciousness. We can rather easily imagine a prewired two-dimensional array for home range information in the brain of the honey bee, and the appropriate programming for filling it in on the basis of foraging experience. Indeed, if the range of potential inputs to the hypothetical grid points on such a map are sufficiently predictable, it might even be possible to imagine a variant of neuroselection which could account for much of the observed behavior of animals. And we do not doubt that suitable programming to enable bees to use landmarks or return home from more or less familiar spots on the basis of such a map would be relatively easy to design.

None of this would, it seems to us, require consciousness, though awareness might on the one hand facilitate it, or, on the other, create behavior which, though as automatic as egg-rolling or stinging, might be easily mistaken for consciousness. The adaptive role of such maps for both long- and short-distance spatial navigators like bees and rats is too obvious to require comment, and the potential advantage of prewiring the predictable aspects of acquisition, transformation, storage, recall, and use of map information is also clear. Whether map abilities are so organized is another question, well worth further exploration.

The analogous ability (obvious at least in our species) to learn by conjuring up and manipulating mental representations is more difficult to study in animals (Lea, this volume). Still more difficult is determining whether such abilities might depend on particular sorts of specialized, preordained wiring. It is tempting to imagine that genetic predispositions and specific neural organization rather than some general, nebulous factor such as "increased size" or "greater complexity" is responsible, and that the many inter- and intrahemispheric processing specializations known or suspected in our species are consonant with the specialization perspective. Piaget's idea that the sort of cognitive abilities we are describing develop in humans in a stereotyped, apparently innately directed way, but require building on information already acquired and formed into "concepts," should remind us that the mechanistic basis of cognitive trial and error in at least our species is probably <u>very</u> complex. Even here, however, there is evidence that some sort of elaborate preordained organization is at work recognizing speech elements (Peterson and Jusczyk, this volume) and generating language (Gleitman, this volume). One evolutionary perspective on human language would see specialized, "dedicated" memory arrays as an essential preadaptation for language. The program for loading, manipulating, reorganizing, and accessing these arrays could be imagined as evolving from the programming of other more modest cognitive abilities.

CONCLUSION

In this paper we have tried to examine laboratory learning studies from an ethological point of view and to restate field observations in terms of the principles uncovered by laboratory work. We have argued that to see learning as for the most part consisting of specialized, dedicated, but well integrated subroutines, based on a small number of general learning strategies which have been "customized" as appropriate for each context and each species, is particularly helpful when considering learning under naturalistic conditions and some apparent laboratory anomalies. We believe that learning theorists made a fundamental mistake when they gave up their attempt to explain all behavior in terms of classical and operant conditioning in favor of a search for a General Theory or Laws from which all learning behavior, conditioning included, is to be explained. We think that both the biological perspective and the evidence argue that associative and trial-and-error learning, along with habituation and sensitization, are the building blocks of learning rather than the consequences. By analogy with the Columbia model discussed above in which sensitization and habituation are wired to generate associative learning, we believe that a relatively straightforward consideration of how natural selection has wired up these various components to create more complex, species-typical behavior is likely to be more fruitful than a search for a General Theory. The latter is no more guaranteed of success than is a general theory of the body in the style of Plato and Aristotle necessarily more productive than the physiological approach of first studying the subunits and then how they interact. For example, imprinting looks very like a two-step, innately triggered associative learning sequence, so that a well informed building-block approach immediately suggests tests not apparent either to ethologists or learning theorists. With this overly theoretical predisposition aside, we believe that a synergistic collaboration of ethology and psychology should be undertaken to understand the mechanistic basis of learning and memory - whether "cognitive" or "noncognitive" (associative and trial-and-error) - with a view to comprehending the evolution of the strategies of specialized learning which play such a crucial role in the adaptive, well-tuned behavior of animals in the natural world.

Acknowledgements. We thank C. Bicker, T.S. Carew, C.G. Gould, W.G. Quinn, and J.E.R. Staddon for especially helpful criticisms. Supported by NSF Grants BNS 82-01004 and 81-18769 and NIH Grant MH 14651.

REFERENCES

(1) Bailey, M.B. 1984. Intelligence and the ecological niche. In Animal Intelligence, ed. R.J. Hoage. Washington, D.C.: Smithsonian Press, in press.

(2) Bateson, P. 1978. Sexual imprinting and optimal outbreeding. Nature 273: 659-660.

(3) Bateson, P. 1982. Preferences for cousins in Japanese quail. Nature 295: 236-237.

(4) Beck, B.B. 1984. Tools and intelligence. In Animal Intelligence, ed. R.J. Hoage. Washington, D.C.: Smithsonian Press, in press.

(5) Birkhead, R.T. 1978. Behavioral adaptations to high-density nesting in the common guillemot. Anim. Behav. 26: 321-331.

(6) Boesch, C., and Boesch, H. 1983. Optimisation of nutcracking with natural hammers by wild chimpanzees. Behaviour 83: 265-286.

(7) Bonner, J.T. 1980. The Evolution of Culture in Animals. Princeton: Princeton University Press.

(8) Brzoska, J., and Obert, H.-J. 1980. Acoustic signals influencing hormone production of the testes in the grass frog. J. Comp. Physiol. 140: 25-29.

(9) Burghardt, G.M., and Hess, E.H. 1966. Food imprinting in the snapping turtle. Science 151: 108-109.

(10) Burtt, E.H. 1977. Some factors in the timing of parent-chick recognition in swallows. Anim. Behav. 25: 231-239.

(11) Changeux, J.-P., and Danchin, A. 1976. Selective stabilization of developing synapses as a mechanism for the specification of neuronal networks. Nature 264: 705-712.

(12) Collias, N.E., and Collias, E.C. 1962. An experimental study of the mechanisms of nest building in a weaver bird. Auk 79: 568-595.

(13) Cowie, R.J.; Krebs, J.R.; and Sherry, D.F. 1981. Food storing by marsh tits. Anim. Behav. 29: 1252-1259.

(14) Cullen, E. 1957. Adaptations in the kittiwake to cliff-nesting. Ibis 99: 275-302.

(15) Curio, E., and Vieth, W. 1978. Cultural transmission of enemy recognition. Science 202: 899-901.

(16) Dyer, F.C., and Gould, J.L. 1983. Honey bee navigation. Am. Sci. 71: 587-597.

(17) Edelman, G., and Mountcastle, V. 1978. The Mindful Brain: Cortical Organization and the Group-Selective Theory of Higher Brain Function. Cambridge, MA: MIT Press.

(18) Emlen, S. 1975. The stellar-orientation system of migratory birds. Sci. Am. 233 (2): 102-111.

(19) Ewer, R.F. 1965. Food burying in the African ground squirrel. Z. Tierpsych. 22: 321-327.

(20) Ewert, J.-P. 1981. Neuro-ethology. Berlin: Springer-Verlag.

(21) Forse, D.D., and LoLordo, V.M. 1973. Attention in the pigeon: differential effects of food-getting vs. shock-avoidance procedures. J. Comp. Physiol. Psychol. 88: 551-558.

(22) Gallup, G.G. 1977. Tonic immobility: the role of fear and predation. Psychol. Rec. 27: 41-61.

(23) Gould, J.L. 1982. Ethology: The Mechanisms and Evolution of Behavior. New York: Nortons.

(24) Gould, J.L. 1982. The map sense of pigeons. Nature 296: 205-211.

(25) Griffin, D.R. 1981. The Question of Animal Awareness, revised ed. New York: Rockefeller University Press.

(26) Hailman, J. 1967. The ontogeny of an instinct. Behav. Suppl. 15: 1-159.

(27) Harding, C.F., and Follett, B.K. 1979. Hormone changes triggered by aggression in a natural population of blackbirds. Science 203: 918-920.

(28) Heiligenberg, W. 1974. Processes governing behavioral states of readiness. Adv. Study Behav. 5: 173-200.

(29) Kandel, E.R. 1976. Cellular Basis of Behavior. San Francisco: Freeman.

(30) Keeton, W.T.; Gould, J.L.; and Gould, C.G. 1984. Biological Sciences, 4th ed. New York: Nortons, in press.

(31) Klinghammer, E., and Hess, E.H. 1964. Parental feeding in ring doves: innate or learned? Z. Tierpsych. 21: 338-347.

(32) Lashley, K.S. 1938. Conditional reactions in the rat. J. Psychol. 6: 311-324.

(33) Leger, D.W.; Owings, D.H.; and Gelfand, D.L. 1980. Single-note vocalizations of California ground squirrels: graded signals and situation specificity of predator and socially evoked calls. Z. Tierpsych. 52: 227-246.

(34) Lindauer, M. 1954. Temperaturregulierung und Wasserhaushalt im Bienenstaat. Z. vgl. Physiol. 36: 391-432.

(35) Lindauer, M. 1957. Sonnenorientierung der Bienen unter der Aequatorsonne und zur Nachtzeit. Naturwissenschaften 44: 1-6.

(36) Lorenz, K.Z., and Tinbergen, N. 1938. Taxis und Instinktbegriffe in der Eirollbewegung der Graugans. Z. Tierpsych. 2: 1-29.

(37) Marler, P. 1970. Song development in white-crowned sparrows. J. Comp. Physiol. Psychol. 71(2): 1-25 (Monogr.).

(38) Mason, W.A. 1968. Early social deprivation in the non-human primates: implications for human behavior. In Environmental Influences, ed. D.C. Glass, pp. 70-100. New York: Rockefeller University Press.

(39) McFarland, D.J., and Lloyd, I.H. 1973. Time-shared feeding and drinking. Q. J. Exp. Psychol. 25: 48-61.

(40) McGrew, W.C., and Tutin, C.E.G. 1973. Chimpanzee tool use in dental grooming. Nature 241: 477-478.

(41) Mishkin, M. 1982. A memory system in the monkey. Phil. Trans. Roy. Soc. Lond. B 298: 85-95.

(42) Norton-Griffiths, M.N. 1969. Organization, control, and development of parental feeding in the oystercatcher. Behaviour 34: 55-114.

(43) Nottebohm, F. 1971. Neural lateralization of vocal control in a passerine bird. I. Song. J. Exp. Zool. 177: 229-262.

(44) Olton, D.S. 1978. Characteristics of spatial memory. In Cognitive Processes in Animal Behavior, eds. S.H. Hulse, H. Fowler, and W.K. Honig, pp. 341-373. Hillsdale, NJ: Erlbaum.

(45) Rothstein, S.I. 1978. Mechanisms of avian egg-recognition: additional evidence for learned components. Anim. Behav. 26: 671-677.

(46) Rothstein, S.I. 1982. Successes and failures in avian egg and nestling recognition. Am. Zool. 22: 547-560.

(47) Sackett, G.P. 1966. Monkeys reared in visual isolation with pictures as visual input: evidence for an innate releasing mechanism. Science 154: 1468-1472.

(48) Schwartz, B. 1978. Psychology of Learning and Behavior. New York: Nortons.

(49) Seyfarth, R.M.; Cheney, D.L.; and Marler, P. 1980. Vervet monkey alarm calls: semantic communication in a free-ranging primate. Anim. Behav. 28: 1070-1094.

(50) Smith, A.W. 1978. Investigation of the mechanisms underlying nest construction in a mud wasp. Anim. Behav. 26: 232-240.

(51) Smith, S.M. 1975. Innate recognition of a coral snake by a possible avian predator. Science 187: 759-760.

(52) Staddon, J.E.R. 1983. Adaptive Behavior and Learning. New York: Cambridge University Press.

(53) Thorpe, W. 1950. The concepts of learning and their relationship to those of instinct. Symp. Soc. Exp. Biol. 4: 387-408.

(54) Tinbergen, N., and Perdeck, A.C. 1950. On the stimulus situation releasing the begging response in the newly hatched herring gull chick. Behaviour 3: 1-39.

The Biology of Learning, eds. P. Marler and H.S. Terrace, pp. 75-88. Dahlem Konferenzen 1984. Berlin, Heidelberg, New York, Tokyo: Springer-Verlag.

Genes, Evolution, and Learning

P.P.G. Bateson
Sub-Dept. of Animal Behavior, University of Cambridge
Madingley, Cambridge, CB3 8AA, England

Abstract. Evolutionary approaches to the study of learning imply genetic influence but not rigid determination because of the interplay between internal and external factors during an individual's development. Unlearned and inherited processes are plausibly involved in learning, but these processes are subject to change as the individual gathers experience. The biologists' functional view of learning is valuable, I argue, because it focusses attention on the particular job that has to be done. This in turn raises the issue of what are likely to be critical controlling variables. A related optimal design approach assumes knowledge of what the job is but, nonetheless, provides an attractive interface between the studies of experimental psychologists and those of biologists interested in learning.

INTRODUCTION

My intention in this paper is to show how biological concerns with evolution and adaptation are relevant to an understanding of learning. I believe these biological interests can actively nourish the study of individual development (3). Nonetheless, enthusiastically encouraging a fusion between two separate styles of thought has its dangers which should not be ignored. The distinctions which Tinbergen (37) drew between the four problems of ethology included separating why an individual develops in the way it does from why its ancestors evolved in the way that they did. The logical point remains as true as ever. Knowledge of how a particular car has been assembled does not tell us much about the history of automobile design, and the same is true for living organisms.

The muddling of different categories is probably easier than the perception of their differences. Certainly, studies of behavior have been replete with examples of the confusion of evolutionary and developmental arguments. For instance, much of the dispute generated by the sociobiology movement arose because it was thought that a conventional proposition about changes in gene frequency in the course of phylogeny implied genetic determination in ontogeny. The genes must influence the outcome of development, if the postulated process of evolutionary change has worked at all, but an influence does not imply a one-to-one correspondence between genes and behavior. The confusion between influence and isomorphism is apparent in the metaphors which are commonly applied such as "genetic blueprint" (see discussion in (8)). A more subtle programming metaphor contains elements of the same muddle when it is used to refer to "genetically programmed behavior." The difficulty here is that the "program" that runs gene expression and the assembly of bodies has been conflated with the "program" that runs the expression of behavior by fully developed nervous systems.

The notion of an isomorphic relationship between developmental influence and behavior is found in another metaphor used by Konrad Lorenz. This is the image of "intercalation" in which learned components of behavior are inserted among the instinctive components and, supposedly, can be recognized as such in adult behavior (11). The empirical grounds for doubting the value of this model are strong (4, 21). After Lehrman's (see (25)) critique of classical ethology, many people working in the subject abandoned the old learning/instinct classification of behavior but continued to use circumlocutions which were effectively equivalent. Recently, a number of authors (e.g., (13)) have grown impatient with these devices and have reverted to an explicit distinction between learnt and innate behavior. Before proceeding further, it is as well to be clear about meanings, since the term "innate" is not used in the same way by everybody, and even the issue of definition raises some pertinent general problems about behavioral development.

THE PROBLEMS OF "INNATENESS"

The term "innate" most commonly and most controversially means "unlearned" and is used (by those who use it) for behavior that develops without the individual experiencing stimuli to which it will respond, or without practice of the motor patterns that it will perform. A second usage for innate (introduced by Lorenz (26)) is for phylogenetically adapted behavior. This refers to behavior that has been adapted to its present use by the process of natural selection during the course of evolution. The development of such behavior may involve learning (e.g., imprinting)

(4). Third, a non-controversial usage stems from the well tried method of analyzing sources of differences (17). If animals that are known to differ genetically are reared in identical environments, any systematic differences in their behavior must ultimately have genetic origins and are sometimes spoken of as being "innate." This third sense is the one that Gould and Marler wish to use in their joint and separate papers in this volume. I am not convinced that the full flavor of their argument is retained if "innate" is replaced by "genetically inherited" and suspect they often pun, using "innate" in the sense of "unlearned." A difficulty arises because the development of a genetically inherited difference in behavior may often involve learning, as with kin recognition and song in birds.

The difficulties generated by the notion of unlearned behavior arise because it is not operational. That does not mean it is implausible. When a wasp performs a highly organized sequence of movements which results in an intricate nest (34) or when a hand-reared garden warbler, kept in a laboratory cage from when it was a nestling, attempts to fly in a southerly direction in the autumn and in a northerly direction in the spring (14), such patterns of behavior are not plausibly learned. Nonetheless, defining behavior as unlearned requires us to prove a negative, and in practice the definition is fraught with many difficulties which become especially serious when dealing with complex, long-lived animals (14, 19). In the end, all the characterizations of unlearned behavior rest on plausibility, but people have very different views about what constitutes a convincing case. Where one person is well satisfied that a given behavior pattern is "truly innate," another will feel that alternative explanations have been shut out prematurely. Much blood has been spilt and a great deal of confusion generated. My feeling is that in these circumstances it would be better to abandon the term innate, using the terms "unlearned," "genetically inherited," or "phylogenetically adapted" as required, while recognizing that they are not interchangeable.

Three other difficulties arise in the old style thinking about development. A behavior pattern first expressed without opportunities for learning is not necessarily resistant to modification by learning. Indeed, we know of many examples in which "unlearned" behavior is affected in this way. For instance, a recently hatched laughing gull will peck at models of adults' bills (in the natural world such an activity would evoke regurgitation of food by a parent). As the chick profits from its experience after hatching, the accuracy of its pecking improves and the kinds of bill-like objects it will peck at are increasingly restricted (15). Behavior patterns range from examples that are not influenced at all by learning (or so it plausibly seems!) to examples that are greatly influenced (1); the distribution is not bimodal.

The second difficulty relates to a common opposition between "maturation" and "experience." Maturational processes depend on conditions within the organism and are not influenced by extra-organismic factors insofar as the intra-organismic environment is maintained constant by homeostatic mechanisms. However, the course of tissue differentiation and its effects on behavior may, nonetheless, vary with the experience of the individual. "Experience" is often wrongly equated with "learning," giving the false implication that developmental changes in the central nervous system are unaffected by external conditions, such as temperature or food availability (25). An excellent modern discussion of this issue is provided by Oyama (30).

A remaining confusion is over the relationship between the heritability of a character and how little it has been influenced by learning. Heritability is the variance in a character in a population due to the variation in the genotypes divided by the total variance in the character. Many problems are associated with its use (12), but if and when estimates of heritability can be legitimately employed, they bear no correspondence to the extent to which behavior has been influenced by learning (21). When an unlearned behavior pattern has been under strong selection pressure in the course of evolution, all the variation in the genes influencing behavior could easily be lost. In such a case, any variation in the behavior would be due to environmental factors and the heritability would be zero. The naive assumption that unlearned behavior is necessarily highly heritable is false.

My purpose in raking over these old wrangles and confusions is to point to the inadequacies of a conceptual framework which so many people have used when thinking about development. The major conclusion, which is commanding an increasing amount of support in many areas of biology (19, 29, 35, 39), is that the genetic consequences of evolutionary change bear no straightforward relationship to the outcome of developmental processes.

Backsliding into misleading but comfortable learning/instinct dichotomies probably occurs when the complexities of behavioral development have not been captured by a vivid enough image. A cake-cooking metaphor goes a long way in the right direction (2, 7, 8). Despite the recognizable raisins (if it were that kind of cake), no one could be expected to identify each of the ingredients and each of the actions involved in cooking as separate components in a slice of cake. What is lacking here, though, is a sense of how the individual may influence its environment. A more useful image may be one in which the developing animal is seen as playing a game with its environment. It comes equipped with a set of rules for the game, one or more opening moves, and some conditional instructions about what must be done in particular circumstances. Beyond that, nobody can predict how it will develop without knowing the external conditions through which it will pass. This game-playing metaphor is helpful because it suggests that each player may in part generate the conditions to which it will then respond. Also, its conditional instructions can be refined or embellished as it gathers experience. Finally, the complex nature of the play in no ways denigrates the importance of the instructions with which each animal enters its lifetime game.

RULES FOR LEARNING

The biologists with their interest in evolution have clear-headed ways of talking about genetic influences on phenotypic behavioral characters, such as the way in which an animal might learn to solve a problem. If the expression of a particular allele means that its possessor behaves in a different way from another individual that does not have the allele, then the first condition for evolution by natural selection is met. If performing a pattern of behavior gives the actor a competitive edge over others that do not behave in the same way (in the sense that it leaves more of its genes behind in subsequent generations), then that form of behavior will become more common in the population and, necessarily, so will the particular genes that were required for its expression.

The learning theorists, for their part, have started to develop a coherent way of dealing with the diversity found in learning processes. The view that neutral and significant events must be close together in time if an association is to be formed between them was clearly disproved by the work of Garcia and his colleagues on taste aversion learning (see review in (10)). The long delay allowed between taking novel food and falling ill made functional sense, but it looked as though the prospects for a general theory of associative learning had sunk to zero. However, common features are found whether a rat learns to avoid shock or poison (36). If two neutral events of the same class are presented in turn prior to the noxious event, the second one interferes with learning about the first. For instance, rats were given novel saccharin solution and 15 minutes later novel vinegar solution, and finally were made ill with lithium chloride. Subsequently, they were less likely to avoid the saccharin solution than the rats which had been given water instead of vinegar (32). The vinegar had overshadowed saccharin just as a buzz followed by a tone and then shock would subsequently be ignored. As Dickinson ((9), p. 7) put it: "Rather then overthrowing the notion of a general learning process, taste-aversion research has served to change and enrich our conception of these general mechanisms." The common underlying principle can be represented as a conditional instruction. If a neutral event was of a certain category and if another of the same category had not been interposed between it and the important event, then that event itself acquires significance for the animal (6). This approach has considerable promise, even though the instructions are inferences and not observed directly.

It is not difficult to imagine how functional instructions for changing behavior could have evolved. Consider two individuals, both of which are able to associate a painful experience with events that preceded the pain. The two animals differ, though, in how selective they are about the associations they make, and the source of the difference is genetic. One animal freezes in response to every novel sound occurring up to 12 hours before the painful experience. The other is equipped with an overshadowing rule and only responds to the unfamiliar sound that occurred closest in time prior to the painful experience. Which animal is the more likely to survive and propagate its genes into the next generation? The answer depends a lot on how costly it is to freeze and thereby suppress all active movements - and on how beneficial it is to take no chances on novel sounds heard 12 hours before the painful experience. As in other evolutionary problems, the stable solution is likely to be a trade-off between conflicting requirements: in this case, remaining safe and remaining active. The optimum, in terms of the extent to which events preceding painful experience are overshadowed, will depend on conditions. However, in the course of evolution, animals that show greater selectivity will probably be favored over those that respond indiscriminately to every novel sound that preceded a painful experience. In other words, the overshadowing rule is usually going to be "evolutionarily stable" in Maynard Smith's (27) sense that, given the circumstances, a better solution cannot be produced.

In brief, we can begin to perceive how the instructions that greatly influence the outcome of learning can be represented by parameters in a formal mathematical model, and these same instructions could have been subject to natural selection so that they are now closely tuned to the physical and social environment in which members of a given species are likely to find themselves. This point of contact between psychology and biology is valuable and likely to be increasingly important in years to come. The psychologists' search for common underlying principles can be at last reconciled with the biologists' emphasis on adaptiveness to particular conditions.

FUNCTIONAL APPROACHES

Konrad Lorenz (26) coined a memorable phrase when he referred to the "innate school marm" who, he imagined, was busy directing the course of learning. What Lorenz meant was not that the instructions for learning were unlearned, but that the instructions were adapted for their present use by natural selection. This point about the adaptedness of rules for learning has been emphasized many times in recent years (e.g., (8, 23, 31, 33)). I shall now take a closer look at the functional approach to learning.

In coffee-time discussions among biologists about what such-and-such is for, factual knowledge is used to decide between competing hypotheses. On moving into the natural environment of an animal, some favorite coffee-time explanations are hastily abandoned and other functional hypotheses leap out of the bushes, so to speak. This is hardly an impressive claim for the rigor of such thinking. However, functional explanations are discarded in the face of contrary evidence. To go further requires an approach such as that pioneered by Tinbergen and his colleagues (38). In their classic study of nesting black-headed gulls, they noted that, after a chick had hatched and dried, the parents picked the eggshell up in their beaks and flew away with it. Why should the black-headed gull do this? The sharp edges of the shell might injure the chicks; the shell might tend to slip over an unhatched egg thus trapping the chick in a double shell; it might interfere in some way with brooding; it might provide a breeding ground for bacteria or moulds; finally, it might attract the attention of predators and thus endanger the brood. In developing these ideas, Tinbergen and his colleagues not only relied on good knowledge of the ecological conditions in which a black-headed gull lives but also drew on comparative evidence. They knew, for example, that the kittiwake gull does not remove its eggshell from the nest, and that the kittiwake, nesting on steep cliffs as it does, is exposed to far less predation of its eggs and chicks than the black-headed gull. So they concluded that the most likely functional explanation for removing eggshells was that the white inside of the eggshell betrayed the cryptically marked chicks and unhatched eggs to predators. They then did an experiment to examine the effects on predation of leaving open eggshells alongside intact eggs and found that the level of predation was much higher when shells where close to eggs than when they were further away. The conclusion was that the adult gulls' tendency to remove shells after hatching functions to decrease predation and thus increases the chances of the brood surviving.

As far as I know, nobody has tried to study the particular consequences of learning that might influence survival and future gene propagation. A start might be to relate natural variation in the ways animals learn to solve a problem with the fitness of each individual concerned. However, any correlations that emerged could doubtless be explained in many ways. And the same would be true for results obtained from direct manipulations of behavior. It would not be easy work.

Tinbergen's studies were based on many years of field experience with the animals he was interested in. In this respect, at least, he could be emulated. Indeed, the general point that studies of learning should be grounded in strong knowledge of the animal's natural behavior and living conditions has frequently been made (e.g., (18, 22, 23)). The biologists' emphasis on the comparison and the contrast of diverse species is relevant to this grounding process. For anyone who has generalized about the behavior of animals (or even all organisms!), differences between closely related species in the way they learn are salutary. Comparative studies may also draw attention to aspects of behavior which had previously been ignored in the familiar species in much the same way that a lesioned or drugged animal can highlight facets of a normal animal's behavior which had previously seemed unimportant.

A different point is that, in examining what Johnston (22) calls "the task description," we are alerted to the conditions in which the character of behavior changes. Knowledge of those conditions is crucial when we start to design experiments in which, inevitably, only a small number of independent variables are actually manipulated while the others are held constant or randomized. The experiment is a waste of time if the parameter values that are not subject to explicit manipulation are badly chosen. This is the relevance of the frequently stated point that we need to know our animals before we can do anything really interesting with them. The case for an open-ended approach to the study of learning is not that it immediately produces a counter-theory to established ideas, but that it provides a much more secure basis on which to formulate general principles (18). The case against such an approach is that its practitioners can miss the common functional principles which may apply in a great many different contexts (7).

OPTIMAL DESIGN

The classic ethological approach to function has been to ask what a given behavior pattern or process might be for. A different approach which is exciting a lot of attention at present is to beg the functional question and ask how a particular problem might be solved. Optimal foraging is a clear case where the skills of biologists and experimental psychologists have converged in productive ways (for a recent review see (23)). In general terms, the method is to calculate the best theoretical solution and then to examine the match between what the animal does and what it is expected to do. If the animal does more or less what is predicted, it is thought to be using the best engineering solution to the problem that the investigator believes it faces. The implication is, then, that the investigator's identification of the animal's problem has been vindicated. An objection to this line of argument is that another possible engineering solution might provide an even better predictor. Moreover, finding an optimal solution is by no means straightforward in studies of learning (20).

Even when the optimal solution is obviously to maximize the payoffs, the best way of achieving the ideal state is still open to considerable argument. For instance, Harley (16) suggested that when the payoffs from alternative actions influence the likelihood of each of these actions being performed on a future occasion, a simple rule is the following:

$$\text{Make the probability of doing A} = \frac{\text{Total payoff for action A}}{\text{Total payoff for all actions}} .$$

This rule, called Relative Payoff Sum, also provides a developmental explanation for a "mixed evolutionarily stable strategy" (27). In such a strategy, the payoff from a stable mix of alternative actions (such as behaving in hawk-like or dove-like ways) cannot be improved upon, or to be more accurate, it cannot be bettered without adding further actions to the repertoire. Accepting the qualification, Harley feels justified in regarding his learning rule as "adaptive" and hints that it may be universal (as does Maynard Smith (28)). Excessive confidence in the correctness of this conclusion would be unwarranted. Harley's model takes no account of the "time horizon," which is the session length in a laboratory experiment, and incorrectly predicts what happens in extinction after partial reinforcement (24). Furthermore, Houston et al. (20) have found better rules in their computer simulations of great tits foraging at two sites where the probabilities of reward are different - the "two-armed bandit" problem. One good, simple rule is: Sample both sites and then stay in the best one. The decision to work one site (or one arm of the bandit) is taken when the cumulative differences in rewards obtained from each of two sites reaches a predetermined value.

It is possible to argue about how generally useful a Cumulative Difference rule would be, and formidable complications face the theorist when conditions vary over time (20). The problem becomes even worse when dealing with foraging for intelligent prey at many different sites (5). Not only must the predator sort out the regularities of its environment, but it also must avoid becoming too regular itself, lest its own movements be predicted by its prey. It is no accident that in a Skinner box a cat does not behave like a pigeon. When one intelligent animal engages another, the business of uncovering rules for learning is extremely difficult. These considerations raise, once again, the need to be sensitive to the actual world in which an animal lives and to which it presumably is well adapted. In conclusion, it is worth emphasizing that, despite the difficulties, the modern approaches to optimal design are making this area theoretically exciting and well worth watching.

CONCLUSIONS

The dissonance between the notes struck by biologists and psychologists is still pronounced (and painful). But the promise of good music to come is also considerable and abundantly justifies the attempts to work towards closer collaboration. In this paper I have made two major points which I hope will help to remove some of the difficulties that still hamper fruitful dialogue. The first is that an interest in evolutionary process and long-term changes in gene frequencies does not commit anybody to genetic determinism. The old style simplistic notions of development are being rapidly replaced by much more rewarding ways of thinking about the interactions of genetic influences and experience. It is not good enough, though, to utter piously and repeatedly that we are all interactionists these days. We need to translate a general and valid point about ontogeny into particular examples of how the developmental processes actually work.

The second point is that functional explanations have widespread uses and are not merely the playthings of biologists. Many outsiders want to penetrate the arcane abstractions of learning theory by being told what the processes are for. Insights into function do not immediately tell anybody how a learning process works. However, behind the apparent confusion of ends with means lies a cogent fact about human minds. Even a guess about the likely end helps us to think about the ways of getting there.

If the functional approach is adopted, it has to be recognized that common design requirements may apply in a great many different contexts. Causal structure (9) has much the same properties whether predicting threats to life or controlling necessary resources. The functions of all associative processes are broadly the same, namely, to predict and to control the external world. Similarly, the role of becoming familiar by mere exposure with inanimate and animate features of the environment may be general. The implication is that a functional classification of learning would cut across the familiar behavioral categories such as feeding, mating, and so forth.

Acknowledgements. I thank the following for their comments: A. Dickinson, R.A. Hinde, A. Kacelnik, P. Marler, J. Stevenson-Hinde, and N. Thompson.

REFERENCES

(1) Alcock, J. 1979. Animal Behavior: An Evolutionary Approach, 2nd ed. Sunderland, MA: Sinauer.

(2) Bateson, P.P.G. 1976. Specificity and the origins of behavior. Adv. Stud. Behav. 6: 1-20.

(3) Bateson, P[P.G.]. 1982. Behavioural development and evolutionary processes. In Current Problems in Sociobiology, eds. King's College Sociobiology Group, pp. 133-151. Cambridge: Cambridge University Press.

(4) Bateson, P[P.G.]. 1983. Genes, environment and the development of behaviour. In Animal Behaviour. Genetics and Development, eds. T.R. Halliday and P.J.B. Slater, pp. 52-81. Oxford: Blackwell.

(5) Bateson, P[P.G.]. 1983. Review of Functional Ontogeny. Anim. Behav. 31: 634-635.

(6) Bateson, P[P.G.]. 1983. Rules for changing the rules. In Evolution from Molecules to Men, ed. D.S. Bendall, pp. 483-507. Cambridge: Cambridge University Press.

(7) Dawkins, R. 1982. The Extended Phenotype. Oxford: Freeman.

(8) Dawkins, R. 1983. Universal Darwinism. In Evolution from Molecules to Men, ed. D.S. Bendall, pp. 403-425. Cambridge: Cambridge University Press.

(9) Dickinson, A. 1980. Contemporary Animal Learning Theory. Cambridge: Cambridge University Press.

(10) Domjan, M. 1980. Ingestional aversion learning: Unique and general processes. Adv. Stud. Behav. 11: 275-336.

(11) Eibl-Eibesfeldt, I. 1970. Ethology: The Biology of Behavior. New York: Holt, Rinehart and Winston.

(12) Feldman, M.W., and Lewontin, R.C. 1975. The heritability hang-up. Science 190: 1163-1168.

(13) Gould, J.L. 1982. Ethology: The Mechanisms and Evolution of Behavior. New York: Norton.

(14) Gwinner, E., and Wiltschko, W. 1980. Circannual changes in migratory orientation of the Garden Warbler, Sylvia borin. Behav. Ecol. Sociobiol. 7: 73-78.

(15) Hailman, J.P. 1967. The ontogeny of an instinct. Behavior (Suppl. 15). Leiden: Brill.

(16) Harley, C.B. 1981. Learning the evolutionarily stable strategy. J. Theoret. Biol. 89: 611-633.

(17) Hinde, R.A. 1968. Dichotomies in the study of development. In Genetical and Environmental Influences on Behaviour, pp. 3-14. Edinburgh: Oliver & Boyd.

(18) Hinde, R.A. 1973. Constraints on learning: An introduction to the problems. In Constraints on Learning: Limitations and Predispositions, eds. R.A. Hinde and J. Stevenson-Hinde, pp. 1-19. London: Academic Press.

(19) Hinde, R.A. 1982. Ethology. London: Fontana.

(20) Houston, A.; Kacelnik, A.; and McNamara, J. 1982. Some learning rules for acquiring information. In Functional Ontogeny, ed. D. McFarland, pp. 140-191. Boston: Pitman.

(21) Jacobs, J. 1981. How heritable is innate behaviour? Z. Tierpsychol. 55: 1-18.

(22) Johnston, T.D. 1981. Contrasting approaches to a theory of learning. Behav. Brain Sci. 4: 125-173.

(23) Kamil, A.C., and Yoerg, S.I. 1982. Learning and foraging behavior. In Perspectives in Ethology: Ontogeny, eds. P.P.G. Bateson and P.H. Klopfer, vol. 5, pp. 325-364. New York: Plenum.

(24) Krebs, J.R., and Kacelnick, A. 1984. Optimal learning rules. Commentary on Maynard Smith. Behav. Brain Sci. 7: 109-110.

(25) Lehrman, D.S. 1970. Semantic and conceptual issues in the nature-nurture problem. In Development and Evolution of Behavior, eds. L.R. Aronson, E. Tobach, D.S. Lehrman, and J.S. Rosenblatt, pp. 17-52. San Francisco: Freeman.

(26) Lorenz, K. 1965. Evolution and Modification of Behavior. Chicago, IL: University of Chicago Press.

(27) Maynard Smith, J. 1982. Evolution and the Theory of Games. Cambridge: Cambridge University Press.

(28) Maynard Smith, J. 1984. Game theory and the evolution of behaviour. Behav. Brain Sci. 7: 95-125.

(29) Oster, G., and Alberch, P. 1982. Evolution and bifurcation of developmental programs. Evol. 36: 444-459.

(30) Oyama, S. 1982. A reformulation of the idea of maturation. In Perspectives in Ethology: Ontogeny, eds. P.P.G. Bateson and P.H. Klopfer, vol. 5, pp. 101-131. New York: Plenum.

(31) Pulliam, H.R., and Dunford, C. 1980. Programmed to Learn. New York: Columbia University Press.

(32) Revusky, S. 1971. The role of interference in association over a delay. In Animal Memory, eds. W.K. Honig and P.H.R. James, pp. 155-213. New York: Academic Press.

(33) Shettleworth, S.J. 1982. Function and mechanism in learning. In Advances in Analysis of Behavior: Biological Factors in Learning, eds. M. Zeiler and P. Harzen, vol. 3. New York: Wiley.

(34) Smith, A.P. 1978. An investigation of the mechanisms underlying nest construction in the mud wasp, Paralastor sp. (Hymenoptera: Eumenidae). Anim. Behav. 26: 232-240.

(35) Stent, G.S. 1981. Strength and weakness of the genetic approach to the development of the nervous system. Ann. Rev. Neurosci. 4: 163-194.

(36) Testa, T.J. 1974. Causal relationships and the acquisition of avoidance response. Psychol. Rev. 81: 491-505.

(37) Tinbergen, N. 1963. On aims and methods of ethology. Z. Tierpsych. 20: 410-433.

(38) Tinbergen, N.; Broekhuysen, G.J.; Feeks, F.; Houghton, J.C.W.; Kruuk, H.; and Szulc, E. 1962. Eggshell removal by the black-headed gull, Larus ridibundus L.; a behaviour component of camouflage. Behaviour 19: 74-117.

(39) Wright, S. 1980. Genic and organismic selection. Evol. 34: 825-843.

The Biology of Learning, eds. P. Marler and H.S. Terrace, pp. 89-114. Dahlem Konferenzen 1984. Berlin, Heidelberg, New York, Tokyo: Springer-Verlag.

The Study of Animal Learning in the Tradition of Pavlov and Thorndike

H.M. Jenkins
McMaster University
Hamilton, Ontario L8S 4K1, Canada

INTRODUCTION

The tradition founded by Pavlov and Thorndike is motivated by the conviction that certain pervasive phenomena of learning are captured in two paradigms; the one exemplified by Pavlov's experiment in salivary conditioning, the other by Thorndike's experiments with cats in a puzzle box. The Pavlovian experiment arranges a relation between two events, a CS and US, independently of behavior. It is directed toward understanding what Pavlov referred to as the process of interchangeable signification which he believed rested equally on the conditioning of inhibition and excitation. The Thorndikian or operant conditioning experiment arranges a relation between behavior and an event, or a response and reinforcer. It is directed toward understanding the shaping and strengthening of action through reward, the suppression and redirection of action through punishment, and the development of stimulus control over punished and reinforced actions.

The immodest purpose of this paper is to explore the following questions: Is the level of abstraction inherent in this approach workable? Does it capture enough of whatever is important for learning to justify the intensive investigation of a relatively small set of preparations viewed as exemplifications of basic paradigms? Is there sufficient independence of learning processes from the fleshed-out circumstances in which men and animals behave to make the enterprise productive? If so, we should after eighty years or so of work be able to identify the involvement of these processes in significant behavior outside of the laboratory. Can that be done?

OVERVIEW

The title of the present paper reflects appropriately, I believe, the continuity of studies of animal learning today with the tradition begun by Pavlov and Thorndike. But animal learning in the 1980s is not the enterprise it was in the first half of the century nor in the heyday of neobehaviorism. Perhaps most important is that the objective has changed. It is not to see how much of complex human behavior can be explained by using elementary conditioning phenomena as building blocks. The elementary phenomena have turned out to be anything but elementary. The paradigms and the phenomena they engage are now regarded as sources of problems for the understanding of learning processes rather than building blocks with which to synthesize complex behavior. In particular, classical and operant conditioning are not viewed as fundamental and distinct forms of learning but as phenomena which themselves require a theoretical analysis. I hope that this will be evident in the body of the present paper.

Since the time of Thorndike and Pavlov there has been a major expansion in the paradigms used for the study of learning. It is now well recognized that some of the effects due to the repeated presentation of a single stimulus, especially long-term habituation, involve associative processes (55, 58). The demonstration that habituation is affected by an association between the habituating stimulus and the context of its presentation (57) has brought habituation into the field of animal learning.

The definition of a classical conditioning experiment has become highly generalized, and new varieties of classical conditioning preparations have played an increasingly important role. The new varieties, going back quite a way, include fear conditioning studied through the conditioned emotional response procedure (CER) developed by Estes and Skinner (11). In the CER procedure, the conditioned change in behavior is suppression of operantly reinforced behavior (e.g., food-reinforced barpressing) rather than a phasic reaction resembling the reaction to the US as in salivary or eye-blink conditioning. The conditioning of food aversions through induced illness has become a major focus of study since the classic experiments of Garcia and his co-workers in the 1960s (13, 14). The dependent measure is consumption, or relative consumption. As in the CER the conditioned change is related only indirectly to the behavior elicited by the US. The appetitive conditioning of movements directed to a localized stimulus was brought to the fore by the autoshaping or signtracking experiment (3). It has expanded the scope of classical conditioning by showing its relevance to the initiation and form of overt skeletal actions directed toward and away from signaling objects (15). The conditioning of reactions to drugs has assumed increasing importance because of the recognition that learning plays a major role in the development of tolerance and the effects of drug withdrawal (48, 49).

Linked to the increased scope of classical conditioning is the tendency to characterize the domain of classical conditioning in more general and abstract terms than those suggested by Pavlov's treatment. Instead of thinking of the experiment as one in which the CS comes to evoke a US-related response, it can be viewed as an arrangement in which the animal learns about a relation between events. What it learns can later be probed in various ways by performance tests. Rescorla and Holland ((42), p. 172) have expressed this orientation toward conditioning as follows: "In this procedure (classical conditioning) ... some relation is arranged between S_1 and another stimulus S_2, at t_1, without regard to the organism's behavior. The consequences for that arrangement are then assessed at t_2. If arrangement of that relation between stimuli can be identified as responsible for differential t_2 behavior, one may say that Pavlovian conditioning has occurred." This characterization of classical conditioning implies a firm distinction between learning and performance. It is congenial to the development of theories of conditioning using concepts of stimulus processing and memory that have developed in the study of human cognition in recent years. Wagner and his students (59) have produced a well developed theory of conditioning, habituation, and other effects of simple exposure to a stimulus within such a framework.

Many of the basic questions about learning in classical conditioning have a close parallel in operant conditioning. Here, too, there has been a growing emphasis on the learning-performance distinction. Instead of conceiving of the reinforcer as acting directly on the response or on S-R connections, animals are viewed as learning about response-reinforcer and stimulus-response-reinforcer relations (6).

There are, however, two aspects of operant conditioning studies that have no close parallel in classical conditioning: the shaping of responses through selective reinforcement and the allocation of behavior to alternative sources of reinforcement. Neither of these phenomena are readily accommodated to the learning-performance distinction because behaving and learning are so highly interdependent in shaping and response allocation. I begin my discussion of current issues in animal learning with these phenomena.

SHAPING AND RESPONSE ALLOCATION IN OPERANT CONDITIONING

Response shaping or differentiation occurs when reinforcement is made to depend on the form of a response, or on its temporal, intensive, or directional character. We owe to Skinner (52) the first and still widely accepted account of how shaping occurs. According to Skinner, the strengthening effect of a reinforcer is always specific in some degree to properties of the reinforced response. Therefore, all that is required to shape behavior is to arrange for the reinforcer to occur only after responses which exhibit the form or property to be shaped. Suppose, for example, that we wish to shape a rat to barpress with an unusually strong force. Reinforcement would be made to depend on a force value in the upper end of the distribution of naturally occurring values. The first reinforcement of the required value will tend to shift the distribution upwards, simply because reinforcing action is to some degree specific to the properties of the reinforced response. When repeated reinforcements have pulled the center of the distribution toward the criterion value, the criterion may be shifted to a still higher value which will appear as the result of response generalization. The procedure may be continued until a limit is reached. Skinner's treatment of shaping could be viewed as a molecular theory of the motor side of trial-and-error learning.

Despite the centrality of shaping to operant conditioning, there has been relatively little research directed toward an analysis of the process. Perhaps that is because the process is not easy to control experimentally. As learning progresses, the criterion for reinforcement is met more and more frequently. This results in complex changes in several variables including the overall rate of reinforcement, the proportion of responses reinforced, and the size of the discrepancy between reinforced response values and the overall distribution of response values.

Platt and his co-workers (2, 30, 36) devised the percentile schedule of reinforcement in order to gain better control over these variables. The percentile schedule continually adjusts the criterion of reinforcement to keep it at the same relative position in the distribution of current response values. This allows one to fix the proportion of response values reinforced (e.g., the upper decile). With certain modifications of the percentile schedule one can also examine the shaping process with the overall rate of reinforcement fixed. Using these techniques Platt and his co-workers were able to show that reinforcement can exert a selective shaping action on response rate and that this effect is separable from other effects of the probability of reinforcement per response, and of the rate of reinforcement on the strength of the tendency to respond. These techniques are also beginning to provide information on the parameters of reinforcement which drive the shaping process. An extremely important parameter is the extremity of differential reinforcement, as controlled by the percentile criterion, independent of the overall rate of reinforcement.

The problem that has attracted the greatest share of experimentation within the operant framework in recent years is that of choice: the allocation of responses to separate sources of reinforcement. Herrnstein's paper on the matching law in 1961 (16) launched the operant study of choice by uncovering an empirical generalization that has proved to be very robust. The matching law states that responses are allocated to a source of reinforcement in the ratio of reinforcements received from that source to the total reinforcements received.

By far the most commonly studied experimental arrangement for the study of operant choice uses the pigeon facing two keys, each of which reinforces pecking with food on some variable interval schedule. The independent variable is the mean rate at which reinforcements become available on each key. The dependent variable is the allocation of responding to alternatives. The matching law relates response rates to reinforcements actually received, not to the programmed rates of reinforcement. If all responses go to one alternative, the law is satisfied in a trivial sense. But interval schedules of reinforcement return approximately the same rate of reinforcement despite wide variations in rate of response. This affords ample room for the matching law to be violated. In fact, under certain conditions, results do depart systematically from matching. Nevertheless, a good approximation to the matching law has been found in an impressive variety of experiments in which not only the frequency of reinforcement is manipulated, but also the amount or delay of reinforcement. The important question is, What lies behind this regularity?

Herrnstein (17) has argued that the matching law provides the basis for measuring reinforcing value. He thinks of the occurrence of any operant as the outcome of a choice among alternative responses, each with its own source of reinforcement. Viewed in this way, the matching law becomes applicable in principle to the prediction of response rates when there is only one explicit source of reinforcement, as in the traditional study of single schedules of reinforcement. Herrnstein was able to show that the matching law does provide a reasonable account of response rate as a function of reinforcement rate under single schedules of reinforcement as well as in the explicit choice experiment. He concluded that the relative response rate is the proper measure of response strength and that it provides the yardstick for the measurement of reinforcing value; in other words, the matching of relative response rates to relative reinforcement rates constitutes the quantitative law of effect.

The matching law is expressed at a molar level since it states a relation between reinforcements and responses aggregated over large blocks of time. But the process that generates the data that go into the matching law can also be thought of as a series of moment-to-moment choices of when and where to respond. Can the aggregated results be understood as the consequence of a series of moment-to-moment choices governed by local changes in the likelihood of reinforcement, or do they reflect an adjustment process which, like the law itself, is based on aggregated data?

It has become apparent that great care is required in the analysis of the sequence of response choices in order to reach the right answer to the question of whether or not in a particular arrangement of concurrent schedules the sequence of responses reflects a process of choosing the alternative that has the higher probability of reinforcement at the moment of choice (53). With the use of increasingly powerful analytical methods, evidence is accumulating that animals make choices in accord with local differentials in the probability of reinforcement and that matching is a by-product of short-term maximizing rather than a result of adjustments to highly aggregated outcomes (19, 20, 50, 51, 53). Local maximizing does not always lead to the optimal overall harvesting of reinforcements, nor does the matching of relative rates (18), but only under special conditions does local maximizing produce an end result that falls substantially short of the overall optimum (20).

The operant allocation problem can be coordinated to certain features of food foraging in the wild. Searching for food and the capturing and handling of food objects or prey can be viewed as operants with functional properties that might be shared by keypecking or barpressing. The locales or patches in which food objects, or even different types of food objects, are concentrated can be likened to separate sources of reinforcement. Optimal foraging theory has raised questions about foraging behavior that are paralleled by questions about optimizing and matching in the operant allocation problem. The ecological analysis of foraging suggests new variables for study within the operant laboratory while the analytical techniques and theoretical options that have been developed for the operant allocation problem will, I believe, also contribute to the study of foraging (see, for example, (4)).

ISSUES IN THE STUDY OF CLASSICAL CONDITIONING

I turn now to a brief review of the current status of work on certain basic issues in classical conditioning.

Is the CS–US Relation Responsible for the CR?

This question arises because, the efforts of experimenters notwithstanding, the classical conditioning experiment is unavoidably impure. The experimenter arranges a CS-US relation but cannot prevent a response-US relation from intruding. Conceivably, the acquired changes that appear to be due to the explicitly arranged CS-US relation are in fact due to the implicit relation between the response and the US. One approach to the problem is the omission procedure developed by Sheffield (46), which he used in the analysis of salivary conditioning. In this procedure the US is omitted whenever the CR occurs. If conditioning develops and is sustained despite response-contingent omission of the US, the CS-US relation must alone be sufficient.

Consider the application of this procedure to autoshaped keypecking in pigeons. The lighting of the pecking disc is followed by a feeding (CS-US relation) unless the key is pecked. Birds begin pecking and continue to peck at a substantial level, despite this omission contingency (60). A spatial version of the omission experiment places the source of the food signal far enough from the food site itself so that if the bird approaches the signal and remains there too long, the opportunity to eat will have passed (15). Approach develops and persists on a substantial fraction of trials even though it causes the loss of feeding. The sufficiency of the CS-US relation has been shown in other forms of appetitive conditioning (7).

There is a lesson in the omission experiment. If we looked in at a signtracking experiment we might observe keypecking followed directly by food. We know from other experiments that keypecking can be strengthened through a response-reinforcer contingency alone (23). One might be tempted to assume that a reinforceable response followed by a reinforcer is in fact being supported by that reinforcer. The omission experiment tells us that we could be mistaken.

What Is the Critical Property of the CS-US Relation?

What is the critical relation between a CS and US which causes a stimulus, to which the animal is at first relatively indifferent, to act as a signal of another event? Pavlov's experiments always confounded the relative temporal proximity of CS and US with a correlation or contingency between CS and US. A now classic set of experiments by Rescorla (38, 39) separated correlation from proximity, or contiguity. The experiments showed that proximity was not sufficient and that correlation was critical. CS-US correlation was removed, without reducing CS-US contiguity, by presenting the US with sufficient frequency between the CS-US pairs to render the CS nonpredictive of the US. That this prevented conditioning to the CS is perhaps not surprising, but it raises basic questions about the process. Why is CS-US correlation critical?

A brilliant series of experiments carried out by Kamin in CER fear conditioning (29) greatly furthered our understanding of the question. The experiments showed that what is learned about the relation of any one CS to a US depends critically on the context provided by other stimuli and, especially, on the status of those stimuli as signals for the same US. The phenomena of overshadowing and blocking exemplify this principle. If two CSs, A and B, are presented together as a signal of shock, each acquires less signal value than it would have acquired had it been the only signal of shock. The signals overshadow each other. Blocking is closely related. If A is first made to signal the US and B is later added, A blocks the acquisition of signal value by B. One might say that pretraining A increases its ability to overshadow B. The added stimulus in blocking is temporally contiguous with the US and is, moreover, in a contingent relation to the US. Correlation plus contiguity is, accordingly, not sufficient for the development of signal value. If a stimulus contributes nothing further to the prediction of US, it may not condition. Predictive value plays an important role.

Kamin's explanation was as follows. A US produces new learning about a potential signaling stimulus only to the extent that the US is unpredictable or surprising within the context provided by all the stimuli that are currently available as predictors of the US. It is as though, Kamin suggested, a surprising event makes the rat search recent memory for its cause. A fully predicted shock, on the other hand, is attributed to whatever already predicts it, and therefore nothing may be learned about other potential predictors.

In three simple equations, now well-known by students of learning, Rescorla and Wagner (44) captured the fundamental proposition that learning occurs to the extent that the total signal value based on all concurrently active stimuli falls short of the maximum possible signal value, or the level reached when conditioning is complete. The formulation not only yielded overshadowing and blocking, it gave a remarkably satisfactory account of certain basic phenomena of inhibitory conditioning. Furthermore, it made some nonobvious predictions which proved correct. For example, if a stimulus, A, is first made inhibitory (signals the absence of a reinforcer in a context which otherwise signals the presence of the reinforcer), A can serve to induce especially rapid acquisition, and unusually strong positive signal value, in another stimulus, B, if B is reinforced in compound with A rather than alone (44).

With certain quite natural extensions the theory provides an account of Rescorla's finding that when the CS-US correlation is removed by presentations of US alone, conditioning may not occur. The setting of an experiment provides a background or contextual stimulus. The particular CSs presented by the experimenter are always in compound with this contextual stimulus which is treated as acquiring signal value like any other stimulus. When USs occur at the same rate between CSs as they do within CSs (the noncontingent case), the background stimulus will, according to the equations of the theory, eventually acquire all of the available signal value and the CS will be neutralized. The effect of contingency is reduced to the same processes of competition among stimuli for signal value that occurs in overshadowing and blocking.

In summary, the answer to the question of the necessary and sufficient conditions for the acquisition of signal value in the Rescorla-Wagner theory is a) some degree of temporal proximity of CS and US, and b) the absence of a fully conditioned competing stimulus. You will not be surprised to learn that not every investigator concerned with this question believes that this is the final answer (26). But anyone who looks closely at these developments will see, I believe, that they constitute a substantial advance on a very fundamental question.

How Is Selectivity in Conditioning Achieved?

The developments just reviewed are part of a continuing effort to understand how, out of the profusion of stimuli that inevitably accompany every biologically significant happening in an animal's life, only certain stimuli emerge as effective signals. The phenomena of selectivity are more varied than can be indicated here, and some cannot be treated by the conception of stimulus interactions embodied in the Rescorla-Wagner theory. Rudy and Wagner (45) provide an incisive review of selectivity in learning and the theoretical strategies that have been developed to account for them.

Two recent approaches to selectivity (32, 35) can be traced to the theory developed by Sutherland and Mackintosh (54), which puts the locus of selectivity in the front-end processing of stimuli rather than in the loss of surprise value for the US. Dickinson (6) provides an introduction to these ideas. Wagner (59) has developed a comprehensive theory of phenomena generated by compound presentations and by presentations of CS and US singly, and in various temporal relations to CS-US pairings. The theory employs concepts of stimulus processing and memory also found in the human literature. It provides another approach to understanding selectivity.

What Is Learned in Classical Conditioning?

I want to point to certain techniques for exploring this question in order to show that in contemporary work the older controversy between cognitive and S-R theories is now seen as part of a broader question about what animals come to know, encode, or represent about the CS-US relation.

My discussion follows Dickinson (6). Consider excitatory conditioning. Does the animal encode the experience as though it were a one-step procedural rule: "respond to CS with CR" or is encoding in the nature of a declarative rule such as: "CS leads to US," which requires a second rule such as: "If CS leads to US, make the CR to CS"? What are the testable implications of these hypothetical encodings? Suppose, following the development of a CR to a food signal, the food is paired with poisoning so that it becomes aversive. It is then presented again with the CS that signals food. If the encoding of original conditioning was procedural, there would be no basis for the integration of the later-acquired food aversion with the previously acquired response to CS. When tested in extinction with CS alone, the animal should respond as though the food had not been made aversive. In contrast, declarative encoding provides the possibility of integration through an inference: "CS leads to US, US is now aversive, hence CS leads to aversive US." A reduced tendency to respond to CS on an extinction test after food had been made aversive would be consistent with this more cognitive form of declarative encoding and inconsistent with the more response-centered, one-step, procedural encoding.

Both integration and lack of integration have been obtained in experiments on this plan, and it is not yet known what determines the outcome. An experiment by Adams and Dickinson (1) on this issue in the context of operant conditioning suggests that the amount of conditioning prior to the change in the status of the US is one important factor. Rescorla and his co-workers (40) found integration in first-order classical conditioning but not in second-order conditioning. However, integration in second-order conditioning has also been reported (37). The experiments demonstrate that an either-or, cognitive vs. S-R, account is untenable. The tactic of making a post-conditioning test after altering the status of the event being signaled may help to identify the nature of the reorganization of behavior brought about by conditioning.

In some very recent work a related technique points to a form of reorganization often speculated about but rarely pinned down. Classical conditioning can result in one stimulus acting as a modulator of the response-initiating function of another stimulus. This has been shown in the case of excitatory conditioning (41) as well as inhibitory conditioning (21, 24). Consider the inhibitory case. The stimulus A alone is reinforced and AB is non-reinforced with the result that responding is suppressed when AB is presented. What is B's mode of action? It could, and sometimes does, act as though it signaled no food (43). But in other preparations, one of which is signtracking, it acts more directly on the response-initiating function of A. The evidence in support is that B may continue to inhibit the response to A even after it has been converted by separate conditioning into a strong signal for the same reinforcer signaled by A. The finding is important because it demonstrates a second-order or modulating effect of B on A that is independent of the first-order status of B as a signal of the reinforcer. It may be that the modulating function of stimuli on the action of other stimuli is a very common product of learning in animal behavior.

What Is the Role of Intrinsic Variables?

In any species, the outcome of conditioning depends on the selection, as well as the arrangement, of CS and US. Present to a hungry pigeon a localized visual stimulus closely followed by food and, if the parameters are right, vigorous autoshaped keypecking directed at the stimulus is the almost certain result. Carry out exactly this procedure with a diffuse auditory stimulus and, although conditioned activity will develop, it will almost never involve keypecking. The form of conditioned behavior clearly depends on the CS as well as the US. This is but one example of the role of intrinsic variables (the effect of choosing particular CSs and USs) as distinct from arrangement variables (42). To what extent do intrinsic variables determine the effects of arrangement variables?

The asceptic phrasing is not meant to conceal the fact that the question should be considered in the context of the ethological conception of learning as a genetically programmed process, evolved under selection pressures characteristic of a species in its niche, and therefore showing a high degree of adaptive specialization. Lorenz (31) conceived of genetically determined plans for adaptation as relatively open to modification through environmental encounters or relatively closed (see also (33)). An adaptive function carried by a phylogenetically closed program in one species may be left more open, hence more flexibly set by environmental interactions, in another. The same can be said of different adaptive functions within a species such as food getting and courtship.

The theme of specialization in learning had little impact on animal learning until Garcia and his co-workers (13, 14) published on the learning of taste aversions in rats. These experiments led the authors to two major conclusions. First, a conditioned aversion to taste as a CS when poison or radiation-induced illness is the US can be acquired despite an interval of many hours between consumption of a distinctively flavored substance and the onset of illness. Moreover, aversions were acquired after a single trial even with a substantial delay. This contrasts sharply with results found with such other CSs and USs as tone and shock in which conditioning becomes marginal with delays on the order of minutes. Second, the choice of US, induced illness vs. electric shock, exerted a strong selective effect on the type of CS that was conditioned. When a novel taste and a compound auditory-visual stimulus were concurrently available to rats while drinking, taste was associated to illness, but the auditory-visual compound was associated to shock.

What was the force of these findings? That some CSs and some USs were more effective than others had long been recognized. What was new was the compelling demonstration of particularly effective and ineffective combinations of CS and US; in other words, strong interactions. Here was a basis for selective association apparently beyond the reach of arrangement variables. Moreover, the nature of the selective conditioning pointed towards adaptive specialization in the program for learning. Perhaps the role of arrangement variables is strongly contingent on intrinsic variables.

My discussion of these issues follows a review of this literature, now enormous, by Domjan (8). His principal conclusions may be summarized as follows: a) At first disputed, it is now clear that conditioning is capable of inducing taste aversions when a long delay is interposed between the exposure to a novel-tasting substance and the onset of illness. b) The CS-US interaction in taste-aversion conditioning is not an experimental artifact, but evidence that the process responsible for the interaction is a genetic disposition to develop associations selectively between certain types of CSs and USs is indirect. Other interpretations of the interaction have been suggested, but none as yet give a satisfactory account of long delay, selectivity, and very rapid learning. Genetic predisposition remains as a probable basis. c) CS-US interactions and conditioning with long delays are found in systems other than the one involving taste and illness. d) There is evidence that associability of different CS-US combinations may be governed in part by nonspecific stimulus variables such as the similarity of CS and US, their spatial proximity, and the temporal and intensive characteristics of their onset and decay. e) Arrangement variables, such as the CS-US interval, have qualitatively similar effects on taste-aversion conditioning and on other more conventional conditioning preparations.

The last point in particular needs elaboration. Single-trial long-delay conditioning may be special to illness-induced flavor aversions. Nevertheless, it shares with other forms of conditioning a similar dependence on arrangement variables. Acquisition is a decreasing function of CS-US interval; the acquired aversion is subject to extinction and conditioned inhibition; acquisition is impaired by pre-exposure to CS or US, and it is subject to blocking and overshadowing and to other selective effects dependent on the relative predictive value of a CS. Although powerful, intrinsic variables do not appear to override the regularities established for arrangment variables in less favored CS-US combinations.

Despite the generality of qualitative effects of conditioning variables, the power of intrinsic variables and the possibility of interpreting their role within an ethological framework continue to pose important issues for the study of animal learning. I return to the matter in the final section.

APPLICATIONS

The study of learning in conditioning experiments is not an end in itself. Its purpose is to further understanding of behavior at large and to increase effectiveness in bringing about desired changes in behavior by design. I allude briefly to just two examples of the application of knowledge gained by the laboratory study of conditioning.

Applications to Animal Behavior

Hollis (22) provides a useful review and analysis of the role of classical conditioning in foraging, food procuring, ingestion and rejection of food, defense, and reproduction. Her view is that classical conditioning serves the adaptive function, which she calls prefiguring, of optimizing the animal's interaction with biologically important events. The adaptation lies in behavioral and physiological changes induced by the CS as a consequence of its relation to the US.

The process by which prefiguring actions are conditioned in the individual is not, however, governed by their adaptive consequences within the learning episode itself. Evolution results in nervous systems that respond to event sequences with activity that anticipates and enhances the animal's interaction with the important biological event, or US, that is being signaled. But adaptive value on an evolutionary time scale does not imply that the conditioning process is driven in the individual by local adaptive consequences. Hollis supports this point with the example of pigeons which continue to approach and peck at a visual signal of food even under the omission procedure in which pecks cause the animal to lose the feedings that would otherwise become available.

The prefiguring hypothesis also implies a role for the CS in determining the form of the CR. There are constraints on the broad classes of stimuli that function as natural signals for different types of USs. Correlation of the form of conditioned activity with form of the signal could therefore be an adaptive constraint on conditioning. For example, as was noted above, a small visual stimulus is a natural signal of food objects for pigeons, and when used as a CS for food it will be approached and pecked. In contrast, a sound, even when localized, is not a natural signal for food, and when used as a CS it will rarely be approached and pecked. The rule suggested by the concept of prefiguring is that conditioned activity will resemble the response to a class of stimuli that function in nature as learned signals for actions that anticipate and enhance interaction with a biologically important event.

In nature, where omission contingencies are not often found, signtracking leads the animal toward signals that have reliably accompanied food in the past, and away from those that predict the absence of food. An automatic process, not necessarily involving the reinforcement of action, can direct the animal's search for food to especially productive parts of its territory. The same conditioning processes generate anticipatory food-procuring and consummatory responses specific to the item which the signaling stimulus predicts. A number of the phenomena of parent-young feeding interactions appear to rest on signaling effects of the kind seen in the signtracking experiment, and evidence is accumulating for the operation of conditioning in many other vital behavior systems.

The adaptive advantage of processes that accomplish a sorting-out of the more predictive covariations of events from less predictive and accidental covariations seems evident. Adaptive conditioning should reflect the causal texture of the environment. Overshadowing, blocking, and more generally, selectivity in signaling are almost certainly functional in the natural habitat. Less has been done, however, in pinpointing their operation in natural behaviors than has been done for the simple conditioning of action patterns and physiological states. It is in the latter connection that the adaptive function of prefiguring has so far been tested at several points and supported by the evidence.

Applications to Drug Tolerance

The evidence that classical conditioning plays a significant role in the development of drug tolerance has been reviewed by Siegel (48, 49). He presents a conditioning theory of tolerance which rests on the following ideas. The pharmacological effects of drug administration can function as unconditioned stimuli. When reliably signaled by the drug administration procedure, or by other environmental cues, conditioned drug reactions occur. For several classes of drugs these reactions are opposite in direction to the unconditioned changes induced by the drug itself. For example, unconditioned responses to morphine include bradycardia, hyperthermia, and analgesia, whereas the conditioned responses include tachycardia, hypothermia, and hyperalgesia. Conditioned responses of this form will tend to counteract the unconditioned responses and can thereby serve to diminish the total responsiveness to the drug whenever repeated, signaled drug administration provides an opportunity for conditioning. According to this view, tolerance is not the inevitable result of repeated doses but depends on learning as well as on nonassociative physiological processes.

A puzzling fact about drug conditioning is that, although for many drugs the conditioned reaction is opposite to the unconditioned reaction, for other drugs the conditioned reaction mimics the unconditioned reaction. Eikelboom and Stewart (10) have recently published a promising explanation of the form of conditioned reactions. It is based on an analysis of the site of drug action (efferent or afferent), and it has implications for the question of why repeated administration of some drugs leads to sensitization rather than habituation.

Siegel (47) and others have shown that the development of morphine tolerance in rats does depend to a remarkable degree on the action of conditioned cues. Tolerance to the analgesic effect of morphine was assessed by means of the rat's paw-lick response elicited by a hot surface. Tolerance developed with repeated doses when the drug injections were given in the same environment used to test for the analgesic effect. In contrast, when the final injection in a series was given in an environment which had not previously been the site of drug administration, the analgesic effect remained very close to its original strength.

Manipulations of the CS-US relation suggested by the conditioning model of tolerance produce their expected effects. Specifically, tolerance is attenuated by extinction of the drug-signaling stimuli, by repeated pre-exposure of the CS prior to its pairing with drugs, and by reductions in the correlation between a CS and drug administration. The acquisition of tolerance can also be retarded by giving the drug in the presence of a stimulus that has served to signal the absence of the drug, i.e., in the presence of a conditioned inhibitor. The theory predicts that hyperalgesia will occur when pain sensitivity is tested in the presence of a drug signal followed by an inert substance in place of morphine, because in this case only the compensatory effects will be evoked. That prediction has been confirmed. Parallel effects have been shown in experiments on ethanol-induced hypothermia in which purely physiological measures, rather than behavioral measures, are used. The role of drug signals in evoking withdrawal reactions and in influencing the rate of readdiction following withdrawal has also been shown. Thus, the evidence supports the view that learning plays an important role in drug dependence and addiction. Experimental, clinical, and epidemiological evidence strongly points to significant parallels in the role of conditioning in human drug reactions.

IS THE STUDY OF ANIMAL LEARNING TOO ABSTRACT?

The study of animal learning has the special objective of trying to understand processes that underlie signaling, shaping, and the stimulus control of reinforced action. Its more general objective, which it shares with other branches of psychological science, is to improve understanding of the nature of animal and human behavior. The effort is directed not only towards behavior as it is found, but also to the potential for behavior to take on new forms by design. It is, therefore, concerned with the capacity for change as well as with learning processes in behavior at large. I comment first on the issue of level of abstraction as it bears on the more special objective.

I hope that I have been able to convey something of the basis for my belief that a) substantial progress has been made toward answering those questions that are native to the framework of animal learning, and b) there is much yet to be learned by continued efforts of the same kind. The tradition has not run out of gas and there are many more hills to be climbed. Nevertheless, the level of abstraction inherent in the very idea of experimental preparations has liabilities as well as advantages.

Although arrangement variables are not washed out by intrinsic variables, intrinsic variables exert a powerful influence on the outcome of conditioning experiments. They appear to operate on at least four levels:

1. They can bring about profound shifts in the magnitude of the effects of arrangement variables as seen, for example, in the very different effect of CS-US interval on different varieties of classical conditioning.

2. They appear to influence the mode of action of stimulus control processes. For example, conditioned inhibition of the pigeon's keypecking in signtracking appears to act quite specifically on the peck-initiating stimulus (24). On the other hand, in the typical fear-conditioning experiment its action is far less specific to the fear-initiating stimulus (21, 43).

3. Intrinsic variables appear to have an important influence on selective processes in conditioning as exemplified in Garcia's classic experiments.

Although not yet firmly established, there may well be genetic dispositions for certain types of CS-US associations. To take another example, overshadowing is a process widely distributed across phyla (see, for example, (5)), yet potentiation of one stimulus by a concurrent stimulus is found in arrangements that are, abstractly, the same as those which yield overshadowing (8).

4. The forms of conditioned action patterns that emerge in conditioning appear not to be intelligible without reference to naturally occurring action patterns. How else to understand, for example, why certain signal-food arrangements condition scratching in chickens, soliciting behavior in dogs, and social responses in rats (22, 25, 56)?

In my view it is unlikely that arrangement variables, although they are part of the story, will prove sufficient to predict these effects, or even to provide an account of them after the fact. There was, for example, no principle of classical conditioning available that could have predicted that in signtracking experiments pigeons pursue the signal of food rather than approaching the food site. Although signtracking has become a common preparation, we still do not have a principle that explains the form of the conditioned behavior.

I believe that students of learning and of ethology should work toward a classification of behavioral systems that could provide a systematic framework for understanding the operation of intrinsic variables in learning. Mayr (33) suggests that the degree of openness in a genetic program varies according to whether the behavior is intraspecific as in courtship and parenting, interspecific as in prey-predator relations, or noncommunicative (involving no interorganismic interaction) as in habitat and food selection. Johnston (28) has suggested that we begin with functional task descriptions of activities such as feeding, grooming, parenting, and so on. It is difficult to know at this time what level of classification of behavior will shed light on the role of intrinsic variables in learning, but it does seem clear that close attention must be paid to the question of what function learning serves in the natural contexts in which it occurs.

Domjan and Galef (9) outline a search strategy for exceptions to what is often called general process learning theory. The strategy involves a comparison of learning in closely related species subject to different selection pressures that could be expected to create particular kinds of adaptive specializations in learning. Although I do not know exactly what is supposed to be asserted by general process learning theory, I can well imagine that this strategy could produce some valuable insights. It might contribute to development of a classificatory system of the kind that could rationalize the effect of intrinsic variables on learning. It would not, however, obviate the need for such a system, nor do the authors suggest that it would.

I return now to the question of whether the study of learning is too abstract to contribute significantly to our understanding of the nature of animal and human behavior. I believe that the study of learning has gone through a period since the 1930s in which it was conceived too abstractly - at too great a distance from the fleshed-out circumstances in which learning occurs. Greater contact has been made with biology in recent years and that is a good thing for the study of learning. Yet, the knowledge that has been gained from the learning theorist's abstractly conceived preparations does have power, and it has made significant contributions to the understanding of behavior at large. The two applications of learning principles which I have chosen to review - applications to animal behavior and to drug conditioning - could have been multiplied, but I believe that they are of sufficient importance in themselves to make a case for the significance of the study of learning in the tradition of Pavlov and Thorndike.

Current thinking about learning processes in relation to behavior reflects changes which I believe will facilitate productive contact with the broader objective. It is now clear that there are important learning processes not engaged by the conditioning experiment. Song learning is but one example of a process that seems well outside of the range of associative processes engaged by conditioning. I also believe that it is broadly recognized that learning processes do not account for behavior, rather, they enter into the determination of almost every complex behavior system. The recent work of Hoffmann and his associates (12) has, for example, shown an important role for reinforcement in imprinting, but I trust that today no informed psychologist would assert that reinforcement explains imprinting. Recognizing boundaries on the province of knowledge derived from the study of animal learning encourages intelligent use of that knowledge.

Although I think that the record is cause for optimism, I want to raise questions about the nature of the contact between the knowledge gained from studies of animal learning and behavior at large. The applications I chose to cite in some detail are ones in which the behavioral phenomena, as they appear within laboratory experiments, are easily recognized as substantively related to the phenomena referenced in the application. It is, for example, not a very large step from signtracking to search and procurement of food in nature, nor from drug tolerance in a conditioning experiment to drug tolerance outside of the laboratory. Are instructive applications only to be found when the step is based on common forms of behavior? Not entirely, I believe.

There is a kind of application at a very abstract level. Perhaps because of the relative simplicity of the arrangements one thinks to try in studies of animal learning, these studies seem propitious for uncovering basic structures of environment-behavior relations that are then readily seen to recur in different contexts. I take partial reinforcement to be a case in point. Similarly, the so-called feature-positive and feature-negative arrangements in discrimination learning (27) seem to reflect an inescapable logic of signaling structures. In the feature-positive case the distinctive feature accompanies the reinforced occasions while in the feature-negative case it accompanies the non-reinforced occasions. In nature, only the feature-positive structure is capable of signaling precisely the time of occurrence of a reinforcer or other momentary event. The feature-positive signal has higher information value. In the laboratory, where one can arrange for the presence or absence of a feature to be equally informative, the feature-positive arrangement leads to the required performance more readily than does the feature-negative arrangement both for pigeons discriminating between visual displays and for college students learning simple concepts (34). The actual learning processes are almost certainly different, but identifying common structures, and parallel performances, suggests the relevance of an ecological explanation.

In addition to short-step applications and more abstract, common-structure applications, many students of animal learning envision applications of another kind. They argue that evolution is conservative; it builds and elaborates on processes that have evolved in simpler organisms and does not discard them. A number of writers allude to the possibility that in animal learning we deal with relatively automatic associative processes of the kind that underlie lower-level habits in human behavior, as distinct from more deliberative processes that characterize some human actions. In a similar vein, some believe that studies of animal learning offer knowledge about nonverbal learning in humans. The cognitive orientation of much recent work in animal learning appears in part to be motivated by these hopes. Dickinson ((6), p. 12), for example, writes: "What is clear ... is that the study of animal cognition and learning may well illuminate the nature of a general, internal 'language of thought'." I will admit to skepticism about the fulfillment of this promise. My skepticism arises from several considerations, but their common ground is captured in a comment made to me in a note from Dick Millward: "No animal is a subset of any other animal" (personal communication, December 2, 1981).

Whether or not my skepticism about applications at this level proves justified, I believe that the study of animal learning has much to offer and that it can be brought into far more productive contact with biologically centered approaches to learning than it now enjoys. There is a need for scientists who are willing to work towards that goal.

Acknowledgements. The author's research is supported by a grant from the National Science and Engineering Research Council of Canada. I am grateful to B.G. Galef for help in thinking about evolution and animal behavior in relation to learning.

REFERENCES

(1) Adams, C., and Dickinson, A. 1981. Actions and habits: Variations in associative representations during instrumental learning. In Information Processing in Animals: Memory Mechanisms, eds. N.E. Spear and R.R. Miller. Hillsdale, NJ: Erlbaum.

(2) Alleman, H.D., and Platt, J.R. 1973. Differential reinforcement of interresponse times with controlled probability of reinforcement per response. Learn. Motiv. 4: 40-73.

(3) Brown, P.L., and Jenkins, H.M. 1968. Auto-shaping of the pigeon's key-peck. J. Exp. Anal. Behav. 11: 1-8.

(4) Commons, M.L.; Herrnstein, R.J.; and Rachlin, H., eds. 1982. Quantitative Analyses of Behavior: Matching and Maximizing Accounts, vol. 2. Cambridge, MA: Ballinger.

(5) Couvillon, P.A., and Bitterman, M.E. 1982. Some phenomena of associative learning in honeybees. J. Comp. Physiol. Psychol. 96: 192-199.

(6) Dickinson, A. 1980. Contemporary Animal Learning Theory. Cambridge, England: Cambridge University Press.

(7) Dickinson, A., and Mackintosh, N.J. 1978. Classical conditioning in animals. Ann. Rev. Psychol. 29: 587-612.

(8) Domjan, M. 1980. Ingestional aversion learning: Unique and general processes. In Advances in the Study of Behavior, eds. J.S. Rosenblatt, R.A. Hinde, C. Beer, and M.C. Busnel, vol. 11. New York: Academic Press.

(9) Domjan, M., and Galef, B.G. 1983. Biological constraints on instrumental and classical conditioning: Retrospect and prospect. Anim. Learn. Behav. 11: 151-161.

(10) Eikelboom, R., and Stewart, J. 1982. Conditioning of drug-induced physiological responses. Psychol. Rev. 89: 507-528.

(11) Estes, W.K., and Skinner, B.F. 1941. Some quantitative properties of anxiety. J. Exp. Psychol. 29: 390-400.

(12) Gaioni, S.J.; Hoffman, H.S.; and DePapulo, P. 1978. Imprinting in older ducklings: Some tests of a reinforcement model. Anim. Learn. Behav. 6: 19-26.

(13) Garcia, J.; Ervin, F.R.; and Koelling, R.A. 1966. Learning with prolonged delay of reinforcement. Psychonom. Sci. 5: 121-122.

(14) Garcia, J., and Koelling, R.A. 1966. The relation of cue to consequence in avoidance learning. Psychonom. Sci. 4: 123-124.

(15) Hearst, E., and Jenkins, H.M. 1974. Sign-tracking: The stimulus-reinforcer relation and directed action. Austin, TX: The Psychonomic Society.

(16) Herrnstein, R.J. 1961. Relative and absolute strength of response as a function of frequency of reinforcement. J. Exp. Anal. Behav. 4: 267-272.

(17) Herrnstein, R.J. 1970. On the law of effect. J. Exp. Anal. Behav. 13: 243-266.

(18) Herrnstein, R.J. 1982. Melioration as behavioral dynamism. In Quantitative Analyses of Behavior: Matching and Maximizing Accounts, eds. M.L. Commons, R.J. Herrnstein, and H. Rachlin, vol. 2. Cambridge, MA: Ballinger.

(19) Hinson, J.M., and Staddon, J.E.R. 1983. Hill-climbing by pigeons. J. Exp. Anal. Behav. 39: 25-47.

(20) Hinson, J.M., and Staddon, J.E.R. 1983. Matching, maximizing, and hill-climbing. J. Exp. Anal. Behav. 40: 321-331.

(21) Holland, P.C. 1983. The nature of inhibition in serial and simultaneous feature-negative discrimination. Paper presented at the conference on Information Processing in Animals: Inhibition and Contingencies. State University of New York at Binghamton, June 14-16, 1983. Hillsdale, NJ: Erlbaum, in press.

(22) Hollis, K.L. 1982. Pavlovian conditioning of signal-centered action patterns and autonomic behavior: A biological analysis of function. In Advances in the Study of Behavior, eds. J.S. Rosenblatt, R.A. Hinde, C. Beer, and M.C. Busnel, vol. 12. New York: Academic Press.

(23) Jenkins, H.M. 1979. Sensitivity of different response systems to stimulus-reinforcer and response-reinforcer relations. In Operant-Pavlovian Interactions, eds. H. Davis and H.M.B. Hurwitz. Hillsdale, NJ: Erlbaum.

(24) Jenkins, H.M. 1983. Conditioned inhibition of keypecking in the pigeon. Paper presented at the conference on Information Processing in Animals: Inhibition and Contingencies. State University of New York at Binghamton, June 14-16, 1983. Hillsdale, NJ: Erlbaum, in press.

(25) Jenkins, H.M.; Barrera, F.J.; Ireland, C.; and Woodside, B. 1978. Signal-centered action patterns of dogs in appetitive classical conditioning. Learn. Motiv. 9: 272-296.

(26) Jenkins, H.M., and Lambos, W.A. 1983. Tests of two explanations of response elimination by noncontingent reinforcement. Anim. Learn. Behav. 11: 302-308.

(27) Jenkins, H.M., and Sainsbury, R.S. 1970. Discrimination learning with the distinctive feature on positive or negative trials. In Attention: Contemporary Theory and Analysis, ed. D. Mostofsky. New York: Appleton-Century-Crofts.

(28) Johnston, T.D. 1981. Contrasting approaches to a theory of learning. Behav. Brain Sci. 4: 125-173.

(29) Kamin, L.J. 1969. Predictability, surprise, attention, and conditioning. In Punishment and Aversive Behavior, eds. B.A. Campbell and R.M. Church. New York: Appleton-Century-Crofts.

(30) Kuch, D.O., and Platt, J.R. 1976. Reinforcement rate and interresponse time differentiation. J. Exp. Anal. Behav. 26: 471-486.

(31) Lorenz, K.Z. 1965. Evolution and Modification of Behavior. Chicago: University of Chicago Press.

(32) Mackintosh, N.J. 1975. A theory of attention: Variations in the associability of stimuli with reinforcement. Psychol. Rev. 82: 276-298.

(33) Mayr, E. 1974. Behavior programs and evolutionary strategies. Am. Sci. 62: 650-659.

(34) Newman, J.; Wolff, W.T.; and Hearst, E. 1980. The feature-positive effect in adult human subjects. J. Exp. Psychol.: Human Learn. Mem. 6: 630-650.

(35) Pearce, J.M., and Hall, G. 1980. A model for Pavlovian learning: Variations in the effectiveness of conditioned but not of unconditioned stimuli. Psychol. Rev. 87: 532-552.

(36) Platt, J.R. 1973. Percentile reinforcement: Paradigms for experimental analysis of response shaping. In The Psychology of Learning and Motivation, ed. G.H. Bower, vol. 7. New York: Academic Press.

(37) Rashotte, M.E. 1981. Second-order autoshaping: Contributions to the research and theory of Pavlovian reinforcement by conditioned stimuli. In Autoshaping and Conditioning Theory, eds. C.M. Locurto, H.S. Terrace, and J. Gibbon. New York: Academic Press.

(38) Rescorla, R.A. 1967. Pavlovian conditioning and its proper control procedures. Psychol. Rev. 74: 71-80.

(39) Rescorla, R.A. 1968. Probability of shock in the presence and absence of CS in fear conditioning. J. Comp. Physiol. Psychol. 66: 1-5.

(40) Rescorla, R.A. 1980. Pavlovian Second-order Conditioning: Studies in Associative Learning. Hillsdale, NJ: Erlbaum.

(41) Rescorla, R.A.; Durlach, P.J.; and Grau, J. 1983. Contextual learning in Pavlovian conditioning. In Context and Learning, eds. P.D. Balsam and A. Tomie. Hillsdale, NJ: Erlbaum, in press.

(42) Rescorla, R.A., and Holland, P.C. 1976. Some behavioral approaches to the study of learning. In Neural Mechanisms of Learning and Memory, eds. M.R. Rosenzweig and E.L. Bennett. Cambridge, MA: The MIT Press.

(43) Rescorla, R.A., and Holland, P.C. 1977. Associations in Pavlovian conditioned inhibition. Learn. Motiv. 8: 429-447.

(44) Rescorla, R.A., and Wagner, A.R. 1972. A theory of Pavlovian conditioning: Variations in the effectiveness of reinforcement and nonreinforcement. In Classical Conditioning II: Current Research and Theory, eds. A. Black and W.F. Prokasy. New York: Appleton-Century-Crofts.

(45) Rudy, J.W., and Wagner, A.R. 1975. Stimulus selection in associative learning. In Handbook of Learning and Cognitive Processes, ed. W.K. Estes. Hillsdale, NJ: Erlbaum.

(46) Sheffield, F.D. 1965. Relation between classical conditioning and instrumental learning. In Classical Conditioning: A Symposium, ed. W.F. Prokasy. New York: Appleton-Century-Crofts.

(47) Siegel, S. 1977. Morphine tolerance acquisition as an associative process. J. Exp. Psychol.: Anim. Behav. Proc. 3: 1-13.

(48) Siegel, S. 1979. The role of conditioning in drug tolerance and Addiction. In Psychopathology in Animals: Research and Clinical Applications, ed. J.D. Keehn. New York: Academic Press.

(49) Siegel, S. 1983. Classical conditioning, drug tolerance, and drug dependence. In Research Advances in Alcohol and Drug Problems, eds. Y. Israel, F.B. Glaser, H. Kalant, R.E. Popham, W. Schmidt, and R.G. Smart, vol. 7. New York: Plenum, in press.

(50) Shimp, C.P. 1969. Optimal behavior in free-operant experiments. Psychol. Rev. 76: 97-112.

(51) Silberberg, A.; Hamilton, B.; Ziriax, J.M.; and Casey, J. 1978. The structure of choice. J. Exp. Psychol.: Anim. Behav. Proc. 4: 368-398.

(52) Skinner, B.F. 1938. The Behavior of Organisms: An Experimental Analysis. New York: Appleton-Century-Crofts.

(53) Staddon, J.E.R.; Hinson, J.M.; and Kram, R. 1981. Optimal choice. J. Exp. Anal. Behav. 35: 397-412.

(54) Sutherland, N.S., and Mackintosh, N.J. 1971. Mechanisms of Animal Discrimination Learning. New York: Academic Press.

(55) Thompson, R.F.; Groves, P.M.; Teyler, T.J.; and Roemer, R.A. 1978. A dual-process theory of habituation: Theory and behavior. In Habituation: Behavioral Studies, eds. H.V.S. Peeke, and M.J. Herz, vol. 1. New York: Academic Press.

(56) Timberlake, W. 1983. The functional organization of appetitive behavior: Behavior systems and learning. In Advances in Analysis of Behavior: Biological Factors in Learning, eds. M.D. Zeiler and P. Harzem, vol. 3. Chichester, England: Wiley, in press.

(57) Wagner, A.R. 1976. Priming in STM: An information processing mechanism for self-generated or retrieval-generated depression in performance. In Habituation: Perspectives from Child Development, Animal Behavior, and Neurophysiology, eds. T.J. Tighe and R.N. Leaton. Hillsdale, NJ: Lawrence Erlbaum.

(58) Wagner, A.R. 1979. Habituation and memory. In Mechanisms of Learning and Motivation: A Memorial Volume to Jerzy Konorski, eds. A. Dickinson and R.A. Boakes. Hillsdale, NJ: Erlbaum.

(59) Wagner, A.R. 1981. SOP: A model of automatic memory processing in animal behavior. In Information Processing in Animals: Memory Mechanisms, eds. N.E. Spear and R.R. Miller. Hillsdale, NJ: Erlbaum.

(60) Williams, D.R., and Williams, H. 1969. Auto-maintenance in the pigeon: Sustained pecking despite contingent non-reinforcement. J. Exp. Anal. Behav. 12: 511-520.

The Biology of Learning, eds. P. Marler and H.S. Terrace, pp. 115-133. Dahlem Konferenzen 1984. Berlin, Heidelberg, New York, Tokyo: Springer-Verlag.

Learning by Selection

J.-P. Changeux, T. Heidmann, and P. Patte
Neurobiologie Moléculaire, Institut Pasteur
75 015 Paris, France

INTRODUCTION

Living organisms are "open" thermodynamic systems (19) that possess an internal structure and thus correspond to a privileged state of organization of matter in both space and time. The question then arises: where does this order come from? From inside the organism, from the outside world, or from both? To simplify, two extreme views can be put forward to account for this higher internal order, placing the emphasis either outside or within the biological system with respect to its relationships with the environment.

1) Instructive mechanism. The environment imposes an order which is transferred directly into the organism.

2) Selective (Darwinian) mechanism. The increase of internal order is indirect. The organism generates spontaneously a multiplicity of internal "variations" in organization which exists prior to interaction with the environment. This interaction merely selects, or selectively stabilizes, some of these endogenous variations.

To operate, a selective machine must thus contain two basic devices: a) a generator of internal diversity utilizing a combinatorial process, and b) a mechanism for selection of privileged combinations (and/or elimination or rejection of the others) associated with the exchange of signals with the outside world.

In the history of biological thinking, many instructive theories have been suggested and then, more often than not, rapidly abandoned. On the other hand, selective models have encountered success in several major domains of biological sciences and, at least in one case, reached the explanatory level.

1) Evolution of species. Darwin's "variations" which precede selection are the well identified mutations of DNA and/or larger-scale chromosomic changes (such as translocations, transpositions, inversions, deletions, changes in chromosome number...). DNA becomes the "generator of internal diversity." On the other hand, the mechanisms for selection, selective stabilization, or segregation of the variations remain controversial (see (4)).

2) Antibody synthesis. The great capacity of the vertebrate immune system to synthesize antibody proteins directed against foreign "non-self" antigenic molecules is accounted for by a selective mechanism which is almost completely elucidated at the molecular level (see (39, 56)). The genes that specify each antibody are not present as integral units in the DNA of the fertilized egg but, instead, as several sets of genetic segments coding for different regions of the antibody light and heavy chains. The segments of DNA, and of the transcribed messenger RNA (which make a repertoire of about 300 coding units), are shuffled and joined together by an internal combinatorial mechanism to specify billions of antibody molecules with different amino acid sequences. Each particular species of antibody then becomes exposed singly at the surface of a given lymphocyte, and among the large population of lymphocytes carrying different antibody specificities, only the ones which bind the antibody complementary to the foreign antigen proliferate in the presence of the antigen and are thus selected.

3) Neurosciences. Early references to selective theories can be found before Darwin, in the work of the British associationists such as Locke or Hume. Mill (59) even explicitly mentions "elimination" as part of the process of "induction." Taine (59) in "De l'intelligence," quoting Darwin, writes: "in the struggle for life (Darwin) which, at any given time, takes place between all our images, the one which, at the origin, possesses a higher energy, keeps at each conflict, by the law of repetition upon which it is based, the capacity to repel its rivals." Also, James (31) states explicitly that "to think is to make selections."

The recent revival of selective theories in the neurosciences arose from the extension of the antibody synthesis model to the nervous system (see (16-18, 32, 58)) or from the application of a simple Darwinist scheme to the growth of synaptic connections during development (5, 8-10, 12).

In this review of learning by selection, models are discussed as possible contributions to learning theory in two situations: a) in the adult, and b) during postnatal development.

LEARNING BY SELECTION IN THE ADULT BRAIN

Levels of Organization and Relevant Experimental Approaches to Learning

Learning theory basically deals with behavior, under the standard laboratory conditions of classical and operant conditioning (see Jenkins, Terrace, and Lea, all this volume) or under more natural surroundings in the case of the ethological approach (see Gould and Marler, this volume). In general, little reference is made to the neural basis of these behaviors even though their actual determinants lie within the central nervous system (see, however, Bateson, Konishi, and Singer, all this volume). Looking for the mechanisms of learning actually requires the deciphering of the anatomy (the neuronal network concerned) and of the activity (the trains of impulses) traveling in the particular network whose "actualization" generates the particular behavior or "mental" process investigated. In principle, one may infer behavior from anatomy + activity + stimulus. On the other hand, to move in the opposite direction, i.e., to infer the rules of anatomy and/or activity from behavior, hinges upon millions of years of brain evolution which did not result in any simple logic of brain organization (see (6)). A reasonable theory of learning must therefore include anatomy + activity, in addition to behavior.

In the past few decades, knowledge of the elementary electrophysiological (see (35)) and molecular (see (7)) mechanisms of synaptic transmission and neuronal integration has progressed considerably. Attempts have been made to relate data obtained with single synapses or neurons to elementary learning phenomena (see (34) and Quinn, this volume). This information, of course, must be included in any general learning theory but does not suffice by itself to account for larger-scale learning processes present in higher vertebrates and humans. Additional principles are needed. The cognitivist approach to behavior has indeed led to the postulation of "molar" units, commonly referred to as cognitive maps ((55), and see Menzel, this volume), search or mental images (36) or, in a general manner, internal or neural "representations" (30, 50, 51, 59). Such all-or-none global entities which characterize highly evolved vertebrate brains, and possibly some invertebrate ones (see Gould, this volume) cannot simply be accounted for by single cells or small ensembles of cells. This is why (following Hebb (23)), the concept of a large "assembly" of neurons "which can act briefly as a closed system" has been proposed to "constitute the simplest instance of a representative process (image or idea)." The level of organization encompassed by a learning theory which would account for these molar units should thus be that of large populations of cells, at the scale of the brain, not of single neurons or synapses. However, models of such "assemblies" should be based on knowledge presently available on these elementary components.

Neural Representations

Neural representations or "mental objects" (9) are postulated to be the basic units that the brain uses for computational tasks (9, 23, 28, 41-43, 48, 57). They are identified as the physical state created by the correlated or "concerted" transitory activity (both electrical and chemical) of a large population of neurons. This assembly is a discrete, closed, and autonomous unit. It can be described by a mathematical structure called a "graph" (see (10) for an extensive definition). Such a neuronal graph provides a model of the connective organization of the neuronal network and of its geometry. It involves large numbers of neurons possessing different connectivities or "singularities" (8) laid down in the course of development (see below). They may belong to different cortical (or subcortical) areas and thus have a distributed topology. A single neuron may contribute to different representations; on the other hand, the elimination of some neurons from the assembly, at least to some degree, does not interfere with its information content (robustness). The identity of a neural representation is thus defined by a spatio-temporal firing map. Depending on the area of the cortex involved, whether a primary or secondary sensory area or an associative area (such as the prefrontal cortex), the modality of the representations will be more concrete (images) or more abstract (concepts).

On the basis of the relationship of neural representations to the outside world, one may first distinguish two basic classes of mental objects:

1) The primary percept is a labile unit. Its neuronal graph and activity are determined by and dependent upon the direct interaction with the outside world via the sensory organs. The concerted character of the firings results from the simultaneous stimulation of the sensory cells.

2) The stored representation is a memory object. Its evocation is not necessarily linked with the presentation of the stimulus which serves for the storage. It results from the spontaneous and concerted firing of the neurons from the graph. Its autonomy can only be conceived if there exists a "coupling" between neurons stable in time and present as a latent physical trace in the network. The cooperativity of this coupling is responsible for the invasive all-or-none character of the firing of the assembly when the stored representation is evoked. Learning can thus be viewed as the process of establishing this latent physical coupling.

Learning As the Selective Stabilization of Pre-representations by Resonance

At this stage, at least, two basic mechanisms can be envisioned: a) with the instructive model, the percept is directly stabilized into a stored representation; in other words, the neurons of the graph in the evoked representation are identical to those of the primary percept; b) with selective models, the storage is indirect and results from the selection of "pre-representations" that already exist in the brain.

The basic postulate of the proposed selective model (see (9)) is that prior to the contact with the outside world, or concomitant with it, the brain spontaneously generates a third class of mental object which is neither a percept nor a stored (latent or evoked) representation. Such pre-representations result under certain conditions, for instance, following focussing of the attention from the activity of neurons or groups of neurons which spontaneously fire in concert but in a labile and transient manner. The neuronal graph of the active neurons varies with time. At any given time it is composed of sets of cells taken among a large population of uncoupled and inactive cells. As a consequence, firing of various combinations of neurons or groups of neurons (or even already stored assemblies) takes place and a wide variety of active neuronal graphs form successively and transiently.

The selection of a given pre-representation and its storage in memory requires, in the simplest case, the interaction with the environment, i.e., the formation of a percept. The postulated mechanism for the selection is that a "resonance" occurs between the externally evoked percept and the internally generated pre-representation. Resonance will take place if: a) the encounter between percept and pre-representation happens within a given length of time and the firings originating from the two sources are "in phase" or have a definite time relationship; b) spatially, their neuronal graphs overlap and the degree of overlap exceeds a critical size.

As a consequence, a stable, latent, and cooperative coupling becomes established between the neurons of the particular graph fired both by the percept and the pre-representation during the encounter period. The neuronal graph of the stored representation thus differs from that of the pre-representation and from that of the percept.

The evocation of a stored representation consists of the concerted firing of the whole graph of neurons cooperatively linked in a stable and latent manner as a consequence of the resonance step. This concerted firing can be elicited by the activation of a small set of neurons from any part of the stored assembly. As a consequence, the firing of the whole assembly can be triggered by various routes, which include different sensory or "internal" modalities, as long as they are connected with a minimal number of neurons from the graph of the stored representation. Evocation can thus be viewed as a multimodal transient "germination-invasion" of a stored representation.

Synaptic Mechanisms for the Selective Stabilization of Cooperative Coupling

In his discussion of the "growth of the assembly" of neurons, Hebb ((23), p. 62) proposed his now classical "neurophysiological postulate" that "when an axon of cell A is near enough to excite a cell B and repeatedly and persistently takes part in firing it, some growth process or metabolic change takes place in one or both cells such that A's efficiency as one of the cells firing B is increased." But in the same chapter, he also states that "the general idea is an old one, that any two cells or systems of cells which are repeatedly active at the same time will tend to become "associated" so that activity in one facilitates activity in the other. The details of speculation . . . are intended to show how this old idea might be put to work again, with the equally old idea of a lowered synaptic resistance..." Since then, the relevance of the so-called "Hebb synapse" to synaptic modifications observed experimentally in simple systems has been extensively discussed and even sometimes challenged on the basis of the pre- vs. postsynaptic site of the modification (see Quinn, this volume). In fact, Hebb's most general formulation does not mention, or even require, such distinction.

Stent (52) has proposed a molecular mechanism of the Hebb synapse based on the electric field-induced metabolic degradation of the acetylcholine receptor. This mechanism has yet to receive experimental support. Recently (24), we have proposed a mechanism for short-term regulation of synapse efficiency at the postsynaptic level based on the classical phenomenon of "receptor desensitization," which in the case of the acetylcholine receptor consists in a now well understood cascade of molecular transitions of the receptor protein (see (7, 25)).

The central postulate is that the postsynaptic receptor for neurotransmitter spontaneously exists under, at least, two states in reversible equilibrium: one (e.g., R) susceptible to activation by agonists (e.g., R → A) and another one refractory to activation or desensitized (e.g., D). If the amplitude of the postsynaptic response is an increasing function of the fraction of receptor in the R state, then the R/D ratio will simply determine the efficiency of the synapse and, of course, the ability of the postsynaptic cell to fire when a threshold value of the membrane potential is reached.

The ratio of the two states might itself be regulated by "allosteric effectors" such as the membrane potential (since a difference of a dipolar moment may exist between the two states) or ligands (when the affinity for the ligand considered differs in the two states). The regulatory ligands might be the neurotransmitters but might also include "internal" effectors such as Ca^{++} or cyclic nucleotides. The binding site(s) for the regulatory ligands would then either be the receptor site for the neurotransmitter or allosteric sites.

Detailed schemes (24) have been derived on this basis for the regulation of the efficiency of a given synapse by its own state of postsynaptic activity (homosynaptic regulation) or by that of neighboring synapses (heterosynaptic regulation). The rate constants of the R → D transition would then determine the duration of the change of synapse efficiency. Timing relationships between the activity of neighboring synapses may create changes of synapse efficiency (via the shift of the R → D equilibrium) which are those expected in the case of the simple scheme of "classical conditioning." Thus, this molecular model accounts for the Hebb synapse in its general and restricted sense. It can be used to create the latent trace which cooperatively couples the neurons in an assembly. It also offers a simple mechanism for the "resonance" step.

The activation of a pre-representation can be viewed as an increased spontaneous firing rate of neurons scattered among a large "initial population" as a consequence of, for instance, a general activation (or released inhibition) of the population by divergent regulatory neurons. These divergent neurons might, for instance, be those from the mesencephalon which control the focussing of the attention in the alert subject. Concomitant activation by the percept of the spontaneously firing neurons of a pre-representation then makes possible changes of synapse efficiency (via the $R \rightleftharpoons D$ transition) that neither the spontaneous firing along nor the percept associated activity would create independently (Heidmann and Changeux, in preparation).

Receptor desensitization has been found in central synapses and with neurotransmitters such as acetylcholine, glutamate, and γ-aminobutyric acid. The allosteric effects of electric field, Ca^{++} ions, and of various pharmacological agents (such as the noncompetitive blockers histrionicotoxin, phencyclidine, local anesthetics . . .) are well documented in the case of the acetylcholine receptor. The duration of the long lasting changes in synapse efficiency might be several minutes when one takes the values of the rate constants of the $R \rightarrow D$ transition determined for T. marmorata acetylcholine receptor. They could be made longer by the covalent modification of the receptor protein. Such modifications (e.g., phosphorylation-dephosphorylation reactions) are already known to play a fundamental role in the regulation of metabolic pathways (see (37)) and might possibly contribute to the regulation of synaptic properties via similar molecular mechanisms.

The possible contributions of such receptor transitions to actual synapse modifications under "learning" conditions in vivo, however, is still entirely hypothetical.

The synaptic scheme of classical conditioning derived from Aplysia work (see Quinn, Fig. 2, this volume) is centered around an observed presynaptic change of efficiency of sensory/motor neuron synapses. It can be noted, however, that according to this scheme, the primary target of regulation by the conditioned stimulus (expressed as Ca^{++} bound to calmodulin) is the adenylate cyclase-serotonin receptor complex which, within the nerve ending, is primarily postsynaptic to a hypothetical serotonin interneuron. This regulation secondarily manifests itself by a change in transmitter release by the nerve ending. This scheme can thus be reinterpreted in terms of a strictly postsynaptic mechanism.

In a general manner, postsynaptic regulation of synapse efficacy takes advantage of the postsynaptic soma (or dendrite) as being the site of convergence of multiple nerve endings. On the other hand, a presynaptic regulation, in order to be "associative," would require a trans-synaptic transfer of signal. Despite its "indirect" nature, such regulation may nevertheless be involved in the coupling between neurons (see section on SELECTIVE STABILIZATION OF SYNAPSES DURING DEVELOPMENT).

Consequences of the Theory

The proposed theory for adult learning falls into the category of typical "selective" (rather than instructive) mechanisms. The "generator of internal diversity" could be identified with the brain's production of pre-representations. The "mechanism for selection" would result from the resonance between the environmentally-induced percept and the pre-representation. Several points in this scheme are distinct from Edelman's (16) model, in particular, the postulates of "cooperative" assemblies of neurons and resonance as a mechanism of matching between percept and pre-representations. It also differs from those of Little and Shaw (43) and Hopfield (28), which postulate cooperative assemblies of neurons but without selection.

Some of the postulates are reminiscent of old notions from psychological literature. For instance, the concept of pre-representation has some analogies with Hume's "chimera" (59), Helmholtz's "unconscious inferences" (59), or Neisser's "schemata" (46). The possibility of combinations (or association) of images (or representations) is, of course, present in early associationist work (Locke, Herbart, Taine (59)). Finally, the general notion of resonance is found, in a different context, in the work of Helmholtz (59) and already explicitly discussed as a basis for learning by Loeb (44). It is also discussed in more recent work by Greene (21), Thom (53), Grossberg (22), and Shepard (51).

The proposed theory might be of some use for the following reasons:

1) It is based on a simple set of assumptions with the aim of bridging the gap between behavioral and neurobiological approaches. The concepts of pre-representation and stored representation are common to both fields and have experimental implications on both sides. The theory accounts, of course, for associative learning. Storage will take place when the animal produces a pre-representation which links conditioned and unconditioned stimulus before concomitant stimulation. It also fits

with the cognitivist approach to behavior with, in addition, a neural hypothesis for the "molar units" formerly assumed on a behavioral basis. The postulate that selection operates on these "molar units" might be of some use for the interpretation of more complex computations and "mental" performances which are out of the scope of this review.

2) The cellular and molecular mechanisms postulated are simple and based on known molecular properties of, for instance, receptors for neurotransmitters. The production of pre-representation, in addition, makes use of the spontaneous activity of neurons and of their widespread oscillatory behavior (3).

3) The theory predicts a characteristic "variability" of the end product of the selection (see also (10, 18)) which, indirectly, reflects the variability of the pre-representation. Such a variability would be manifested by differences in the neuronal graphs stored after presentation of the same external stimulus in different "isogenic" individuals (or in the same organism at different moments) and by differences in the memorized features of the stored object accessible to introspection from one experience to another.

4) The direct "print" of a percept into the neuronal network according to an instructive mechanism would lead to a rapid saturation of the system. A selective storage mechanism, on the other hand, results in the "printing" of only a few features of the percept. Despite the intrinsic arbitrary character of these features, such a "fragmentary" storage might increase the memory capacity of the system, and this would result in a significantly slower saturation. Also, the selective mechanism leads to the rejection of percepts which do not match any pre-representation. The nervous system then stores representations which match its own organization (9, 53, 54).

5) Direct experimental tests of the theory might result from both the analysis of the cooperative coupling of neurons at the cellular and molecular level and from the development of methods for measuring the concerted firing of large but dispersed assemblies of neurons.

6) Finally, since many speculations have been made about the significance of dream sleep (see (11, 14, 33)), additional ones would not hurt! For instance, dreams could result from the spontaneous genesis of pre-representations under conditions (sleep) where the selection by resonance does <u>not</u> operate.

THE SELECTIVE STABILIZATION OF SYNAPSES DURING DEVELOPMENT

Coding the Complexity of Adult Brain Functional Organization

If learning in the adult results from the selection of "variations" from prewired labile assemblies of nerve cells, the question then is where do these prewired neuronal graphs come from? Since they are laid down in the course of development, two main hypotheses can be envisioned: a) the establishment of the neuronal networks during embryonic and postnatal development is autonomous and results from strictly genetic mechanisms, b) an interaction with the environment takes place and contributes, directly or indirectly, to the development of the adult connectivity.

The main features of the anatomy of the nervous system, and in particular that of the brain, remain invariant from individual to individual for a given species and appear independent of the environment in which the organism developed. On the other hand, gene mutations and/or chromosome rearrangements drastically alter this organization down to the cellular and even synaptic level. Also, structural genes for neurotransmitter biosynthetic enzymes (38) or receptors (15, 47) have been cloned. Thus, genes code for the invariance of the anatomy and for the basic molecular components responsible for nerve activity.

Estimation of the total number of genes present in the genome of the fertilized egg, however, raises several questions. First of all, on a weight basis, the total DNA of the fertilized egg in man cannot contain more than two million genes (coding themselves for proteins of approx. 40,000 mol. wt.). In reality, most of this DNA is either non-coding or repetitive. There are thus no more (perhaps less) than 200,000 structural genes present in the human genome. This number, compared to that of cortical neurons (about 30 billions in humans), looks exceedingly small. On the other hand, the number of combinations of sets of these genes can be significantly larger. In addition, most of the genes are expressed in the brain (13). On strictly theoretical grounds, no opposition thus exists to a full genetic coding of brain organization.

A more serious question is raised by comparing the evolution of brain organization and performances with that of gene numbers. Grossly, the total amount of DNA per cell does not change from mouse to man, despite major differences in brain complexity. The complexity of the anatomy increases much faster than that of the genome in a nonlinear manner. Also, comparison on quantitative grounds of the detailed synaptic organization of identified neurons in genetically identical (isogenic) individuals (Daphnia, Poecilostoma) reveals a significant variability from one individual to the other (40). Such a phenotypic variance is not expected from a strict, genetically coded organization.

Finally, following ethological work on imprinting (see Bateson, this volume), song learning (see Marler and Gould, this volume), and, of course, Hubel and Wiesel's (29) work on the visual cortex, a large body of experimental evidence indicates that the interaction with the outside world regulates, or is even required, for a full development of the adult brain's connectivity and performance.

These few observations and remarks, as well as many others, suggest that epigenetic mechanisms associated with an interaction with the outside world contribute to the development of brain organization, particularly in higher vertebrates and especially in humans.

Epigenesis by Selective Stabilization of Synapses

In the course of the development of the nervous system, cell proliferation is followed to an even larger extent by synapse proliferation. Since cell division mostly takes place in mammals before birth (with the major exception of the cerebellum), the theory proposed deals with the outgrowth and stability of synapses (5, 10, 12). It postulates that, to some extent, a Darwinian selection of synapses contributes to the final establishment of the adult organization.

The initial state is taken as the "critical" stage of development where the connectivity between neurons reaches a maximum. For a given set of neurons many more connections are present at this stage than in the adult. At the cellular level, there may be redundancy of the innervation. At the network level, the diversity of possible connections between neurons is at its maximum.

At this stage, any given synapse may exist under at least three states: labile (L), stable (S), and degenerated (D). Only the L and S states transmit nerve impulses, and the acceptable transitions between states are L→S, L→D, and S→L. According to the theory (10) the evolution of the connective state of any given synapse is governed, in a retrograde manner, by the total message of activity afferent to the postsynaptic soma during a given time interval. As a consequence, according to rules analogous to those discussed in the section on LEARNING BY SELECTION IN THE ADULT BRAIN a given afferent multimessage will cause the stabilization of a matching set of synapses from the maximal neuronal graph, while the other will regress.

Consequences of the Theory and Test

The theory is a selective one. The "generator of internal diversity" lies in the mechanics of neurite extension, in the motion of the growth cone, and in the invasion of the target by the exploratory axons and dendrites. Despite its genetic determinism, this development takes place with important fluctuations. The "mechanism of selection" associated with the interaction with the environment consists in the stabilization of a preexisting pattern of connections matching the afferent message. The spontaneous activity of the neurons, known to exist very early in development, may contribute to a "resonance" with percepts, as in the adult case (see section on LEARNING BY SELECTION IN THE ADULT BRAIN) but may also directly participate in a strictly internal selection, in particular during embryonic development when the embryo is isolated from the outside world.

1) This epigenetic step creates an increased internal order as a consequence of selection. To operate, it does not need much additional genetic information. Moreover, the genetic information can be shared between different systems of neurons. The cost in genes is low. The theory thus offers one plausible mechanism among many for coding complexity in organization with a small set of genes. By the same token, it accounts for the paradoxical nonlinear increase of complexity of the functional organization of the nervous system compared with that of the genome in the course of mammalian evolution.

As a consequence, diversification of neurons belonging to the same category occurs: each one acquires its individuality or "singularity" (8) identified as the precise pattern of connections it establishes. A major consequence of the theory is that the distribution of these singularities may vary significantly from one individual to the next. Indeed, it can be demonstrated rigorously that the same afferent multi-message may stabilize different connective organizations that nevertheless result in the same input-output relationship (see also section on LEARNING BY SELECTION IN THE ADULT BRAIN). The "variability" referred to in the theory may account for the phenotypic variance observed between different isogenic individuals.

2) Still, only fragmentary experimental data are available as <u>tests</u> of the theory. They have been obtained mainly with simple systems: the

developing neuromuscular junction (see (8)), the climbing fiber-Purkinje cell synapse in the cerebellum (see (45)), the autonomic ganglion (see (49)) from mammals. In all these instances, a stage of transient multiple innervation of the target muscle cell or neuron has been identified followed by the elimination of a large contingent of synapses (from 1/2 to 4/5). In a few cases (see for the neuromuscular junction (1, 2) activity has been shown to regulate this evolution. Systematically, in agreement with the theory, blocking slows and enhanced firing speeds up the elimination of supernumerary synapses. In the case of the neuromuscular junction, a detailed molecular mechanism has been proposed for the competition of several motor nerve endings (20) on the basis of a limited stock of a retrograde factor, μ , produced by the muscle cell and actively taken up and transformed in a stabilization factor by the active nerve endings. Attempts are presently being made to isolate such factors (26, 27).

3) The theory also accounts for the so-called critical periods or sensitive phases of learning and/or imprinting. They may correspond to the transient stage of maximal innervation (or diversity) where the synaptic contacts are still in a labile state. This stage is well-defined in the case of a single category of synapses. In the case of complex systems, such as the cerebral cortex, multiple categories of circuits become successively established, and, accordingly, many outgrowth and regression steps successively take place. In this sense the whole period of postnatal development becomes "critical"! It is worth recalling that in humans this period is exceptionally long. This prolonged epigenesis of the cerebral cortex would not cost many genes but has a considerable impact on the increased complexity and performances of the adult brain.

CONCLUSIONS

These two contributions to learning theory, in the adult and during development, are based upon Darwinian selection mechanisms. The "variations" upon which the selection operates are caused by the concerted spontaneous firing of labile assemblies of neurons or pre-representations in the adult and associated with the three-dimensional patterns of labile connections which form at the critical stage of maximal connectivity in the developing organism. Selection results from the resonance or matching with environment-induced percepts or stimuli which stabilize either a conformational state of a regulatory protein (such as a receptor for a neurotransmitter) in an already-wired network or a growing nerve connection.

The number of "combinations" internally generated by the pre-representations or the developing network of connections might be large enough to offer, at critical times, a pattern sufficiently close to the externally evoked percept so that it becomes stabilized. An isomorphism with elements or features of the outside world may thus, indirectly, develop within the organism. The genesis of pre-representations would make use of such patterns of connections themselves selected during the epigenetic steps of development. Also, the epigenesis operates upon a general organization of the nervous system, which has been selected in the course of the evolution of species and might itself constitute a representation of the world.

REFERENCES

(1) Benoît, P., and Changeux, J.-P. 1975. Consequences of tenotomy on the evolution of multi-innervation in developing rat soleus muscle. Brain Res. 99: 354-358.

(2) Benoît, P., and Changeux, J.-P. 1978. Consequences of blocking nerve activity on the evolution of multi-innervation in the regenerating neuromuscular junction of the rat. Brain. Res. 149: 89-96.

(3) Berridge, M., and Rapp, P. 1979. A comparative survey of the function, mechanism and control of cellular oscillations. J. Exp. Biol. 81: 217-280.

(4) Bodmer, W., and Cavalli-Sforza, L. 1976. Genetics, evolution and man. San Francisco: W. Freeman.

(5) Changeux, J.-P. 1972. Le Cerveau et l'évènement. Communications 18: 37-47.

(6) Changeux, J.-P. 1980. Genetic determinism and epigenesis of the neuronal network; Is there a compromise between Chomsky and Piaget? In Language and Learning, ed. M. Piattelli, pp. 184-202. Cambridge, MA: Harvard University Press.

(7) Changeux, J.-P. 1981. The acetylcholine receptor: an allosteric membrane protein. Harvey Lect. 75: 85-254.

(8) Changeux, J.-P. 1983. Concluding remarks: about the "singularity" of nerve cells and its ontogenesis. Progr. Brain Res. 58: 465-478.

(9) Changeux, J.-P. 1983. L'Homme neuronal. Paris: Fayard.

(10) Changeux, J.-P.; Courrège, P.; and Danchin, A. 1973. A theory of the epigenesis of neural networks by selective stabilization of synapses. Proc. Natl. Acad. Sci. USA 70: 2974-2978.

(11) Changeux, J.-P., and Danchin, A. 1974. Apprendre par stabilisation sélective de synapses en cours de développement. In L'unité de l'homme, eds. E. Morin and M. Piatteli, pp. 320-357. Paris: Le Seuil.

(12) Changeux, J.-P., and Danchin, A. 1976. Selective stabilization of developing synapses as a mechanism for the specification of neuronal networks. Nature 264: 705-712.

(13) Chaudhari, N., and Hahn, W. 1983. Genetic expression in the developing brain. Science 220: 924-928.

(14) Crick, F., and Mitchison, G. 1983. The function of dream sleep. Nature 304: 111-114.

(15) Devillers-Thierry, A.; Giraudat, J.; Bentaboulet, M.; Changeux, J.-P. 1983. Complete mRNA coding sequences of the acetylcholine binding α-subunit of Torpedo marmorata acetylcholine receptor: A model for the transmembrane organization of the polypeptide chain. Proc. Natl. Acad. Sci. USA 80: 2067-2071.

(16) Edelman, G. 1978. The Mindful Brain. Cortical Organization and the Group-selective Theory of Higher Brain Functions. Cambridge, MA: MIT Press.

(17) Edelman, G. 1981. Group selection as the basis for higher brain function. In The Organization of the Cerebral Cortex, eds. F. Schmitt et al. Cambridge, MA: MIT Press.

(18) Edelman, G., and Finkel, L. 1984. Neuronal group selection in the cerebral cortex. In Dynamic Aspects of Neocortical Function, eds. G. Edelman et al. New York: John Wiley, in press.

(19) Glansdorff, P., and Prigogine, I. 1971. Structure, stabilité et fluctuations. Paris: Masson.

(20) Gouzé, J.-L.; Lasry, J.-M.; and Changeux, J.-P. 1983. Selective stabilization of muscle innervation during development: a mathematical model. Biol. Cybern. 46: 207-215.

(21) Greene, P. 1962. On looking for neuronal networks and "cell assemblies" that underlie behavior. Bull. Math. Biophys. 24: 247-275; 395-411.

(22) Grossberg, S. 1980. How does the brain build a cognitive code? Psych. Rev. 87: 1-51.

(23) Hebb, D. 1949. The Organization of Behavior. New York: Wiley.

(24) Heidmann, T., and Changeux, J.-P. 1982. Un modèle moléculaire de régulation d'efficacité au niveau postsynaptique d'une synapse chimique. C.R. Acad. Sc. Paris 295: 665-670.

(25) Heidmann, T.; Oswald, R.; and Changeux, J.-P. 1983. Multiple sites of action for non competitive blockers on acetylcholine receptor rich membrane fragments from Torpedo marmorata. Biochemistry 22: 3112-3127.

(26) Henderson, C.E.; Huchet, M.; and Changeux, J.-P. 1981. Neurite outgrowth from embryonic chicken spinal neurons is promoted by media conditioned by muscle cells. Proc. Nat. Acad. Sci. USA 78: 2625-2629.

(27) Henderson, C.E.; Huchet, M.; and Changeux, J.-P. 1983. Denervation increases the neurite-promoting activity in extracts of skeletal muscle. Nature 302: 609-611.

(28) Hopfield, J. 1982. Neural networks and physical systems with emergent collective computational abilities. Proc. Nat. Acad. Sci. USA 79: 2554-2558.

(29) Hubel, P., and Wiesel, T. 1977. Functional architecture of macaque monkey visual cortex. Ferrier Lecture. Proc. Roy. Soc. Lond. B 198: 1-59.

(30) Hunter, W.S. 1913. The delayed reaction in animals. Behav. Monogr. 2: 6.

(31) James, W. 1909. Précis de psychologie. Paris: Marcel Rivière.

(32) Jerne, N. 1967. Antibodies and learning: selection versus instruction. In The Neurosciences, eds. G. Quarton et al., pp. 200-205. New York: Rockefeller University Press.

(33) Jouvet, M. 1974. Neurobiologie de rêve. In L'Unité de l'Homme, eds. E. Morin and M. Piattelli, pp. 354-392. Paris: Ed. Le Seuil.

(34) Kandel, E. 1979. Cellular insights into behavior and learning. Harvey Lect. 73: 19-92.

(35) Katz, B. 1966. Nerve Muscle and Synapse. New York: McGraw-Hill.

(36) Kosslyn, S. 1980. Images and Mind. Cambridge, MA: Harvard University Press.

(37) Krebs, G., and Beavo, J. 1979. Phosphorylation, dephosphorylation of enzymes. Ann. Rev. Biochem. 48: 923-960.

(38) Lamouroux, A.; Biguet, N.; Samolyk, D.; Privat, A.; Salomon, J.-C.; Pujol, F.; and Mallet, J. 1982. Identification of cDNA clones coding for rat tyrosine hydroxylase antigen. Proc. Natl. Acad. Sci. USA 79: 3881-3885.

(39) Leder, P. 1981. The genetics of antibody diversity. Sci. Am. 246 No. 5: 72-83.

(40) Levinthal, F.; Macagno, E.; and Levinthal, C. 1976. Anatomy and development of identified cells in isogenic organisms. Cold Spring Harbor. Symp. Quant. Biol. 40: 321-331.

(41) Little, W. 1974. Existence of persistent states in the brain. Math. Biosci. 19: 101-120.

(42) Little, W., and Shaw, G. 1975. A statistical theory of short and long term memory. Behav. Biol. 14: 115.

(43) Little, W., and Shaw, G. 1978. Analytic study of the memory storage capacity of neural network. Math. Biosci. 39: 281-290.

(44) Loeb, J. 1900. Comparative Physiology of the Brain and Comparative Psychology. New York: Putnam.

(45) Mariani, J. 1983. Elimination of synapses during the development of the central nervous system. Progr. Brain Res. 58: 383-392.

(46) Neisser, U. 1976. Cognition and reality. San Francisco: Freeman.

(47) Noda, M.; Takahashi, H.; Tanabe, T.; Toyosato, M.; Kikyotani, S.; Furutani, Y.; Horose, T.; Takashima, H.; Inayama, S.; Miyata, T.; and Numa, S. 1983. Structural homology of Torpedo californica AchR subunits. Nature 302: 528-532.

(48) Peretto, P. 1983. Collective properties of neural networks: a statistical physics approach. Biol. Cybern., in press.

(49) Purves, D., and Lichtman, J. 1980. Elimination of synapses in the developing nervous system. Science 210: 158-157.

(50) Shepard, R. 1975. Form, formation and transformation of internal representations. In Information Processing and Cognition, ed. R. Solso. Hillsdale, NJ: Erlbaum.

(51) Shepard, R. 1984. Ecological constraints on internal representation. Third J. Gibson Memorial Lecture, Cornell University, in press.

(52) Stent, G. 1973. A physiological mechanism for Hebb's postulate of learning. Proc. Natl. Acad. Sci. USA 70: 997-1001.

(53) Thom, R. 1968. Topologie et signification in "l'Age de la Science" n° 4. In Modèles mathématiques de la morphogénèse, ed. R. Thom. Paris: Bourgeois.

(54) Thom, R. 1980. Modèles mathématiques de la morphogénèse. Paris: Bourgeois.

(55) Tolman, E.C. 1948. Cognitive maps in rats and men. Psychol. Rev. 55: 189-208.

(56) Urbain. 1981. Le réseau immuniaire. La Recherche 126: 1056-1066.

(57) Von der Malsburg, C. 1981. The correlation theory of brain function. Internal report 81-2, July 1981. Göttingen: Department of Neurobiology, Max Planck Institute for Biophysical Chemistry.

(58) Young, J.Z. 1973. Memory as a selective process. Australian Academy of Science Report: Symposium on Biological Memory, pp. 25-45.

(59) Note: For references before 1900, see: Bercherie, P. 1983. Génèse des concepts freudiens. Paris: Navarin.

The Biology of Learning, eds. P. Marler and H.S. Terrace,pp. 135-147. Dahlem Konferenzen 1984. Berlin, Heidelberg, New York, Tokyo: Springer-Verlag.

Learning in Invertebrates

B. Heinrich
Dept. of Zoology, University of Vermont
Burlington, VT 05405, USA

Abstract. Most of our knowledge of learning in the hymenoptera with its hundreds of thousands of species is so far restricted to one species, the honey bee. Learning mechanisms in this animal as it may relate to ecological requirements will be discussed. Furthermore, I will explore ecological adaptations of behavior and learning drawn from the literature and make predictions about the types of learning expected in some other hymenoptera that face entirely different learning problems than those faced by honey bees. Finally, I suggest that a comparative approach to a study of learning in the hymenoptera may enlarge our insights into constraints in the biology of learning.

INTRODUCTION

The diversity within the hymenoptera, with its hundreds of thousands of species of contrasting life strategies, is based primarily on specific behavioral adaptations. The behavioral specificity is in large part based on hard-wired responses, but specificity of response in some species is also achieved through learning. Details of learning have been explored in only one species, the honey bee, which now provides a strong basis not only for studies on learning mechanisms (Gould, this volume) but also for investigations on evolutionary aspects of learning.

An evolutionary approach to a problem is necessarily a comparative approach, and the hymenoptera with its 38 major families offers the biologist an incomparably rich array of diversity for evolutionary and comparative studies. In one family alone, the bees, there are over 20,000 species in about 100 genera. If learning by the honey bee is part of a behavioral repertory fitted to a specific ecological strategy, then it can be expected that learning in other hymenoptera has also been shaped by ecological requirements and constraints. A strong case has been made by Gould and Marler (this volume) for species-specific learning, especially in birds as it relates to ecological requirements. I will try here to identify some of these requirements, particularly in social predatory wasps and in bumblebees, in order to suggest where future experiments might differentiate between rules of learning per se and specific adaptations.

There are solitary, semi-social, eusocial, and parasitic bees, and they are obligatory flower specialists, facultative specialists, and generalists. But almost all of the flowers from which bees forage have evolved to stand out from the environment by means of bright colors, scents, and conspicuous geometric patterns. In contrast, the vespid wasps face a somewhat different world than the bees. In the 3,000 or so vespid species there is also the great gamut of sociability. However, these animals derive the protein needed for larval growth not from pollen, but from live insect prey that has evolved to remain hidden, blending in closely with its environment. Again we find specialists that feed only on one kind of prey, while others are generalists.

HONEY BEES

Honey bees learn to associate specific scents, colors, and geometrical patterns and their movements (5) with food. They learn and remember some colors (19, 20) and scents (13) more easily than others, and the learning tasks are time-linked (14), with the time signal apparently involving a humoral transfer of at least some information in the mushroom bodies of the brain (6, 18). In addition, they learn the solar azimuth (16) through a memory of the sun's course with respect to local landmarks (3). Comb building by young bees is, in part, learned using as a reference the comb they emerged from and walk over (24). Honey bees also learn to recognize their queen by odor (2). Bees learn mechanical tasks (25) such as where to insert their tongue in some flowers (20, 30) and how to avoid the tripping mechanism of others (26). Nevertheless, there is still much of learning in honey bees that remains to be explained (Gould, this volume).

From the classical work on the physiology of learning and memory in honey bees (22), we know that the bees store information on odor and spectral colors first in a short-term sensory integration center and then may transfer information (of odor) in several minutes into a long-term memory. Nevertheless, inheritance plays no small part in behavior (27), including the flowers visited. In several generations, strains of honey bees were produced that collected pollen from alfalfa, Medicago sativa, from which they usually collect no pollen unless no other pollen source is available (23).

The above information suggests that there is no hard and fast distinction between hard-wired behavior and learned behavior, except that one is less flexible than the other. It would be of great interest to know if the specific physical changes of the nervous system that result through development under a genetic program are the same as those that subsequently modify the physical system through experience to affect the changed behavior (Quinn, this volume). Perhaps the "open" program of honey bees is a delayed maturation that allows for environmental factors to affect the course of the maturation.

In a colony of highly social bees with short-lived individuals and a long colony cycle, the late generations in any one season will find entirely different kinds of flowers in the field than those emerging in the spring. Under these conditions any one bee should have a relatively open program when beginning to forage. Nevertheless, some flower signals are more reliably associated with profitable rewards than others, and signal-biased learning in honey bees (22) and bumblebees (12) can thus be explained in terms of its potential ultimate mechanisms. Honey bees learn to associate sugar with violet at 85% probability after one reward (their fill of 2m sucrose), while similar probability of choosing green requires four rewards (22). The bumblebees, Bombus terricola, learn to forage from blue and white flowers (12) after 50 and 250 rewards, respectively (each reward was 1μ of 50% sucrose). To what extent violet and blue flowers are reliable reward indicators in the field has so far, however, not been critically examined. However, the natural history of bee learning, like that of birdsong learning (Immelmann, this volume) can provide an ultimate basis of understanding for the proximal mechanisms.

STM-LTM

It is generally thought that there is a functional similarity between inter-event relationships that influence learning in vertebrates and invertebrates (Sahley, this volume). Honey bees first store information in a short-term memory (STM) and then within minutes transfer it into a long-term memory (LTM) that, at least in some species, is relatively immune to extinction (22). But apparently LTM is sometimes of short duration. Bumblebees that have been rewarded during many foraging trips at artificial blue flowers (12) or from a highly rewarding jewelweed flower in the field (8) are almost immune to switching. However, if they have been foraging from white artificial flowers (12) or from jewelweed for only a few days (9), then they are capable of switching on subsequent days, following distinct "forgetting" overnight. Apparently the time-course of STM and LTM of the two kinds of bees are different, or the two types of memory are not all-or-none phenomena.

Is STM a physiological necessity for LTM to occur? What can the animal do with STM that it could not do if it had LTM without the intervening STM? By making the color signal at artificial flowers available to bees during the approach to the flower, and then eliminating it while the bees fed, Menzel (22) was able to show that honey bees enter the color signal of a flower they have approached into their STM, and only later transfer it into their LTM. Why do they "see" the signal during the approach before they are potentially rewarded and when they will in all likelihood not be rewarded? (In the field where there is competition between foragers, most flowers visited may be empty, and even in the laboratory (12) bees may visit >70 empty artificial flowers in a row.)

In the absence of more comparative data it is probably fruitless to speculate on why honey bees enter information on flower color and scent into their STM on approach to a flower at which they may or may not be rewarded, rather than after being rewarded. However, the second question of why they have an STM, or how long or short it is, might be related to the reward schedules experienced in nature.

Erber (4) showed that short-term memories are quickly lost in honey bees if not reinforced within minutes. He speculated that rapid decay of the short-term memory phase prevents the bee from entering the signal of an uneconomical flower-kind into its long-term memory. Menzel (21) has also concluded that the short-term memory may not necessarily be a physiological precursor to long-term memory. Rather, it may be a mode of information storage that allows for refinement of incoming information. "Refinement of incoming information" could include not only the ability to overshadow, but also to resist overshadowing while encountering non-rewarding flowers until a sufficient sample has been taken of the flower population. Interestingly, bumblebees prefer a constant to a variable reward schedule (29) which should minimize sampling error.

It is possible, however, that there is no ethological significance to the above physiological and behavioral correlates. Future comparative studies might throw light on possible neurophysiological mechanisms and specific adaptations.

If at least some of the bees' search behavior during foraging is genetically predisposed (17, 28), then it should reduce the time in which rewarding flowers are found and committed to memory. Flowers have evolved to be conspicuous to bees by their bright colors, sweet scents, and geometrical patterns in order to secure pollination service. Within any one plant species the flowers are uniform in signal, further aiding learning of that flower kind. In addition, the signals associated with the flowers of concurrently blooming plants tend to contrast strongly with each other, so that the choices a bee has to make are relatively few and easy.

PREDATORY WASPS

Not all hymenoptera face such an "easy" foraging task as bees. The predatory wasps, for example, live from insect prey that has evolved to blend into the background and to resemble common inedible objects so as to avoid detection and to confuse the predators. Only the unpalatable insects have evolved to be colorful and conspicuous so as to be unambiguously identifiable. How have the predatory wasps evolved to solve this problem, and what role does learning play in it?

Many solitary predatory wasps, like some solitary bees, specialize. Different species hunt for cicadas, spiders, flies, caterpillars, or some other specific prey. Having such predictable prey (due to the narrowness of their specialization) they can evolve to be attuned to very specific subtleties. Whether or not learning in recognition of prey types is involved during foraging by the specialists is not known. It is tempting to speculate, however, that something analogous to birdsong learning might be involved (Marler, this volume), where the predators match information with specific templates.

Extreme specialization as seen in most solitary wasps could be a liability in a social predatory wasp for the same reason that it is in social bees: the potential insect prey are nearly as seasonal as are the different kinds of flowers. The prey available to the wasps at the beginning of the colony cycle in the spring will likely be unavailable later in the summer when cohorts of new prey have appeared. Presumably social wasps cannot be hard-wired to find and capture an almost infinite variety of potential prey that closely resembles an even greater variety of unpalatable features, unless they can extract a common feature or features that separate living from nonliving matter. Can the problem faced by these social wasps be solved by capacity for classical conditioning, as it is in bees foraging from flowers? If so, how might the learning differ?

At the present time no data are available to provide an answer. I offer only the following preliminary observations of foraging by the social wasp, Dolichovespula maculata, to suggest that there are potential differences. Dolichovespula maculata forage for live insect prey rather than pollen to feed the larvae in their nests. The insect prey upon which they live - small moths, flies, caterpillars, membracid bugs - are usually cryptic and are available in the field in an almost infinite array of shapes and colors, as are the leaf blemishes, twigs, seeds, buds, bark, and other natural features that they may mimic closely. If a wasp attacks one "green stub" that is a leaf petiole and subsequently rejects all green stubs, it might potentially miss many geometrid caterpillars, membracid bugs, grasshoppers, and some other insects. How does Dolichovespula deal with the dazzling variety of potential prey that often cannot be distinguished from inanimate matter except under detailed inspection?

As a first step it would be predicted that they are highly responsive to movement and possibly to size, rather than to color, shape, or scent. Since the difference in signals between the hundreds of potential prey and inanimate objects are often extremely subtle, it might be too much to expect of the wasps to discriminate on the basis of physical features. Furthermore, prey items will be encountered very infrequently, while non-rewarding inanimate objects will be encountered at very high frequency. This situation may have resulted in the evolution of learning patterns or mechanisms very different from those of honey bees that learning theorists have examined in detail.

I observed D. maculata foraging in the field (11) and these wasps (aside from pursuing all small moving objects) appeared to pounce every few seconds on all sorts of inanimate matter such as leaf petioles, spots on leaves, bird droppings, etc. In a total of 260 pounces I only recorded two food captures, a small moth and a fly. Most wasps could not be followed for long as they hunted in the field, but I observed one for more than an hour as it patrolled at a pine tree where an aphid colony attracted prey. This wasp attacked the same upright dead twig on one branch a total of 32 times. Indeed, it attacked the same twig every time it came near it, unless it had just attacked it within the previous minute. From an evolutionary perspective, and from these initial field observations, it appears that these wasps either do not learn to avoid non-rewarding cues that are already in their inherited "window" or their "motivation" is not decreased when confronted with non-rewarding cues.

OPERANT CONDITIONING?

The association of signals with food is only the first part of a successful foraging where behavior grades from innate to learned patterns of behavior. In bees the motor skills required to handle flowers are an obvious second component.

Leaf-cutter bees place pollen onto the ventrum of the abdomen. Honey bees and bumblebees place pollen into the corbiculae of their hind legs. The motor patterns for these behaviors are presumably inherited. However, some of the other motor skills required for foraging depend on the shape of the flowers and the location within the flower where the pollen and/or nectar is located. These skills appear to be inherited as well in bee species that are associated with specific kinds of flowers, and at least in part learned in other bee species that necessarily depend on many kinds of flowers.

Whether or not the skills are "hard-wired" or learned appears, from an evolutionary perspective, to be related to foraging economics. Solitary bees of desert regions that have flowers available only for a limited time following the seasonal rains are usually species-specific flower specialists (17) that presumably cue to the correct flowers and manipulate them correctly without costly and time-consuming trial-and-error learning. Few data are available to evaluate quantitatively this hypothesis, although the recent study by Strickler (28) shows that the bee species which in the field always specializes on Echium vulgare removes more pollen from each flower per visit and visits more flowers per unit time than nonspecialists that are normally not associated with these flowers.

No extended colony cycle, with emphasis on sociality, would likely be possible if some bees could not harvest from a variety of plants that come into bloom throughout the season. Clearly the social bees cannot be hard-wired for any one specific kind of flower when they necessarily depend on many. Being born with an open program of responses, they adapt to some flower kinds by learning. However, not all flower kinds require learning to be manipulated. When the flowers provide easy access to the food rewards, bumblebees handle the flowers correctly from their first foraging trip from the hive (8), but when foraging from other flowers where the nectar and pollen are hidden or are in uncommon arrangements, the forager's performance improves with experience (8, 15).

The necessary motor patterns for foraging for pollen and nectar differ at different kinds of flowers. To collect pollen from blue bindweed, Solanum dulcamara, bumblebees need to extract the pollen from tubular hanging anthers. Bumblebees do so routinely by grasping the flowers with their mandibles and shaking the flowers. The pollen drops down onto their ventrum, from where it is scraped off into the corbiculae. To collect pollen from wild rose flowers, bumblebees grasp groups of anthers between thorax and abdomen, shaking these anthers to release the pollen. At the horizontal platform flowers of wild carrot, Daucus carota, they press their bodies close to the florets and rapidly crawl over the flower while friction loosens the pollen. The pollen then becomes attached to the body hairs from which it can be transferred to the corbiculae during flight to the next inflorescence. Presumably some of the behavior involved in the mechanical aspects of foraging are part of a "closed" program, while others due to environmental unpredictability (flower variety) are in an "open" program (Terrace, this volume).

Nectar collecting also requires specialized motor patterns. To obtain nectar from turtlehead, Chelone glabra, bumblebees need to pry apart the corolla lips of the flower in order to crawl inside. To collect nectar from jewelweed, Impatiens biflora, long-tongued bumblebees enter these tubular flowers probing deep into the interior. The pollen that inadvertently adheres to the bees as they enter these flowers is discarded. In addition to correct motor patterns for specific kinds of flowers, the bees must know whether to collect nectar, pollen, or both from different kinds of flowers. Bumblebees sometimes attempt to collect nectar from species of flowers that only provide pollen, and among those that do collect pollen, foraging speeds vary severalfold among individual bees (8, 9).

MULTIPLE SKILLS

Although bumblebees tend to specialize on one kind of flower even when several different kinds are simultaneously available at the same site, they generally do not restrict their visits to their specialty. Instead, they have "major" and "minor" specialties (9). A strict major on the best flower kind is the most profitable strategy at any one time, but the reward spectrum keeps changing with time. Minoring, which is observed most at the beginning of a bee's foraging career, or in experienced bees, at times when overall rewards are low, allows the bees to change majors to keep up with changing resources. In honey bees, in contrast, individual foragers do not need to track changing resources individually in the field since they are informed of resources in the field by scouts dancing in the hive. It would be of interest to know if there are different learning/forgetting rates between scouts and recruit honey bees, and whether or not there are different learning capacities between honey bees and bumblebees that relate to their different foraging strategies.

Honey bees learn to associate specific color and scent signals with food in one to several trials, and the learning acquisition is independent of the amount of reward (22). If learning by bumblebees follows similar rules, then it is difficult to explain how they are able to narrow down to one kind of flower after sampling many kinds, each offering different amounts of nectar and/or pollen, especially since some of the least rewarding flowers that will be rejected may be visited at 20-30 times per minute, while some of those offering the highest net rewards can be visited at less than a third of this rate and will still be preferred.

If bees can be trained in the laboratory to one signal in one trial, it begs the question of why in the field they may be attentive to numerous signals, at the same time or alternately. Bumblebees having different majors make consecutive visits to the same flower kinds ("runs") of varying lengths, with runs at their major being interspersed with runs at their minor. The length of the runs at major and minor flowers are relatively independent of flower density. For example, in one field one Bombus vagans worker usually visited 1-5 Leontodon autumnalis flowers and then made runs of 10-15 visits to Daucus carota, occasionally visiting several Trifolium pratense. Another B. vagans in the same locality made runs of 10-20 flower visits to T. pratense, interspersed with 1-3 flower visits to D. carota. Still a third B. vagans at that locality made runs of over fifty flower visits to L. autumnalis with short runs of 1-3 flower visits to T. pratense (8). On what basis do the bees determine lengths of runs and switching? At the present time we have no answer. We do know, however, that the bees are highly sensitive to changes in net food rewards, and the length of the runs at any one flower kind increases when the food rewards increase (9).

Length of training is another important variable. Bumblebees that had been trained in the laboratory to artificial flowers by several hundred visits to "flowers" of one color later refused to sample and switch to another color even though these contained superior rewards (12). Similarly, jewelweed specialists of 3-4 weeks of experience refused to switch to abundantly available aster that other individuals were specializing on, after the jewelweed was almost all experimentally removed (8).

Novel signals, which are often associated with high rewards, may also be a factor in the learning biology of bees. The most novel flowers are often the most mechanically "difficult," which also often offer tremendous potential profits in comparison to those that are available to all bees and require little skill to manipulate.

COST OF LEARNING

The caloric cost of learning to manipulate the difficult flowers can be estimated from the improvement of handling accuracy and rate of flower visitation. For example, with Bombus vagans workers foraging from jewelweed, Impatiens biflora, about 50% landed on and probed in the wrong place for nectar when first starting to forage from these flowers (9). At these and other zygomorphic flowers, the bees' foraging speed improved substantially with experience (9, 15). Experienced bees visited eleven flowers per minute. Bees just starting to forage from these flowers visited only 1-3. Thus, in this case the cost of having to learn rather than handling the flowers correctly from the beginning is 110 cal/min (the reward available from eleven flowers minus 80 cal/min, the minimum reward missed). This cost is well worth it. It would take B. vagans one week of 8 hrs/day foraging to make the same profit at less rich but easy to handle alternative flowers such as Solidago canadensis. However, it only takes the bees one hundred flower visits on the rich flowers such as Impatiens biflora to become proficient at handling them (9).

TEMPERATURE

Under natural conditions nonaquatic invertebrate animals are often subjected to great changes of body temperature, whereas in the laboratory temperature is usually maintained constant. In larvae of the mealworm, Tenebrio molitor, cooling to 2°C following training in a T-maze facilitated retention (1). In adults the results were ambiguous, possibly depending on the duration and timing of the cold. Little is known of the effects of high temperature. In the field foraging bees (10), wasps (11), and bumblebees (7), depending on the species and the conditions, generate thoracic temperatures of 20-46°C, and when inactive, the body temperature of these endothermic insects may reach 0°C in the cold.

Do the body temperatures normally maintained, and the changes of body temperature with time, affect the input and output from the nervous system, including learning? Perhaps temperature is a variable that should also be kept in mind when making direct comparisons of results from different taxa.

A key to the understanding of the evolution of most behavior concerns the balance between potential costs and benefits of various options in the context of physiological constraints or rules. A broad, comparative perspective provides opportunities for distinguishing specific adaptations from physiological rules. Alternately, knowing the physiological rules can give insights into the potential evolutionary options. I have argued that an ethological approach to learning behavior in hymenoptera provides rich opportunities for investigating a diversity of potential learning options. The hymenoptera face a great variety of different ecological and physiological problems, and I conclude that the comparative approach to learning behavior in this group has the potential of enriching our understanding through a synthesis of physiological and ethological insights.

REFERENCES

(1) Alloway, T.M. 1969. Effect of low temperature upon acquisition and retention in the grain beetle (Tenebrio molitor). J. Comp. Physiol. Psychol. 69: 1-8.

(2) Breed, M.D. 1981. Individual recognition and learning of queen odors by worker honey bees. Proc. Natl. Acad. Sci. USA 78: 2635-2637.

(3) Dyer, F.C., and Gould, J.L. 1981. Honey bee orientation: A backdrop system for cloudy days. Science 214: 1041-1042.

(4) Erber, J. 1976. Retrograde amnesia in honey bees (Apis mellifera carnica). J. Comp. Physiol. Psychol. 90: 41-46.

(5) Erber, J. 1982. Movement learning of free flying honey bees. J. Comp. Physiol. A 146: 273-282.

(6) Erber, J.; Masuhr, T.; and Menzel, R. 1980. Localization of short-term memory in the brain of the bee, Apis mellifera. Physiol. Entomol. 5: 343-358.

(7) Heinrich, B. 1975. Thermoregulation in bumblebees II. Energetics of warm-up and free flight. J. Comp. Physiol. 76: 155-166.

(8) Heinrich, B. 1976. The foraging specializations of individual bumblebees. Ecol. Mon. 46: 105-128.

(9) Heinrich, B. 1979. "Majoring" and "Minoring" by foraging bumblebees, Bombus vagans: an experimental analysis. Ecol. 60: 245-255.

(10) Heinrich, B. 1980. Mechanisms of body-temperature regulation in honey bees, Apis mellifera. J. Exp. Biol. 85: 73-87.

(11) Heinrich, B., and Heinrich, M.J. 1983. Strategies of thermoregulation and foraging in two vespid wasps, Dolichovespula masculata and Vespula vulgaris. J. Comp. Physiol. B, in press.

(12) Heinrich, B.; Mudge, P.R.; and Deringis, P.A. 1977. Laboratory analysis of flower constancy in foraging bumblebees: Bombus ternarius and B. terricola. Behav. Ecol. Sociobiol. 2: 247-265.

(13) Koltermann, R. 1973. Rassen - bzw. artspezifische Duftbewertung bei der Honigbiene und ökologische Adaptation. J. Comp. Physiol. 85: 327-360.

(14) Koltermann, R. 1974. Periodicity in the activity and learning performance of the honey bee. In Experimental Analysis of Insect Behavior, ed. L. Barton Browne, pp. 218-227. Berlin, Heidelberg, New York: Springer Verlag.

(15) Laverty, L.M. 1980. The flower-visiting behavior of bumblebees; floral complexity and learning. Can. J. Zool. 58: 1324-1335.

(16) Lindauer, M. 1960. Time compensated sun orientation in bees. Cold Spring Harbor Symp. Q. Biol. 25: 371-377.

(17) MacSwain, J.W.; Raven, P.H.; and Thorpe, R.W. 1973. Comparative behavior of bees and Onagraceae. IV. Clarkia bees of the Western United States. Univ. Calif. Publ. Ent. 70: 1-80.

(18) Martin, V.; Martin, H.; and Lindauer, M. 1978. Transplantation of a time-signal in honey bees. J. Comp. Physiol. A 124: 193-201.

(19) Meineke, H. 1978. Umlernen in der Honigbiene zwischen Gelb- und Blau-belohnung im Dauerversuch. J. Insect Physiol. 24: 155-163.

(20) Menzel, R. 1968. Das Gedächtnis der Honigbiene für Spektralfarben. Z. vergl. Physiol. 60: 82-102.

(21) Menzel, R. 1979. Behavioral access to short-term memory in bees. Nature 281: 368-369.

(22) Menzel, R.; Erber, J.; and Masuhr, T. 1974. Learning and memory in the honey bee. In Experimental Analysis of Insect Behavior, ed. L. Barton Browne, pp. 195-217. Berlin, Heidelberg, New York: Springer Verlag.

(23) Nye, W.P., and Mackenson, O. 1968. Selective breeding of honey bees for alfalfa pollen: fifth generation and backcrosses. J. Apic. Res. 7: 21-27.

(24) Oelsen, von G., and Rademacher, E. 1979. Untersuchungen zum Bauverhalten der Honigbiene (Apis mellifera). Apidol. 10: 175-209.

(25) Pesotti, I. 1972. Dissemination with light stimuli and lever pressing responses in Melipona rufiventris. J. Apic. Res. 11: 89-93.

(26) Reinhardt, J.F. 1952. Some responses of honey bees to alfalfa flowers. Naturalist 86: 257-275.

(27) Rothenbuhler, W.C. 1964. Behavioral genetics of nest cleaning in honey bees. I. Responses of four inbred lines to disease-killed brood. Anim. Behav. 12: 578-583.

(28) Strickler, K. 1979. Specialization and foraging efficiency of solitary bees. Ecol. 60: 998-1009.

(29) Waddington, K.D.; Allen, T.; and Heinrich, B. 1981. Floral preferences of bumblebees (Bombus edwardsii) in relation to intermittent versus continuous rewards. Anim. Behav. 29: 779-784.

(30) Weaver, N. 1957. The foraging behavior of bees on hairy vetch. Insectes Sociaux 4: 43-57.

The Biology of Learning, eds. P. Marler and H.S. Terrace, pp.149-180. Dahlem Konferenzen 1984. Berlin, Heidelberg, New York, Tokyo: Springer-Verlag.

Natural History of Honey Bee Learning

J.L. Gould
Dept. of Biology, Princeton University
Princeton, NJ 08544, USA

Abstract. The life style of honey bees requires them to learn a variety of specific things in order to navigate and forage. The need for learning is sufficiently predictable with regard to the behavioral contexts and useful cues that it is highly structured and relatively easily studied. Although each cue seems to be learned independently, the set of cues memorized in each context appears to be stored as a unit. The exact nature of the storage system for flower shape, landmarks near the food, and navigational landmarks, as well as certain navigational parameters, is not yet perfectly clear. The nature of these ambiguities is explored. The possibility that some form of a synaptic selection strategy might underlie bee learning is also explored in a preliminary way.

INTRODUCTION

Ethology of Learning

The paper by Gould and Marler (this volume) provides much of the conceptual basis for this consideration of insect learning and defines many of the terms used here. In that paper, we suggested that much of the learning in the natural world is organized around innate recognition systems which act as contextual triggers for learning. With regard to associative learning, we argued that the nature and relative importance (salience) of the cues which are subsequently learned are frequently innately preordained. We found it more useful to think of most associative learning as being organized as behavior-specific subroutines, rather than as some undifferentiated general learning process, although we hope that the routines use the sorts of classical, unitary components elucidated by students of learning theory as the building blocks. We concluded that associative learning is adaptive when the information which might otherwise be stored innately either is not sufficiently predictable or is too detailed to be encoded genetically.

In the same paper, we suggested that trial-and-error learning is a strategy for wiring up novel motor behavior into autonomous motor programs. We noted many similarities with the organization of associative learning which suggested that the cueing and the systems for recognizing "progress" the organism makes toward the goal may often be endogenously specified and specific to the particular class of behavior being shaped. And again, as in associative learning, we theorized that trial-and-error learning is adaptive when the motor behavior necessary to accomplish a task is either too unpredictable to be stored innately, or too complex. We also pointed out that the associative and motor-learning routines for a specific behavioral situation frequently appear to be innately linked. We also discussed briefly behavior which appears to depend on learning and mental manipulation of information - what we called cognitive trial and error.

Finally, we considered in passing how the neural substructure for learning might in some cases be organized. Here and in the paper by Marler (this volume), we will consider in a preliminary way the value of classifying behavior into three general levels, a division which can be imagined as cutting orthogonally across the categories of associative, motor, and cognitive learning. These three levels, which probably have at least somewhat different functional bases, relate to the three sorts of behavioral situations animals face: a) a situation sufficiently simple and predictable that recognition or response behavior can be innately specified; b) a simple situation whose details are not sufficiently predictable to preordain recognition or response; and c) a situation too complex to lend itself to completely endogenous recognition of whole patterns or to motor programming. (These levels correspond to the Type 1, 2, and 3 circuits of Marler.) The distinction between b) and c) - between unpredictability and complexity - is that for level b) it might be possible to list and store innately all the likely components of a stimulus to be recognized or a motor behavior to be performed and then select, on the basis of learning, the appropriate elements and arrange them into the proper spatial or temporal relationships. In the context of associative learning, this is something like alpha-conditioning and will be explored here with regard to flower learning, while Marler (this volume) will examine birdsong learning from the same perspective. The motor version of this selection strategy was discussed briefly by Marler and Gould (this volume) for rhesus macaques, squirrels, and oystercatchers.

The third level - the situation for which innate recognition or response is too complex - would require a different strategy for learning and using information. Both this review and the Marler paper will look at this case as well.

Why Bees?

Since this review will focus on honey bees, I feel compelled to explain why other insects do not receive equal treatment. After all, some of the best work on landmark learning was done with digger wasps (77, 78) and hover flies (19). Very elegant experiments have shown how true flies (70) form associative visual or olfactory search images. Some flies and other insects perform reasonably well in laboratory situations in both associative (e.g., (68, 71, 76)) and trial-and-error (e.g., (11, 50)) situations, and both wasps (8) and solitary bees (51) appear to learn and use map-like cognitive lists by which they schedule their day's activity in terms of a series of locations with unique requirements.

The advantage of honey bees is that all these abilities are expressed in one species, and moreover, bees are on the whole far more convenient to work with. As a result, we can hope to see how several different learning programs are integrated together to enhance the survival of a single species with a well-defined niche.

NATURAL HISTORY OF FORAGING

Niche

I will concentrate on the learning which takes place during foraging. In order to see how the learning phenomenology fits the life-style of honey bees, let us look briefly at the foraging cycle (for a detailed review see (37)). Honey bees are somewhere in the middle on the specialist/generalist scale described in the Gould and Marler paper. They gather nectar (carbohydrates) and pollen (protein) from a wide variety of flowers and carry these substances back to the hive. Colonies are perennial and survive the temperate-zone winter by living in tree trunks or other protected cavities and metabolizing their honey stores to keep the queen and many of the worker bees warm all winter. This allows honey bees to exploit spring flowers more effectively - that is, earlier and in greater numbers - than annual species, whose queens overwinter alone. Honey bees are also unusual in that they communicate to one another the coordinates of good food sources.

The Trip Out

We can consider the foraging cycle for a scout bee - a forager who does not know where food is to be found - or a recruit - a bee who has attended a dance and so knows where to look. If it is a bee's first flight of the day, she will hover in front of the hive and "scan" in an apparent attempt to memorize the appearance of the hive entrance and surrounding landmarks. Experiments by von Frisch and his students show that bees readily learn colors in this context (32), and they almost certainly must form a photograph-like image of the hive (as discussed below). The scanning is very similar to that observed when bees "study" a novel food source (46).

Based on rather limited data, we conclude that scout bees fly off to search, by no means restricting their flight paths to a straight line. A scout is able to find food and return home by a direct route after such a search even if the hive has been moved the previous night so that she is unfamiliar with the locale, and even if her route out was, as is likely, highly circuitous. She accomplishes this feat of navigation by keeping track of the distance and direction of each leg of the flight path (review in (24, 37)). Whether this involves remembering each leg separately and integrating over the entire route at the end, or continuously calculating a new set of coordinates for her location relative to the hive, is not known. In either case, learning is an essential element of a specialized processing system.

Recruits are in a different position because they have learned the precise coordinates - distance and direction - of the food source from the dances of experienced bees. They store and use this information to navigate to the food (review in (32, 35)). The lack of reward offered by pollen and nest-site dancers does not reduce the effectiveness of the dance communication. The odor of the food or pollen, present on the hairs of the dancer's body, is also memorized and then used when the recruit begins her search at the location specified by the dance. The way in which the information is stored and processed is not known, but it is clear that the dance attenders average several different dance runs (32, 39). Information from separate dancers is not combined (34), but since bees compare dances (to judge which food source or potential nest site is more attractive), separate dances probably can be remembered.

The bees' navigation and communication systems are referenced to the sun, which they recognize as a sign stimulus (13). Indeed, even after three weeks of active flight with the sun usually in view and serving as the major navigational aid, a forager appears to learn nothing about the bright, white, circular, unpolarized, 0.5° disk we call the sun and is completely taken in by a dim, green, triangular, fully polarized, 10° artificial sun because it satisfies an innate criterion (13). Honey bees are also innately able to substitute polarization patterns and landmarks for the sun when it is not available (review (24)). In a later section, we will look at the role of learning in navigation and how learning is interwoven with complex innate processing.

At the Food Source

A bee searching for a food source, whether she is a scout or a recruit, does not land at random. Instead, she has innate guidance information - what we could also call behavioral biases or spontaneous preferences or, in the language of programming, default values. Default values are the values to be assumed in the absence of more precise information. They are functionally analogous to the sign stimuli animals use to direct behavior before associative learning, which in many cases is triggered by the same cue and replaces the innate initial guidance. Hence it is that bees know innately that flowers usually have dark UV centers and bright surrounds (22), normally possess nectar guides (3), are often finely divided (43, 44), frequently have points (4,5), and so on. Targets with these features attract searching bees. None of these are absolute behavioral criteria any more than they correlate perfectly with flowers. Instead, the bees' choice of targets to land on and explore is a probabilistic combination of variables. This principle of probabilistic control of behavior is particularly obvious in the case of spontaneous color and odor preferences (40, 53). Of all colors, for instance, I have shown that naive recruit bees prefer to land on violet, although it is always possible that bees new to the food source may have different expectations than perfectly naive bees. This preference is probabilistic rather than absolute, so that even in the presence of a violet flower, a naive bee will sometimes choose a blue or yellow or white blossom. This makes sense evolutionarily, because not all flowers are violet. As Maynard Smith (6) and others have shown in the context of evolutionary stable strategies, distributing behavior probabilistically is highly adaptive so long as the probabilities assigned to the various alternatives correspond well to the likely distribution of correct choices in the outside world. In a very real sense, many cases of selective learning should be thought of as mechanisms by which experience serves to tune an animal's behavior from the default distribution of alternatives to the actual odds in the world around it (40).

Once a bee lands on a potential food source and finds a sufficient reward, she makes repeated visits and memorizes its location, odor, color, shape, nearby landmarks, time of nectar production, and how to harvest it. If there are enough flowers with nectar or pollen to allow the bee to exploit the source during several trips, the forager will probably become a temporary specialist on this species of flower. So much is known about the behavioral organization of flower learning that I will treat it in a separate section below.

Returning to the Hive

The flight back can require using the navigational calculations made on the way out (for at least scouts), or perhaps relying on the information originally obtained from the dance (for recruits). If the area is familiar, bees also use prominent landmarks, though landmark use is far more frequently observed on outward flights. Once near the hive, the returning bee searches for the particular tree or hive box whose appearance she has memorized earlier. Moving the hive even a few meters, or rotating it 90°, or changing its color after foragers have made their initial inspection flight can create enormous confusion, and once the bee does find the hive, she reinspects it thoroughly before venturing forth again (31, 41, 80).

A returning forager may perform a communication dance if the food she has gathered is of adequate quality. This requires her to "read out" her navigational coordinates and encode them according to dance conventions. Performance and interpretation of the communication dance is innate (57).

As we can see, a single foraging trip involves a complex and well regulated interplay of learning with innate behavioral programming and information processing. I will now consider the organization of this interaction in three specific cases: navigation, flower learning, and formation of a locale "map."

LEARNING IN NAVIGATION

Sun Orientation

As I described above, scout bees keep track of each leg of their outward search flight. The direction of each leg is determined relative to the sun's azimuth. Since the sun moves from east to west, bees must compensate for this motion if they are to make accurate calculations. There is abundant evidence that they do so (review in (36)). The actual rate of movement depends on the season, latitude, and time of day, and the direction of movement is also an essential variable (left to right in the northern temperate zone, right to left in the southern temperate zone, and either direction in the tropics depending on the time of year). Bees appear to learn the direction of movement when they begin to fly (about 10 days into their life span of 4-6 weeks in summer), and this learning may not be reversible (58). Although Menzel (personal communication) asserts that his bees cannot learn that a food station moves in azimuth, Lindauer (58) demonstrated this ability in bees reared in a basement. In fact, the basic calibration bees perform is to discover that the food source and the world on which they are to be found move systematically relative to the insects' essential landmark - the sun.

The actual rate of movement at each time of day is derived from two alternative systems. In the absence of usable landmarks, bees extrapolate the most recent rate of the sun's azimuth movement (36) - which rate they must be constantly measuring and remembering. (Actually, they use a running average of several measurements, though whether each measurement is remembered separately or weighted and added into the existing running-average value is not yet known (38).)

If useful landmarks are available, however, bees memorize the azimuth movement of the sun over the course of the day (23, 24). This requires some sort of time-linked memory array which must be periodically updated. Under most contemporary situations, this is probably the most heavily used system. However, bees evolved in (and frequently still reside in) forests, where the multiplicity of similar-looking trees, combined with the low visual resolution (1-4° depending on the part of the eye) of these insects, which corresponds to about 20/2000 vision in humans, may severely limit their use of landmarks. Menzel (personal communication), on the other hand, feels this would not limit an animal's ability to distinguish landmarks.

Polarized-light Navigation

If the sun is not visible, bees are able to infer its location on the basis of patterns of UV polarization in patches of blue sky (review in (14, 31)). The exact mechanism by which this is accomplished is not known. One school of thought holds that the bees have a system by which the angle of the polarized light visible to the bee is used to compute the sun's location (e.g., (74)), with the necessary ambiguities in such calculations being resolved by a system of innate rules (13). If this is the case, no learning need be involved, though the navigation would not be particularly accurate under some conditions.

The other general possibility is that bees remember the pattern of polarized light in the form of a photograph, analogous to the system of canopy orientation in ants (49). Several lines of evidence support this notion, including the reduced accuracy with which bees orient their dances to artificial polarization patterns on cloudy days ((12); reviewed in (24)), and the newly discovered ability of at least one species of tropical honey bee (and our temperate-zone bee evolved in the tropics from a common ancestor) to orient dances to canopy information (39). If this photograph strategy is the actual mechanism involved, then bees would need to store a sky picture, perhaps as an array of at least 100 points.

In the absence of both sun and blue sky, bees attempt to use prominent landmarks which they have memorized (23). Indeed, bees experienced with the route to and from a food source prefer prominent landmarks over celestial cues (23, 33). The nature of this landmark learning, as well as its storage and use, will be considered in a later section.

The general pattern of navigational learning is clear: specific information is gathered in specific, innately recognized behavioral contexts, fed into specific information-processing circuits, almost certainly stored in temporally defined memory-arrays, and retrieved and used in an appropriate way at the correct time. All this specialized learning behavior is possible because the processing strategy can be innately programmed and the important variables specified in advance. Neither the pure use of innately stored information nor unguided learning could work in this context.

FLOWER LEARNING

Color

The first aspect of bee learning investigated by von Frisch was the ability of bees to learn the color of a food source. Using simple associative conditioning, he demonstrated that bees could memorize any cue from yellow to ultraviolet (UV) (31). Later, Opfinger (69) showed that the color was learned only during the bees' approach to the flower before the innately recognized unconditioned stimulus of sugar could tell the bees that the color was worth learning.

Menzel and his colleagues have vastly extended our knowledge about this episode (26-28, 30, 61, 63-66). They have shown that this learning is restricted to the color seen in the final two or three seconds before landing and not the color seen while feeding or hovering near the flower before departing. This seems reasonably consistent with the usual pairing relationship seen in associative learning, and Bitterman has demonstrated many of the other phenomena of classical conditioning in color learning (9, 20, 52), though Menzel (personal communication) disputes Bitterman's methods, results, and interpretations. He favors a more naturalistic representation of classical conditioning in bees (63). The Menzel group has also shown that the rate at which bees learn color depends on the color used, so that innately preferred colors are learned more readily (Fig. 1).

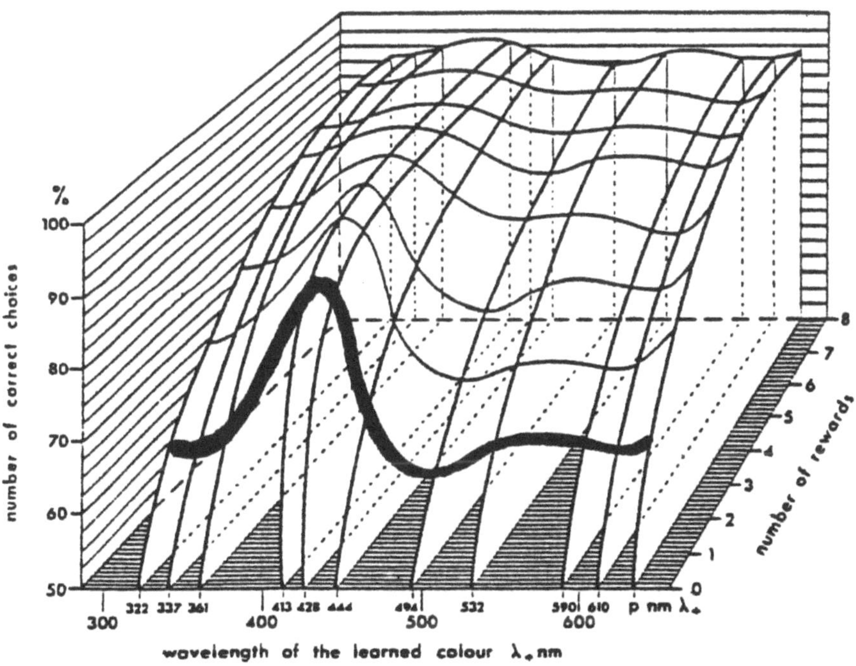

FIG. 1 - Average learning curves in a two-color choice test for eleven colors visible to honey bees. The alternative color was always the complement. Violet (frequently called purple in the English-speaking world) is learned most quickly, while blue green is the most slowly learned color (n = 54,712). The curve for one-trial learning has been emphasized. Data from Menzel et al. (66).

Bees require about three to five trials to select a particular color with 90% certainty in a two-color choice test. Regardless of the amount of training, bees rarely get beyond the 95% level. According to Menzel (personal communication), neither a polarized target nor a flashing light can be remembered, and differential color training - training in the presence of an unrewarded alternative - has no effect. Bees forget colors they have been trained to only once over about a week, but they seem never to forget a food source they have visited three times. Color learning is subject to disruption by electroconvulsive shock, chilling, and CO_2 just after training, which implies to many researchers that there is a short-term memory phase followed by consolidation. Subsequent work on the short-term memory phase of color learning suggests that information from separate trials can interact during the first moments of short-term processing, but that as consolidation begins the input of new information is temporarily blocked ((62); see also (46)). This suggests that a behavioral analysis of short-term memory processing using the neural models suggested by Kandel and his colleagues (Quinn, this volume) might be very rewarding.

Odor

Von Frisch demonstrated over half a century ago that bees learn odors quickly and accurately (review (31)). Again, Menzel and his colleagues have greatly extended our understanding (e.g., (29, 30, 65, 66)). Odor learning is virtually one trial and takes place the instant before the nectar is tasted. Using a supercooled needle to "anesthetize" one locus at a time, Menzel and Erber have traced the site of active processing in the bee brain as a function of the time after the pairing of odor and sugar takes place. The neural activity essential to later storage of the odor association begins in the antennal lobe of the brain and proceeds about two minutes later to the alpha lobe of the mushroom body. Roughly two minutes later still, activity has shifted to the calyx. After another two minutes or so, chilling no longer has any effect anywhere. By training only one antenna, Menzel and his colleagues showed that the processing is initially unilateral, but the information leaving the alpha lobe is sent to both calyxes.

Shape

The exact nature of shape learning is more difficult to describe. Hertz (42-44), for instance, found almost no ability to learn shape, and though later tests have demonstrated that bees have an appreciable aptitude for shape recognition, considerable debate remains about what information is actually stored. For instance, many experimenters report that simple, closed shapes (circles, squares, triangles, and the like) of equal area cannot be diffentiated (review in (80)), but, like Anderson (1), I find that bees acquire the distinction to the 90% level in about fifteen trials. (If an unrewarded copy of the alternative shape is offered next to the training shape, learning takes only about ten trials.) By switching shapes while the forager is feeding, I have shown that shape is learned after feeding.

The most widely held opinion on shape learning from 1930-1970 was that the information was encoded as a single "parameter" such as contour density or flicker rate or the ratio of contour to area. There was no evidence suggesting that specific, photograph-like representations could be stored. Attractive as this very simple system of coding is, however, it cannot be the whole story. Both Anderson (4) and Cruse (21) find that a single parameter is insufficient to account for the distinctions bees can make, and they suggest that two parameters must be necessary. Schnetter (75) thinks that four must be required, and Ronacher (72, 73) also argues for the existence of several. None of these models is wholly satisfactory, however. To make matters worse, virtually all these hypotheses are based on tests with flat, horizontal shapes. Wehner (review in (80)) and Anderson (1) demonstrated very clearly that the orientation of a flat vertical shape can also be learned. As this figure-orientation constraint is missing from horizontal tests, at least one additional parameter must be necessary.

The obvious alternative is that bees take and store some sort of mental picture (that is, eidetic image) of the food source. Wehner (79) argues for this model on the basis of his data since the alternative shapes he offered were identical in terms of all the previously proposed parameters. More recently, however, he appears to have retreated to a neutral position (80). The major argument against the photograph alternative seems to be the apparent inability of bees to "generalize" shapes. Both Anderson (1) and Ronacher (72, 73) found that honey bees could not transfer training experience from a figure with one contour density to a figure of the same shape with another density, implying that density is more important than a generalized concept of shape. This is not a fully convincing argument since storage of a picture could be independent of the ability to perform particular transforms on it.

There are two points that may help clarify both the questions here and the more serious problem that will arise when we consider landmark learning. The first is that, given the observation that sign stimuli may be equivalent to neural feature detectors, we might get further by substituting "features" for parameters. (I refer here to a feature detector as a higher level visual cell with properties which result from integrating information from lower level cells.) The term "feature" arose on the basis of early studies in which the cell preferences were for logical, unitary shapes such as bars which in a behavioral specialist could be associated with particular functions. Any arbitrary shape will excite a different pattern of activity among an ensemble of feature detectors with differing receptive field preferences, color preferences, inhibitory and excitatory surrounds, and so on, in the manner of olfactory codings (cf. Gould and Marler paper, this volume). The visual experience of animals including both bees and mammals is necessarily built up out of a variety of simple and complex feature detectors, a subset of which can serve as specific releasers. An orthogonal pair of spatial frequency detectors would probably be sufficient to account for the contour density parameter, and Wehner's bees could easily have distinguished the figures he used (disks with one half painted black, differing in the orientation of the black/white boundary) on the basis of other well-known feature detectors. Perhaps thinking in more physiological terms would help define the parameters of flower learning more precisely. Certainly it could not hurt, and it seems obvious that the nature and distribution of various feature detectors in a species will greatly color its perception of the world (37).

Second, we have the problem posed by Lashley's rats (Gould and Marler, this volume): even though these animals can remember photographic images, they frequently rely on diagnostic features, probably equivalent to sign stimuli, for making these distinctions. Hence, demonstrating that an animal is using features (Lashley's "part figures") does not prove that it is unable to use pictures when necessary. The training must involve a series of different unrewarded alternatives.

As a result, it is not obvious to me that the distinction between photographs (eidetic images) and parameters (neural abstractions probably based on feature detectors) has been clearly drawn and put to the test. Both schools of thought seem to make assumptions about visual resolution, feature detection, and neural processing. The essential difference I see is that a photograph strategy preserves the spatial relationship between the components of the figure, whereas a parameter approach only encodes abstractions about the elements. The reason I said earlier that bees probably form an eidetic image of their hive, for instance, is based on this line of reasoning. Among other things, honey bees are able to remember which hole out of an array of at least sixteen identical holes in a featureless wall leads to their hive. Incidentally, this ability to remember the special spatial relationship between holes which defines the entrance to their colony breaks down when the array becomes too large (see also Wehner's work on nest location by bees and ants (80)).

Although I do not believe that the decisive experiment on flower learning has yet been done (and distinguishing in practice between what is an element and what is the true figure is not simple), my tentative vote is with the parameter hypothesis. It is striking, of course, that a creature which in other contexts probably can store and use images cannot remember flower shapes in this way. Perhaps the advantages of neural economy more than offset any ambiguity a parameter strategy might generate. The crucial experiment is to compare a vertical target consisting of four circles arranged in a square with one edge of the imaginary square horizontal with the same target whose edge is oriented 45° from the horizontal.

Time

Von Frisch demonstrated the link between time and food location in the first half of the century and pointed out the adaptive nature of this learning: particular species of flowers produce nectar only during particular parts of the day (31). Individual bees, then, will specialize on different flowers at different times of day and maintain their periodicity indefinitely. In exploring this invertebrate version of the Kamin effect - the spontaneous assumption that a particular food source will reappear roughly 24 hours later - Koltermann (54) succeeded in training bees to remember nine different times of day. I do not know of anyone who has attempted to push the temporal memory of bees further. Koltermann (54) reports that he could show temporal discrimination of two separate food sources down to an interval of 20 minutes (this required the temporal equivalent of differential conditioning). The discrimination was sharp only when nearby times had different food cues. In theory, bees might be capable of storing 40 or even 50 different times. It would be extraordinarily difficult to test the limits of the time learning of bees by the usual free-flying training techniques. It may be that a laboratory approach would succeed in defining the size and resolution of this system.

Landmarks Near the Food

Anyone who has ever trained bees knows that three-dimensional landmarks near a food source strongly affect where bees search for the target (reviewed in (31, 80)). This learning takes place during departure (69), thereby violating the normal US/CS relationship. The question, of course, is how bees remember the relationship of the landmark or landmarks to the food. When there is only one landmark, bees must know its distance and bearing to the food. The sun seems their obvious cue, and in its absence, the pattern of polarized light overhead (47). Under overcast, bees cannot determine this bearing. Bees determine their distance from a landmark by its angular size: they move along the correct bearing until the size of the image on their retina matches what they experienced during training (16). Hence, bees will search nearer a test landmark narrower than the one they were trained on, and further from a larger one, so long as the landmark is visible along the bee's visual horizon as it searches.

So far, so good: bees can remember something about vertical landmarks and use that information to locate food. What, however, do they remember? An analogy with shape learning gives us two general possibilities to work with: parameters and pictures. Either could account for the single-landmark data. Anderson (2) showed that if he trained bees to a target surrounded by landmarks and then removed some of them on one side, bees would not fly to the former target location. In general, they searched closer to the remaining landmarks. Hence, Anderson concluded, the bees were not behaving as if they had stored an image to which they were trying to find the best match (Fig. 2A). Instead, he suggested that they probably broke the landmark data down into parameters (such as the bearing of the sun to the array, and whether or not there was a nearby landmark of some sort within each quadrant of the horizon) and did their matching on an abstract level.

More recently, Cartwright and Collett (15, 18) have argued that bees store an eidetic image of nearby landmarks against the horizon and orient subsequently by a matching process, taking into account bearing and angular width (Fig. 2B). They do not address the apparent discrepancy with Anderson's interpretation in any detail. There are several differences between the two sets of experiments. Anderson always surrounded the target with landmarks, always used at least eight markers, and usually tested with an array in which landmarks had been both moved out of place and removed. Cartwright and Collett, by contrast, always used

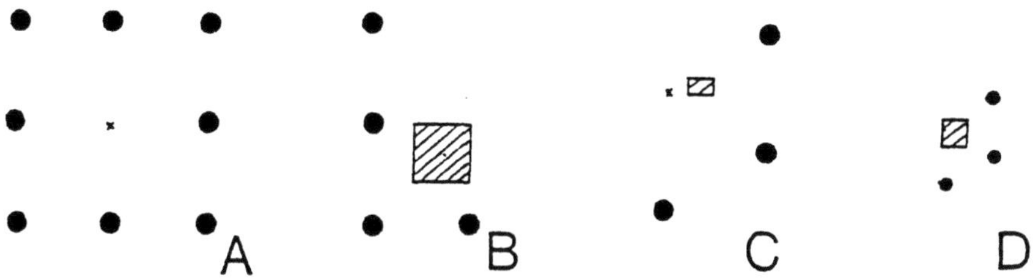

FIG. 2 - A) Anderson (2) trained bees to the center of a landmark array and then B) tested with an incomplete or rearranged set of markers. The foragers searched nearer the remaining landmarks than they had previously (hatched area). C) Cartwright and Collett (15, 18) used three offset landmarks and then D) tested with markers at the same bearing whose angular size was altered to compensate for distance. The foragers searched just where they had before.

three or fewer landmarks, placed them to one side, and moved them so that a perfect visual match was usually possible - that is, all were usually kept at the same relative bearing and the same apparent angular width from the "correct" location.

To explain the latter results by Anderson's parameter scheme, we would probably only have to assume that the bees subdivided the horizon into at least eight or more sectors. I am mildly attracted by this alternative since I see no evidence that bees can distinguish the shape of food landmarks, and they even seem to be willing to trade width and height as though they employ a degree-of-sector-occupancy coding scheme. On the other hand, I imagine Anderson's results could be explained along eidetic lines if we were to make the appropriate assumptions about visual resolution, kinds of features being encoded, and the neural processing underlying the bees' attempts to find the best match. In short, though I am partial to the photograph alternative, I am not convinced that the parameter/photograph dichotomy has yet been clearly defined and tested. The critical experiment is to vary landmark shape without altering its effective visual area.

Cue Hierarchies

As we saw in the case of navigation, bees have a hierarchy of cues:

recruits navigating to a food source prefer the sun to polarized light, and polarized light to landmarks. A similar pattern of preferences is clear in flower learning. After bees are trained to a food source until their accuracy in choice tests reaches its maximum level, the component cues can be separated to see which are more potent. Hence, a purple, orange-scented, triangular target can be removed and the returning bee offered a target with the correct scent but the wrong color and shape, one with the training color but a different odor and shape, and so on. The preference for a higher ranking cue is very strong, and the rank order is odor, color, and then shape (45-47, 53, 55, 56, 59, 60). Oddly enough, the position of landmark cues in this hierarchy varies between races. For the German race, Apis mellifera carnica, landmarks are more important than shape, whereas for the Italian race, A.m. ligustica, shapes outweigh landmarks (42, 56, 60). Moreover, the learning curves for odor, color, shape, and landmarks, though quite consistent among bees of a particular race, are distinctly different between races (review in (59)), as are some elements in the spontaneous preference behavior (53). These observations underscore how evolution has acted to orchestrate and fine tune virtually every aspect of the learning behavior of bees to meet the challenges posed by their niche.

Organization of Memory

In the process of investigating how time is linked to other cues, Bogdany (10) has uncovered a fascinating pattern of organization in flower learning. Consider the following situation: we train a group of foragers to a blue, peppermint-scented, triangular target from 9:00-10:00, and then to a yellow, orange-scented, star-shaped target from 10:00-11:00. We repeat this sequence for a few days. On the next day, we set out both targets together, each supplied with sugar solution. The foragers appear at about 8:45 and land almost exclusively on the triangle. At about 9:45, they begin to switch from the triangular target, which still has food, to the star-shaped feeder. By 10:00, all the trained bees have abandoned the perfectly good triangle for the star. It is as though the bees have organized their daily activities by means of a mental appointment book listing all of the relevant cues for each time of day (37, 40).

Bogdany also showed that if one component of an expected combination were changed, all of the components had to be relearned. For example, if we were to substitute rose for peppermint in the blue triangular target to which bees had been trained at 9:00, we would find that after landing, feeding, and returning to the hive, the forager on her next visit would be able to choose rose odor over a new odor at the 90% level but would be very poor at selecting the color blue or the triangle shape in a similar choice test. The bee also loses and must relearn the precise beginning and ending times for the food source. Apparently, the bee must start over and relearn each cue if any one is changed, each cue at its own specific rate.

This picture of storage in terms of sets of independently acquired cues is reinforced by the observation that if a cue - color or odor, for instance - is omitted initially but then added later, the set does not have to be relearned. Apparently the blank space for, say, odor can be filled in after the time-linked slots for color and shape have been filled. Clearly there is no hint of classical "blocking" here. It is tempting to think of flower memory organized as a matrix. I will explore this line of thinking below.

Flower Handling

In addition to all of the associative tasks confronting a bee as it memorizes the information necessary for efficient foraging, bees also face an operant task: how best to land on the flower and extract the nectar or pollen. There are many different flower morphologies, and most require very different strategies for harvesting the reward. Recruits to an artificial food source are remarkably incompetent at getting the food but soon perfect a technique. This technique can differ between bees, so that one individual may always land at the periphery of the feeder and walk directly up to one of the feeding grooves where she feeds until full; another will always land on the food reservoir, feeding in an awkward head-down posture; still another will land directly on the grooves and move counterclockwise from groove to groove around the circular feeder, and so on. At flowers a similar pattern is clear, though the range of behavioral choices is much restricted. Through experimentation, then, bees discover the motor behavior necessary to exploit blossoms, and so handling time goes down (e.g., (48)). This is probably one of the reasons bees tend so strongly to specialize.

Honey bee flower handling falls neatly into the category of learned motor programs described by Gould and Marler (this volume). So far as I know, no one has studied the exploratory behavior of naive bees at a food source to see if there might be an innate repertoire of motor subunits which bees bring to bear on a new flower. If so, we might think of the behavior as a process of learning the proper order of motor acts. One interesting observation is that a really hungry bee will (inefficiently) chew through the side of a flower whose morphology prevents her from getting her head within proboscis-length of the nectar. Unless bees reason out the problem for themselves, this behavior is probably an innate backup routine which, because of its inefficiency, is called into play only in extreme circumstances.

Since honey bees probably store information "bits" about flowers in sets, we must wonder whether their harvesting behavior is also time-linked. Although this would certainly make sense, I know of no experimental test of this idea.

LOCALE MAPS

There is a modest literature dealing with the ability of insects to find their way home after being displaced (review in (80)). The phenomenon is equally apparent in bees. A particularly apt set of examples is described by Brines (12). He would capture departing foragers at the hive entrance and carry them in total darkness several hundred meters to locations well out of sight of the hive and its nearby landmarks. These displaced bees would be permitted to feed and then depart. Two general patterns of behavior were commonly observed: young foragers (recognizable by their furry backs and undamaged wings) would usually circle and either land back on the feeder or drift off in random directions; many older bees, however, would circle and then depart directly toward the unseen hive. Flight times indicated that the return path must have been virtually direct, and because of the serious need for food in the hive, the returning bees would sometimes perform dances. These dances were quite accurate.

I assume that the experienced bees knew where they were, based on the landmarks they saw upon departure. It is unlikely that the transported foragers were operating on the basis of any memory of a specific food source in the close vicinity since most of the releases were from a large, empty, perfectly barren parking lot at least 50 meters from the nearest vegetation (grassy playing fields). Reviewing the hymenoptera literature in general, Wehner (80) concludes that insects store a set of "snapshots" from along familiar foraging paths and consult this neural "album" when lost. He also believes that the pictures are arranged in serial order along the flight paths. Bees trained along a dog-leg to a food source, for instance, frequently continue to navigate out day after day along the indirect route if there is a prominent landmark or set of landmarks along the first leg or at the turn (e.g., (31)). Navigation back, by contrast, is usually direct.

In honey bees (and perhaps other insects) the possibility exists that there may even be something more involved. We once trained bees along a lake shore ((40), and Dyer, in preparation) (Fig. 3) and used the redirection technique (34) to cause their dances to indicate a spot in the lake. (Redirection causes recruits to interpret a dance as though a bright light is the sun, while the dancers, who have been treated to make them less sensitive to light, are actually orienting their dances to gravity.) Recruits refused to explore the indicated location in the lake, though they arrived in large numbers at the training station if the redirection was set to zero (39, 40). Thinking that site-specific odors might be involved, we

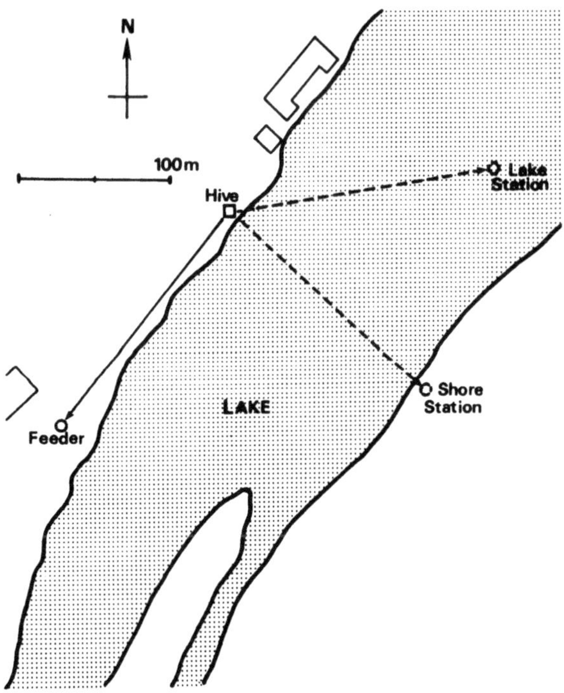

FIG. 3 - When foragers were trained along a lake shore to a distance of 175 meters, and their dances "redirected" to the spot indicated in the lake, recruits would not fly there. Even when the foragers' station was in a boat at the same spot, recruits would not come. When the boat reached the far shore, however, recruits arrived in large numbers. Apparently the recruits could use the dance coordinates to evaluate the indicated site without actually flying there ((33, 38, 39), and Dyer, in preparation).

trained the foragers in a boat to a feeder at the spot their dances had formerly indicated, but no recruits came. And yet, when we moved the boat and feeder to the far shore of the lake, recruits arrived in considerable numbers.

The most obvious interpretation is that the recruit bees were able to "place" the location indicated by the dance, either on a mental map or by calling up the "snapshot" nearest the dance coordinates. Presumably, they would see that the target was in the middle of the lake and look for a different dance to follow. When the dance indicated the far shore, however, the recruits could place the target in a plausible locale and so decide to fly across the lake. The crucial experiment (suggested by Quinn) is to train the shore bees far enough so that their redirected dances indicate the far shore.

The possibility exists, then, that information about navigational landmarks may be stored not just in route-specific series, but in true map-like fashion, each landmark in correct spatial relation to the other. If this were the case, the storage could even be strictly map-like, with landmarks encoded on Cartesian arrays. The nature of the storage is not known, though it cannot be of very high resolution. Dyer (in preparation), for instance, finds that bees readily confuse road edges with field boundaries, substituting corn for asphalt (see also (33)). Obviously, a simple horizon model would not account for the ability of bees to use flat landmarks on the ground (e.g., (31)). The opposite is true during navigation.

One curious set of observations may relate to the possible existence and use of locale maps. Von Frisch (31), Markl (personal communication), and many people in my lab (40) have observed that bees trained away from the hive according to the strict schedule described by von Frisch (31) begin to anticipate movements of the feeding station. In the course of my thesis work (reviewed in (35)), I even once observed the ability to anticipate regular movements toward the hive. On the other hand, Menzel asserts that his bees cannot learn that a station is moving either away from or toward the hive. Clearly, there remains a great deal to be learned about the way in which bees "map" their environment.

SYNAPTIC SELECTION

As I suggested at the onset, enough is known about the way bees learn and remember what they need to know that we can ask whether the neuronal and neural group selection hypotheses of Changeux ((17), and this volume) and Edelman (25) can account for some of the observed behavior. The gist of this line of reasoning is that, by analogy with the immune system, all the possible answers to a question are wired up in advance, and learning merely selects the correct one(s). Conditioning biases would then be a consequence of the preexisting wiring. The selection itself could be at the cellular or the synaptic level.

This scheme is attractive in the context of honey bee flower learning because the behavioral situation is relatively predictable: flowers have a limited set of parameters to be learned - color, odor, shape, etc. - and the range of variables to be encoded can be specified in advance. Hence, it would be possible to construct the matrix of "answer" cells required by neuronal selection, and the correspondence between spontaneous preferences and learning rate (or conditioning bias) suggests that the two are part of the same system. Menzel's report cited above that perceivable cues (UV and flashing lights) cannot be learned is consistent with the idea that some answers are excluded because the proper connections are not wired in advance. Erber's report (29) that only cells which displayed a weak preexisting response to the stimuli he was going to train the bees to (sugar, the US, and a particular odor as the CS) subsequently "learned" strengthens the case and suggests that the model of alpha-conditioning may apply even when there is no overt behavioral response in a naive animal.

Bogdany's data (10) on learning and forgetting by sets is also suggestive of a selectionist organization and provide a starting point for asking how such a system would have to be structured in bees. To make any plausibility analysis it will be necessary to assume a processing strategy here, a coding system there, and so on. Let me begin by making it clear that as of now no one knows which systems or strategies are involved, though perhaps this exercise will indicate the importance of finding out. In any case, the plausibility analysis does not depend on these details so long as we are conservative in our assumptions. If the selection strategy turns out to be plausible, then we can begin to worry about such details and try to make a prediction.

We can probably begin by dismissing the idea that the selection is at the cellular level, because the minimum number of cells needed would be too great if each were specific to a single combination - a blue, 6-pointed, orange-scented flower located 120 miles SSE of the hive providing food at 11:15, for instance. If there are even just twenty times of day possible (and as with all these numbers, I am using a low estimate), ten colors to distinguish, only one hundred odors, only ten shapes, only ten landmark combinations, and only one hundred possible location coordinates, we would need one or two orders-of-magnitude more cells for this one purpose than bees have in their entire nervous system.

Synaptic selection is more reasonable, and Murphey's work (67) on the cercal interneuron demonstrates that presynaptic phenomena of the sort discussed in Singer's paper (this volume) are found in invertebrates as well. Let us assume for the moment that the sign stimulus alerts the system that a winning combination is being sensed, and that the various characteristics of current synaptic activity are worth remembering, at least provisionally. Let us also assume for the sake of simplicity that the synaptic input is organized at the dendritic level (as seems to be the case for the interneurons in crickets which receive information from the cerci, for instance (7)). In vertebrate color learning, the hue of a target is defined by the ratio of activity of the three classes (colors) of photoreceptors (review in (37)), which is probably actually reported by three classes of color-opponent neurons. (This is a useful refinement since the ratios reported by these sorts of cells are relatively intensity-independent. Moreover, the total activity of the three classes taken together is relatively constant for any color, although the distribution of that activity among the three sorts is unique for each color. I refer to this organization as "reciprocal coding." (As of now, Menzel and his colleagues (personal communication) have only succeeded in finding two classes of color-opponent cells in bees.) At this point, we have two general ways of organizing our model: we could suppose that the precise color is determined by a separate circuit, categorized, and fed to the memory cell through the appropriate member of a set of color-specific "lines." Alternatively (and I have chosen this possibility to discuss), a memory cell must itself remember the ratio experienced just before the US. I use the term "memory cell" in a manner analogous to Konishi's song-recognition cell (this volume) which responds only to the simultaneous presence of all the learned criteria which define the song. Needless to say, the memory proper is more likely stored by the ensemble of active synapses on the cell. The sign stimulus could act presynaptically on the incoming data to provide filtering at that level, or postsynaptically to change which (or how many) synapses are listened to. The former alternative seems more attractive to me, but either would probably work. Once a dendrite adjusted to the CS so that it "listened" to each class of input only to the degree that they were active during the critical period for color learning, no other color, regardless of brightness, would stimulate the dendrite as much. The different learning rates for individual colors would then be in part a consequence of the amount of change necessary to alter the cell from the innate or default-value wiring to a state which matches the CS. We could think of this hypothetical dendritic learning process as analogous to habituation, with a sensitization-like event serving to trigger the adaptive forgetting which is evident in Bogdany's data. This is very like the Hawkins and Kandel scheme described in Quinn's paper (this volume) and the group report by Menzel et al. (this volume). The dendritic-level selectionist model is also consistent with the observation that a missing cue can be added without creating problems.

Carrying this exercise a bit further, we could imagine an odor dendrite listening to the output of something like odor-opponent cells in the olfactory lobe. Again, the sign stimulus would have to trigger a memorization of the pattern of input. The ability of bees to use a pheromone to mark a food source has been viewed as ruling out synaptic selection (Menzel, personal communication), but, although pheromones are detected by specialized receptors with private lines to the brain where they help prime behavior, they are detected by the many classes of generalist odor receptors as well. It is difficult to imagine how the odor-learning system could be denied access to any particular odor. Nor, as should be obvious from the Gould and Marler paper (this volume), is it the case that the absence of an overt feeding response to a particular learnable odor is a problem: the essence of the idea is that no overt response is necessary so long as relatively weak connections exist in the association centers. A long list of such quibbles can be drawn up without thoughtful consideration of the sufficient and likely patterns of initial connectivity in a generalist association area.

As another aspect of this intriguing hypothesis, it would obviously be a great convenience if shape were represented as a set of discrete parameters, rather than as a photograph. Again, if the parameters are based on groups of feature detectors with compensating outputs (as, for instance, a bright-center/dark-surround cell paired with a dark-center/bright-surround cell), the invariance of total stimulation of this reciprocal coding, combined with the unique pattern of activation characteristic of each distinguishable shape, would make the encoding of shape memory relatively straightforward.

It seems clear that both distance and direction could be encoded reciprocally. For instance, a set of widely tuned direction cells centered on, say, 0°, 90°, 180°, and 270° could have the same total activity regardless of direction but a unique pattern corresponding to direction. Landmarks could be encoded either as parameters or a photograph. For the moment, I shall assume the former.

The result of encoding each of these cues in terms of patterns of synaptic activation provides, at least potentially, an enormous neural economy. Instead of 10^8 cells, a bee might, if we ignore eidetic images for now, need fewer than one hundred. I have not mentioned time as one of the variables since I think that time of day is the most likely way in which the cells in a selectionist array would be defined initially. This is because although two behaviorally different food sources can share the same color or odor or location and so on, time cannot be shared: reusing the same time wipes out the previous memory, so long as care is taken that the overlap is complete. Paradoxically, time is one of the most slowly learned cues, but this is probably an adaptive ambiguity: flowers are not terribly punctual, and if the bee has no other "appointment," she might as well check on a temporally "nearby" source. Other hypotheses of flower storage must account for the robust observation that, when it needs to be, time is a unique parameter to bees which cannot be shared with other learned flower "sets."

As mentioned earlier, differential time training serves to eliminate temporal sloppiness from the outset: when we train a bee to two sources, one of which begins to be available just as the other ceases, the bees learn the start-up time of the second source almost immediately while the start-up time for the first one is learned at the usual, slow rate. I can easily imagine a series of time-specific cells, set at perhaps fifteen minute intervals, with a relatively wide "gating" signal from the endogenous clock which easily activates many temporally adjacent "empty" or "naive" cells but triggers "defined" or "experienced" cells only at the peak of the gating. (The intercell interval would have to be about five minutes if we assume that the navigation/dance system shares this array, which is an attractive possibility.) This "core" time-defined matrix would have to contain a minimum of only one hundred cells to cover the entire day (or about three hundred if a 5-minute resolution is assumed.) If incoming sensory information is categorized before being sent to the dendrites, another 200-500 cells might be necessary.

The real pressure for space comes when the animal begins to encode pictures. A selectionist view of eidetic image coding might imagine that one cell records each spot on the visual field, and that the input to the cell (again, perhaps segregated onto dedicated dendrites) might be in the form of several sets of reciprocally coding inputs from feature detectors.

We must consider the problems both of storage and of access. The number of cells required for storage depends on the resolution of the system. The bees' polarized light navigation system, for instance, integrates over 15° segments of the sky (31), suggesting that a 100-element array would be sufficient for it. The resolution of other systems has not been defined: the horizon-image matching data would require no more than twenty cells on the basis of present data, and shape-image matching could probably be done with fewer than one hundred (assuming, in both cases, a photograph is even involved.)

Access presents another set of relevant considerations. For example, if images are used, is there an array dedicated to each time-defined storage cell, or might there be some smaller number of random-access arrays linked as necessary to one or more time-defined cells when learning takes place? The former is simpler, the latter more economical and might permit time-independent generalization. Assuming that bees can remember ten sources simultaneously (the maximum ever demonstrated being eight), the dedicated array alternative might require 12,000 cells, while the random-access strategy would perhaps need only 1,200. Neither number is out of the question, but it would be nice to know just how many different food sources bees can remember.

From a selectionist's perspective, the bee's locale map might be organized along one of several lines. If a snapshot system is involved (as opposed to or in addition to a Cartesian map), we must wonder about the resolution of the photograph and the number of images required. Present behavioral data are not too helpful, but let us suppose the number lies between 100 and 1000. How many images would be needed? Again, this is hard to know. The evidence suggests that experienced bees know the 1000 m around their hive pretty well, but no experiments addressing the necessary inter-image spacing have been done. Let us assume that there is one for every 100 m, meaning that in theory, if the images were being stored by their coordinates in preaddressed neural locations, about sixty image arrays would be needed. Conservatively, then, a low-resolution snapshot map of the first 1 km of the 12 km flight range would require 6000-60,000 cells. This begins to sound like a lot, unless we can suppose that a cell can store points in more than one array (superimposed storage). Another interesting possibility is that a reverse strategy might be used: that is, instead of organizing snapshots into a map, perhaps a higher resolution map encoding the location, size, color, orientation, and so on, of landmarks might be stored, so that when the bee "consulted" any coordinate, a synthetic image of the surroundings could be generated based on the map information.

I find this line of thinking about the possible organization of learning and memory provocative but not compelling. Selectionist models are intriguing for a variety of reasons, and they can suggest explicit behavioral experiments which ought to be practicable with honey bees. Certainly the selectionist interpretation of the highly ordered and adaptive process of flower learning by bees is strikingly consistent with the behavioral data. In any case, it should be clear that bees are highly specialized learning machines which illustrate well the evolutionary potential for "customizing" learning behavior to suit a species' needs, and that this specialization provides an excellent opportunity for studying the natural history of learning.

Acknowledgements. I thank T.S. Collett, C.G. Gould, R. Menzel, W.G. Quinn, J.E.R. Staddon, and W. Towne for helpful criticisms. Supported by NSF Grant BNS 82-01004.

REFERENCES

(1) Anderson, A.M. 1972. The ability of honey bees to generalize visual stimuli. In Information Processing in the Visual System of Arthropods, ed. R. Wehner, pp. 207-212. Berlin: Springer-Verlag.

(2) Anderson, A.M. 1977. A model for landmark learning in the honey bee. J. Comp. Physiol. 114: 335-355.

(3) Anderson, A.M. 1977. Parameters determining the attractiveness of stripe patterns in the honey bee. Anim. Behav. 25: 80-87.

(4) Anderson, A.M. 1977. Shape perception in the honey bee. Anim. Behav. 25: 67-69.

(5) Anderson, A.M. 1977. The influence of pointed regions on the shape preference of honey bees. Anim. Behav. 25: 88-94.

(6) Anderson, A.M. 1979. Visual scanning in the honey bee. J. Comp. Physiol. 130: 173-182.

(7) Bacon, J., and Murphey, R.K. 1984. Receptive fields of cricket giant interneurones are determined by their dendric structure. J. Physiol., in press.

(8) Baerends, G.P. 1941. Fortpflanzungsverhalten und Orientierung der Gradwespe. Tijdschrift Entomol. 84: 68-275.

(9) Bitterman, M.E., and Couvillon, P.A. 1982. Compound conditioning in honey bees. J. Comp. Physiol. Psychol. 96: 192-199.

(10) Bogdany, F.J. 1978. Linking of learning signals in honey bee orientation. Behav. Ecol. Sociobiol. 3: 323-336.

(11) Booker, R., and Quinn, W.G. 1981. Conditioning of leg position in normal and mutant Drosophila. Proc. Natl. Acad. Sci. 78: 3940-3944.

(12) Brines, M. 1978. Skylight polarization patterns as cues for honey bee orientation: physical measurements and behavioral experiments. Ph. D. Thesis, Rockefeller University, New York.

(13) Brines, M.L., and Gould, J.L. 1979. Bees have rules. Science 206: 571-573.

(14) Brines, M.L., and Gould, J.L. 1982. Skylight polarization patterns and animal orientation. J. Exp. Biol. 96: 69-91.

(15) Cartwright, B.A., and Collett, T.S. 1982. How honey bees use landmarks to guide their return to a food source. Nature 295: 560-564.

(16) Cartwright, B.A., and Collett, T.S. 1983. Landmark learning in bees. J. Comp. Physiol. 151: 521-543.

(17) Changeux, J.-P., and Danchin, A. 1976. Selective stabilization of developing synapses as a mechanism for the specification of neuronal networks. Nature 264: 705-712.

(18) Collett, T.S., and Cartwright, B.A. 1983. Eidetic images in insects: their role in navigation. Trends Neurosci. 5: 101-105.

(19) Collett, T.S., and Land, M.F. 1975. Visual spatial memory in a hoverfly. J. Comp. Physiol. 100: 59-84.

(20) Couvillon, P.A., and Bitterman, M.E. 1980. Some phenomena of associative learning in honey bees. J. Comp. Physiol. Psychol. 94: 878-885.

(21) Cruse, H. 1972. A qualitative model for pattern discrimination in the honey bee. In Information Processing in the Visual System of Arthropods, ed. R. Wehner, pp. 201-206. Berlin: Springer-Verlag.

(22) Daumer, K. 1958. Blumenfarben, wie sie die Bienen sehen. Z. vgl. Physiol. 41: 49-110.

(23) Dyer, F.C., and Gould, J.L. 1981. Honey bee orientation: a backup system for cloudy days. Science 214: 1041-1042.

(24) Dyer, F.C., and Gould J.L. 1983. Honey bee navigation. Am. Sci. 71: 587-597.

(25) Edelman, G., and Mountcastle, V. 1978. The Mindful Brain Cortical Organization and the Group-Selective Theory of Higher Brain Function. Cambridge, MA: MIT Press.

(26) Erber, J. 1972. The time-dependent sorting of optical information in the honey bee. In Information Processing in the Visual System of Arthropods, ed. R. Wehner, pp. 309-314. Berlin: Springer-Verlag.

(27) Erber, J. 1975. The dynamics of learning in the honey bee. I. The time dependence of the choice reaction. J. Comp. Physiol. 99: 231-242.

(28) Erber, J. 1976. Retrograde amnesia in honey bees. J. Comp. Physiol. Psychol. 90: 41-46.

(29) Erber, J. 1981. Neural correlates of learning in the honey bee. Trends Neurosci. 4: 270-273.

(30) Erber, J., and Schildberger, K. 1980. Conditioning of an antennal reflex to visual stimuli in bees. J. Comp. Physiol. 135: 217-225.

(31) Frisch, K.V. 1967. The Dance Language and Orientation of Bees. Cambridge, MA: Harvard University Press.

(32) Frisch, K.V., and Jander, R. 1957. Über den Schwanzeltanz der Bienen. Z. vgl. Physiol. 40: 239-263.

(33) Frisch, K.V., and Lindauer, M. 1954. Himmel und Erde in Konkurrenz bei der Orientierung der Bienen. Naturwissenschaften 41: 245-253.

(34) Gould, J.L. 1975. Honey bee communication: the dance-language controversy. Science 189: 685-692.

(35) Gould, J.L. 1976. The dance-language controversy. Q. Rev. Biol. 51: 211-244.

(36) Gould, J.L. 1980. Sun compensation by bees. Science 207: 545-547.

(37) Gould, J.L. 1982. Ethology: The Mechanisms and Evolution of Behavior. New York: W.W. Norton.

(38) Gould, J.L. 1984. Processing of sun-azimuth information by bees. Anim. Behav. 32, in press.

(39) Gould, J.L.; Dyer, F.C.; and Towne, W.T. 1984. Recent progress in understanding the honey bee dance language. Fortsch. Zool., in press.

(40) Gould, J.L., and Gould, C.G. 1982. The insect mind: physics or metaphysics? In Animal Mind - Human Mind, ed. D.R. Griffin, pp. 269-298. Dahlem Konferenzen. Berlin: Springer-Verlag.

(41) Heran, H. 1958. Die Orientierung der Bienen im Flug. Ergebn. Biol. 20: 199-239.

(42) Hertz, M. 1929a. Die Organisation des optischen Feldes bei der Biene, I. Z. vgl. Physiol. 8: 693-748.

(43) Hertz, M. 1929b. Die Organisation des optischen Feldes bei der Biene, II. Z. vgl. Physiol. 11: 107-145.

(44) Hertz, M. 1931. Die Organisation des optischen Feldes bei der Biene, III. Z. vgl. Physiol. 14: 629-674.

(45) Hoefer, I., and Lindauer, M. 1975. Das Lernverhalten zweier Bienenrassen unter veränderten Orientierungsbedingungen. J. Comp. Physiol. 99: 119-138.

(46) Hoefer, I., and Lindauer, M. 1976. Der Einfluß einer Vordressur auf das Lernverhalten der Honigbiene. J. Comp. Physiol. 109: 249-264.

(47) Hoefer, I., and Lindauer, M. 1976. Der Schatten als Hilfsmarke bei der Orientierung der Honigbiene. J. Comp. Physiol. 112: 5-18.

(48) Heinrich, B. 1979. Bumble Bee Economics. Cambridge, MA: Harvard University Press.

(49) Holldobler, B. 1980. Canopy orientation: a new kind of orientation in ants. Science 210: 86-88.

(50) Horridge, G.A. 1962. Learning of leg position by headless insects. Nature 193: 697-698.

(51) Janzen, D. 1974. The deflowering of Central America. Nat. Hist. 83(4): 48-53.

(52) Klosterhalfen, S.; Fischer, W.; and Bitterman, M.E. 1978. Modification of attention in honey bees. Science 201: 1241-1243.

(53) Koltermann, R. 1973. Rassen- bzw. artspezifische Duftbewertung bei der Honigbiene und ökologische Adaption. J. Comp. Physiol. 85: 327-360.

(54) Koltermann, R. 1974. Periodicity in the activity and learning performance of the honey bee. In Experimental Analysis of Insect Behavior, ed. L.B. Brown, pp. 218-227. Berlin: Springer-Verlag.

(55) Kriston, I. 1973. Die Bewertung von Duft- und Farbsignalen als Orientierungshilfen an der Futterquelle durch Apis mellifera. J. Comp. Physiol. 84: 77-94.

(56) Lauer, J., and Lindauer, M. 1971. Genetische fixierte Lerndisposition bei der Honigbiene. In Informationsaufnahme und Informationsverarbeitung im lebenden Organismus 1, pp. 1-87. Wiesbaden: Franz Steiner Verlag.

(57) Lindauer, M. 1959. Angeborene und erlernte Komponenten in der Sonnenorientierung der Bienen. Z. vgl. Physiol. 42: 43-62.

(58) Lindauer, M. 1963. Kompaßorientierung. Ergebn. Biol. 26: 158-181.

(59) Lindauer, M. 1969. Lernen und Vergessen bei der Honigbiene. Proceedings of the VI Congress IUSSI, Bern.

(60) Lindauer, M. 1976. Recent advances in the orientation and learning of honey bees. Proceedings of the XV International Congress on Entomology, pp. 450-460, Washington, D.C.

(61) Maynard-Smith, J. 1976. Evolution and the theory of games. Am. Sci. 64: 41-45.

(62) Menzel, R. 1979. Behavioral access to short-term memory in bees. Nature 281: 368-369.

(63) Menzel, R. 1983. Neurobiology of learning and memory: the honey bee as a model system. Naturwissenschaften, in press.

(64) Menzel, R., and Erber, J. 1972. Influence of the quantity of reward on the learning performance in honey bees. Behav. 41: 27-42.

(65) Menzel, R., and Erber, J. 1978. Learning and memory in bees. Sci. Am. 239(1): 102-110.

(66) Menzel, R.; Erber, J.; and Masuhr, J. 1974. Learning and memory in the honey bee. In Experimental Analysis of Insect Behavior, ed. L.B. Browne, pp. 195-217. Berlin: Springer-Verlag.

(67) Murphey, R.K., and Matsumoto, S.G. 1976. Experience modifies the plastic properties of identified neurons. Science 191: 564-566.

(68) Nelson, M.C. 1971. Classical conditioning in the blowfly. J. Comp. Physiol. Psychol. 77: 353-368.

(69) Opfinger, E. 1931. Über die Orientierung der Biene an der Futterquelle. Z. vgl. Physiol. 15: 431-487.

(70) Prokopy, R.J.; Averill, A.L.; Cooley, S.S.; and Roitberg, C.A. 1982. Associative learning in egglaying site selection by apple maggot flies. Science 218: 76-77.

(71) Quinn, W.G.; Harris, W.A.; and Benzer, S. 1975. Conditioned behavior in Drosophila melanogaster. Proc. Natl. Acad. Sci. 71: 708-712.

(72) Ronacher, B. 1979a. Äquivalenz zwischen Groß- und Helligkeitsunterschieden im Rahmen der visuellen Wahrnehmung der Honigbiene. Biol. Cybernetics 32: 63-75.

(73) Ronacher, B. 1979b. Beitrag einzelner Parameter zum wahrnehmungsgemäßen Unterschied von zusammengesetzten Reizen bei der Honigbiene. Biol. Cybernetics 32: 77-83.

(74) Rossell, S., and Wehner, R. 1982. The bee's map of the e-vector pattern in the sky. Proc. Natl. Acad. Sci. 79: 4451-4455.

(75) Schnetter, B. 1972. Experiments on pattern discrimination in honey bees. In Information Processing in the Visual System of Arthropods, ed. R. Wehner, pp. 195-200. Berlin: Springer-Verlag.

(76) Templeton, B.L.; Bonini, N.; Dawson, D.R.; and Quinn, W.G. 1983. Reward learning in normal and mutant Drosophila. Proc. Natl. Acad. Sci. 80: 1482-1486.

(77) Tinbergen, N., and Kruyt, W. 1938. Ueber die Orientierung des Bienenwolfes III. Die Bevorzugung bestimmter Wegmarken. Z. vgl. Physiol. 25: 292-334.

(78) Van Beusekom, G. 1948. Some experiments on the optical orientation in Philanthus triangulum. Behav. 1: 195-225.

(79) Wehner, R. 1972. Pattern modulation and pattern detection in the visual system of hymenoptera. In Information Processing in the Visual System of Arthropods, ed. R. Wehner, pp. 183-194. Berlin: Springer-Verlag.

(80) Wehner, R. 1981. Spatial vision in arthropods. In Handbook of Sensory Physiology, ed. H. Autrum, pp. 287-616. Berlin: Springer-Verlag.

The Biology of Learning, eds. P. Marler and H.S. Terrace, pp. 181-196. Dahlem Konferenzen 1984. Berlin, Heidelberg, New York, Tokyo: Springer-Verlag.

Behavior Theory and Invertebrate Learning

C.L. Sahley
Dept. of Psychology, Yale University
New Haven, CT 06520, USA

INTRODUCTION

Historically, the study of animal learning has been the domain of the psychologist. Researchers within the "learning theory" tradition have been concerned with three defined problems: the necessary and sufficient conditions for learning to occur, the nature of what is learned, and the ways in which learning influences performance (cf. (25)). Within the field, a particular theory of learning is judged by how well it deals with these problems. Many different theoretical approaches to these issues have emerged, and it will avail us little to discuss any particular one in detail. On the other hand, we do need to consider the general approach.

To a behaviorist, learning is a theoretical construct, an inference made on the basis of an observed relationship between an animal's behavior and its known past experience. To conclude that learning has occurred is to say that an animal's behavior is a product of its prior experience with one or more stimulus events and of the relationship between these events (57, 59). The problem for the learning psychologist is to demonstrate that it is the organism's experience with the inter-event relationship per se and not some other variable confounded with that relationship that produced the observed change in behavior. This sounds abstract and grim. However, within this general definition one can begin to identify the kinds of organism-environment interactions that result in a learned change in behavior. The task of the learning theorist is then to derive a set of organizing principles that explain the existing data and generate ideas for future research.

As neurobiology has developed in the last decade, researchers have become increasingly interested in describing the cellular and molecular events that mediate the learned changes in behavior. This work has proceeded most successfully in invertebrates. The rationale for this approach has been described by Kandel (31). Briefly, the argument is that to elucidate the cellular basis of learning one must be able to identify the specific cells in the circuit and the cellular and molecular events in the cells that mediate the behavioral changes that result from learning. The virtue of certain invertebrates is their extreme relative simplicity, with fewer and larger cells, than those of any intact vertebrate. Consequently, one's chances for identifying individual cells which are relevant to learning, and the cellular and molecular events within the cells that produce the relevant changes, are increased enormously.

The invertebrate approach to understanding learning is now a vigorous and exciting area. Research at both the behavioral and cellular level is thriving. In this paper the relation of the behavioral aspects of the study of learning in invertebrates to the traditional psychological approach to the study of learning will be considered.

The study of learning in invertebrates is in its infancy. In less than a decade we have progressed from wondering if invertebrates can learn, to identifying the necessary and sufficient conditions that produce learning, and now we can begin to ask the nature of what is learned. In this progress invertebrate researchers have adopted from the "learning theory" tradition the well described operational criteria for identifying what constitutes learned changes in behavior and have adapted many paradigms known to produce learning in vertebrates. Psychologists studying vertebrates have examined in detail variations of the paradigms known to produce learning (57). From this work has emerged substantial knowledge about how a range of organism-environment interactions influence learned changes in behavior. What is now emerging in the invertebrate literature is a similar collection of data derived from experiments that employ methods (experimental operations) that are functionally equivalent to those that have been employed to study vertebrates. Consequently, there is now a basis for determining if there may be some general behavioral principles that are shared by vertebrate and invertebrate species. The strategy is to determine if there is a "functional similarity" (cf. (62)) in vertebrate and invertebrate learning processes by comparing how they respond to similar learning experiences.

In the remainder of this paper I will discuss recent experiments on learning in invertebrates to allow comparison with vertebrates subjected to similarly arranged environmental experiences. Specifically, I will consider the influence of inter-stimulus event relationships that result in associative learning on the two species. As a reflection of the bias of contemporary research on learning in invertebrates, the scope of this paper will be restricted to Pavlovian conditioning.

ASSOCIATIVE LEARNING: CONTINGENT INTER-STIMULUS EVENT RELATIONSHIPS

Associative learning in vertebrates has largely been defined by Pavlovian conditioning experiment procedures. The animal is exposed to a CS-US pairing, and associative learning is indexed by a change in the animal's behavior to the CS (the conditioned response) that can be uniquely attributed to the CS-US pairing operation and reflects the learned inter-event relationship between the CS and the US. Here I will describe several types of inter-event relations known to influence associative learning in vertebrates and compare the effects of these relationships in vertebrates to their effects on learning in invertebrates.

Temporal Relationships

One of the first inter-event relationships identified as being important for conditioning is the temporal relation of the CS and US (cf. (40)). In any vertebrate, regardless of species, the time interval separating the CS and US exerts a major influence on the probability that the CS and US will become associated, and thus, the CS will acquire conditioned responding. Depending on particular associations to be learned and the response that is measured, the optimal time interval for learning varies from milliseconds, as with the rabbit's nictitating membrane response (22), to hours, as with conditioning a taste aversion in rats (e.g., (18, 19, 59)). Nevertheless, there is a general rule in vertebrates: as this interval increases, the likelihood of successful conditioning decreases. This principle holds for invertebrates as well. The list of invertebrates that show a learned change in behavior as a result of experiencing two stimuli presented contiguously in time is rapidly increasing and includes examples from almost all invertebrate species. These include planarian flatworms (4, 29, 32), bees (5, 33, 35, 39, 44, 65, 67), hermit crabs (41), crayfish (4, 12, 72), fruit flies (6, 16, 46, 63), leeches (24), starfish (36), octopuses (71), and many other molluscs (1, 2, 7, 13-15, 20, 42, 43, 61, 62, 69, 70). In several preparations there has been an attempt to determine the temporal CS-US interval, the time between the onset of the CS and the onset of the US. In bees this interval was first examined by Menzel (39). He found that free-flying bees associate a color with food only when it is closely contiguous with the initiation of feeding, the time window ranging from approximately +3 to -2s.

More recently, the CS-US interval has been examined in restrained bees using the proboscis-extension procedure which allows much better control of the interval (5). In a trace conditioning procedure, Bitterman et al. (5) were able to vary the CS-US interval from +15s to -15s in different groups. They found that forward pairings result in better conditioning than backward pairings and that the optimal CS-US interval is from +1 to -3s.

Perhaps the most rigorous analysis of inter-stimulus interval in invertebrates is a recent study on Aplysia (7). Carew et al. examined the effect of inter-stimulus interval using differential conditioning of the gill-withdrawal reflex system. Using different animals, they systematically varied the interval between the presentation of the CS and the US. They found a sharp decline in the conditioned response at intervals greater than 2s. In addition, no significant learning was observed when the CS and US were presented simultaneously or when the US onset preceded the CS by 0.5, 1.0, or 1.5s.

Thus, as is the case for vertebrates (22), temporal CS-US relationships are critical for learning in invertebrates.

Stimulus Relationships

When two or more stimuli are presented together as a compound and paired with a US, it is often the case that only one component of the compound will acquire conditioned properties, even though both stimuli have been paired with the US. This phenomenon is known as stimulus selection. It is important because it indicates that temporal pairing of the CS and US is not the sole determinant of conditioning. Two important factors which influence the CS's ability to acquire conditioned responding are a) the predictive value of the CS, as originally revealed by Kamin's blocking experiments (30), and b) the salience of the CS, as originally revealed by Pavlov's overshadowing experiments (45).

Predictive value of the CS. Kamin (30) was the first to describe the importance of the predictive value of the CS for determining which stimuli would acquire conditioned responding. In his experiment, animals were first conditioned to one CS, S1. Once S1 reliably evoked the conditioned response, another CS, S2, was then presented in compound with the S2, and this compound, S1S2, was paired with the US (S1S2-US). He found that the added CS, S2, failed to elicit conditioned responding, even though it had been repeatedly paired with the US. The prior training to the S1 blocked conditioning to the S2.

Kamin concluded that the added stimulus, S2, failed to acquire conditioned properties because the information it provided the animal, with respect to the US occurrence, was redundant with that already provided by S1. He suggested that when the US is well predicted, by the presence of a stimulus with which it had been previously paired, that stimulus will interfere with or block conditioning to other stimuli. The importance of the predictive inter-event relations in the establishment of associative learning has more recently been demonstrated in a variety of species and tasks (8, 21, 26, 38, 64, 66), including invertebrates (17, 62). However, in the honey bee, Couvillon and colleagues reported that a Kamin-like paradigm revealed no evidence of blocking (11). Two groups of bees were compared in a free-flying procedure. The blocking group was first trained with a single stimulus, S1-US, and then in a second phase with a compound stimulus, S1S2-US. The control group was trained first with the compound and then with the single stimulus. Bees in both groups were then tested for conditioned responding to S2. If predictability of the US were important in stimulus selection, one would expect less response to S2 in the blocking group. In contrast, Couvillon et al. found relatively more response to S2 in the blocking group.

Overshadowing. In addition to the predictive value, the salience of a component in a compound stimulus greatly influences stimulus selection. That is, when two stimuli are presented as a compound, the particular stimulus which is more strongly perceived by the animal will tend to "overshadow" the weaker stimulus (45). In Pavlov's original experiments, one group received compound stimuli consisting of, for example, a tactile and thermal stimulus paired with the US. The second group received each stimulus separately paired with the US. In this case the dogs trained with a compound CS and tested to each component separately failed to show conditioned responding to one of the components of the compound. Pavlov's control experiments in which dogs were conditioned to the individual stimuli showed that the dogs were able to learn either of the stimulus associations, if such were presented separately. Contemporary theorists offer two possible explanations for this phenomenon. Rescorla and Wagner (58, 68) suggest that stimuli compete for a fixed amount of associative strength and the most salient stimulus gets a larger amount. In contrast, Mackintosh (37) suggests that each stimulus competes for the attention that is assumed necessary for learning.

A recent study of overshadowing in honey bees has been substantially guided by these existing theoretical analyses. In several experiments, Couvillon et al.'s experiments patterned largely after experiments with vertebrates, training to compounds stimuli, and then testing to the individual components, demonstrate clear indication of an overshadowing effect. The experiments, however, do not clearly support either competition for associative strength (58) or competition for attention (37). These experiments are difficult to interpret because configurational effects could account entirely for overshadowing (28), and bees are known to be greatly influenced by configurations of stimuli (11).

It appears, then, that predictive relationships are important for learning in molluscs but not for bees.

Inhibitory Relationships

Pavlov was the first to describe learning of inhibitory relationships (45), but recently Rescorla has systematically extended the notion of inhibitory learning (48-50). In his experiments, Rescorla varied the predictive value of the CS by altering the degree to which US occurrence was contingent upon CS occurrence. This was accomplished by giving all animals the same number of CS-US pairings but varying the number of extra (irrelevant) US presentations. The idea behind these experiments is that if the US occurs occasionally in the absence of the CS, then the CS begins to predict the absence of the US. Rescorla found that when the CS began to predict the absence of the US, conditioned inhibition was observed (23, 34, 38, 45, 50). Several other procedures including explicitly unpaired CS-US presentations also produce conditioned inhibition (51, 65).

In invertebrates, the only report of conditioned inhibition is in bees (5, 10). In a two-phase experiment using proboscis extension, Bitterman et al. (5) found that animals experiencing explicitly unpaired presentations of the CS and US could learn to discriminate between the presence and absence of the odor. That is, animals which had previously received explicitly unpaired presentations of the CS and US would immediately extend their proboscises, when placed in the training situation with no odor and would retract them if the odor were presented. In contrast, animals trained with paired CS-US presentations would extend their proboscises only when the odor was presented. The behavior of the first group indicated that some kind of inhibitory relationship had been learned. In another series of experiments by the same group, bees were trained with an odor explicitly unpaired with sugar. They were then subjected to a second conditioning regimen and their subsequent behavior was examined. The bees were presented with a new odor paired with the sugar, while the remaining bees received paired presentations of the odor they had previously experienced unpaired with the US. The results showed that the bees learned a) to respond to the new odor, and b) to cease responding to the placement. In contrast, the bees experiencing CS-US pairings with the previously unpaired odor showed a substantial resistance to conditioning of this previously unpaired odor. Thus, the previous unpaired CS-US pairings had resulted in inhibitory learning to the CS, and this inhibition resulted in a retardation of the learning of the new paired relationship.

Further support in invertebrates of the idea that explicit unpairing of the CS and US inhibits subsequent learning of an association comes from Bitterman et al.'s further finding that explicitly unpaired training facilitated subsequent extinction of the learning (5).

Higher-order Relationships

The associative learning of vertebrates is also influenced by higher-order inter-event relations. The defining feature of these phenomena is that a CS can acquire the ability to elicit conditioned responding without ever being directly paired with the US (see (55)). The most familiar example of this is second-order conditioning (cf. (45, 52)). In this two-phase experiment, first-order conditioned responding is established to one conditioned stimulus, S1, by pairing it with the US (S1-US pairings). In Phase 2 another CS, S2, is introduced and paired with S1 (S2-S1 pairings). The result of this two-phase training procedure is that S2 acquires the ability to evoke conditioned responding even though it was never paired with the US.

In the years since Pavlov (48), psychologists have observed second-order conditioning in a variety of vertebrates (3, 9, 27, 58) and invertebrates (40, 41, 62), including the terrestrial mollusc, Limax maximus.

To produce second-order conditioning in Limax, we (62) exposed slugs to two phases of training. In Phase 1 slugs experienced pairings of carrot odor and quinidine (S1-US pairings), and in Phase 2 they received simultaneous potato odor-carrot odor pairings (S2-S1 pairings). The question was: Would this training reduce the slug's preference for potato odor (S2), even though it was never paired with the quinidine? Since second-order conditioning is dependent on pairings in both phases, we compared learning in this group with the learning in two control groups which did not experience both pairings. We found that the slugs in the experimental group displayed a reduced preference for potato odor in comparison to the slugs in the control groups. The same inter-event relationships that produce second-order conditioning in vertebrates produce second-order conditioning in Limax.

Recently, Rescorla and his colleagues (27, 50, 51, 60) have begun to examine the nature of the associations mediating second-order conditioning in vertebrates. Rescorla (54-56) has discussed the issues involved in considerable detail. During Phase 2 of a second-order conditioning experiment (S2-S1 pairings), there are several associations that the animal can learn that could mediate a conditioned response to S2. One possibility is an S2-S1 association. In this case, it is assumed that the animal associates representations of the two CS events. For example, in our study this would imply that the slugs associated representations activated by the potato odor (S2) and carrot odor (S1) experience. According to this view, the second-order conditioned response to potato odor (S2) would be mediated by a two-link chain of associations. First, potato odor (S2) would activate the slugs' representation of carrot odor (S1) which in turn, due to its first-order association with quinidine, would evoke the conditioned response.

A second alternative is the S2-CR association. This account assumes that during the Phase 2 (S2-S1 pairings), S1, due to its pairings with the quinidine US, evokes a conditioned response (CR). This allows for the possibility that the animal will associate S2 with the CR activated by the S1. For example, in our experiment, this would mean that the slugs associated potato odor (S2) with the conditioned response that carrot odor (S1) activated. The second-order response to potato odor (S2) during testing would then be directly evoked by potato odor and not via carrot odor.

Although this analysis is abstract, there is a way of experimentally distinguishing between these two classes of interpretation. The distinguishing experiment makes use of what Rescorla (53) terms a post-conditioning treatment strategy. The first alternative, the S2-S1 association, implies that conditioned responding to S2 is dependent upon the ability of S1 to evoke a conditioned response. In contrast, the S2-CR association view implies no such dependency. Thus, if the S1 extinction treatment occurs following the two phases of training necessary to produce second-order conditioning, but prior to testing for second-order conditioning to S2, S1's ability to evoke a first-order conditioned response is extinguished. The S1-S2 association view would predict that the S1 extinction treatment would eliminate S2's ability to evoke a second-order conditioned response, because S2's ability depends on the integrity of S1's first-order conditioning properties. In contrast, the S2-CR association view predicts that the S1 extinction treatment would not influence S2's ability to evoke a conditioned response, because S2 can directly evoke conditioned responding.

As Rescorla (55, 56) points out, the implications of these conceptions of the associations involved in second-order conditioning have been widely investigated in vertebrates. Moreover, evidence consistent with both views has been reported. That is, in some instances the S1 extinction procedure has eliminated second-order conditioning (e.g., (9, 58)), and in other instances has not (47).

Recently, some of the variables which give rise to this discrepancy in results have been identified (see (9, 54)). What appears to be especially important is the manner in which S2 and S1 are presented during Phase 2 (second-order training). When they are presented simultaneously (S2 and S1), the acquisition of S2-S1 associations is favored. When they are presented sequentially (S2, S1), acquisition of S2-CR associations is favored.

Working with the invertebrate Limax, we have analyzed the associations that contribute to second-order conditioning using the S1-extinction procedure. We wanted to determine if the vertebrate generalization - that simultaneous S2+S1 presentations promote S2-S1 associations, and if sequential S2, S1 presentations promote S2-CR associations - also applies to Limax (62).

We found that the second-order response to potato odor by slugs exposed to simultaneous Phase 2 pairings of two odors was dependent on the integrity of their first-order responses to carrot odor. That is, extinguishing the first-order response also extinguishes the second-order reaction. This outcome strongly suggests that the second-order response produced by simultaneous S2-S1 pairings was mediated by the slug, in some sense, by associating representations of the two odors, i.e., an inter-odor S2-S1 association.

In contrast, the second-order response to potato odor (S2) by slugs exposed to sequential pairings of potato and carrot odor was independent of their first-order response to carrot odor. The S1 extinction procedure attenuated the first-order response but had no effect on the second-order response. This result provides evidence in support of the S2-CR association view. Thus, slugs associated potato odor with some component of the conditioned response complex evoked by carrot odor (S1) during the Phase 2 potato-carrot odor pairings.

Thus, the second-order conditioned change in Limax's odor preference behavior can be mediated by alternative associative structures, and simultaneous S2-S1 presentations favor the formation of an S2-S1 association, whereas sequential S2, S1 presentations favor formation of an S2-CR association. The similarity of these conclusions to those that apply to second-order conditioned responses of vertebrates is remarkable.

CONCLUSIONS

From the preceding brief review, it is clear that striking commonalities exist between learning in invertebrates and learning in vertebrates. Guided by the operational definitions and paradigms developed by learning psychologists, a set of principles that has considerable generality within and between species has begun to emerge. Over a broad range of learning phenomena, there is a remarkable functional similarity between the inter-event relationships that influence learning in vertebrates and invertebrates.

Whether these common principles are mediated by similar cellular and molecular processes is still an open question. Learning provides animals with the ability to extract causal relationships among biologically significant events in their environment. The emergence of common behavioral principles may have come about as a result of a convergent evolution in response to a common set of selection pressures. That is, to have the ability to learn that events are connected by a causal relationship may give an animal such a selective advantage that this ability may have evolved in parallel in many animals. In this case, these common behavioral principles between vertebrates and invertebrates may be mediated by a multiplicity of physiological and biochemical processes. On the other hand, nature tends to be conservative. The fine-grained phenomenological similarity may reflect an underlying mechanistic similarity.

The exciting prospect for the near future is that the studies in invertebrates, always accessible neurophysiologically, can address the fine-grained issues of what association is learned and can determine how learning influences performance.

REFERENCES

(1) Alkon, D.L. 1974. Associative training of Hermissenda. J. Gen. Physiol. 64: 70-84.

(2) Alkon, D.L. 1980. Cellular analysis of a gastropod mollusc, Hermissenda crassicornis, model of associative learning. Biol. Bull. 159: 505.

(3) Amiro, T.W., and Bitterman, M.E. 1980. Second-order conditioning in goldfish. J. Exp. Psychol.: Anim. Behav. Proc. 6: 41-48.

(4) Applewhite, P., and Morowitz, H.J. 1966. The micrometazoa as model systems for studying the physiology of memory. Yale J. Biol. Med. 39: 60-105.

(5) Bitterman, M.E.; Menzel, R.; Feitz, A.; and Schafer, S. 1983. Classical conditioning of Proboscis Extension in Honeybees (Apis mellifera). J. Comp. Psychol. 97: 107-119.

(6) Booker, R., and Quinn, W.G. 1981. Conditioning of leg position in normal and mutant Drosophila. Proc. Natl. Acad. Sci. USA 78: 3940-3944.

(7) Carew, T.J.; Hawkins, R.D.; and Kandel, E.R. 1983. Differential classical conditioning of a defensive withdrawal reflex in Aplysia californica. Science 219: 397-400.

(8) Cheatle, M.D., and Rudy, J.W. 1978. Analysis of second-order odor-aversion conditioning in neonatal rats: Implications for Kamin's blocking effect. J. Exp. Psychol.: Anim. Behav. Proc. 4: 237-249.

(9) Cheatle, M.D., and Rudy, J.W. 1979. Ontogeny of second-order odor-aversion conditioning in neonatal rats. J. Exp. Psychol.: Anim. Behav. Proc. 5: 142-151.

(10) Couvillon, P.A., and Bitterman, M.E. 1980. Some phenomena of associative learning in honey bees. J. Comp. Physiol. Psych. 94: 878-885.

(11) Couvillon, P.A.; Klosterhalfen, S.; and Bitterman, M.E. 1983. Analysis of overshadowing in honeybees. J. Comp. Psychol., in press.

(12) Cowles, R.P. 1908. Habits, reactions, and associations in Ocypoda arenaria. Papers Tortugas Lab. Carnegie Inst. Wash. 2: 1-41.

(13) Croll, R., and Chase, R. 1980. Plasticity of olfactory orientation to foods in the snail Achtina fulica. J. Comp. Physiol. 136: 266-277.

(14) Crow, T.J., and Alkon, D.L. 1978. Retention of an associative behavioral change in Hermissenda. Science 201: 1239-1241.

(15) Davis, W.J., and Gillette, R. 1978. Neural correlates of behavioral plasticity in command neurons of Pleurobranchaea. Science 199: 801-804.

(16) Dudai, Y. 1980. Properties of learning and memory in Drosophila melanogaster. J. Comp. Physiol. 114: 69-89.

(17) Farley, J., and Alkon, D.L. 1983. Primary Neural Substrates of Learning and Behavioral Change. New York: Cambridge University Press, in press.

(18) Garcia, J.; Ervin, F.R.; and Koelling, R.A. 1966. Learning with a prolonged delay of reinforcement. Psychonom. Sci. 5: 121-122.

(19) Garcia, J., and Koelling, R.A. 1966. Relation of cue to consequence in avoidance learning. Psychonom. Sci. 4: 123-124.

(20) Gelperin, A. 1975. Rapid food aversion learning by a terrestrial mollusc. Science 189: 567-570.

(21) Gillan, D.J., and Domjan, M. 1977. Taste-aversion conditioning with expected versus nonexpected drug treatment. J. Exp. Psychol.: Anim. Behav. Proc. 3: 297-309.

(22) Gormezano, I. 1975. Yoked comparisons of classical and instrumental conditioning of the eyelid response; and an addendum on "voluntary responders". In Classical Conditioning: A Symposium, ed. W.F. Prokasy. New York: Appleton-Century-Crofts.

(23) Hearst, E. 1972. Some persistent problems in the analysis of conditioned inhibition. In Inhibition and Learning, eds. R.A. Boakes and M.S. Halliday, pp. 5-39. London: Academic Press.

(24) Henderson, T., and Strong, T. 1972. Classical conditioning in the leech, Macrobdella dititra as a function of CS and US intensity. Conditioned Reflex 7: 210-215.

(25) Hilgard, E.R., and Bower, G.H. 1966. Theories of Learning. New York: Appleton.

(26) Holland, P.C. 1977. Conditioned stimulus as a determinant of the form of the Pavlovian conditioned response. J. Exp. Psychol.: Anim. Behav. Proc. 3: 77-104.

(27) Holland, P.C., and Rescorla, R.A. 1975. Second-order conditioning with food unconditioned stimulus. J. Comp. Physiol. Psych. 88: 459-467.

(28) Hull, C. 1934. Principles of Behavior. New York: Appleton.

(29) Jacobson, A.L.; Fried, C.; and Horowitz, S.D. 1967. Classical conditioning, pseudoconditioning and sensitization in the planarian. J. Comp. Physiol. Psych. 64: 73-79.

(30) Kamin, L.J. 1969. Predictability, surprise, attention, and conditioning. In Punishment and Aversive Behavior, eds. R. Church and B.A. Campbell, pp. 279-296. New York: Appleton-Century-Crofts.

(31) Kandel, E.R. 1976. Cellular Basis of Behavior, pp. 279-296. San Francisco: Freeman.

(32) Kimmel, H.D., and Garrigan, H.A. 1973. Resistance to extinction in planaria. J. Exp. Psychol. 101: 343-347.

(33) Koltermann, R. 1969. Lern- und Vergessensprozesse bei der Honigbiene aufgezeigt anhand von Duftdressuren. Z. Vgl. Physiol. 5: 762-800.

(34) Konorski, J. 1967. Integrative Activity of the Brain. Chicago: University of Chicago Press.

(35) Kriston, I. 1971. Zum Problem des Lernverhaltens von Apis mellifica L. gegenüber verschiedenen Duftstoffen. Z. Vgl. Physiol. 74: 169-189.

(36) Landenberger, D.E. 1966. Learning in the Pacific starfish, Piaster giganteus. Anim. Behav. 14: 414-418.

(37) Mackintosh, N.J. 1974. The Psychology of Animal Learning. London: Academic Press.

(38) Marchant, H.G., and Moore, J.W. 1973. Blocking of the rabbit's conditioned nictitating membrane response in Kamin's two-stage paradigm. J. Exp. Psychol. 101: 155-159.

(39) Menzel, R. 1968. Das Gedächtnis der Honigbiene für Spektralfarben I. Kurzzeitiges und langzeitiges Behalten. Z. Vgl. Physiol. 60: 82-102.

(40) Menzel, R. 1983. Neurobiology of learning and memory: The honeybee as a model system. Naturwiss. 70: 504-511.

(41) Mikhailoff, S. 1923. Experience reflexologique: Experiences nouvelles sur Pagurus striatus, Leander xiphiaas et treillanus. Bull. Inst. Oceanograph. Monaco 375.

(42) Mpitsos, G.J., and Collins, S.D. 1975. Learning: rapid aversion conditioning in the gastropod mollusc Pleurobranchaea. Science 188: 954-957.

(43) Mpitsos, G.J.; Collins, S.; and McClellan, A.D. 1978. Learning: A model system for physiological studies. Science 199: 497-502.

(44) Opfinger, E. 1931. Über die Orientierung der Biene an der Futterquelle. Z. Vgl. Physiol. 15: 431-487.

(45) Pavlov, I.P. 1927. Conditioned Reflexes, pp. 33-35. London: Oxford University Press.

(46) Quinn, W.G.; Harris, W.A.; and Benzer, S. 1974. Conditioned behavior in Drosophila melanogaster. Proc. Natl. Acad. Sci. USA 71: 708-712.

(47) Rashotte, M.E.; Griffin, R.W.; and Sisk, C.L. 1977. Second-order conditioning of the pigeon's key peck. Anim. Learn. Behav. 5: 25-38.

(48) Rescorla, R.A: 1967. Pavlovian conditioning and its proper control procedures. Psychol. Rev. 74: 71-80.

(49) Rescorla, R.A. 1968. Probability of shock in the presence and absence of the CS in fear conditioning. J. Comp. Physiol. Psych. 66: 1-5.

(50) Rescorla, R.A. 1969. Conditioned inhibition of fear resulting negative CS-US contingencies. J. Comp. Physiol. Psych. 67: 504-509.

(51) Rescorla, R.A. 1973. Effect of US habituation following conditioning. J. Comp. Physiol. Psych. 75: 77-81.

(52) Rescorla, R.A. 1973. Second-order conditioning: implications for theories of learning. In Contemporary Approaches to Conditioning and Learning, eds. F.J. McGuigan and D.B. Lamsden, pp. 7-33. Washington, D.C.: Winston.

(53) Rescorla, R.A. 1975. Pavlovian excitatory and inhibitory conditioning. In Handbook of Learning and Cognitive Processes: Conditioning and Behavior Theory, ed. W.K. Estes, vol. 2, pp. 7-35. Hillsdale, NJ: Erlbaum.

(54) Rescorla, R.A. 1980. Pavlovian Second-order Conditioning: Studies in Associative Learning. Hillsdale, NJ: Erlbaum.

(55) Rescorla, R.A. 1980. Simultaneous and successive associations in sensory preconditioning. J. Exp. Psychol.: Anim. Behav. Proc. 6: 207-216.

(56) Rescorla, R.A. 1982. Simultaneous second-order conditioning produces S-S learning in conditioned suppression. J. Exp. Psychol.: Anim. Behav. Proc. 8: 23-32.

(57) Rescorla, R.A., and Holland, P.C. 1976. Some behavioral approaches to the study of learning. In Neural Mechanisms of Learning and Memory, eds. M.R. Rosenzweig and E.L. Bennett. Cambridge, MA: MIT Press.

(58) Rescorla, R.A., and Wagner, A.R. 1972. A theory of Pavlovian conditioning: Variations in the effectiveness of reinforcement and nonreinforcement. In Classical Conditioning II: Current Research and Theory, eds. A.H. Black and W.F. Prokasy. New York: Appleton-Century-Crofts.

(59) Revusky, S.H., and Garcia, J. 1971. Learned associations over long delays. In The Psychology of Learning and Motivation, ed. G.H. Bower, vol. 4, pp. 1-84. New York: Academic Press.

(60) Rizley, R.C., and Rescorla, R.A. 1972. Associations in second-order conditioning and pre-conditioning. J. Comp. Physiol. Psych. 81: 1-11.

(61) Sahley, C.L.; Gelperin, A.; and Rudy, J.W. 1981. One-trial associative learning in a terrestrial mollusc. Proc. Natl. Acad. Sci. USA 78: 640-642.

(62) Sahley, C.L.; Rudy, J.W.; and Gelperin, A. 1981. An analysis of associative learning in a terrestrial mollusc. I. Higher-order conditioning, blocking, and a US-pre-exposure effect. J. Comp. Physiol. 144: 1-8.

(63) Tempel, B.L.; Bonini, D.R.; Dawson, D.R.; and Quinn, W.G. 1982. Reward learning in normal and mutant Drosophila. Proc. Natl. Acad. Sci. USA 80: 1482-1486.

(64) Tennent, W.A., and Bitterman, M.E. 1975. Blocking and overshadowing in two species of fish. J. Exp. Psychol.: Anim. Behav. Proc. 1: 22-29.

(65) Vareschi, E. 1971. Duftunterscheidung bei der Honigbiene. Einzelzell-Ableitungen und Verhaltensreaktionen. Z. Vgl. Physiol. 75: 143-173.

(66) Vom Saal, W., and Jenkins, H.M. 1970. Blocking the development of stimulus control. Learn. Motiv. 1: 52-62.

(67) Von Frisch, K. 1921. Über den Geruchssinn der Bienen und seine blütenbiologische Bedeutung. Zool. Jahrb. 37: 1-238.

(68) Wagner, A.R. 1969. Stimulus selection and a modified continuity theory. In The Psychology of Learning and Motivation, eds. G.H. Bower and J.T. Spence, vol. 3. New York: Academic Press.

(69) Walters, E.T.; Carew, T.J.; and Kandel, E.R. 1981. Associative learning in Aplysia californica. Proc. Natl. Acad. Sci. USA 76: 6675-6679.

(70) Walters, E.T.; Carew, T.J.; and Kandel, E.R. 1981. Associative learning in Aplysia. Evidence for conditioned fear in an invertebrate. Science 211: 504-506.

(71) Wells, M.J., and Young, J.Z. 1968. Learning with delayed rewards in Octopus. Z. Vgl. Physiol. 61: 103-128.

(72) Yerkes, R.M., and Huggins, G.E. 1903. Habit formation in the crawfish, Cambarus affinis. Harv. Psychol. Stud. I: 565-577.

The Biology of Learning, eds. P. Marler and H.S. Terrace, pp. 197-246. Dahlem Konferenzen 1984. Berlin, Heidelberg, New York, Tokyo: Springer-Verlag.

Work in Invertebrates on the Mechanisms Underlying Learning

W.G. Quinn
Dept. of Biology, Princeton University
Princeton, NJ 08544, USA

INTRODUCTION

The lotus-eaters may be the first amnesiacs in literature. Penelope, at the other extreme of memory retentiveness, personifies the salient ideal of the Odyssey - faith and constancy in the face of time and change. In the twenty-eight centuries since Homer, humanists have speculated occasionally, often eloquently, about the origins and effects of memories. The issue was turned into a scientific one at the beginning of this century by Pavlov, Thorndike, and others, who took a few instances of learning, removed them from the sphere of subjective speculation, and made them accessible to quantitative experiments in the laboratory. In the years following, there was a tendency to emulate the experiments of the pioneers rather than their reductionist spirit. Psychologists have been industrious and successful in cataloguing the phenomenology of learning, but scientists of all stripes have met with real difficulty in pursuing their understanding past overt behavior.

A change in an animal's behavioral propensities must reflect physical or chemical changes in its brain. What is the character of the change? Recent work in invertebrates has begun to crack this problem. Workers on Aplysia have assembled detailed molecular and physiological mechanisms to explain nonassociative learning and classical conditioning in that organism. The picture drawn from the biochemical lesions in learning-deficient Drosophila mutants converges with the Aplysia model. Other work, on a somewhat special type of associative learning in

Hermissenda, suggests a mechanism which is different, but not entirely different. Neurophysiological work in a variety of other preparations suggests that the biochemical machinery implicated in these learning changes may operate in many neural systems, even neuromuscular junctions, and may mediate a variety of behavioral alterations.

The picture will be easiest to follow if we first review Aplysia, the most developed system, and then discuss other work in terms of this scheme. In Aplysia, we will start with the simplest behaviors first - nonassociative learning.

APLYSIA

Aplysia californica is an ugly marine mollusc about 15 cm long. Its behavior consists largely of creeping along the ocean floor and eating seaweed, and it escapes predation by looking unattractive, tasting unpalatable, and exuding foul-smelling purple ink when disturbed. What behavior it has, however, is eminently accessible to analysis, because the animal is built like an old Philco radio, with simple circuits and large, easily identifiable components. Aplysia's nervous system has about 10,000 large neurons organized in ganglia. Many are identifiable from individual to individual by size, color, location, connections to other neurons, and the behavioral patterns they subserve (80). Kandel and his associates have made most of their headway by concentrating on one reflex response - gill or siphon withdrawal in response to siphon stimulation. A wild Aplysia in the ocean ordinarily has its gill expanded and its siphon protruding from its mantle. If it receives a moderate tactile stimulus to its mantle or siphon, it withdraws its siphon and contracts its gill. This reflex, stereotyped and seemingly monotonous, is actually subject to two forms of behavioral modulation. If the sensory stimulus is repeated again and again, the response progressively wanes in intensity, or habituates. If, on the other hand, the animal is nonspecifically aroused by a novel, strong stimulus, such as hitting it over the head or shocking it on the tail, the gill-withdrawal response is accentuated, or sensitized. Both of these behavioral changes are long lasting, both result from alterations in the central nervous system, and both are examples of nonassociative learning. Moreover, the circuit underlying this reflex is a relatively simple one, even for Aplysia. If the siphon is poked or touched, sensory end organs send impulses via axons to sensory cells on the abdominal ganglion. These make monosynaptic connections to motor neurons in the same ganglion, which stimulate the muscles for gill withdrawal. Under the test conditions, the peripheral parts of the reflex - sensory end organs, the neuromuscular junctions, and the muscles - all respond faithfully and repetitively, and modulation of the reflex takes place at central connections in the ganglion. Thus habituation, for the most part, reflects decreasing efficacy of transmission between sensory and motor neurons, and sensitization reflects increasing efficacy of transmission (23, 24, 33, 62, 81, 82).

Actually, the circuitry is not quite as straightforward as this sketch suggests. Elements of a peripheral nerve net can mediate this reflex (92), although they tend to do so at much higher stimulus intensities than normally occur in the lab or the ocean (21). Even within the CNS, some of the reflex and much of the experience-dependent modulation appear to be mediated by four or five interneurons (30, 61, 62) (Fig. 1). Moreover, the parallel often drawn between behavior and physiology is somewhat of an oversimplification. Neurophysiological work has been done mostly with the gill-withdrawal reflex (because it is simpler); behavioral measurements are mostly on the similar siphon-withdrawal reflex (because it is easier to observe in intact animals). Therefore, the parallel often drawn between changes in strength of a behavioral reflex strength and changes in strength of identified synaptic connections represents to some extent a heuristic simplification. Still, we are left with a nice, semiquantitative correlation between behavioral changes and synaptic changes, for both habituation and sensitization.

Given a change in synaptic properties, a logical question is whether the alteration occurs in the presynaptic or the postsynaptic cell. The classical method for deciding this is quantal analysis of transmission in low calcium solution. Castellucci and Kandel (26, 27, 70) did this for the identified synapses. The Aplysia neurons they studied do not naturally lend themselves to quantal analysis as did the classical frog neuromuscular junctions. The Aplysia cells' geometry is irregular, and the relevant synapses are distributed over various cell processes which are far from the best physiological recording site. Because of this unfavorable situation, the data in these analyses look unconvincing at first glance. Nevertheless, the results (particularly the data on frequency of failures) are in reasonable accord with changes at the presynaptic side of the synapses. The number of vesicles released (quanta) decreases during habituation and increases during sensitization, while the postsynaptic response to a vesicle's worth of transmitter stays at least approximately constant. (The corollary conclusion, that the synaptic change is <u>entirely</u> in presynaptic quantal content, does not follow strongly from the data, but it is the simplest interpretation and is supported by later evidence.)

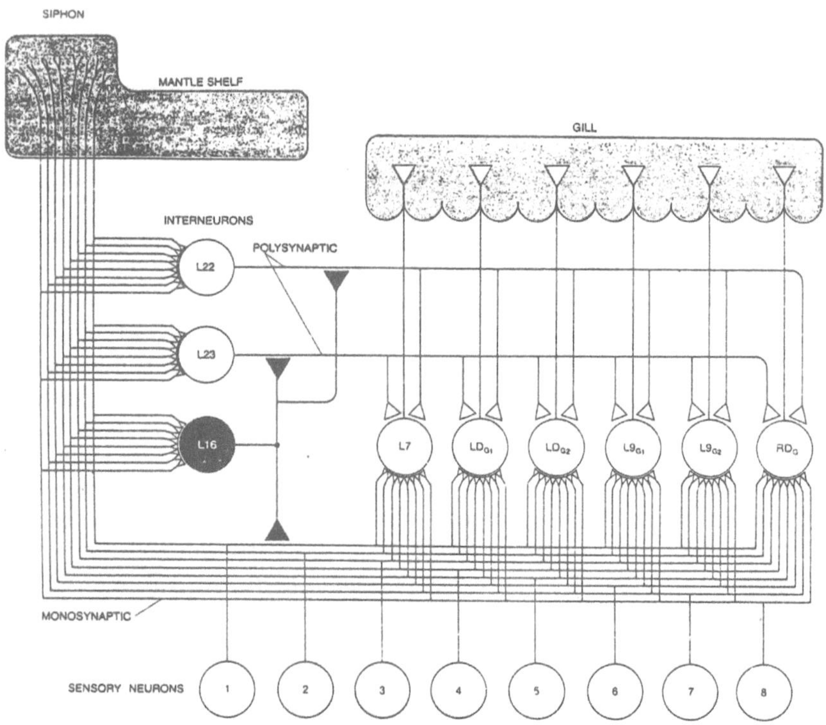

FIG. 1 - Schematic diagram of neuronal connections in the gill-withdrawal circuit. In the reflex action the animal withdraws its gill when the siphon (the mantle shelf) is stimulated in some way. The skin of the siphon is innervated by about twenty-four sensory neurons; the diagram includes only eight of them. The sensory neurons make direct monosynaptic connections to six identified gill motor neurons, which are shown in the row beginning with L7, and to at least one inhibitory cell (L16) and two interposed excitatory interneurons (L22 and L23) which make synapses with motor neurons. Since the diagram was published, other interneurons and facilitatory neurons of the L29 group have been discovered, and the connections between interneurons has been found to be much more complex. (See Fig. 9 in (58)). This figure is from Kandel (68).

Thus Kandel and his colleagues at Columbia, by circuit-tracing methods, went from experience-dependent modulation of a reflex to a change in efficacy of identified, parallel synapses then, by physiological experiments, to a change in the presynaptic half of the synapse - the terminals of sensory neurons. The evidence up to this point, while not compelling, was persuasive, so it made sense to continue by looking for chemical and physiological changes in the sensory cells themselves.

Aplysia can be sensitized in general and its gill-withdrawal response can be enhanced in particular by beating the animal over the head or (more efficaciously) by applying electric shock to its tail. Cedar, Schwartz, and Kandel found that concentrations of biogenic monoamine transmitters (principally dopamine, octopamine, and serotonin) in Aplysia and cyclic AMP were elevated in the entire abdominal ganglion after behavioral sensitization (31, 32). Of the three monoamines, only serotonin altered the efficacy of the synapses being studied, and this compound caused cAMP levels to rise in the (presynaptic) sensory neurons (32, 70). At this point, given these biochemical clues, the logical thing to do was to carry out behavioral and biochemical procedures which might mimic sensitization and to look in the presynaptic terminals for neurophysiological changes which produce enhanced vesicle release.

As it stands, this experiment is technically too difficult. The terminals, small and distant from the cell bodies, are inaccessible to direct study with microelectrodes. Therefore, Klein and Kandel (76, 77) looked instead for changes in the membranes of sensory cell bodies, on the plausible supposition that qualitatively similar events occurred in the terminals. They found that physiological consequences of behavioral sensitization (shocking the animal's tail, stimulating the relevant connective to the abdominal ganglion, exciting identified facilitatory neurons in the ganglion, applying serotonin to a sensory neuron, or injecting cAMP into the neuron) all produced a similar change in a sensory neuron's electrical properties - an alteration in the shape of its action potential. Remember in the classical Hodgkin-Huxley model that membrane depolarization during the rising phase of the action potential results from inward sodium current (and also inward calcium current in some cells). Membrane repolarization during the later falling phase is brought about by outward potassium current. Aplysia sensory neurons "sensitized" by any of the above procedures were slower to repolarize after action potentials (especially in the presence of tetraethylammonium (TEA) (76, 77)). Voltage-clamp studies in different solutions showed that this delay of repolarization results from inactivation of a new potassium channel, apparently different in kind from those previously characterized.

How might raising the level of cyclic AMP in a cell decrease potassium conductance? The action of cAMP itself is apparently straightforward. In all cases so far examined, cAMP activates a specific protein kinase (similar in all cells) which phosphorylates target proteins which are different in different target tissues (79). Castellucci et al. (28) confirmed some of these steps by injecting activated cAMP kinase into sensory neurons and producing a decrease in potassium conductance. How might this effect be brought about? Perhaps the kinase phosphorylates a protein constituent of a potassium channel, thereby blocking the channel. Support for such a mechanism comes from a remarkable series of experiments by Camardo, Siegelbaum, and Kandel (19, 108). With luck and skill one can press a smooth, fire-polished microelectrode against a cell's membrane, apply gentle suction, and retract the electrode, now with a patch of the cell membrane covering it like a drumhead, forming a high resistance (gigaohm) seal (57). Often the membrane patch is so small that it contains just a few channel proteins; in this case, if one maintains a constant voltage across the membrane patch, one can observe quantized increases and decreases in current corresponding to single channels opening and closing. Camardo et al. did all these things, and in good experiments they observed 3 to 7 potassium channels opening and closing in the membrane. If they then added cAMP-activated protein kinase to the solution outside the membrane patch (corresponding to the inside of the biological cell), the channels inactivated; i.e., the quantized conductances and the fluctuations corresponding to single channels opening and closing disappeared one by one. Afterwards, the channel conductances could be reactivated by alkaline phosphatase applied to the patch. The strong conclusion is that cAMP-dependent kinase, acting on proteins in the membrane patch, inactivates potassium conductance channels. This does not strictly imply that the channel itself is phosphorylated; it could be some nearby protein in the membrane which acts like a plug. Nevertheless, the experiments make the simplest picture very likely. They also document the causal connection between cAMP increase and physiological changes in the presynaptic neuron. These experiments, together with earlier voltage-clamp work, also indicate the nature of the potassium conductance channel, I_S, which is altered by phosphorylation (76, 108). It is different in character than three traditional potassium currents found in invertebrates (early, inactivating conductance I_A; the Hodgkin-Huxley current I_K; and the calcium-activated conductance $I_{K(Ca)}$) (5). It is relatively voltage-independent and is not inactivated by depolarization (108). In its properties and particularly in its ability to be inactivated by a second messenger system, it resembles vertebrate

I_M conductances (17) and also a novel potassium conductance found in vertebrate CNS which is also inactivated by cAMP (86).

The logical chain of this explanation has a missing link. Can a fractionally decreased potassium conductance, which under normal conditions results in a relatively slight increase in the duration of an action potential, produce the large increase in synaptic efficacy observed in sensitization? The answer is: probably. The change is in the right direction - decreasing potassium conduction depolarizes the cell for longer periods, increasing the open time of voltage-sensitive calcium channels in the synaptic terminals; more calcium entering the terminal after an action potential should enhance vesicle exocytosis and transmitter release and potentiate transmission. The change observed in the cell body looks too small to produce the effect. However, the relative densities of calcium and potassium channels might be very different in the synaptic terminal. In recent experiments, agents which prevent cyclase and kinase activation (29, 105) have been shown to block all the synaptic potentiation when injected into a sensory cell (29, 105). These experiments indicate that the physiological effects arise primarily or entirely in the biochemical pathway studied.

At this point, the causal chain in the model for sensitization is logically complete. Sensitizing stimuli coming from Aplysia's head or tail are transmitted via nerve fibers in the left pleuroabdominal connective to excite facilitatory interneurons. These cause the release of serotonin (or a serotonin-like agonist) onto sensory neurons, particularly onto their synaptic terminals on motor neurons. Released transmitter activates an adenylate cyclase in the cell membrane and increases cyclic AMP levels in the cytoplasm. Cyclic AMP activates a protein kinase there, which phosphorylates a membrane protein, inactivating one species of potassium channel present in the cell's membrane. Consequently, each action potential which enters the terminal will be broadened, resulting in more calcium entering into the terminals and more neurotransmitter release. Finally, the increase in transmitter release from the sensory neurons to the motor neurons will strengthen the reflex circuit and accentuate the behavior (Fig. 2).

If one can account for the acquisition of a simple behavioral change in biochemical terms, then one can hope similarly to understand the persistence of the change - the memory. Ideally the chemical change should match this learning in duration and kinetics of decay. Short-

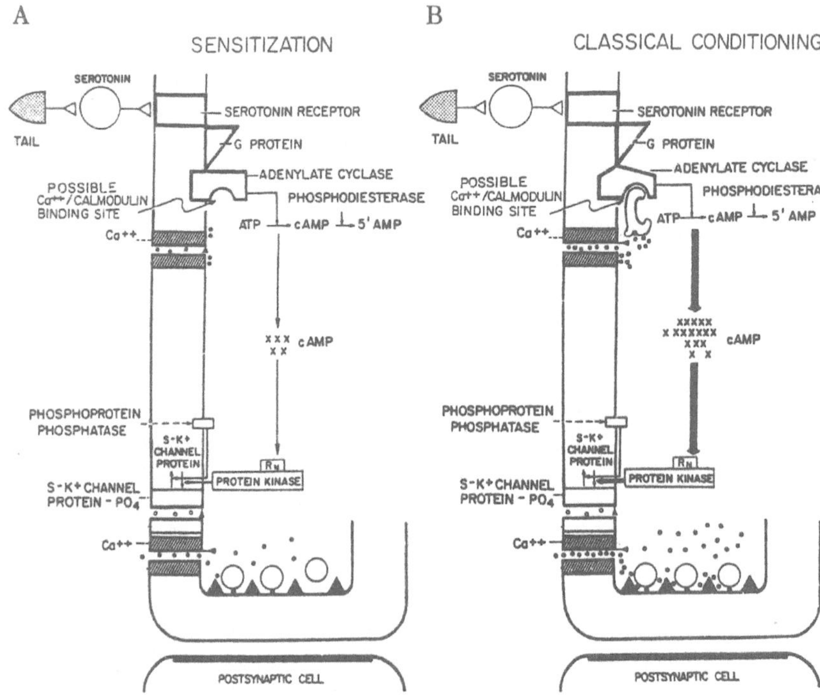

FIG. 2 - Diagrammatic (rectangular) sensory neuron in Aplysia showing schematic sequence of molecular events in facilitation and in classical conditioning. Both types of plasticity use the serotonin and cyclase signaling cascade. In classical conditioning, the cascade is amplified when Ca^{++} ions, entering the cells as a consequence of action potentials, bind to calmodulin which in turn interacts with adenylate cyclase to potentiate its serotonin-induced response (from Kandel et al. (69)).

term memory after sensitization lasts about 30 min, almost exactly as long as the rise, above baseline, of intracellular cAMP levels. Given this nice correspondence, one would next hope to understand the persistence in more detail, to find the most rate-limiting, "timekeeping" step in the metabolic step involved in sensitization. (The cascade can be sketched as: serotonin binding → regulatory (N_S) protein activation → cyclase stimulation → cAMP accumulation → kinase activation → protein phosphorylation. The relevant "timekeeping step" in the memory decay involves a back reaction of one of these activated states or of an accumulated messenger molecule. Two early experiments have shown where this "timekeeping step" is not found. The Walsh inhibitor is a protein which blocks cAMP-dependent kinase action. Injecting this inhibitor into sensory neurons blocks spike broadening and synaptic potentiation, as stated above (29). More to the point, injecting the inhibitor a few minutes after facilitation causes potentiation to disappear rapidly (∿ 5 min, versus ∿ 20 min in uninhibited control cells). This indicates that the change in protein phosphorylation is a relatively transitory one, otherwise the potassium channel activation and consequent synaptic facilitation would have persisted; the timekeeping step must be earlier.

Guanosine 5'-0-thiodiphosphate (GDPβS) inhibits another step in the cascade, the conversion of a regulatory protein called N_s (also G_s or G/F) to an activated form (conjugated with guanosine triphosphate), which in turn activates adenylate cyclase activity. The inhibitor GDPβS injected into sensory cells has physiological effects similar to the Walsh inhibitors - injected before a sensitizing response it prevents synaptic facilitation; injected shortly afterwards it causes rapid decay of the effect (105). The interpretation of this result is complicated. Nevertheless, it is clear that the persistence of the physiological change requires continued activation of some component or components of the receptor-cyclase complex; the most enduring chemical alteration has to be at or before the cyclase step.

All these changes address short-term retention of a sensitized response - memory which lasts about 20 min. Under different training conditions, the behavioral and neurophysiological change (enhanced gill-withdrawal responsiveness and enhanced synaptic transmission) can persist for days or weeks, long after all elements of the cAMP system have returned to baseline level (94). Anatomical concomitants to such long-term changes have been found in the sensory neuron terminals (12). However, there are no current biochemical results which relate directly to long-term memory. One guess about the mechanism is that increased cAMP levels in the sensory cells, in addition to causing immediate physiological change, may cause changes in the cell's nucleus and activate specific structural genes for synaptic components. Whether this idea is true or not, it has initiated an exciting venture in Aplysia gene cloning (see (104)). However, at present, long-term memory is as much a mystery as short-term memory was ten years ago.

Behavioral sensitization, as mentioned before, is nonassociative learning. It is similar to associative, classical conditioning, in that the nervous system uses one sensory input to alter its response to another. However, classical (Pavlovian) conditioning contains an additional requirement - the cue (CS) and the reinforcement (UCS) must occur close together in time, in a sequence compatible with a causal relationship, in order to elicit a response. The Aplysia gill-withdrawal reflex showed behavioral sensitization from the outset, but for years it seemed recalcitrant to associative conditioning. Somewhat serendipitously, Carew et al. (25) found that tail shock was a better sensitizing stimulus than head shock, and with this new reinforcement they found that they could classically condition this reflex, obtaining enhanced siphon withdrawal after training with paired stimuli.

The most recent variation of these behavioral experiments involves differential conditioning. The experimenters (22) gave a light tactile stimulus to two different sites on, say, the siphon, with one stimulus paired with tail shock, the other specifically unpaired. After training, the paired stimulus produced much greater siphon withdrawal than the unpaired stimulus. This procedure, which incorporates a dissociation control run on the same preparation, lends itself particularly well to neurophysiological analysis. Moreover, the fact that the reflex had been extensively analyzed in connection with sensitization meant that the experimenters knew immediately where in the nervous system to look for changes - in the synapses between sensory and motor neurons in the abdominal ganglion.

Hawkins et al. (59) moved from the whole animal to the isolated ganglion, substituted current injection into identified cells for the sensory stimuli used in training, and used excitatory synaptic input into a motor neuron as a measure of the response the animal would have made if it were intact. Instead of touching two sites on the siphon, they stimulated two sensory neurons which had receptive fields in the siphon and which made synaptic connections onto the motor neuron. Instead of shocking the tail, they stimulated the left pleuro-abdominal connective group. "Training" consisted of firing one sensory neuron temporally paired with connective stimulation; firing of the other sensory neuron was specifically unpaired. Afterwards, the synaptic connection from the paired sensory neuron was strongly enhanced both over pre-training levels and over connections from unpaired control cells. The molecular machinery which produced this alteration in training was apparently entirely presynaptic

- preventing the postsynaptic motor neuron from firing did not prevent the change, and firing the motor neuron in synchrony with synaptic stimulation did not accentuate it (20). This last result is congruent with the presynaptic site of the change in sensitization, but it is incompatible with Hebb's postulate for learning (64), an attractive, historically important model proposed thirty-five years before.

One final aspect of these physiological experiments led, almost by chance, to important mechanistic consequences. In these neurophysiological studies above, the sensory neurons to be differentially trained were arbitrarily selected in pairs from a cluster of about twenty small, similar-looking cells. Previous synaptic mechanisms for associative learning, proposed by these workers, involved relatively elaborate facilitatory circuitry. The fact that each of a large number of functionally indistinguishable cells could be individually altered by training (59) made the previous mechanisms unlikely. It was cumbersome to imagine that each cell came with its very own facilitatory circuit. Something simpler had to account for the paired enhancement; the simplest thing imaginable is activity (action potentials) in the neuron itself. Serotonergic input was known to activate adenylate cyclase in the sensory neurons. Perhaps this cyclase response in these cells was amplified by some concomitant of spiking activity. Three ionic consequences for the cell of firing action potentials are: increased extracellular potassium, increased intracellular sodium, and increased intracellular calcium. Of these, the last is the best candidate, because calcium, in combination with calmodulin, is a ubiquitous modulator of cell-biochemical processes and is specifically known to interact with cyclic AMP responses (97).

The fact that the gill-withdrawal reflex showed no enhanced response after backwards conditioning (after UCS-CS pairing) (60) suggested that the calcium (or whatever) interacts with the cAMP metabolic cascade at a step at or before the level of cyclase activation. (The best way to see this is to imagine an alternative - e.g., that calcium inhibits cAMP breakdown - and work out the behavioral consequences.) Moreover, in mammalian systems, there is a separate, brain-specific adenylate cyclase (16) which is specifically stabilized by interaction with calcium calmodulin. Recently, direct but preliminary evidence shows that the cyclic AMP response in the relevant Aplysia sensory neurons is enhanced by spiking activity, and that calcium is important for at least some component of this enhancement ((1, 69); also see Fig. 2B).

Another set of experiments involving differential enhancement of synaptic transmission in another Aplysia ganglion was carried out independently and simultaneously with the studies above by Walters and Byrne (117). These experiments, and the mechanistic conclusion drawn from them, are very similar to those of Hawkins et al. (59) above, except that the neurophysiology was done without the connection established by circuit tracing to a specific, learned behavior.

This is a good place to stop, catch our breath, look around. First of all, we are in the high mountains - the conclusions drawn from the Aplysia work, if true, are centrally important to neurobiology and psychology. Therefore, the stakes are high, and we need to examine the way we have come. In tracing events underlying sensitization along the pathway from behavior through physiology, biochemistry, biophysics, physiology, and back to behavior, we have followed a long sequence of postulated events, all supported by different kinds of experimental evidence. In general, the earliest work - the circuit tracing for the reflex and the localization of the critical change to the presynaptic endings of the motor neuron - is the easiest to question, especially if we ask for a quantitative accounting of the behavior and the changes.

The problem here is not a trivial one. Neural correlates of learning have been found in several vertebrates studies, but in each case the importance of the finding has been clouded by the question of whether observed alterations are primary or are a secondary consequence of other changes in other neurons, upstream, downstream, or in feedback loops in the behavioral circuit, where the important events have occurred (114). In Aplysia the problem is not so dire, but it is not absent. Interneurons (62) may influence behavior profoundly, and the possibility of other important interneurons among the 500 or so little neurons in the abdominal ganglion, or of reverberating loop circuits, has not been excluded beyond doubt.

This problem is particularly important because sensitization is measured neurophysiologically as the height (in mV) of transient postsynaptic potentials, but measured behaviorally as the duration (up to 40 sec) of withdrawal response. How a change in the one leads to a change in the other has not been fully established. It may not be straightforward. If one asks a Kandelian aplysiologist, "What fraction of the observed behavioral change comes from alterations of the synapses you study?", a typical answer is "30-50%," but this is only an educated guess at present.

The hope, which is a reasonable one, is that central synapses in all parts of the behavioral circuit are changed in the same way by the same molecular mechanism. The circuit tracing work is excellent, arguably the best of its kind, but at present it runs up against the complexity of even "simple" nervous systems. A new technique which may help with this issue is the "Miller killer" method (89), in which one ablates an individual cell in its ramified entirety by injecting Lucifer yellow and irradiating it with blue light. Removing a single neuron in this way, followed by behavioral testing, might allow one to assess the neuron's contribution to the animal's behavior. It would also help if one could study sensory neuron-motor neuron synapses in the absence of input from other cells. This also may soon be possible, because pairs of identified cells taken from leech or Aplysia ganglia can now be made to grow, extend processes, and form biologically appropriate synapses in tissue culture dishes (53, 103).

The strength of the Aplysia study is the recent biochemical work. The metabolic cascade from adenylate cyclase stimulation through kinase activation, protein phosphorylation to potassium S channel inactivation is spectacularly well documented. Several neural and biochemical events in the postulated causal pathway from behavioral sensitization - stimulating the left pleuro-abdominal connective (30) firing of cells in the L29 group (62), applying serotonin onto the sensory neurons (70), injecting cyclic AMP (70) or catalytic kinase (28) into these neurons - produce the logical physiological result: enhanced synaptic transmission onto the motor neurons. Intracellular cAMP in the neurons falls after sensitization with the same kinetics as short-term for sensitized response (29, 49). Pharmacological antagonists (GDPβS, Walsh inhibitor) both block the effect (29, 105) - and with a time course consistent with their effect on phosphoprotein levels. Finally, the gigaohm-seal membrane patchwork (19, 108) gives unequivocal evidence of a new variety of potassium channel, one whose properties were predicted by the cAMP model for sensitization (76). The consistency, detail, and compelling nature of the biochemical evidence would be a triumph in experimental neurobiology independent of its relevance to learning, and it does much to shore up earlier aspects of the model.

The recent result on classical conditioning of this reflex is important, and also lucky in its outcome. It makes most of the earlier, detailed work on sensitization directly relevant to higher learning. Apparently neurons involved in associative learning use all the old molecular machinery for sensitization, with a new wrinkle - activity-dependent amplification (59, 118). This is a much happier conclusion than finding all-new machinery, for then one would have had to start over from the beginning on associative learning.

By chance, the new work also strengthens conclusions drawn from the old. The fact that sensory cells arbitrarily selected in pairs can have their synaptic output differentially enhanced reinforces the notion that the circuitry producing the synaptic changes is as simple as the early, neuron-tracing studies indicated. The fact that this differential enhancement is independent of postsynaptic activity (20) reinforces the evidence that the changes really are in the presynaptic terminal. However, it has been less than a year since the neurophysiological studies on associative learning were started, and there are still loose ends, unresolved details, and important unanswered questions. For example, how does one construct an adenylate cyclase system which is susceptible for amplification by calcium only in the brief 1-2 sec interval prior to serotonin activation (the maximum effective CS-US interval in this case), yet which once activated stays activated for $\sim$30 min (the duration of short-term memory)? Fortunately, kinetic experiments of the type required to answer this go very quickly at Columbia, so we will not grow old in ignorance.

This is the state of the Aplysia learning model in September 1983. In the near future we can expect rapid advances, but also some revisions. A case in point is the transmitter which acts on the sensory neuron terminals to produce facilitation. For nearly a decade this was thought to be serotonin, but now the issue is in doubt. On the one hand, serotonin concentrations do rise in the abdominal ganglion after behavioral sensitization, and serotonin does activate adenylate cyclase in the relevant sensory neurons (31, 32, 70). Known pharmacological agonists and antagonists of serotonin produce their expected effects on the response, so serotonin was postulated to be the "sensitizing" transmitter (70). On the other hand, the sensitizing interneurons in the L29 group were recently examined and found to be devoid of serotonin immunoreactivity (75). The resolution of this disparity is not clear. The transmitter in the L29 cells may be a monoamine other than serotonin (an unusual one, since other major monoamines have been experimentally ruled out). The L29 transmitter might be a peptide (although the vesicles in the cell terminals "look" monoaminergic in ultrastructural studies (75)). Alternatively, the L29 neurons may connect the sensory neurons not monosynaptically but indirectly via serotonergic interneurons. At any rate, the Columbia model, like many from Detroit, occasionally has to be recalled for repairs.

Actually, this is a healthy development. To an outside observer, Kandel and his associates often appear to march in lockstep; there is always a Columbia model, and it is usually presented with a clarity which appears to admit no alternatives. Nevertheless, the model has grown continually more detailed, more plausible, and more thoroughly documented. Weak points have been shored up by experiment, and details have been altered. The work has rarely, for whatever reason, been directly tested by workers outside the Columbia group, but it appears to be continuously, honestly scrutinized by those within. The fact that the model is often administered as if it were ultimate truth does not seem to interfere with the search for ultimate truth. For the molecular basis of learning, it is the best approximation to the truth we have.

HERMISSENDA

Hermissenda crassicornis is another marine mollusc, small (7 cm), sylphlike, and tastefully colored, but with a somewhat less blatantly convenient nervous system than Aplysia. Relative aesthetic appeal aside, Hermissenda is emerging as a valuable alternative to Aplysia. Alkon, who christened this animal as a learning preparation (9), concentrated from the outset on associative learning, which was found early and documented convincingly. Workers in this system, particularly Alkon and his immediate associates, appear determined to keep their conceptual framework independent from the Columbia model. This is probably a good thing in the long run, because if new mechanistic principles emerge, this is a likely way to find them; and if mechanistic similarities to Aplysia emerge in the end, they will be constrained by the data. "In the end" may be some time off. Although early and current experimental results insure Hermissenda's high value as a learning preparation, the system is currently less developed than Aplysia, particularly the biochemistry and membrane biophysics. Ionic mechanisms, as formulated in recent models for Hermissenda learning, are unusually interrelated and complex. It may be several years before they are known in enough detail for an outsider to assess the work with genuine understanding.

Hermissenda's nervous system is concentrated in fused ganglia clustered over its esophagus. Situated near the ganglia are four conspicuous sense organs, two eyes and two statocysts, and it happens that these two sensory modalities, light and acceleration, interact to produce associative behavior.

Wild Hermissenda in the ocean crawl toward light because they find their food in brightly lit, shallow water. In captive Hermissenda this behavior is easily quantified by putting an animal in one end of a narrow glass tube filled with seawater and measuring the time it takes to crawl to the lighted other end (about 60 cm). Crow and Alkon (36) trained Hermissenda not to phototax by spinning them on a turntable while simultaneously exposing them to light; afterwards their phototactic response was slower. Training procedures and testing measurements are entirely automated, and the behavioral change is a robust one - at least a twofold increase in response time by trained animals. The controls for associativity in this study were unusually complete - light and rotation must be temporally paired for the behavioral change to occur - and more recent experiments demonstrate other, more fine-grained properties of associative learning: extinguishability and depression of the learned response by noncontingent exposures to light during training (49). Acquisition of the response is slow (50 trials) but once acquired, it persists for several days (36).

The ease of the learning assay and the scrupulousness of the measurement procedure are one of the real strengths of Hermissenda as a learning system. Another strength is the size, simplicity, and accessibility of the relevant sensory receptors. One can record for hours intracellularly from the caudal hair cells of the statocysts, and from the five photoreceptors (two type A cells and three type B cells). Such electrophysiological experiments, augmented by anatomical studies, indicated the major mutual influences between photoreceptor cells and hair cells (6, 9, 50, 56). Among these, the most important influences used in explaining the inter-sensory interactions are the following: a) Type A photoreceptors seem most directly connected to motor neurons involved in orientation toward light and are the best candidates for primary sensory influences on phototaxis. b) Type B photoreceptors are directly inhibitory to type A photoreceptors; stimulating B photoreceptors decreases the phototactic response. c) Stimulating the eyes with light tends to excite the statocyst hair cells, probably by second-order connections through optic ganglion cells. d) Stimulating ipsilateral hair cells tends to immediately hyperpolarize type B photoreceptors. However, inhibition is followed by a later excitatory phase when the rotation stops.

Behavioral training induces changes in all the sensory cells above; the statocyst hair cells become less excitable; type A photoreceptors become less excitable, and type B photoreceptors become generally more excitable, particularly in the 60 sec period after a light pulse. Among all these changes, the changes in type B photoreceptors looked most striking, lasted longest, and, on the basis of electrophysiological interconnections, were the most likely candidates among sensory cells to be the primary source of the changes observed.

Do the type B photoreceptors really change during training? Other, unanalyzed cells in the nervous system probably also change, and at this point in the analysis, analogizing from vertebrate studies, one would guess that the changes in the sensory cells result indirectly from such changes elsewhere. This possibility was excluded in an elegant series of experiments by Crow and Alkon (37). They trained Hermissenda as before, then impaled B cell photoreceptors in trained and unpaired control animals. They observed more positive resting potentials and higher input resistances in the "trained" photoreceptors and found that synaptic input (measured by standard neurophysiological techniques) could not account for the changes. Finally and more decisively, they cut the photoreceptor's axon near the cell body, eliminating detectable synaptic activity to the B cell; after such axotomy, membrane voltage and input resistance of the B cells remained higher in the trained animals than in controls. This is strong evidence that associative pairing of two stimuli (light and synaptic input from the vestibular system) can produce relatively long lasting membrane changes in a single cell. The experiment does not indicate that the network encoding the changes in the B photoreceptors is necessarily as simple as one would like to draw it, or that the B cells are the only cells changed by associative training, or that the B cells give rise to the associatively learned behavioral change.

This last question is an important one - do the B cells have anything to do with behavioral learning? A priori one would guess not, first because only a small fraction of the cells in the nervous system were examined for such changes, and second because photoreceptors are for seeing, not learning. In an attempt to answer this question, Farley et al. (51) have taken naive Hermissenda, surgically exposed the nervous system, impaled type B photoreceptors with intracellular electrodes, and "trained" B cells (one per animal) with paired light and intracellular current injection. They reported that cells receiving paired current injection and light were afterwards more excitable than control cells in other animals, which received the current and light unpaired. Moreover, a day later, when the animals had recovered, individuals that contained an associatively "trained" cell appeared relatively slower to phototax than those which contained an "unpaired control" B cell. The experimental approach here is an incisive one. However, examination of the data show severe mismatches between paired and unpaired control cells ((57), Table 1) in pre-training input resistance, and also apparent mismatches in general (nonphototactic) motility between trained and control whole animals ((57), notes 13, 19). To be conclusive, or even persuasive, the experiments will have to be done over.

In any case, the type B photoreceptors do have permanent changes encoded in their membranes by associative training, and these changes probably induce or help induce learned behavioral changes. What is the ionic nature of the membrane changes in the B cells, and how are they encoded? First of all, both the cells' increased excitability (predisposition to fire action potentials) and their increased input resistance (R_m) could logically result from decreased potassium currents. In fact, voltage-clamp studies (7, 8, 10) indicate that potassium conductance is decreased after training. So far so good, and similar to the model from Aplysia: associative training shuts off a potassium channel. From here on, however, the models diverge. Neuron membranes in invertebrates can have at least four different kinds of potassium currents: I_A, I_K, $I_{K(Ca)}$, and I_S (5, 108). Remember that it is I_S which is inactivated in Aplysia sensory neurons. Voltage-clamp studies done in Hermissenda divide potassium currents into two voltage-dependent types: a) I_A (35), which is rapidly activated by depolarization, rapidly inactivated, and can be blocked by the drug 4-aminopyridine; and b) I_B, which has slower onset, is not inactivated, and can be largely blocked by the drug tetraethylammonium (7, 10). The salient differences measured in "trained" type B cells are in I_A conductance channels, although there is also some difference in I_B (9, 10). So apparently the potassium channel altered by training is a different one from Aplysia. The I_A current in Hermissenda can be partially inactivated by a protein kinase, but by a calcium-calmodulin dependent kinase (one of four now reported) (100) rather than the cyclic AMP-dependent kinase implicated in Aplysia learning (9).

The ionic and synaptic mechanism, as currently modelled from experiments with Hermissenda, is also completely different from Aplysia. Enhanced responsiveness to paired stimuli comes not from converging biochemical pathways in a cell but from the nonlinear interactions of membrane conductances in the B cells and from the network properties of an ensemble of cells. What is found after behavioral training, and after "B cell training" with light and depolarization, is increased calcium accumulation in the type B photoreceptors (34). The problem is how to account for this. Two relatively independent, mutually reinforcing mechanisms are envisioned. During light and rotation, the B photoreceptor is depolarized by direct response to illumination, with an increase in inward calcium currents. However, the direct input of statocyst hair cells on B cells is inhibitory (hyperpolarizing) (6). As it stands, this looks like the opposite of what we want, since it is paired light and depolarization which mimics the effects of behavioral training. However, at the termination of rotation, there is an excitatory (depolarizing) rebound in the synaptic input to the B cells (110), and events after stimulus offset may be important. As it happens, there is a long "tail" of depolarization in the photoreceptors which persists for about 60 sec after the termination of a light pulse. This tail current results, in part, from voltage-dependent inactivation of both potassium I_A current and potassium $I_{K(Ca)}$ current. So during this period the photoreceptor's membrane is depolarized, and its membrane resistance is high because of potassium current inactivation. This increased resistance makes it more responsive to synaptic input, in particular to the excitatory input from statocyst hair cells after the termination of rotation. Therefore, the "tail" of depolarization after the paired light and rotation is increased and prolonged over the tail after light alone.

Another mechanism that accentuates the tail current involves the properties of a neuronal network, particularly of an interneuron in the sensory pathway. An identified cell in the optic ganglion, called "E," gives excitatory input to type B photoreceptors. Both B cells and statocysts' cells give inhibitory input to the E cell, and during paired light and rotation, the inhibition is more than doubly strong. At the offset of paired stimulation, there is excitatory rebound from inhibition in the E cell and consequent excitation of the type B photoreceptors (9, 110). Note that the synaptic excitation also occurs during the tail after the light stimulus, so it should reinforce the rebound excitation from direct vestibular input, and it should be accentuated by the inactivation of potassium channels in the B cell membrane.

This bipartite explanation for the effects of paired stimuli during training has conceptual advantages and disadvantages. Since it relies on the time constants of cells in microcircuits, the time constraints for pairing could be carefully tailored by natural selection to fit the exigencies of the task to be learned. This is convenient, because pairing specificity apparently varies for different learning tasks even in the same organism, a phenomenon most dramatically demonstrated by the studies of Garcia and others on toxophobic conditioning in rats (54). The apparent disadvantages of the system as modelled is that it relies on the near simultaneity of the offset rather than the onset of the stimuli to be associated in the CS and the UCS. This predicted dependence on offset has not yet been directly looked for in behavioral conditioning experiments in Hermissenda. If found, it would make learning in this organism quite unlike that in vertebrates, where the rules for stimulus association have been carefully examined and found to depend on onset.

The mechanism for encoding associative change in Hermissenda may or may not be exactly correct as currently modelled. In any case, it is a fact that behavioral training or paired light and B photoreceptor depolarization both increase the photoreceptor's internal calcium concentration, as measured with indicator dye arsenazo III (34). How does increased intracellular calcium account for long-lasting changes? At least part of the answer comes from a positive feedback loop involving calcium and potassium conductances. Remember that increased intracellular calcium can activate a calcium-calmodulin dependent kinase which partially blocks early I_A (9). The resulting decrease in the outward potassium current should depolarize the cell slightly, producing increased inward voltage-dependent calcium conductance, further depolarizing the cell, increasing intracellular calcium, and initiating another round of the feedback cycle. Given this feedback mechanism, an elevation in intracellular calcium might persist for minutes to hours, depending on the time constants of the feedback loop and the kinase system. Longer-term memory, persisting over hours or days, is currently unexplained in this system, as in all others.

This is a sketch of the processes currently believed to underlie acquisition and retention in Hermissenda. The bipartite mechanism proposed to explain acquisition is very ingenious, but also very complicated, depending on the subtleties of ionic conductances and synaptic couplings between neurons in a network. Whether the model as it stands can account quantitatively for the effects of paired stimuli in the calcium accumulation observed is uncertain without sophisticated modelling and may always be difficult to convey to outsiders in intuitable form. However, long-lasting calcium changes are found in some vertebrate neural assemblies implicated in learning. Moreover, the calcium current-potassium current feedback cycle - enhancement by higher voltage-dependent calcium channels (Fig. 3) (9, 10) - is not only clever but has a simplicity and elegance that suggest that it is biologically true. Whether the cycle operates exactly as currently formulated for Hermissenda seems debatable and will also require quantitative modelling to demonstrate that it works.

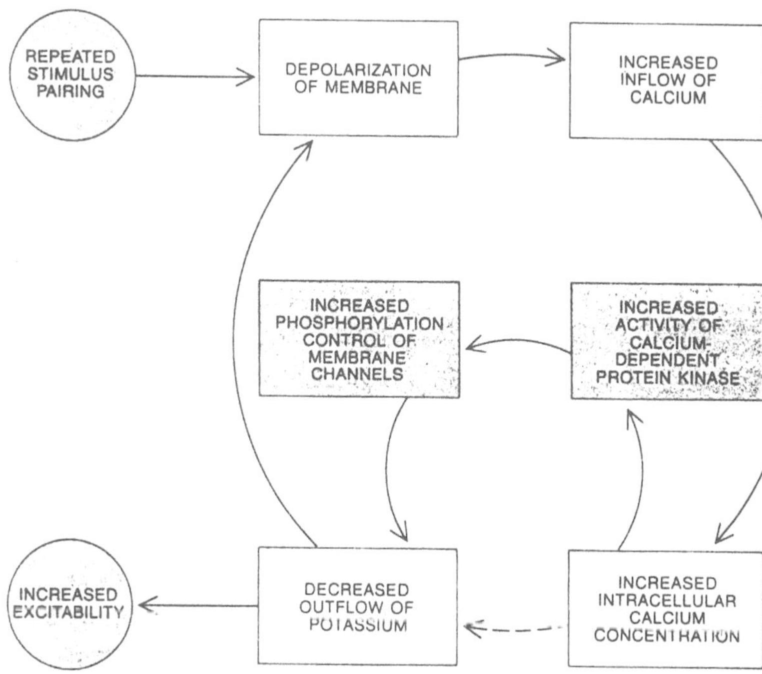

FIG. 3 - The calcium-potassium positive feedback loop, as modelled from Hermissenda, to account for long-term potentiation (stably increased membrane excitability). Strong or long membrane depolarization produces elevated intracellular calcium concentration. This activates a calcium-calmodulin dependent kinase, which shuts down some potassium channels, including I_A channels. Decreased potassium conductance results in membrane depolarization and increased excitability, ensuring that more calcium will enter and completing the cycle. Note that this is an inhibition-of-inhibition, positive feedback cycle, of a type especially predisposed to two-state ("flip-flop") behavior. Note also that this is only part of the picture in real neurons. Other potassium channels under other conditions are <u>opened</u> by intracellular calcium (from Alkon (9)).

For an outsider, assessing the plausibility of individual details is very difficult because of the numerous complications in the Hermissenda system. The late potassium current, called I_B in Hermissenda work, is probably subdivided, as in other invertebrates into I_K, $I_{K(Ca)}$, and perhaps I_S; apparently training has effects on late potassium currents in the type photoreceptor as well as on I_A, and cyclic AMP-dependent kinase also affects I_B currents (10). The neurophysiological effects of training, already unwieldly to explain, are liable to grow more complicated with further work.

Some definitive experiments cry out to be done on this system: a) behavioral experiments, testing for pairing specificity for stimulus offset; b) training experiments (on animals in which type B photoreceptors or optic ganglion E cells have been killed (89)), to assess the relative contributions of these cells to the behavioral effects observed; c) single-channel recordings from type B membrane photoreceptor patches to resolve all the types of calcium and potassium channels present and to see how they change during training. Single-channel recordings, in particular, would give solidity to those aspects of the model which are most ephemeral in the mind. The specific second messenger system and the specific ionic conductances implicated in Hermissenda learning have both been changed in the past, and they could easily change again. This said, we are left with several excellent conclusions from the work: a) Hermissenda show very strong, well characterized associative learning. b) Training induces changes in type B photoreceptors which are intrinsic; they persist in the absence of outside synaptic input. c) In the type B photoreceptor, two consequences of associative training are increased intracellular free calcium and a decrease in potassium conductance. Given this experimental foundation, the Hermissenda work is bound to grow in value as it continues.

DROSOPHILA

Drosophila melanogaster is tiny, but it is not a simple system in the conventional sense. A fruit fly has about 100,000 very small neurons in its brain and thoracic ganglia; some peripheral neuromuscular junctions are very accessible, but detailed CNS physiology looks fairly hopeless. If Aplysia is like an old Philco standby, then Drosophila is like some miniaturized wonder from Panasonic, packed with inscrutable silicon and germanium. Nevertheless, Drosophila has a hidden simplicity. It comes with a relatively small genome (five E. coli equivalents of structural genes) and with well developed genetic methodology, both classical and molecular. The fly's domesticated genetics allow an alternative approach to learning mechanisms, which uses different methods and often provides complementary answers to neurophysiological work - studying the learning machinery by deranging its components one by one with single gene lesions.

Normal flies can be trained in an associative, olfactory discrimination task using electric shock reinforcement. Chemically induced mutants, altered in one gene, which fail to learn in this situation can be selected, maintained as true-breeding populations, and characterized with whatever techniques come to hand. All this has been done, and the most productive technique has been old-fashioned biochemistry. Comparisons with other systems will be easiest if we take an antichronological approach, discussing biochemical abnormalities in the mutants first, then describing how they change behavior.

Four mutations - dunce (44), turnip, rutabaga, and cabbage (4) - were isolated on the basis of their deficient learning in the original olfactory test. A fifth mutant, amnesiac (40), learned normally but forgot rapidly after training - within one hour instead of the normal 4-6 h. Dopa decarboxylase, a mutation previously isolated in another lab (121), turned out to be deficient in neurotransmitter synthesis (39, 85) and also deficient in learning (112, 113).

The metabolic abnormalities in four of the learning mutations are known or partly known. These fall into a striking pattern, affecting sequential steps in one-cell signalling pathway. The dopa decarboxylase mutation blocks synthesis of the monoamine neurotransmitters dopamine and serotonin in the brain (39, 86). The turnip mutation, directly or indirectly, affects a serotonin receptor in neuron membranes, altering its binding affinity for this neurotransmitter ((3) and Smith, unpublished). Rutabaga reduces adenylate cyclase activity (3, 83), and dunce alters or eliminates a cyclic AMP phosphodiesterase isozyme (18, 72, 106). These alterations (all of which affect the monoamine-induced cAMP response of cells) are so self-consistent, and so consistent with other work, that it is worth examining the evidence that the metabolic lesions are specific. We will start with the best understood lesions.

Dunce appears to be the structural gene for cAMP phosphodiesterase II (PdE II), one of two enzymes in Drosophila which hydrolyze cyclic AMP (18, 73, 74). Six separate mutations in this gene have been isolated. All reduce or abolish activity of PdE II. Moreover, two of the mutations alter the enzyme activity in ways which indicate a structural alteration in the enzyme protein: dunce^1 increases its temperature sensitivity measured in vitro, and dunce$_2$ increases its K_m (reduces its apparent affinity for its substrate (72)). Still more to the point, in extracts from dunce/+ heterozygotes (flies with one normal and one mutant gene), half the PdE II activity is perfectly normal, and half has clearly mutant properties (72). Another, independent line of evidence indicates coding of PdE II by dunce. In Drosophila, using mild genetic trickery, one can vary the number of copies of a given gene from one to three per cell. This was done with dunce, and in all cases the total amount of PdE II (V_{max}) is proportional to the number of dunce gene copies present (74, 106). By molecular genetic standards, the two arguments above, taken together, provide good evidence that the dunce gene codes for part or all of the PdE II protein. The dunce gene has recently been cloned (Davis and Davidson, unpublished), raising the hope that coding of the enzyme can be directly demonstrated with expression vectors, and that regions of active dunce gene expression can be localized within the fly brain.

The enzyme dopa decarboxylase is present in many fly tissues and used for several purposes (121, 122). In the brain it is necessary for the synthesis of dopamine and serotonin (39, 85) (but not for octopamine, the other principal monoamine present in invertebrates) (85). Wright and his colleagues have isolated temperature-sensitive dopa decarboxylase mutant alleles, as well as chromosomal deficiencies that remove the gene altogether (122), and used them to provide evidence similar to that for dunce above, that the mutants are in the structural gene for the dopa decarboxylase enzyme. The dopa decarboxylase gene, like dunce, has been cloned (65), raising similar hopes.

The rutabaga mutation decreases cyclic AMP levels in the flies and decreases adenylate cyclase levels in vitro (3,75). Adenylate cyclase, with its attendant modulatory proteins, is a complex system; the biochemical work (done by Livingstone et al. (83, 84), and outlined in (3)) to determine which subunit is affected by the mutation has been difficult and still contains minor inconsistencies. The major conclusions are a) rutabaga cyclase activity is <u>not</u> altered in either of the known cyclase regulatory proteins, N_s (also called G_s) or N_i. b) The rutabaga mutation appears to affect adenylate cyclase catalytic activity but not all activity. For example, in Drosophila, only 20% of brain cyclase is affected. Apparently there are two or more cyclase isozymes, only one of which is affected by the mutation. c) The particular cyclase enzymatic activity affected by the mutation most resembles cyclase activity newly found in vertebrates, which is brain-specific and enhanced by calcium-calmodulin (16, 102). Livingstone has recently found that cyclase catalytic activity from rutabaga membrane extracts is specifically deficient in its ability to be activated by exogenous calmodulin added to the extract.

Mutant turnip flies have normal cyclase activity measured in vitro. However, along with its behavioral defects, the mutant showed a striking physiological abnormality. Superfused serotonin increased the normal heartbeat rate of normal larval hearts but not of turnip hearts (2). This abnormality has a molecular correlate. Membrane extracts from normal fly heads show high-affinity serotonin binding, with at least two dissociation constants ((3, 45) and Smith, unpublished). Smith has found that mutant turnip extracts specifically lack the highest-affinity serotonin binding component ((3) and Smith, unpublished). The simplest explanation (93) is that the turnip mutation alters the synaptic receptor for serotonin, decreasing its affinity for the ligand.

We do not really know whether the defect is in the receptor molecule itself. Typically, in cyclase-coupled receptor systems the highest-affinity binding comes not from the receptor alone, but from the receptor coupled to the GTP-binding regulatory protein N_S (99). So the mutant's observed defect in high-affinity binding could result from abnormal receptor or from abnormal N_S protein. In fact, Choi in our laboratory has observed abnormalities in membrane GTP-binding proteins and in GTP hydrolysis, in turnip mutants. Therefore, at present, our best bet is that the primary lesion caused by turnip is in the N_S protein. Nevertheless, at present this is just a bet; we need to do more work on this question.

Now it is time to see how biochemistry affects behavior. It is worth worrying a little at the outset about the original olfactory learning test (95), because it was used to isolate the mutants and also because more recent tests tend to be variations on this theme. Typically, a group of about forty flies is placed in an apparatus with tubes and trained by being exposed alternately to two chemical odorants (denoted "A" and "B"), one of which is coupled to 90 V electric shock. They are then tested most simply (44) by being transported to a choice point between tubes with odors A and B and seeing which way they run. Odor concentrations are arranged so that naive flies distribute themselves 50-50 between the two odorants. If the flies were shocked during training in the presence of odor A, they now tend to run toward odor B (35%-65%). On the other hand, if they were shocked on odor B, they now run towards A (65%-35%). The relevant results of this type of experiment can be expressed as a simpler number, i.e., the fraction of the population avoiding the shock-associated odor minus the fraction avoiding the control odor (averaged for groups of flies trained in opposite directions). This numerical training index "Λ," normally can vary between 1 (perfect learning) and 0 (no learning). In the fly learning experiment outlined above, Λ would be 0.30. A typical set of ten real experiments gives $\Lambda = 0.34 \pm 0.03$ (95).

This type of test has advantages and disadvantages. The learning effect is easily demonstrated and easily quantified, because working with populations gives instant statistics. A disadvantage of the test in its original form is that the learning effect was relatively weak. A Λ value of 0.34 means that only a third of the flies express learning in this test. This made isolating mutants laborious, since one had to breed up groups of identical, mutagenized flies, then measure their average score. Nevertheless, the prize seemed worth the game. The first four mutants selected in this way were severely but not completely deficient in learning: dunce, $\Lambda = 0.04 \pm 0.02$ (40); rutabaga, $\Lambda = 0.02 \pm 0.03$; cabbage, $\Lambda = 0.02 \pm 0.01$; and turnip, $\Lambda = 0.06 \pm 0.03$ (4). A fifth mutation, amnesiac, was separately selected in a deliberate screen from memory mutants. It learns normally but subsequently forgets in about an hour vs. 4-6 h for normal flies (96).

Recently, the training procedure has been improved to give much stronger learning effects. Jellies at Illinois State (67) and later Tully at Princeton (3) have made several modifications of the original shock avoidance olfactory learning paradigm. The most important of these are: a) eliminating disturbances to the flies, such as mechanical shaking, b) presenting odor stimuli at carefully controlled concentrations in laminar air currents, and c) sequestering flies in a chamber and conditioning them classically so that they are made to encounter odorants and shock inevitably for several training cycles. When Tully tests his flies in an odor T maze now, he obtains Λ values of 0.90 (corresponding to 95% correct choice by the trained flies). So almost all wild-type flies can learn, which removes some of the labor and many of the uncertainties in using a "group" learning test. Memory in wild-type flies also lasts longer after this type of training. Tully has tested the mutants dunce, rutabaga, and amnesiac in this test (unpublished experiments). These mutants all show appreciable initial learning, but all have unusually rapid memory decay for the first hour afterwards. The simplest explanation for these findings is that dunce and rutabaga are actually memory mutants - i.e., that they can learn even in shock-avoidance training, but that their memory span under those conditions is so short and labile that it is virtually undetectable. In fact, experiments with positively reinforced training by Tempel et al. (111) had already given this conclusion, as had suggestive experiments by Dudai (41, 42).

Interestingly, in Tully's test, memory decay for all three mutants slows at about one hour, and the remaining memory persists for at least six times longer. These mutants appear to be affected in only short-term retention. Memory which happens to survive the initial period is stored normally as long-term memory. Savings-type experiments on the mutants suggest a similar conclusion (42, 96).

One would like similarly to study memory in dopa decarboxylase mutants, which have altered neurotransmitter levels. Flies with severe lesion in gene never learn, so studying their retention is impossible. However, one can breed flies with moderate lesions in this gene; these have reduced (but not absent) dopa decarboxylase enzyme levels in their brain and reduced (but not absent) initial learning levels (112). Memory retention for these "leaky" mutants is similar to wild-type flies, after both positively and negatively reinforced training (113). If dunce, rutabaga, and amnesiac are primarily affected in short-term memory, then dopa decarboxylase is primarily affected in learning acquisition. Naively extrapolating from the Aplysia model, this all makes sense. Remember that in Aplysia, the kinetics of short-term memory decay corresponded to the decay kinetics of intracellular cyclic AMP concentrations in the sensory neurons (105). Crudely put, short-term memory may be elevated cyclic AMP in the relevant cells. If so, mutations like dunce and rutabaga, which alter cyclic AMP metabolism, should also affect short-term memory. In contrast, a mutation like dopa decarboxylase, which causes a block well ahead of cyclase activation in the metabolic cascade, should logically interfere with the acquisition of a behavioral change without altering its subsequent retention. This correlation is nice, and also informative, because what is measured in Aplysia is memory after sensitization; what is measured in Drosophila is memory after associative learning.

Wild-type and mutant Drosophila have been examined in a number of other learning situations. These are useful as a sort of obstacle course, broadly delimiting the range and severity of the mutants' deficits and, in one case, suggesting how the ability to learn might be useful to flies in the wild.

Tempel et al. (111) trained flies to discriminate between odors as above but substituted reward for punishment. They found that hungry flies would specifically migrate towards odorants previously associated with the opportunity to feed on sucrose. These studies indicated, among other things, that the mutations affected positively reinforced learning. The mutations also affect learning throughout development. Normal Drosophila larvae, even a day after hatching, can learn an olfactory discrimination task similar to that above for adults, and dunce, rutabaga, cabbage, and turnip larvae showed learning deficiencies similar to their parents (4).

Individual Drosophila can be conditioned in the leg flexion task developed by Horridge for cockroaches (66). If a tethered animal is shocked whenever it extends one of its legs, it rapidly comes to maintain that leg in a flexed position. Wild-type flies, tested in this way, learned about as well as cockroaches. The learning mutants did worse, although some individuals of all genotypes did learn (13, 14). This test is useful because the behavior shown by the flies is unambiguously operant learning; it is nice to know that the mutations affect this as well as classical conditioning.

Drosophila can learn to visual cues. Menne and Spatz at Freiburg trained flies to discriminate between different colored lights, avoiding a color that was associated with severe mechanical shaking (87). Dudai and Bicker (43) and also Folkers (52) tested dunce, rutabaga, turnip, and amnesiac in this visual test. They found that the mutants could eventually learn this task, although it took them many more trials to do so. However, of all the tests developed so far, visual learning is the test on which the mutants' performance is the least altered.

Drosophila courtship and mating behavior was long considered an unalterable sequence of programmed fixed-action patterns. Recently, several components of this pas de deux have been shown to be influenced by experiences in both male and female flies. The best-characterized example of such modifiable behavior is the courtship-depression effect of Siegel and Hall (107). Male flies placed with sexually unreceptive females become, as it were, discouraged and show markedly less ardor in courtship for about three hours afterwards. Some recent behavioral experiments (115) suggest that this behavior represents associative conditioning. Nevertheless, the first evidence for this was the fact that several of the learning mutants showed less of it (107). In the past, behavior has been used to test the mutants. This use of the mutants to test behavior (for modifiability) is novel and may be generally useful in discovering the importance of learning for natural flies.

In all the fly-learning tests described above, the experimenters have gone to some pains to ensure that the behavioral changes were associative. Nevertheless, Drosophila also show nonassociative forms of learning - habituation and sensitization - just like Aplysia. These are most easily measured as modulations of the Drosophila's proboscis extension response to sugar, a reflex behavior studied extensively in larger flies (38). Duerr (46) examined habituation and sensitization of this response in Drosophila and found that all the associative learning mutants tested showed abnormalities: dunce and turnip had low habituation; dunce, rutabaga, and amnesiac showed unusually brief sensitization. Her results indicated that sensitization and associative learning, for example, require some of the same gene products and so suggested that they share common mechanistic features. This conclusion is deeply unsurprising, given the recent results on classical conditioning in Aplysia. However, Duerr's work with Drosophila was actually the first experiment suggesting such mechanistic similarities and was one of the principal inducements for us to examine our mutants for Kandelian abnormalities. So, antichronologically, the behavior towards the end of this discussion of flies leads us to the biochemistry at the beginning.

Flies have more to contribute than convergent evidence. The mutants were isolated for deficiencies in a different form of learning, using a different sensory modality, in a very different organism from Aplysia. This suggests that the monoamine-cyclic AMP signalling system is centrally involved in diverse learning behaviors ranging far beyond one reflex in one particular organism. Drawing big conclusions from little flies makes us pleased with ourselves. However, Drosophila, like other invertebrate learning systems, has exoskeletons in its closet. The genetic approach has intrinsic limitations. We can get well-defined pieces of information about the biochemical machinery underlying learning behavior, but we depend on others for the intervening physiological processes. More specifically, one can raise legitimate questions about the nature of the mutants. I will try to deal with four that have bothered me.

1) Why are the mutants not dead? Four of the learning mutations affect important transmitters or metabolites. How do they manage to stay alive, let alone perform reasonably in non-learning behaviors? Severe dopa decarboxylase mutations are in fact lethal (121, 122) if expressed early in development. Tempel and Livingstone (85, 112) produced normal flies by using temperature-sensitive mutants, raising them at permissive temperature until adulthood, then shifting them to restrictive temperature

for a few days, shutting off the enzyme. Even so, one wonders how the flies could function without dopamine or serotonin. We can only surmise that in Drosophila the monoamine systems are purely modulatory and do not disrupt vital circuits, an idea consistent with the small amounts of the relevant enzymes present in the fly's brain (39).

Dunce, rutabaga, and turnip may be alive because the metabolic lesions are partial, not total. The dunce mutation removes only one of two cAMP hydrolyzing enzymes in Drosophila. Turnip and rutabaga may be "leaky" mutations, altering a protein's function without destroying it entirely.

2) Are the metabolic lesions really what we say they are? The molecular alterations observed in dopa decarboxylase, turnip, rutabaga, and dunce are so consistent with each other and fit so well with other work that one naturally wonders whether they are primary effects of the mutations or incidental changes wishfully dredged from a welter of secondary metabolic alterations in sick mutants and seized upon because they happened to make sense. This seems not to be the case. In two instances, for dunce and dopa decarboxylase, there is excellent evidence that the mutated genes directly code for the enzymes they are supposed to affect. For the other two mutations, there is no such direct evidence. However, the biochemical lesions in rutabaga and turnip are too well-defined to result from general metabolic malaise. With these, we could be wrong about exactly what they are, but we are on the right track.

3) How general is the block in learning? Early on, the question arose as to whether these were "shallow" mutations, affecting learning only in the avoidance task and used to isolate them, or "deep" ones, disrupting learning in a variety of situations. The answer seems to be that the mutations are "deep," but not perfectly so. The flies have been tested in a number of labs, in associative and nonassociative tests using different cues (odors, colors, presence of other flies, limb position) and different reinforcement (electric shock, mechanical shaking, bitter quinine, tasty sucrose). In all cases, the mutants' performance was deficient. However, in no case was the block in learning complete. Even in the test used to isolate them, the mutants showed a little learning (.02 vs. 0.34). In other cases, as the tasks were made easier for wild-type flies, the mutants tended to show appreciable learning. The genetically induced learning deficiencies, though severe, are relative, not absolute.

4) Are the mutants specifically affected in learning, or are all behaviors deranged? Again, the answer to this question is reassuring, but not

perfectly so. Dunce, rutabaga, cabbage, and amnesiac are all normal in development, external morphology, and neuroanatomy (2, 44). (Severe dunce alleles are almost completely female sterile (7), an interesting and unexplained result.) The mutants were apparently normal in non-learning behaviors: locomotion, phototaxis, negative geotaxis, and flying. In what is probably the most stringent test of fly normalcy, the mutants, male and female both, perform all the fixed-action steps of courtship and mating properly, except for those distinct components explicitly shown to be experience-dependent (107). The mutants can all respond to the sensory cues and reinforcements used to train them. They detect the odors and electric shocks with normal or near-normal thresholds. (44, 111, 112).

Turnip is different. Turnip is a somewhat sick fly, small, slow to phototax, 30% slower to develop, and with reduced brain size (2). Nevertheless, we think that the mutation is relevant to learning, because turnip/+ heterozygotes are indistinguishable from normal flies in all our tests but forget very rapidly after training (30 min vs. 4-6 h) (96).

In summary, the learning mutants seem to be widely affected in learning behaviors and relatively normal in non-learning ones. They appear to represent a "special" subset of possible Drosophila mutants. Dudai (43) tested a number of standard Drosophila morphological and behavioral mutants to see if they had learning deficiencies. He found in general that the "sicker" a mutant appeared in overt behavior or phototaxis, the worse it learned. However, none of these "off-the-shelf" mutants was as severely learning-deficient as those discussed above, whereas most looked and behaved sicker. Still, perhaps best evidence for the specialness of the learning mutants is that they were isolated "blind" in a behavioral screen, before there were any specific models for learning and before we really knew what we were doing. The pattern of metabolic lesions in monoamines and cyclic AMP metabolism (Fig. 4) emerged several years later. We would have been silly to hope for such luck.

OTHER SYSTEMS

Higher invertebrates - opisthobranch molluscs, cephalopod molluscs, and arthropods - may all be able to learn, provided one knows what to teach them. Pleurobranchia can be classically conditioned to suppress its feeding response (90), and Limax maximus, a garden variety land slug, can be taught food aversion tasks (33, 55, 101) of the type pioneered in rats by Garcia. These molluscs, like Aplysia and Hermissenda, have ganglia full of large, accessible neurons, and neurophysiological correlates

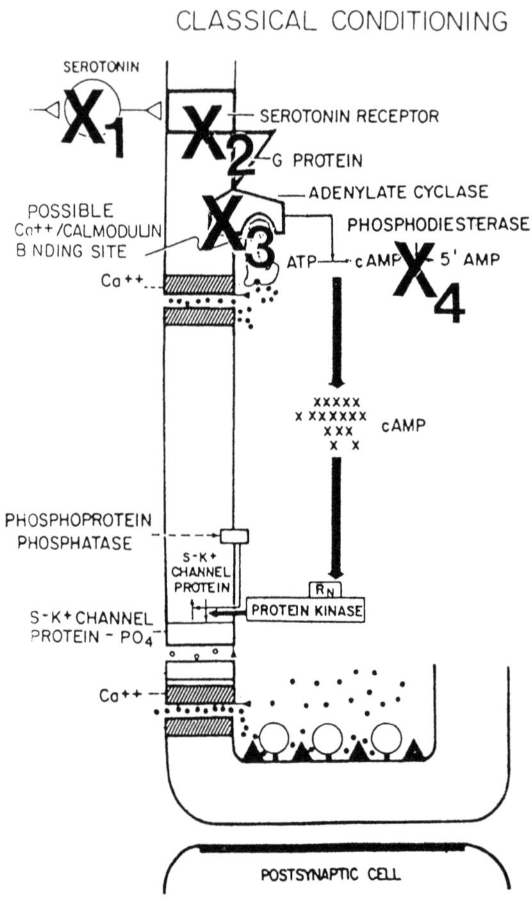

FIG. 4 - Diagrammatic synapse modelled from Aplysia work (Fig. 2B) with additions showing where the identified lesions from Drosophila learning mutants disrupt the cascade. 1) dopa decarboxylase - blocks serotonin synthesis; 2) turnip - alters serotonin receptor - N_S protein interaction; 3) rutabaga - alters cyclase interaction with calcium-calmodulin; 4) dunce - deletes a cyclic AMP phosphodiesterase.

of training have been found in both. However, learning in both Pleurobranchia and Limax involves modulation of a feeding response based on chemosensory input. In both animals the relevant sensory system consists of multitudes of small cells, and the behavioral output - a feeding motor program - is generated by a complicated neuronal network. Therefore, mechanistic work on learning in these animals has been difficult and slow. The current value of these systems is in pure behavior, an area covered by Sahley (this volume).

Octopuses make intelligent and empathetic pets (E.B. Lewis, personal communication). They can readily learn visual and tactile discrimination tasks (122). This capability has been used to measure their sensory capabilities and, in combination with lesion studies, to study the functional organization of brain centers involved in learning (119, 123). The trouble with octopuses is that they are too intellectually well endowed, with large brains composed of millions of small neurons. They are arguably no more simple and much less relevant to us than rats.

Bees suffer similar problems of over-intelligence. They are the heroes of centrally important studies on adaptive constraints on learning (reviewed by Gould, this volume) and on time-dependent processes in learning (reviewed by Sahley, this volume). However, cellular work with bees is a relatively grisly enterprise because their brains consist of millions of tiny, inconvenient neurons. Nevertheless, there are intriguing recordings by Erber (48) and Menzel and Erber (88) of "conditionable cells." Although these cells are exceptionally miniscule, the recordings are intracellular, so that prospect for mechanistic experiments on them is grim but not hopeless.

Work with plague locusts has yielded one genuine mechanistic insight. Horridge (66) had trained decapitated cockroaches to flex their legs to avoid electric shock, and Eisenstein and Cohen (47) had shown that this behavior can be mediated by a single segmental ganglion. The leg flexion is genuine operant conditioning (see above). Tosney and Hoyle (116), working with isolated ganglia in the locust, developed a plausible neurophysiological analogue of this behavior. They recorded continuously from a motor neuron involved in leg flexion, and they were able to condition the motor neuron to increase its spontaneous firing rate by applying electric shock to the ganglion via a connective whenever the motor neuron firing rate fell below a criterion value. (They could similarly depress the mechanism's firing rate by reversing the shock contingency.) Woolacott and Hoyle (120) "trained" motor neurons in isolated ganglia as above. If they then bathed the ganglion in high Mg^{+2} - low Ca^{2+} solution, eliminating detectable synaptic input to the motor neuron, the cell's firing rate remained higher than untrained controls. This experiment is formally similar to the Crow and Alkon experiment on Hermissenda photoreceptor cells; it indicates that some of the training-induced difference observed in the cell is intrinsic. (However, procedures for producing and demonstrating synaptic isolation in the locust are more difficult and less rigorous.) These results obtained with the locust are, on the face of it, quite different from those with other systems. Conditioned changes in the present case are encoded in a motor neuron rather than a sensory neuron. Moreover, the change is manifest as a change in "pacemaker" spiking rhythm rather than in synaptic facilitation (as in Aplysia) or prolonged depolarization and excitability (as in Hermissenda). Nevertheless, there may be less to this difference than meets the eye. The logical way to increase pacemaker rhythm, as Woolacott and Hoyle pointed out, is by blocking a potassium conductance, thereby increasing the cell's input resistance. If this is the case, the ionic mechanisms may be more generally similar in the three cases than the relevant cells' "overt behavior."

If receptor cells, interneurons, and motor cells can all show somewhat similar changes with training, then perhaps the molecular machinery for encoding the changes is widespread among neurons, provided one knows how to look for it. In fact, a number of vertebrate and invertebrate neuromuscular junctions can undergo facilitation, and the compounds most frequently implicated in these modulatory changes are monoamines - the class of neurotransmitters implicated in sensitization and conditioning in Aplysia and Drosophila. Specific neuromuscular junctions in cockroaches and lobsters are potentiated by octopamine and serotonin (15, 78, 80, 91). A recent study of a crayfish neuromuscular junction (12) indicates that applied octopamine can induce presynaptic facilitation and, more excitingly, that the facilitatory effect is amplified by activity (action potentials) in the presynaptic neuron. The parallel between these observations and the recent classical conditioning studies in Aplysia is striking and leads one to wonder: is there pairing specificity in the crayfish case? Can a neuromuscular junction show the neural properties of associative learning? The idea sounds preposterous. Alternatively, the machinery might be ubiquitous because learning-like mechanisms may play a role in selecting appropriate connections in normal development.

RÜCKBLICK UND ÜBERBLICK

If we want biochemically detailed mechanisms for learning, we have two choices. The first, formulated in Aplysia, involves a monoamine-activated cyclase and a cyclic AMP-dependent kinase which activates a newly discovered potassium conductance channel, I_S. Associativity in this case is conferred by activity-dependent amplification of the cyclase response, probably by increased intracellular calcium. The other model, formulated in Hermissenda, involves a calcium-calmodulin dependent kinase which inactivates a previously characterized potassium conductance, I_A. Associativity in this model comes from properties of a neural network, and from nonlinear responsiveness to excitation in both a photoreceptor cell and an interneuron. The systems look mutually exclusive. Which is correct?

Before we wheel out the artillery on this issue, we should note the points of convergence. In both instances, the critical alterations are in sensory receptor cells. Both instances involve cellular second-messenger signalling systems and kinases; both involve alterations in potassium conductance. Common sense, as well as this convergence, suggests that these general principles are liable to be true and universal. Intracellular second messengers have the right kinetic "feel." They can respond to neurotransmitters as conventional channel-coupled receptors do, but the response endures for minutes - about as long as short-term memory. Similarly, it makes sense that these second messenger systems produce their effects by altering potassium conductance. The fact that a neuron has only one kind of sodium conductance channel but may have up to five kinds of potassium channels indicates that the cell goes to a great deal of care to regulate and modulate this ion's conductance, both near resting potential where it affects membrane excitability (as in Hermissenda or locusts) and during depolarization, where it affects the informational consequences of an action potential (as in Aplysia).

The existence of the five potassium channels suggests that ionic events during learning may be complex. Therefore, we may never have a strict choice between the Aplysia and the Hermissenda models. Some learning systems may work like the one, and others like the other. More likely, neuronal ensembles that learn may use some of both mechanisms. In their understandable haste to establish basic principles, each group may have overlooked subtleties in their system that resemble the other's. The role of cAMP and of the I_S channel is firmly established as necessary for sensitization and classical conditioning in Aplysia. Remember, though, that the best bet for associative effects in this system is amplification of the cyclase response by calcium. Could not the amplification be made more efficacious by blending in a little of the calcium-potassium positive feedback loop found in Hermissenda? This is an idle bystander speculating. Recently, however, Walters and Byrne, working with their Aplysia tail preparation, reported increased calcium conductance and altered membrane voltage after associative entrainment of their neuronal system in Aplysia (118).

Learning studies in Hermissenda have concentrated on one photoreceptor cell type and, more recently, have focussed on a single second messenger system and a single potassium channel type. Again, these are judicious choices; experimental results indicate that the observed changes are important in the learned response. Results do <u>not</u>, however, rule out the possibility that the monoamine-cyclase-potassium I_S channel cascade is also involved. Spinning a Hermissenda around in its training machine probably sensitizes it, and this almost certainly alters circulating monoamine levels. The direct synaptic influences on type B photoreceptors have been carefully measured and catalogued both during and after training, but the possibility of circulating, neurohormonal influences has not been excluded. Moreover, the type B photoreceptors are known to have cAMP-dependent kinase, with affects on non-I_A potassium channels (10). If associative learning in Aplysia goes a little like that in Hermissenda, then the reverse is also likely to be true.

All this said, one is still left with the question of which features of the two systems are likely to have more widespread importance. In this respect, a betting man would go with Aplysia, because the proposed mechanism is biochemical. Neuronal networks, like the one in Hermissenda, tend to be individually evolved and tuned for the purposes at hand; biochemical principles, in contrast, tend to be conserved across cell types and across phyla. An exception to this generalization, and a contribution from Hermissenda that looks fundamental, is the potassium-calcium positive feedback cycle (9, 10).

The work with Drosophila mutants reinforces one's intuition and goes, by and large, with Aplysia. In particular, the fact that the mutations rutabaga (adenylate cyclase) and dunce (cyclic AMP phosphodiesterase) affect sensitization, classical conditioning, and also positively and negatively reinforced operant learning, strongly suggests that cAMP is involved in many kinds of learning in a variety of species. It is the best present evidence for underlying mechanistic species unity. However, the fact that no fly mutation blocks learning completely suggests that cAMP changes are not the whole story. Remember also that not all Drosophila learning mutants have been biochemically deciphered. A metabolic block monoamine synthesis or cyclic AMP metabolism is relatively easy to detect with the present methods. A mutation with a lesion more in line with the Hermissenda story - in potassium or calcium channels, say, or in a calcium-calmodulin dependent kinase - would go unexplained for the present, as have cabbage and amnesiac.

In our search here for universal learning mechanisms, we have worried only about invertebrates. Is any of this work relevant to vertebrates like us? On this issue, the neuroanatomical arrangement of monoaminergic systems in mammalian brains is quite provocative. Noradrenergic and serotonergic systems originate from cells in discrete brain stem nuclei to ramify widely and diffusely in the cerebrum and cerebellum. Such cortical "sprinkler systems" seem tailor-made for altering the attentiveness or retentiveness of large regions of the brain at once. Moreover, the mammalian cortex is full of the biochemical machinery found in Aplysia - cyclase-coupled monoamine receptors, a special calcium-enhanced cyclase, a kinase, even a potassium conductance which is inactivated by cyclic AMP (86). Having evolved this neuroanatomy and packed itself with these molecules, the mammalian brain would be foolish not to indulge in learning along Kandelian lines. Along with this appeal to reason and parsimony go a large number of drug studies (reviewed by Squire in (109)) which, taken in toto, persuasively implicate monoamines in learning. Nevertheless, given the nature of the mammalian brain, it may be years before we have a direct test of all the principles from Aplysia.

The cortex is spacious and diverse. If much of it is biochemically reminiscent of Aplysia, then one piece of it is neurophysiologically suggestive of Hermissenda. The hippocampus is a specialized, highly organized structure which is necessary for most types of learning in humans (Squire, this volume). This role, and its regular anatomy, have made it the subject of extensive neurophysiological studies, particularly in the CA_2 and CA_3 cells. Two unusual features of these cells have emerged: a) after-potentials that last long after excitation (11) and b) large increases in intracellular calcium (Krinjevic, unpublished). The mechanistic basis for these events is not at all clear, but the phenomenological similarity to events in Hermissenda photoreceptors is intriguing. Whether the analogy is a detailed one, and whether the calcium-potassium feedback cycle functions as in Fig. 3, with a calcium-calmodulin dependent kinase acting on potassium I_A channels, are speculative questions at present. Cell-biological experiments to resolve them (injecting calcium-calmodulin dependent kinase into hippocampal cells, for example) are probably within present technical capacity and would be very valuable. However, even with a mechanism for long-term potentiation in hand, one would still be faced with the question of how the hippocampal circuit functions in learning.

So ... molecular and cell-biological events underlying learning are being elucidated in Hermissenda. They are fairly well in hand in Aplysia, and some work with flies suggests that principles from Aplysia have widespread importance. The relevance of this work to mammals is not established but is an excellent bet; nature would be unusually perverse if she stockpiled the mammalian brain with the relevant molecules and then used them for some unrelated purpose. What good does all this information do us, and what relevance does it have to ethologists and psychologists?

First of all, large areas of traditional American psychology - schedules of reinforcement, interstimulus intervals, memory phases - may turn into chemistry (see also (63, 98)). Here are some naive extrapolations from the Aplysia model: the maximum CS-US interval may reflect the perdurance of intracellular calcium in the relevant cells; cells that carry out Garcia learning (54) may have a mechanism of prolonging this, perhaps with a feedback cycle as in Fig. 3. The sensitivity of short-term memory to electroconvulsive shock, long modelled in terms of reverberating circuits for want of a better idea, now seems more explicable in terms of general disruption of monoamine and second-messenger levels by abnormal activity patterns - in effect turning a wild-type rat into a dunce phenocopy. U-shaped memory retention curves, studied by Kamin and many others, may simply reflect a time lag between the decay of short-term cAMP effects in the cells that learn and the onset of long-term gene activation. These extrapolations are simpleminded and probably wrong, but I think that they suggest the way the field will go.

The invertebrate work speaks much less directly to issues in classical ethology, cognitive learning, and language learning. Lorenz-style imprinting is most analogous to long-term memory. Although present work provides clues as to how this process might take place (cAMP-dependent gene activation is a good bet), these ideas are speculative and likely to remain so for some time. Questions such as: what stimulus features of parents are imprinted in greylag geese; what visual features of landscapes and flowers are remembered by bees; and what features of learned language and recollected experience are stored in humans; all have little to do with the problem of how one physically changes the properties of a unitary synapse, and more to do with how one assembles a network of synapses to read in, store, and read out ordered patterns of information. In the near future, local (synapse-mechanistic) studies may help global (network) learning studies only a little, by indicating which cortical areas, layers, or cell types in the brain happen to contain the molecular learning machinery - monoamine receptors, kinases, and S-type potassium channels, if you will.

For assessing the long-term effect of the current work, a historical analogy may help. Many experimenters, including some great ones, spent years rearranging genes on chromosomes and altering them with chemicals, establishing a catalogue of effects in hopes of deducing universal properties of the genetic material. They had no success, although they created an intellectual structure, genetics, which was valuable for other reasons. At length, the genetic material was purified by straightforward biochemical techniques and its structure elucidated. Within ten years the old phenomenology was reinterpreted in ways that were simple, rememberable, and generative of new experimental areas. The Aplysia gill is not the double helix. Learning, even in its cellular aspects, does not look as simple or as unitary as heredity, and pieces of the puzzle are still missing. Nevertheless, elements of the Aplysia model, with elements of related work, will have a lasting and centrally important effect on the field, and areas of behavior that are abstract and formalistic will have molecular palpability. People who recoil at the prospect of a humanistically interesting problem turning into squalid chemistry should remember that Penelope's steadfastness, for example, depended on long-term memory, a problem only lightly touched by experiment, and on the network properties of the human cortex, a tangle unlikely to be completely unravelled, ever.

Acknowledgements. I thank D. Alkon, T. Carew, T. Crow, J. Farley, and E. Kandel for preprints, for helpful conversations, and for patient answers to questions. I thank J. Gould, D. Ready, and especially R. Smith for improving the manuscript, and J. Nielsen for typing it.

REFERENCES

(1) Abrams, T.W.; Carew, T.J.; Hawkins, R.D.; and Kandel, E.R. 1983. Aspects of the cellular mechanism of temporal specificity in conditioning in Aplysia: Preliminary evidence for Ca^{+2} influx as a signal of activity. Soc. Neurosci. Abstr. 9: 168.

(2) Aceves-Pina, E.O. 1982. Behavioral, anatomical and physiological analysis of normal D. melanogaster and learning-deficient mutants. Ph.D. Thesis, Princeton University, Princeton, NJ.

(3) Aceves-Pina, E.O.; Booker, R.; Duerr, J.S.; Livingstone, M.S.; Quinn, W.G.; Smith, R.F.; Sziber, P.P.; Tempel, B.L.; and Tully, T.P. 1983. Learning and memory in Drosophila, studied with mutants. Cold Spring Harbor Symp. 48: 831.

(4) Aceves-Pina, E.O., and Quinn, W.G. 1979. Learning in normal and mutant Drosophila larvae. Science 206: 93.

(5) Adams, D.J.; Smith, S.J.; and Thompson, S.H. 1980. Ionic currents in molluscan soma. Ann. Rev. Neurosci. 3: 141.

(6) Alkon, D.L. 1973. Intersensory interactions in Hermissenda. J. Gen. Physiol. 62: 185.

(7) Alkon, D.L. 1979. Voltage-dependent calcium and potassium ion conductances: A contingency mechanism for an associative learning model. Science 205: 810.

(8) Alkon, D.L. 1980. Cellular analysis of a gastropod (Hermissenda crassicornis) model of associative learning. Biol. Bull. 159: 505.

(9) Alkon, D.L. 1983. Learning in a marine snail. Sci. Am. 249(1): 70.

(10) Alkon, D.L.; Lederhendler, I.; and Shoukimas, J.J. 1982. Primary changes of membrane currents during retention of associative learning. Science 215: 693.

(11) Andersen, P.S.; Sundberg, S.H.; Sveen, O.; and Wigstrom, H. 1977. Specific long-lasting potentiation of synaptic transmission in hippocampal slices. Nature 266: 736.

(12) Bailey, C.H., and Chen, M. 1983. Morphological basis of long term habituation and sensitization in Aplysia. Science 220: 61.

(13) Booker, R. 1982. A behavioral-genetic analysis of learning in Drosophila melanogaster. Ph. D. Thesis, Princeton University, Princeton, NJ.

(14) Booker, R., and Quinn, W.G. 1981. Conditioning of leg position in normal and mutant Drosophila. Proc. Natl. Acad. Sci. USA 78: 3940.

(15) Breen, C.A., and Atwood, H.L. 1983. Octopamine - a neurohormone with presynaptic activity-dependent effects at crayfish neuromuscular junctions. Nature 303: 716.

(16) Brostrom, M.A.; Brostrom, C.O.; and Wolff, D.J. 1978. Calcium dependent adenylate cyclase from cerebral cortex: Activation by guanine nucleotides. Arch. Biochem. Biophys. 191: 341.

(17) Brown, D.A., and Adams, P.R. 1980. Muscarinic suppression of a novel voltage-sensitive K^+ current in a vertebrate neurone. Nature 183: 672.

(18) Byers, D.; Davis, R.L.; and Kiger, J.A. 1981. Defect in cyclic AMP phosphodiesterase due to the dunce mutation of learning in Drosophila melanogaster. Nature 289: 79.

(19) Camardo, J.S.; Shuster, M.S.; Siegelbaum, S.A.; and Kandel, E.R. 1983. Modulation of a specific K^+ channel in sensory neurons of Aplysia by serotonin and cyclic AMP dependent protein phosphorylation. Cold Spring Harbor Symp. 48: 213.

(20) Carew, T.J.; Abrams, T.W.; Hawkins, R.D.; and Kandel, E.R. 1983. A test of Hebb's postulate of identified synapses which mediate classical conditioning in Aplysia. Soc. Neurosci. Abstr. 9: 168.

(21) Carew, T.J.; Castellucci, V.F.; Byrne, J.; and Kandel, E.R. 1979. Quantitative analysis of relative contribution of central and peripheral neurons to gill-withdrawal reflex in Aplysia californica. J. Neurophys. 42: 497.

(22) Carew, T.J.; Hawkins, R.D.; and Kandel, E.R. 1983. Differential classical conditioning of a defensive withdrawal reflex in Aplysia californica. Science 219: 397.

(23) Carew, T.J., and Kandel, E.R. 1973. Acquisition and retention of long-term habituation in Aplysia: correlation of behavioral and cellular processes. Science 182: 1158.

(24) Carew, T.J.; Pinsker, H.M.; and Kandel, E.R. 1972. Long-term habituation of a defensive withdrawal reflex in Aplysia. Science 175: 451.

(25) Carew, T.J.; Walters, E.T.; and Kandel, E.R. 1981. Classical conditioning in a simple withdrawal reflex in Aplysia californica. J. Neurosci. 1: 1426.

(26) Castellucci, V.F., and Kandel, E.R. 1974. A quantal analysis of the synaptic depression underlying habituation of the gill-withdrawal reflex in Aplysia. Proc. Natl. Acad. Sci. USA 71: 5004.

(27) Castellucci, V.F., and Kandel, E.R. 1976. Presynaptic facilitation as a mechanism for behavioral sensitization in Aplysia. Science 194: 1176.

(28) Castellucci, V.F.; Kandel, E.R.; Schwartz, J.H.; Wilson, F.D.; Nairn, A.C.; and Greengard, P. 1980. Intracellular injection of the catalytic subunit of cyclic AMP-dependent protein kinase simulates facilitation of transmitter release underlying behavioral sensitization in Aplysia. Proc. Natl. Acad. Sci. USA 77: 7492.

(29) Castellucci, V.F.; Nairn, A.; Greengard, P.; Schwartz, J.H.; and Kandel, E.R. 1982. Inhibitor of adenosine 3':5'-monophosphate-dependent protein kinase blocks presynaptic facilitation in Aplysia. J. Neurosci. 2: 1673.

(30) Castellucci, V.F.; Pinsker, H.; Kupfermann, I.; and Kandel, E.R. 1970. Neuronal mechanisms of habituation and dishabituation of the gill-withdrawal reflex in Aplysia. Science 167: 1745.

(31) Cedar, H.; Kandel, E.R.; and Schwartz, J.H. 1972. Cyclic adenosine monophosphate in the nervous system of Aplysia californica I. Increased synthesis in response to synaptic stimulation. J. Gen. Physiol. 60: 558.

(32) Cedar, H., and Schwartz, J.H. 1972. Cyclic adenosine monophosphate in the nervous system of Aplysia californica. II. Effect of serotonin and dopamine. J. Gen. Physiol. 60: 870.

(33) Chang, J.J., and Gelperin, A. 1980. Rapid taste aversion learning by an isolated molluscan central nervous system. Proc. Natl. Acad. Sci. USA 77: 6204.

(34) Connor, J., and Alkon, D.L. 1982. Light-induced changes of intracellular Ca^{2+} in Hermissenda photoreceptors measured with Arsenazo III. Soc. Neurosci. Abstr. 8: 944.

(35) Connor, J.A., and Stevens, C.F. 1971. Voltage clamp studies of a transient outward current in gastropod neural somata. J. Physiol. Lond. 213: 21.

(36) Crow, T.J., and Alkon, D.L. 1978. Retention of an associative behavioral change in Hermissenda. Science 201: 1239.

(37) Crow, T.J., and Alkon, D.L. 1980. Associative behavioral modification in Hermissenda: Cellular correlates. Science 209: 412.

(38) Dethier, V.G. 1976. The Hungry Fly. Cambridge, MA: Harvard University Press.

(39) Dewhurst, S.A.; Croker, S.G.; Ikeda, K.; and McCaman, R.E. 1972. Metabolism of biogenic amines in Drosophila nervous tissue. Comp. Biochem. Physiol. 43B: 975.

(40) Dudai, Y. 1977. Properties of learning and memory in Drosophila melanogaster. J. Comp. Physiol. 114: 69.

(41) Dudai, Y. 1979. Behavioral plasticity in a Drosophila mutant, dunce. J. Comp. Physiol. 130: 271.

(42) Dudai, Y. 1981. Olfactory choice behavior of normal and mutant Drosophila in a conflict situation in a successive conditioning paradigm. Soc. Neurosci. Abstr. 7: 643.

(43) Dudai, Y., and Bicker, G. 1978. Comparison of visual and olfactory learning in Drosophila. Naturwissen. 65: 495.

(44) Dudai, Y.; Jan, Y.N.; Byers, D.; Quinn, W.G.; and Benzer, S. 1976. Dunce, a mutant of Drosophila deficient in learning. Proc. Natl. Acad. Sci. USA 73: 1684.

(45) Dudai, Y., and Zvi, S. 1982. Heterogeneity of serotonin receptors in Drosophila melanogaster. Soc. Neurosci. Abstr. 8: 989.

(46) Duerr, J.S., and Quinn, W.G. 1982. Three Drosophila mutations which block associative learning also affect habituation and sensitization. Proc. Natl. Acad. Sci. USA 79: 3646.

(47) Eisenstein, E.M., and Cohen, M.J. 1965. Learning in an isolated prothoracic ganglion. Anim. Behav. 13: 304.

(48) Erbur, J. 1981. Neural correlation of learning in the honey bee. Trends Neurosci. 4: 270.

(49) Farley, J. 1983. Contingency-sensitive neural and behavioral change in Hermissenda. J. Neurophys., in press.

(50) Farley, J., and Alkon, D.L. 1980. Neural organization predicts stimulus specificity for a retained associative behavioral change. Science 210: 1373.

(51) Farley, J.; Richards, W.; Ling, L.; Liman, E.; and Alkon, D.L. 1983. Membrane changes in a single photoreceptor cause associative learning in Hermissenda. Science 221: 1201.

(52) Folkers, E. 1982. Visual learning and memory of Drosophila melanogaster wild-type C-S and the mutants dunce, amnesiac, turnip and rutabaga. J. Insect. Physiol. 28: 535.

(53) Fuchs, P.A.; Nicholls, J.G.; and Ready, D.F. 1981. Membrane properties and selective connexions of identified leech neurones in culture. J. Physiol. 316: 203.

(54) Garcia, J.; McGowan, B.K.; and Green, K.F. 1972. Biological constraints on learning. In Classical Conditioning II: Current Research and Theory. New York: Appleton-Century-Crofts.

(55) Gelperin, A. 1983. Neuroethological studies of associative learning in feeding control systems. In Neuroethology and Behavioral Physiology, eds. F. Huber and H. Markl. Berlin: Springer-Verlag.

(56) Goh, Y., and Alkon, D.L. 1982. Convergence of visual and statocyst inputs on interneurons and motoneurons of Hermissenda: A network design for associative conditioning. Soc. Neurosci. Abstr. 8: 825.

(57) Hamill, O.; Marty, A.; Neher, E.; Sakmann, B.; and Sigworth, F.J. 1981. Improved patch-clamp techniques for high resolution current recording from cells and cell-free membrane patches. Pflug. Arch. 391: 85.

(58) Hawkins, R.D. 1981. Interneurons involved in mediation and modulation of gill-withdrawal reflex in Aplysia. III. Identified facilitating neurons increase Ca^{2+} current in sensory neurons. J. Neurophys. 45: 327.

(59) Hawkins, R.D.; Abrams, S.W.; Carew, T.J.; and Kandel, E.R. 1983. A cellular mechanism of classical conditioning in Aplysia: activity-dependent enhancement of presynaptic facilitation. Science 219: 400.

(60) Hawkins, R.D.; Carew, T.J.; and Kandel, E.R. 1983. Effects of interstimulus interval and contingency on classical conditioning in Aplysia. Soc. Neurosci. Abstr. 9: 168.

(61) Hawkins, R.D.; Castellucci, V.F.; and Kandel, E.R. 1981a. Interneurons involved in mediation and modulation of gill-withdrawal reflex in Aplysia. I. Identification and characterization. J. Neurophysiol. 45: 304.

(62) Hawkins, R.D.; Castellucci, V.F.; and Kandel, E.R. 1981b. Interneurons involved in mediation and modulation of gill-withdrawal reflex in Aplysia. II. Identified neurons produce heterosynaptic facilitation contributing to behavioral sensitization. J. Neurophysiol. 45: 315.

(63) Hawkins, R.D., and Kandel, E.R. 1984. Is there a cell-biological alphabet for learning? Psychol. Rev., in press.

(64) Hebb, D.O. 1949. The Organization of Behavior. New York: J. Wiley and Sons.

(65) Hirsh, J., and Davidson, N. 1981. Isolation and characterization of the Dopa decarboxylase gene of Drosophila. Mol. C. Bol. 1: 475.

(66) Horridge, G.A. 1962. Learning leg position by the ventral nerve cord in headless insects. Proc. Roy. Soc. Lond. B. 157: 33.

(67) Jellies, J.A. 1981. Associative olfactory conditioning in Drosophila melanogaster and memory retention through metamorphosis. Master's Thesis, Illinois State University, Normal, IL.

(68) Kandel, E.R. 1979. Small systems of neurons. Sci. Am. 241(2): 66.

(69) Kandel, E.R.; Abrams, T.; Bernier, L.; Carew, T.J.; Hawkins, R.D.; and Schwartz, J.H. 1983. Classical conditioning and sensitization share aspects of the same molecular cascade in Aplysia. Cold Spring Harbor Symp. 48: 821.

(70) Kandel, E.R.; Brunelli, M.; Byrne, J.; and Castellucci, V. 1975. A common presynaptic locus for the synaptic changes underlying short-term habituation and sensitization of the gill-withdrawal reflex in Aplysia. Cold Spring Harbor Symp. 40: 465.

(71) Kandel, E.R.; Frazier, W.T.; Waziri, R.; and Coggeshall, R.E. 1967. Direct and common connections among identified neurons in Aplysia. J. Neurophysiol. 30: 1352.

(72) Kauvar, L.M. 1982. Defective cyclic adenosine 3'5' monophosphate phosphodiesterase in the Drosophila memory mutant dunce. J. Neurosci. 2: 1347.

(73) Kiger, J.A. 1979. A genetically distinct form of cyclic AMP phosphodiesterase associated with chromomere 3D4 in Drosophila melanogaster. Genetics 91: 521.

(74) Kiger, J.A., and Golanty, E. 1977. A cytogenetic analysis of cyclic nucleotide phosphodiesterase in Drosophila. Genetics 85: 609.

(75) Kistler, H.B.; Hawkins, R.D.; Koester, J.; Kandel, E.R.; and Schwartz, J.H. 1983. Immunochemical studies of neurons producing presynaptic facilitation in the abdominal ganglion of Aplysia californica. Soc. Neurosci. Abstr. 9: 915.

(76) Klein, M.; Camardo, J.; and Kandel, E.R. 1982. Serotonin modulates a specific potassium current in the sensory neurons that show presynaptic facilitation in Aplysia. Proc. Natl. Acad. Sci. USA 79: 5713.

(77) Klein, M., and Kandel, E.R. 1978. Presynaptic modulation of voltage dependent Ca^{2+} current: mechanism for behavioral sensitization in Aplysia californica. P.N.A.S. 75: 3512-3516.

(78) Kravitz, E.A.; Glusman, S.; Harris-Warick, R.M.; Livingstone, M.S.; Schwarz, T.; and Goy, M.F. 1979. Amines and peptides as neurohormones in lobsters: actions on neuromuscular preparations and preliminary behavioral studies. J. Exp. Biol. 89: 159.

(79) Kuo, J.F., and Greengard, P. 1969. Cyclic nucleotide-dependent protein kinases. IV. Widespread occurrence of adenosine 3'-5': monophosphate-dependent protein kinase in various tissues and phyla of the animal kingdom. Proc. Natl. Acad. Sci. USA 64: 1349.

(80) Kupfermann, I. 1981. Modulatory action of neurotransmitters. Ann. Rev. Neurosci. 2: 447.

(81) Kupfermann, I.; Carew, T.J.; and Kandel, E.R. 1974. Local, reflex and central commands controlling gill and siphon movements in Aplysia. J. Neurophysiol. 37: 996.

(82) Kupfermann, I., and Kandel, E.R. 1969. Neuronal controls of a behavioral response mediated by the abdominal ganglion of Aplysia. Science 164: 847.

(83) Livingstone, M.S.; Sziber, P.P.; and Quinn, W.G. 1982. Defective adenylate cyclase in the Drosophila learning mutant rutabaga. Soc. Neurosci. Abstr. 8: 384.

(84) Livingstone, M.S.; Sziber, P.O.; and Quinn, W.G. 1984. Loss of calcium/calmodulin sensitivity of adenylate cyclase in rutabaga, a Drosophila learning mutant. Cell 37: 205.

(85) Livingstone, M.S., and Tempel, B.L. 1983. Genetic dissection of monoamine transmitter synthesis in Drosophila. Nature 303: 67.

(86) Madison, D.V., and Nicoll, R.A. 1982. Noradrenaline blocks accommodation of pyramidal cell discharge in the hippocampus. Nature 299: 636.

(87) Menne, D., and Spatz, H.C. 1977. Colour learning in Drosophila. J. Comp. Physiol. 114: 301.

(88) Menzel, R.; Erbur, J.; and Masuhr, J. 1974. Learning and memory in the honey bee. In Experimental Analysis of Insect Behavior, ed. L. Barton-Browne, pp. 195-217. Berlin: Springer-Verlag.

(89) Miller, V.P., and Selverston, A.I. 1979. Rapid killing of single neurons by irradiation of intracellularly injected dye. Science 206: 702.

(90) Mpitsos, G.J., and Davis, W.J. 1973. Learning: Classical and avoidance conditioning in the mollusk Pleurobranchaea. Science 180: 317.

(91) O'Shea, M., and Evans, P.D. 1979. Potentiation of neuromuscular transmission by an actopaminergic neuron in the locust. J. Exp. Biol. 79: 169.

(92) Peretz, B.; Jacklet, J.W.; and Lukowiak, K. 1976. Habituation of reflexes in Aplysia: Contribution of the peripheral and central nervous systems. Science 191: 396.

(93) Pert, C.B., and Snyder, S.H. 1973. Opiate receptor: demonstration in nervous tissue. Science 179: 1011.

(94) Prinsker, H.M.; Henning, W.A.; Carew, T.J.; and Kandel, E.R. 1973. Long term sensitization of the defensive gill withdrawal reflex in Aplysia. Science 182: 1039.

(95) Quinn, W.G.; Harris, W.A.; and Benzer, S. 1974. Conditioned behavior in Drosophila melanogaster. Proc. Natl. Acad. Sci. USA 71: 708.

(96) Quinn, W.G.; Sziber, P.P.; and Booker, R. 1979. The Drosophila memory mutant amnesiac. Nature 76: 3430.

(97) Rasmussen, H. 1981. Calcium and Cyclic AMP as Synarchic Messengers. New York: Wiley & Sons.

(98) Rescorla, R.A. 1980. Pavlovian Second-Order Conditioning: Studies in Associative Learning. Hillsdale, NJ: Erlbaum.

(99) Rodbell, M. 1980. The role of hormone-receptors and GTP-regulatory proteins in membrane transduction. Nature 284: 17.

(100) Reichert, L.F., and Kelly, R.B. 1983. A molecular description of nerve terminal function. Ann. Rev. Biochem. 59: 871.

(101) Sahley, C.; Gelperin, A.; and Rudy, J.W. 1981. One-trial associative learning modifies food odor preferences of a terrestrial mollusk. Proc. Natl. Acad. Sci. USA 78: 640.

(102) Salter, R.S.; Krinks, M.H.; Klee, C.B.; and Neer, E.J. 1981. Calmodulin activates the isolated catalytic unit of brain adenylate cyclase. J. Biol. Chem. 256: 9830.

(103) Schacher, S., and Camardo, J.S. 1983. Properties of a specific transmitter-mediated connection formed between identified neurons of Aplysia in dissociated cell cultures. Soc. Neurosci. Abstr. 9, in press.

(104) Scheller, R.H.; Jackson, J.F.; McAllister, L.B.; Schwartz, J.H.; Kandel, E.R.; and Axel, R. 1982. A family of genes that codes for ELH, a neuropeptide eliciting a sterotyped pattern of behavior in Aplysia. Cell 32: 7.

(105) Schwartz, J.H.; Bernier, L.; Castellucci, V.F.; Palazzolo, M.; Saitoh, T.; Stapleton, A.; and Kandel, E.R. 1983. What molecular steps determine the time course of the memory for short-term sensitization in Aplysia? Cold Spring Harbor Symp. 48, in press.

(106) Shotwell, S.L. 1983. Cyclic adenosine 3' 5' monophosphate phosphodiesterase and its role in learning in Drosophila. J. Neurosci. 3: 739.

(107) Siegel, R.W., and Hall, J.C. 1979. Conditioned responses in courtship behavior of normal and mutant Drosophila. Proc. Natl. Acad. Sci. USA 76: 3430.

(108) Siegelbaum, S.A.; Camardo, J.S.; and Kandel, E.R. 1982. Serotonin and cyclic AMP close single K^+ channels in Aplysia sensory neurones. Nature 299: 413.

(109) Squire, L.R., and Davis, H.P. 1981. The pharmacology of memory, a neurobiological perspective. Ann. Rev. Pharmacol. Toxicol. 21: 237.

(110) Tabata, M., and Alkon, D.L. 1982. Positive synaptic feedback in visual system of nudibranch mollusk Hermissenda crassicornis. J. Neurophys. 48: 174.

(111) Tempel, B.L.; Bonini, N.; Dawson, D.R.; and Quinn, W.G. 1983. Reward learning in normal and mutant Drosophila. Proc. Natl. Acad. Sci. USA 80: 1482.

(112) Tempel, B.L.; Livingstone, M.S.; and Quinn, W.G. 1984. A mutation in Drosophila that reduces dopamine and serotonin synthesis abolishes associative learning. Proc. Natl. Acad. Sci. USA, in press.

(113) Tempel, B.L., and Quinn, W.G. 1982. Mutations in the dopadecarboxylase gene affect learning but not memory in Drosophila. Soc. Neurosci. Abstr. 8: 385.

(114) Thompson, R.F.; Berger, T.W.; and Madden, J. 1983. Cellular processes of learning and memory in the mammalian CNS. Ann. Rev. Neurosci. 6: 447.

(115) Tompkins, L.; Siegel, R.W.; Gailey, D.A.; and Hall, J.C. 1983. Conditioned courtship in Drosophila and its mediation by association of chemical cues. Behav. Genet., in press.

(116) Tosney, B., and Hoyle, G. 1977. Computer-controlled learning in a simple system. Proc. Roy. Soc. Lond. B. 195: 365.

(117) Walters, E.T., and Byrne, J.H. 1983. Associative conditioning of single sensory neurons suggests a cellular mechanism for learning. Science 219: 405.

(118) Walters, E.T., and Byrne, J.H. 1983. Slow depolarization produced by associative conditioning may enhance Ca^{++} entry into Aplysia sensory neurons. Brain Res., in press.

(119) Wells, M.J. 1962. Brain and Behavior in Cephalopods. Stanford, CA: Stanford University Press.

(120) Woolacott, M., and Hoyle, G. 1977. Neural events underlying learning in insects: changes in pacemaker. Proc. Roy. Soc. London B. 195: 395.

(121) Wright, T.R.F. 1977. The genetics of dopa decarboxylase and methyl dopa sensitivity in Drosophila melanogaster. Am. Zool. 17: 707.

(122) Wright, T.R.F.; Steward, R.; Bentley, K.W.; and Adler, P.N. 1981. The genetics of dopa decarboxylase in Drosophila melanogaster III. Effects of a temperature sensitive dopa decarboxylase deficient mutation in female fertility. Dev. Genet. 2: 223.

(123) Young, J.Z. 1961. Learning and discrimination in the octopus. Biol. Rev. 36: 52.

Standing, left to right:
Martin Heisenberg, Karl Fischbach, Chip Quinn, Randolf Menzel, Tom Carew, Martin Lindauer, Gerd Bicker.

Seated, left to right:
Bernd Heinrich, Jim Markl, Chris Sahley, Jim Gould, Allan Wagner.

The Biology of Learning, eds. P. Marler and H.S. Terrace, pp. 249-270. Dahlem Konferenzen 1984. Berlin, Heidelberg, New York, Tokyo: Springer-Verlag.

Biology of Invertebrate Learning
Group Report

R. Menzel, Rapporteur
G. Bicker
T.J. Carew
K.-F. Fischbach
J.L. Gould
B. Heinrich
M.A. Heisenberg
M. Lindauer
H.S. Markl
W.G. Quinn
C.L. Sahley
A.R. Wagner

INTRODUCTION

Invertebrates offer many opportunities to study several of the fundamental questions about learning posed by experimental psychology, ethology, and neurobiology. Invertebrates range in neurological complexity from jellyfish and primitive worms to insects, crustaceans, and cephalopods. Comparative learning studies should help us decide whether learning follows certain rules, whether these rules are the same as in vertebrates, and whether learning has independently evolved in different phyla. A comparison with well-studied vertebrate species such as rats, pigeons, even man, may inform us about alternative strategies of learning or about similarities in coping with change and causality in the environments. This comparison should also help us to understand whether the rules have a common physiological basis or represent convergent evolution.

In contrast to vertebrates, many invertebrate species have small brains, often with only a few hundred neurons in each ganglion. These neurons can be gigantic when compared with neurons in the vertebrate brain, facilitating intracellular studies of the neural effects of plastic changes of behavior. Understanding all the interconnections and monitoring all the neurons in a circuit is no longer a pipe dream, but a reality in some rhythm-generating invertebrate circuits and a near-reality in some learning circuits. But as we have learned in this workshop, although unprecedented progress has been made in identifying and measuring neuronal or even molecular events during a learning process, the crucial questions are not definitively answered. However, they can be phrased in a more precise way. We think this is already a big step on the way towards understanding the neuronal basis of learning.

Neurophysiological techniques are still somewhat outmatched by the complexity of the nervous system. But with large enough neurons and simple enough circuits, as in some invertebrates, electrodes become less intrusive and more revealing. What does an electrode see when a circuit is changing? Where does the change occur? What actually changes?

The fruit fly Drosophila melanogaster has, since the turn of the century, been the favorite object of geneticists, who profit from the quick succession of generations and from the small genome. With that creature, some scientific fantasies have become reality, allowing the genetic and biochemical analysis of neural plasticity, including learning. As Quinn points out (this volume), the genetic analysis of learning in Drosophila can be done on the level of single gene mutations, allowing analysis of the biochemical basis of learning with respect to identifiable enzymes and offering the prospect of tracing the causal chain to the information stored in DNA. In addition, another old question of learning psychology and a traditional battlefield between experimental psychologists and ethologists can be scrutinized: How do genetic programming and individual experience interact to produce real behavior?

Ecological adaptations in invertebrates seem infinitely varied. Animals without backbones come small and large, fast and slow. They walk, swim, or fly, they are blind or vision-dominated, live in small microhabitats or range over the oceans or continents for their resources. Invertebrates communicate intensely in societies or they live as individuals and never interact, not even with their sexual partners. All the various adaptations shape the behavior of the members of each species and set the framework for the species-specific solutions of behavioral adaptivity. Are these solutions independent discoveries of each species, or are there some common features of learning applicable to every species and independent of the vast differences in species-specific ecological adaptations? This question has been of central interest throughout the workshop. Are there rules and/or common physiological mechanisms of learning applicable despite all the differences in the structure of the nervous systems, in sensory and motor capacities, and in the organization of behavior? If so, are these patterns of learning behavior the result of interaction with very specific environments, which simply allow the animal to develop only these patterns to adjust to the causal relationships of the outside world through experience, or are these patterns an indication for a phylogenetic history of learning mechanisms? We have not been able to solve these questions, but we are convinced that they can be tackled by focussing on comparative studies including invertebrates.

Learning is a facet of cognition and intelligence. If learning really exists in invertebrates, and not only in the simplest forms, then would it not force us to think about what we actually mean by "cognitive" and "intelligent" functions?

WHAT INVERTEBRATES LEARN

Before we come to the specific points, let us first examine whether invertebrates can really learn. This was not seriously discussed in our group because everyone agreed that many invertebrate species have been found to learn, often in complex fashion.

A marine slug, Aplysia, habituates its reflexes to repeated stimulation. After habituation, a strong stimulus sensitizes the reflex again, causing almost an equally strong reflex response before habituation (17). Aplysia is also capable of associative learning. An elegant series of experiments (4) involves differential conditioning. "The experimenters gave a light tactile stimulus to two different sites on, say, the siphon, with one stimulus paired with tail shock, the other specifically unpaired. After training, the paired stimulus produced much greater siphon withdrawal than the unpaired stimulus" (Quinn, this volume). Another marine slug (Hermissenda) can be trained to reduce its tendency to crawl towards light, if it has been rotated in the presence of light (1). A similar exposure to the same stimuli, but unpaired, does not produce this behavioral change. The garden slug (Limax maximus) likes to eat mushrooms, carrots, and potatoes. But if the smell of one of these vegetables is associated with bitter-tasting substances (quinine) and another not associated with the bitter taste, the slug avoids the smell paired with the taste but not an unpaired smell (see Sahley, this volume). Pleurobranchea is another marine mollusc which has been found to change its behavior as a consequence of unpleasant experience. The taste of squid meat triggers feeding, but pairing the taste with an electroshock decreases feeding (5).

Drosophila can also be conditioned to avoid a smell if it experiences an electric shock through its tarsi. Proper experimentation causes a behavioral change of up to 90% (Quinn, this volume). This is reassuring, since the former value of only 30% behavioral change was always disturbingly low. Furthermore, appetitive conditioning experiments with sucrose reward can also change the behavior of the animal effectively, and with longer memory from fewer trials.

Hymenoptera, in particular the social groups (bees, wasps, ants) control many of their behaviors through learning. Forel (10), in his pioneering work, recognized that bees and wasps find their way around by individual learning, and von Frisch (11) proved for the first time that honey bees can be trained to visual and olfactory cues by food reward. This learning is fast and effective, and memory is long-lasting. Learning in honey bees is one of the major subjects of this report and is also discussed in great detail by Gould (this volume).

These few examples were chosen from among many established cases because we shall return to them in detail. There is ample evidence for learning phenomena such as habituation, sensitization, classical conditioning, operant conditioning, latent learning, and even observational learning and memory can last for a lifetime in many invertebrates (see, for example, (2, 3, 7, 27, 30)). Little is known, however, at which evolutionary level the nervous system develops the potential of learning. All species with neurons fused in ganglia, with well-developed senses and a rich repertoire of movements are capable of both nonassociative and associative learning. This applies to all phyla above and including flatworms. Coelenterates (jellyfish, sea anemones), which lack ganglia and control their simple movements by nerve nets, habituate to repeated stimuli, but associative learning has not yet been demonstrated unambiguously (12, 25). However, there is no reason to assume that a nerve net is too primitive for associative learning. Experiments with the proper controls are urgently needed.

LEARNING THEORY AND INVERTEBRATE LEARNING

Learning theory has not always had a congenial relationship to phenomena and interpretations of invertebrate learning. A common metatheoretical view among learning theorists (e.g., (26)) was that the empirical laws of Pavlovian and Thorndikian conditioning as discovered in laboratory investigations with such conventional subjects as dog and rat, could serve as the axioms of a theory, capable of deducing all instances of learning as properly defined. One consequence of this view was that an important instance of behavioral plasticity investigated in invertebrates, that is, habituation, was not admitted as true learning (e.g., (19)). Another consequence was a prejudice against interpretations of learning that made use of another vocabulary (e.g., that of physiology or ethology) than that involving the "conditioned response." While learning theory still places strong emphasis upon Pavlovian and Thorndikian conditioning, in recent years there has been a notable change in the presumed objectives of this emphasis. As Jenkins (this volume) observes, the phenomena of classical and operant conditioning are no longer viewed as elementary theoretical "building blocks." Rather, the research paradigms are simply regarded as especially useful techniques for isolating and evaluating the processes that may be involved in associative learning. The paradigms are thus assigned a place (albeit a presumably most strategic one) alongside of other paradigms, e.g., habituation, song learning, problem solving, etc., in the attempted development of theories that may be phrased in physiological, information-processing, or any potentially useful language.

With this change in attitude has come a much greater interest on the part of learning theorists in the phenomena of invertebrate learning as potential sources of theoretical principles and as important tests of the generality of available theories. The extensive investigation of habituation in Aplysia (18) has had substantial impact upon current learning theory approaches to this phenomenon. And substantial investigations have begun to evaluate the applicability of such theories of associative learning as the Rescorla-Wagner model (24) to invertebrates (see Sahley, this volume).

Although few invertebrate species (Limax, Hermissenda, bees, and to some extent, Aplysia) have been studied carefully enough to comment on the generality of the principles emphasized by current learning theories, the available evidence indicates more similarities between species than differences. As in vertebrates, the temporal relationship between CS and US appears highly critical for the development of associations; the predictability of the US is important in determining the degree of conditioned responding, as shown in blocking* and overshadowing†,

---

*In a "blocking experiment" the animal is first conditioned to respond to stimulus S_1 and then to a combination of S_1 and S_2. The animal is then tested as to whether S_2 has been learned. In the case of blocking by prior conditioning to S_1, the animal does not respond to S_2 alone.

conditioned inhibition training can produce an inhibitory CS; and higher-order conditioning phenomena have also been documented in invertebrates. In the honey bee, several additional phenomena have been observed which attest to the similarities in invertebrate and vertebrate learning (e.g., in regard to reversal learning, performance on fixed ratio and fixed interval schedules of reinforcement, matching to reward probability and the distinction between long-term and short-term memory (5, 23). There are also reported differences in the learning observed between invertebrate species and between certain well-studied invertebrates and mammals (e.g., overshadowing in bees (5)). But these reports are harder to interpret at the present time since they essentially stem from "negative results," i.e., failure to observe some phenomenon with a particular species under given experimental circumstances. They may be due to lack of appreciation of the animals' sensory and/or motor capacities or to liabilities in experimental design. As Sahley (this volume) points out, we cannot yet decide what degree of functional similarity obtains between invertebrate and vertebrate learning, or what the similarities and differences may mean with respect to underlying mechanisms.

NEURAL MODELS OF LEARNING

Learning is the adaptive modification of behavior through experience with the environment. Since behavior is the product of neural processes, there must be a change in the neural events underlying these processes. What are these changes? During the past decade there has been substantial progress in identifying the cellular mechanisms of these changes. The heroes of the story are two molluscs, Aplysia and Hermissenda, which supply us with the necessary information to speculate on elementary learning processes. In particular, studies on Aplysia have reached the point where the questions can even be addressed at the molecular level (see Quinn, this volume). From studying the cellular processes underlying habituation, sensitization, and classical conditioning, the most important result appears to be that classical conditioning is an augmented or amplified form of sensitization: specifically, it appears that when a sensory neuron is active just prior to input from a sensitizing stimulus

---

†In an "overshadowing experiment" the animal is conditioned initially to a combination of two stimuli (S_1, S_2) and is then tested for its separate responses to each of the stimuli S_1 and S_2. In the case of overshadowing, the animal responds less to one stimulus, as opposed to its responses after conditioning to that stimulus alone.

which produces presynaptic facilitation at the sensory neuron terminals, the sensory neuron enjoys a long-lasting facilitation. This persistent "activity-dependent presynaptic facilitation" appears to resemble the presynaptic facilitation seen in cases of sensitization only in an amplified form which might be viewed as a "conditioned sensitization" effect.

Is such a cellular mechanism a viable model of elementary learning processes? Let us first consider the limitation of the model and ask in which direction future research should go: a) Thus far only the sensory neurons have been studied in detail. Although the cellular events in the presynaptic terminals of the sensory neurons parallels behavioral observations, future research must also study the interneurons in the reflex. One may predict that associative changes might occur there as well, perhaps using the same mechanisms which have been seen in the sensory neurons. b) Long-term memory is not yet understood in this system. It might well be that it involves a selective gene activation for long-lasting memory which is initiated parallel to the shorter-lasting effect of activation of a specific phosphodiesterase (see Quinn, this volume). c) A few potentially important technical details have to be kept in mind. All the recordings which characterize the membrane-bound changes in the ion flux come from the soma; the effective changes occur in the synaptic terminals. Kandel and his colleagues have good reason to believe that the soma membrane has properties in common with the terminal membrane and may be electrically close to the terminals, but still the process at the synapses is inferred from the events in the soma. Furthermore, one should keep in mind that the behavioral measure for conditioning is the duration of the response, whereas the sizes of the PSP's are compared in neurophysiological studies. The relationship between these measures remains to be elucidated.

The Aplysia model is not the only one on the market, but it is the one most convincingly documented. Comparing it with the studies on Hermissenda and Drosophila (see Quinn, this volume), one might conclude that the acquisition processes are different in Aplysia, Hermissenda, and Drosophila, but that the memory process may have features in common which involve second messengers (cAMP) and protein phosphorylation.

Now let us ask what kind of behavioral plasticity might be plausibly modelled by synaptic events in Aplysia. "Alpha conditioning" is defined in learning psychology as a behavioral product of CS-US pairings in which a) the CS prior to training has some tendency to produce the same response produced by the US, and b) the response becomes more probable or stronger with training. This is the case in the demonstration of associative learning in Aplysia as demonstrated by Carew et al. (4). The theory developed to account for the phenomenon, however, does not depend on this characteristic. Figure 1 shows the minimal circuit necessary for associative learning. With pairing of the CS (S.N. in Fig. 1) and US (tail S.N. in Fig. 1) there will be facilitative input to the terminals of the S.N. after the latter has been activated, resulting in a subsequent enhanced (sensitization-like) effectiveness of the S.N. It could be called "conditioned sensitization."

What should be appreciated is that in the simple circuit (Fig. 1) the CS prior to conditioning would tend to produce the eventual CR, but it would not necessarily resemble any UR as required for "alpha conditioning."

If the figure included a synapse of the facilitatory interneuron to the motor neuron, it could mediate alpha conditioning. But this is not necessary and is misleading in terms of the theoretical possibilities of the model. It is therefore more appropriate to talk about "conditioned sensitization" if one wants to refer to the Aplysia model as a building block of an elementary learning process.

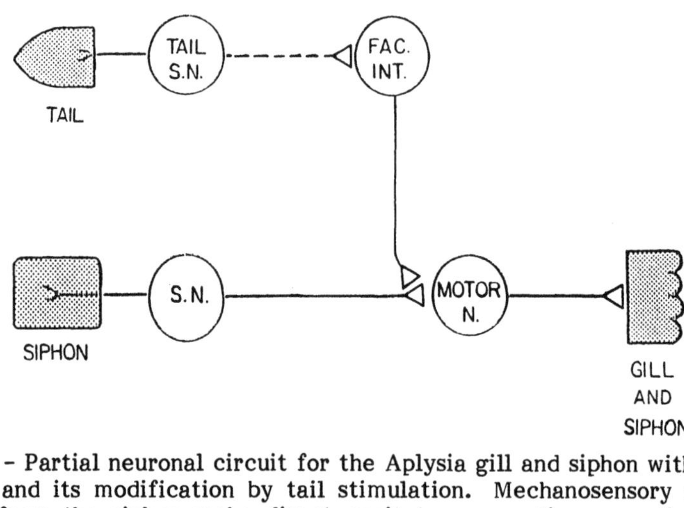

FIG. 1 - Partial neuronal circuit for the Aplysia gill and siphon withdrawal reflex and its modification by tail stimulation. Mechanosensory neurons (S.N.) from the siphon make direct excitatory synaptic connections onto gill and siphon motor neurons. Tail sensory neurons (S.N. tail) excite facilitator interneurons, which produce presynaptic facilitation of the siphon sensory neurons (from (13)).

One might ask whether the Aplysia model of conditioned sensitization may account for most types of conditioning. At first glance this seems unlikely since in many experiments on classical conditioning the CS is a neutral stimulus and does not provoke a response. However, there are cases where at close inspection small responses, e.g., in the motor neuron, are found prior to conditioning. One example is the rabbit nictating membrane, but here again only certain CS's (tone) cause a subthreshhold response in the motor neuron prior to conditioning, as opposed to other stimuli, which can be conditioned but do not produce any noticeable response in the motor neuron (e.g., light). It is an interesting but unresolved question as to whether most (or all?) potential CS's are prewired to be associated with most responses. What may happen during learning is the strengthening (or weakening) of previously existing connections. There was lively discussion in the group as to how general such a concept of conditioning can be.

We may also question whether the Aplysia model is too special because the plastic properties due to paired stimuli alter a sensory synapse. Although it seems more likely that learning is predominantly a change in the synaptic properties of interneurons, there is no substantive reason why this sensory-motor synapse should not be a good model for interneuronal synapses. Finally, while the Aplysia model is quite attractive, it is wise not to overemphasize it as the only possible elementary building block for associative learning. As Hawkins and Kandel (13) point out, it does not account for every type of learning, and polysynaptic components are probably of equal importance as monosynaptic effects.

It has been pointed out by Quinn (this volume) that the analysis of single gene mutants in Drosophila provide a first test for the applicability of the Aplysia model in a different species. Such a test is facilitated by the fact that the mutants best analysed mainly display short-term memory defects. It is important to note that mutants behave differently in visual and olfactory learning tasks. This is not clearly understood, and it is contraintuitive if one assumes a general applicability of the model. However, if one considers the mechanisms of the model as applying only to a certain part of the fly brain (e.g., mushroom bodies) and only for the chemosensory modality, then neither this effect nor the other performances of the mutants are surprising.

The brain is a network of neurons whose pattern of connections are genetically organized according to function. It is conceivable that the genetic program includes the information of a specific pattern of Aplysia-type modifiable synapses as building blocks in the network. Unfortunately, there are practically no physiological experiments to test this idea, but one can use behavioral experiments to speculate along this line. We realize that the Aplysia model does not account for conditioned inhibition unless additional circuit elements are added and that it explains extinction as habituation to CS, a view not in line with the traditional Pavlovian way of thinking (see (13, 24)). Among invertebrates conditioned inhibition has so far only been reported for the bee (see (23)). So it might well be that the way out of this dilemma is the additional action of an inhibitory circuit. But if this is true, then why should extinction not be hooked up to the inhibitory circuit, too? The complexity of the bee's CNS would certainly allow this. Other arguments along this line can be found in Gould (this volume). By defending the concept of conditioned sensitization as a basic strategy of simple forms of learning, Gould suggests that the rapidity of selective learning and the associational cue biases are in favor of this concept. How could such a hypothesis be tested? Does the existence of cue biases already set a limit on the mechanisms of conditioned sensitization? Would one expect the biases to show up in spontaneous preferences? Are different acquisition rates indications for prewired sensory-motor circuits?

One should be cautious in correlating the speed of synaptic change with the neuronal selection of biased circuits. It is easily conceivable that quickly modified synapses are just the least strictly selected ones. Thus, if a quantitative measure is of little help in deciding on prewired circuits, is there a qualitative measure? If the concept should be more than just the trivial statement that the nervous system has to be wired up for sensory and motor integration, one has to conclude that there is a finite number of cues to be learned by a given animal. If the number of cues to be learned turns out to be practically unlimited, the concept is of little value, except insofar as the ease of learning correlates well with the spontaneous choice behavior. In the end, most members in our and other groups (see also Kroodsma et al. and Holland et al., both this volume) agreed that the mechanism of synaptic selection may be the most attractive hypothesis, which at least allows the formulation of predictions and helps to design tests for both behavioral and neurophysiological experiments.

By discussing the network properties we realized how urgently neurophysiological data are needed. Again very useful animals will be the molluscs (Aplysia, Hermissenda, Pleurobranchea, Limax), which have been analyzed to some degree with respect to adaptive processes, but other invertebrate species offer promising preparations for future work, too (3, 6, 7, 22, 29). Studies on operant conditioning of the neural activity in a motor neuron of an insect (Locusta) add additional insight in plastic changes on the level of single neurons, but network properties are unknown in the most interesting preparation (2, 16). Neurons in the brain of the honey bee were found to undergo changes of their responsiveness during repeated unpaired or paired stimulations (9), but here again much more work is needed to define the changes on the neuronal level. The phenomena of associative learning are well documented in these animals and preparations for neurophysiological recordings have been developed. Now painstaking work is needed to collect the necessary data for an analysis which looks at both the cellular and network properties.

WHAT INVERTEBRATES CAN AND CANNOT LEARN

A priori, it seems plausible that invertebrate species with very small or simple brains should behave stereotypically and exhibit only rudimentary plasticity, but that "higher" forms with brains containing millions of neurons should be marvels of learning. Indeed, at the one extreme there is no firm evidence for associative learning in unicellular organisms such as protozoans or "brainless" multicellular organisms such as the coelenterates, whereas nonassociative plasticity is found in both phyla. But this is all that can be said for the idea, although the problems inherent in negative results and the generally inferior level of experiments in very primitive invertebrates indicate the need for better studies. Associative plasticity may exist even in unicellular organisms, although if it does, the whole issue of neural plasticity mediated by synapses would then appear in a new light. The perspectives related to these questions are so fascinating that well controlled experiments are strongly called for.

Learning studies in invertebrates have often been hindered by superficial extrapolations from experiments on mammals and humans. We have learned to be especially cautious with negative outcomes of such experiments. In general, learning behavior may be more strictly constrained by preexisting neural connections in invertebrates than in vertebrates. An earthworm responds and habituates to a vibrational stimulus but cannot associate such a stimulus with electric shock as a reinforcer. If, however, the taste of the substrate is a predictor for a hot and dry place (which the animal tries to avoid) or for a dark, humid, and cool place (which it tends to reach), the animal will learn either to avoid or to select this taste (21). Many experiments of these kinds have been reported in invertebrates (see, e.g., (7, 27)), but it is unclear whether the experiments have been appropriately designed to allow identification of these prepared associations.

The honey bee provides many examples of what an invertebrate can and cannot learn. Colors, odors, black and white patterns, landmarks, time of day, or any direction with respect to the sun or the sky's polarized light pattern can all be learned either as near or far cues at the feeding place. The orientation of the substrate relative to gravity, the fine structure of the substrate, and the spatial and temporal sequence of odors are learned very effectively and retained in a long-lasting memory. However, not every perceived cue is learned as a food signal. Neither a polarized light pattern marking the feeding place nor a flashing light can be associated with food; time schedules other than a 24 hour rhythm (under natural light conditions) or day-night rhythms outside the range of 22-25 hours per light cycle are learned. In proboscis reflex conditioning, bees have difficulty in associating visual cues - colors or moving stripe patterns - whereas they very quickly associate an odor stimulus with sucrose solution. If one takes the rate of acquisition as a measure for selectivity in associations, one finds a gradation which relates very nicely to naturally occurring flower cues (violet is learned more quickly than other colors and flower-like odors are learned faster than unnatural odors.). With bees there is a clearly defined hierarchy of stimuli as potential CS's, and apparently there is a difference in the proclivity of bees to establish different associations (see (23)).

The evidence indicates that each animal species possesses a set of specific learning dispositions and is thus species-specifically prepared for combining hard-wired with modifiable neural circuits. Ethologists have used various measures to characterize such preparedness, e.g., rate of acquisitions, stability of memory, precision of controlling behavior, sensitive periods for learning. These measures are indeed useful, if the gradation of learning is not simply the result of perceptual or motor limitations.

The picture emerging from these experiments envisages species-specific preparedness as neural processes which favor certain stimuli-response or response-reward associations in a graded way, ranging from exclusive, specialized, and selective associations to wide and free associations of "neutral" stimuli with very little prewired tuning. The one extreme - selective association - which we called "preparedness in the strict sense," allows only certain stimuli-response associations and excludes others. So far there is no experiment on invertebrates which demonstrates preparedness in the strict sense using the 2 x 2 design known from the taste aversion and electroshock aversion learning of mammals (see Revusky, this volume, for more information on the taste aversion experiments).* A 2 x 2 experimental array is helpful in deciding between a capacity to respond to certain stimuli and a preparedness for selective associations of stimuli in a learning paradigm. Selective associations in a less strict sense, however, have been demonstrated in several invertebrate species. Most instructive are the results on honey bees: high conditionability of the proboscis extension reflex to odors and very low conditionability of the same response to color stimuli; high conditionability of that reflex to movements of stripes from front and back and hardly any conditionability to moving stripes from back to front. Even in these cases, though, selectivity is not exclusive. Rather, the overwhelming evidence in honey bees suggests a great freedom for associations of cues with reward, particularly under natural conditions.

So far we have restricted ourselves to the learning phenomena of classical conditioning. There is ample evidence of other forms of learning in invertebrates. Operant conditioning and latent learning are traditional types of learning occurring in freely behaving animals. Motor programs underlying walking and stridulating in arthropods, for example, adjust adaptively to manipulations of legs and wings (8, 16). Amputation of a leg causes changes in the walking rhythm of the remaining legs. A forced change in the position of stridulating wings in crickets results in an adjustment to the new wing position. Locusts and cockroaches can be trained to maintain a certain leg position if any change of the position results in strong electrical stimulation to the tarsi. The neural analysis of this preparation has already provided insight into the plasticity at the level of the motor neuron (3, 16). Flies learn to reverse their visually guided flight control system if the feedback from the movement of objects is artificially reversed with respect to the intended movement of the stationarily flying animals (Heisenberg, personal communication). Bumblebees and honey bees learn to manipulate flowers (see Heinrich, this volume), and honey bees show many properties of schedule performance in appetitive learning known from operant conditioning

---

*The selective learning test following the 2 x 2 design consists of two CS's and two US's for which association is tested in all combinations, and it has been found that only a certain CS is associated with a certain US.

in mammals and birds (see above). Learning without an obvious relationship to positively or negatively reinforcing stimuli (latent learning) is common to invertebrates and to vertebrates during orientation in space and time. Communication in the dance behavior of bees may include a component of observational or instructional learning. The recruited bee learns the distance and direction of a food source from the dancing bee by following the turns during the dance performance, and it transposes this information into the control of the flight path outside the hive.

We have used many examples from honey bees to illustrate learning in an invertebrate. This may give a wrong impression because the bee's life strategy is rather versatile (see Gould and Heinrich, both this volume) as compared to most invertebrate species. A more typical case for an invertebrate would be an animal specifically adapted to the ecological niche which controls its life-style to a greater extent by prewired circuits. But is there anything else different in honey bees which makes them so capable of learning?

Ever since von Frisch began working with honey bees, they have been known to be effective learners, which makes them most suitable animals for behavioral studies (see (11)). Using the training technique, von Frisch and hundreds following him have examined in great detail the sensory capacity of the bee. The results of these studies provide a firm knowledge about what the bee sees, what it smells, how well it orients in time and space, and for what purpose all these perceptual capacities are used in its natural life. Since the bee communicates odor, nectar, taste, distance, direction, and profitability of a food source to its hivemates, one can also ask the bee what it knows about the food source even when it is not actually approaching it. More details of these fascinating capabilities can be found in the paper by Gould (this volume).

Is the honey bee a unique species in this sense? Other hymenopteran insects are social, too; they also have good vision including color vision, a good sense of smell, and a detailed knowledge of landmarks; in addition, they orient themselves to the sun and the polarized light pattern of the sky. What makes the bee so special and so useful for learning experiments is the combination of a number of sociobiological factors. The bee colony is perennial. Outside the tropical areas this means that bees have to collect enough food to enable them to survive a sometimes long winter in a hive which always has to be kept at a constant high temperature. Foraging efficiency is enormously increased by the dance communication.

Each individual foraging bee is informed about its foraging effectiveness by three means: a) the vigor with which the young hive bees collect the nectar from the returning forager, b) the observation of other dancing bees, and c) tasting their nectar samples. Thus, the individual bee can afford to be very precise during its foraging flight and stick to the previously rewarding food source. As opposed to bumblebees, wasps, ants, and solitary bees, most honey bee workers (except for the scout bees) have no need to probe around to find out whether they are still foraging on the most profitable food source. This behavior makes the bee an extremely useful animal for learning experiments, because it will display precise choice behavior under both experimental and natural conditions.

There may be nothing more that is special about the honey bee. The bee may even be inferior to certain bumblebee species with respect to motor learning. In particular, there is no indication whatsoever that specific functions of brain structures are different in the bee as compared with other insects.

ARE THERE HIGHER COGNITIVE PERFORMANCES IN INVERTEBRATES?

When an animal's behavior is attributable to learning, it must be assumed that the animal possesses some representation of past experience upon which performance is dependent. From such an abstract starting point one is led to ask for, among other things, a characterization of the basic representational entities (what is encoded), a description of the manner in which representations are organized (what the structure of memory is), and specification of the operations that are performed on them (how intelligent performance is generated). It might be thought, however, that the learned behavior of invertebrates would provide little challenge in relationship to such questions, perhaps being easily addressed by some relatively simple theory of knowledge. This is not so, at least in the case of the honey bee. At issue is how the bee is able to orient in space and time in relationship to nest sites and food sources, often appearing to perform mental operations on acquired information in the course of adaptive problem solving.

The honey bee locates the nest site and food sources very effectively by using both landmarks and celestial cues for orientation. The sensory capacities involved in this orientation have been well documented in von Frisch's famous detour, displacement, and competition experiments

(11). The results of these experiments and others (see Gould, this volume) have led to the suppositions that the bee has a representational system that is map-like and that it can carry out cognitive operations that are akin to geometric calculations. But, as in the case of Tolman et al.'s (28) supposition that rats' ability to negotiate a maze is best understood in terms of a cognitive map, there are alternative interpretations, e.g., that the bees' representations consist of a series of "snapshots."

Further research on this question promises to be informative. For example, we discussed an extension of Gould's lake experiment (see Gould, Fig. 3, this volume) which rests on the assumption that bees leave a vacant space in their cognitive map for an area of which they have no experience (e.g., on a lake). Bees with such a hole in their map should not fly to a location mapped within that hole when a redirected dancer is recruiting for it as a feeding place. In contrast, unexperienced bees without a relevant map should be ready to fly to such an area. It is hard to design a decisive experiment to demonstrate bees' possession of a map-like representational system. Furthermore, it could be taken as an oversimplification to assume that a single orientation system is involved. It is well-known from studies of humans that orientation is possible without a mental map. Comparably, the bee may sometimes behave as though it holds and is dependent upon a map, and in other circumstances it may not.

ECOLOGICAL ASPECTS OF LEARNING

The ecological approach to learning is, of course, an evolutionary approach. Specific learning capacities are studied best under controlled field or even semi-laboratory conditions in order to make comparisons between different species and to correlate them with the environmental conditions in which these species live. From this one tries to derive hypotheses about possible adaptive functions of these specific learning performances, which then have to be tested again in controlled experiments. Ultimately, we thus try to estimate the fitness benefits derived from possessing these abilities. Attempts to explain why specific learning capacities have evolved in a given species lead in turn to efforts to derive more general rules, e.g., relating particular types of life-styles, life cycle strategies, ecological conditions, social organizations, reproductive modes, foraging types, etc., with particular characteristics of preprogrammed vs. learned performances.

What these rules could be is only slowly emerging (see papers by Gould and Marler, Heinrich, and Shettleworth, all this volume). On a very generalized level (e.g., comparing generalist vs. specialist life strategies, short- vs. long-lived species, etc.) rules may be too "soft" or even trivial to be of much use. They have to be formulated in much more specific terms to be useful, i.e., to allow testable predictions.

So far we can see that the common denominator for the solution to the question, "under which conditions should a behavior be genetically closed, and when and in what way should provision be made for learning?" seems to be related mainly to the probability with which an object will be accessible when it is needed ("ecological predictability"). The more certain the occurrence of the goal of a behavior (e.g., food, mate, etc.) is, the less plastic the behavior may be, and the more likely it is that a genetically controlled subroutine can be used to reach the goal. Few comparative studies have been carried out so far, and only a few assumptions can be made which need experimental testing. For example, communication between the sexes of a species during courtship should be highly genetically controlled, whereas individual recognition within species has to be learned, even more so with increasing intraspecific competition for limited resources. Animals living in societies can make much better use of learning than can solitary animals, e.g., since the food supply changes unpredictably during the lifetime of the members of a colony and since nest mates can communicate information about newly discovered food sources.

Learning about the environment can be particularly useful for the central place foragers such as social wasps, bees, or ants, which collect and store as much food as possible when it is available and then live with their brood on the harvest at less favorable times. Many solitary wasps and bees can be genetically highly specialized with respect to their prey items or the flowers they visit, but they are still remarkable in their ability to learn the features of their nest site and of the flight routes leading to them. Therefore, the relevant issues may be the specific form of brood care, the length of seasonal activity, foraging, etc. For example, with localized nest sites where the brood has to be continuously supplied with food and with the problem of the mother having to discriminate between her own brood and those of other females in the same breeding colony, ecological conditions and necessities for learning - about a profitable crop site, the way home, the breeding site, and the identification of brood, etc. - will be quite distinctly and predictably different from species with a different life-style. Of course, this is no one-way relationship: just as the social life provides for many opportunities that make learning adaptive, superior learning capacities also allow advanced forms of social behavior.

Other factors that determine the balance of learning vs. innate control of behavior seem to be of much less importance than determinants of whether and when behavior shall be closed or open to learning (21). Brain size was traditionally thought to be a limiting factor for learning. But just consider what a bee learns with its 1 mm^3 large (small) brain, and you may wonder what you are doing with the rest of your 999,999 mm^3. There is evidence for a correlation between the volume of mushroom bodies - the highest-order neuropile in the insect brain - of various insect species and the number of different food sources on which each species lives (15). The discussion, along with one of the exceptions to this rule, exemplifies how difficult the interpretation of such correlations can be. Social roaches and termites eat "only" wood but have very large mushroom bodies. First, this observation supports the old findings that there is a correlation between mushroom body size and social living, but this correlation has not led us very far - the determining factors are unknown. Second, we may be very much biased in considering "wood" as a restricted menu. Third, perhaps termites must learn the elaborate mazes of their nests, as do rats.

Thus, to regard "predictability" of the environment as a precondition for making learning both more beneficial and less costly (see (17)) than genetic programming may be necessary but insufficient to explain the evolution of specific learning adaptations. The predictability of the environment may be just the same for the honey bee as for the fruit fly; however, a number of life-style adaptations with respect to brood care, foraging, nesting, food storage, etc., allow and even force the bee to make much more effective use of learning (i.e., making learning more fitness-relevant) and thus make the bee quite different from the fly as a learner. The bee may well be supplied by the same basic learning mechanisms as the fly, but depending on many details of life strategies, it makes use of learning in much more diversified ways than the fly. So it may be that life strategy adaptations, as it were, set the stage for making the bee adaptive to open up genetic programs at well-defined steps and in certain behavioral conditions allowing learning to take over adaptive modification.

The matter of ecological predictability appears in another context, e.g., in the genetic characteristics of one's own offspring or of one's genetically related co-members in a social group of sexually reproducing species. Brood care is a heavy investment and should preferably be directed to closest kin. Kin recognition necessarily involves learning because of the similarities of all offspring of the same species and the basic unpredictability of those features which make one individual's offspring different from those of members of the same species due to sexual recombination (14). In desert wood lice, for example, learning of the individual odor marks has been found to be most important for kin recognition and organization of social behvior (20).

Therefore, what the ecological-evolutionary comparative approach to learning tries to do and can do is to relate causally the specific learning dispositions of different groups and species with their particular life strategy adaptations, thus answering the question as to why they are as we find them to be. It is very much hoped that ethology and sociobiology will address these questions in the future, since as yet very little empirical material is available. Again, it is our belief that invertebrates, specifically the hymenopteran insects, will be the most rewarding study objects due to their rich diversity of life-styles.

CONCLUSION

Three major themes guided this discussion on the biology of invertebrate learning: physiology of elementary learning processes, invertebrate learning and learning theory, and ecology and invertebrate learning. It is superfluous to mention that each of these research areas has its own history of ideas, its own concepts, and its own vocabulary, all of which sometimes made understanding difficult. The discovery we made was that common ground is emerging for all three disciplines. This common ground is a firm basis for future work, building especially on neurobiological model systems of learning, such as that of Aplysia. Cellular analyses of learning phenomena force us to reconsider our terminology and theories. They offer the chance for the reconciliation of otherwise diverging disciplines. Many controversies and battles over meanings disappear when we have the neural substrate in our hands. We can then focus our efforts on the really important questions. So far we have not reached this firm ground, but we already sense an improvement. Learning studies in invertebrates are moving in the right direction.

REFERENCES

(1) Alkon, D. 1983. Learning in a marine snail. Sci. Am. 249(1): 71-84.

(2) Alloway, T.M. 1972. Learning and memory in insects. Ann. Rev. Entomol. 17: 43-56.

(3) Bullock, T.H., and Quarton, C.C., eds. 1967. Simple Systems for the Study of Learning Mechanisms. Cambridge, MA: MIT Press.

(4) Carew, T.J.; Hawkins, R.D.; and Kandel, E.R. 1983. Differential classical conditioning of a defense withdrawal reflex in Aplysia californica. Science 219: 397-400.

(5) Couvillon, P.A.; Klosterhalfen, S.; and Bitterman, M.E. 1983. Analysis of overshadowing in honeybees. J. Comp. Psychol. 97: 154-166.

(6) Davis, W.J., and Gilette, R. 1978. Neural correlate of behavioral plasticity in command neurons of Pleuro-branchaea. Science 199: 801-804.

(7) Eisenstein, E.M. 1967. The use of invertebrate systems for studies on the basis of learning and memory. In The Neurosciences, A Study Program, eds. G.C. Quarton, T. Melneshuk, and F.O. Schmitt, pp. 653-665. New York: Rockefeller University Press.

(8) Elsner, N. 1983. Neuroethological approach to the phylogeny of leg stridulation. In Neuroethology and Behavioral Physiology, eds. F. Huber and H. Markl, pp. 54-68. Heidelberg, New York, Tokyo: Springer Verlag.

(9) Erber, J. 1981. Neural correlates of learning in the honeybee. TINS 4: 270-273.

(10) Forel, A. 1910. Das Sinnesleben der Insekten. München: Reinhardt.

(11) Frisch, K. von. 1967. The dance language and orientation of bees. Cambridge: Harvard University Press.

(12) Haralson, J.V., and Groff, C.I. 1975. Classical conditioning in the sea anemone, Cribrina xanthogrammica. Physiol. Behav. 15: 455-460.

(13) Hawkins, R.D., and Kandel, E.R. 1984. Is there a cell biological alphabet for learning? Psychobiol. Rev., in press.

(14) Hölldobler, B., and Michener, C.D. 1980. Mechanisms of identification in social hymenoptera. In Evolution of Social Behavior: Hypotheses and Empirical Tests, ed. H. Markl, pp. 35-58. Dahlem Konferenzen. Weinheim: Verlag Chemie.

(15) Howse, P.E. 1974. Design and function in the insect brain. In Experimental Analyses of Insect Behavior, ed. L. Barton-Browne, pp. 180-194. Berlin, Heidelberg, New York: Springer Verlag.

(16) Hoyle, G. 1979. Mechanisms of simple motor learning. TINS 2: 153-155.

(17) Johnston, T.D. 1982. Selective costs and benefits in the evolution of learning. Adv. Study Behav. 12: 65-106.

(18) Kandel, E. 1978. A Cell-biological Approach to Learning. Bethesda, MD: Society for Neuroscience.

(19) Kimble, G. 1961. Hilgard and Marguis' Conditioning and Learning, 2nd ed. New York: Appleton-Century-Crofts.

(20) Linsenmair, K. 1984. Individual and family recognition in subsocial arthropods. In Experimental Behavior Ecology and Sociobiology, eds. B. Hölldobler and M. Lindauer. Stuttgart, New York: G. Fischer Verlag.

(21) Mayr, E. 1974. Behavior programs and evolutionary strategies. Am. Sci. 62: 650-659.

(22) McManus, F.E., and Wyers, E.J. 1979. Olfaction and selective association in the earthworm, Lumbricus terrestris. Behav. Neur. 25: 39-57.

(23) Menzel, R. 1983. Neurobiology of learning and memory: The honeybee as a model system. Naturwiss. 70: 504-511.

(24) Rescorla, R.A., and Wagner, A.R. 1972. A theory of Pavlovian conditioning: Variations in the effectiveness of reinforcement and non-reinforcement. In Classical Conditioning II, eds. A.H. Black and W.F. Pokasy, pp. 64-99. New York: Appleton-Century-Crofts.

(25) Ross, D.M. 1965. The behavior of sessile coelenterates in relation to some conditioning experiments. Anim. Behav. 13(Suppl.1): 43-57.

(26) Skinner, B.F. 1938. The Behavior of Organisms. New York: Appleton-Century-Crofts.

(27) Thorpe, W.H. 1963. Learning and Instruction in Animals. London: Methuen and Co.

(28) Tolman, E.C.; Ritchie, B.F.; and Kalish, D. 1947. Studies in spatial learning (I). J. Exp. Psychol. 36: 221-229.

(29) Tolman, E.C.; Ritchie, B.F.; and Kalish, D. 1947. Studies in spatial learning (V). J. Exp. Psychol. 37: 285-292.

(30) Young, J.Z. 1966. The Memory System of the Brain. Berkeley, CA: University California Press.

The Biology of Learning, eds. P. Marler and H.S. Terrace, pp. 271-288. Dahlem Konferenzen 1984. Berlin, Heidelberg, New York, Tokyo: Springer-Verlag.

The Natural History of Bird Learning

K. Immelmann
Lehrstuhl für Verhaltensphysiologie, Fakultät für Biologie
Universität Bielefeld
4800 Bielefeld 1, F.R. Germany

Abstract. Birds are characterized by the rapidity of their ontogenetic development as compared, for example, to mammals. Many of their learning processes, therefore, proceed rather quickly. They are called imprinting or imprinting-like processes. Birds also possess advanced learning and memory capacities in areas such as orientation learning, song learning, food caching, and individual recognition. The occurrence of sensitive phases in bird learning may serve to concentrate learning capacities to periods of optimal learning opportunity. Mechanisms of early learning include hormonal control and a phase-specificity in the neuroanatomical development of certain areas in the brain. Behavioral data on sensitive phases and neuroanatomical data on brain development suggest that the great stability found as a result of imprinting may be due to morphological alterations. After the end of the sensitive phase, morphological plasticity is reduced and environmental stimulation only leads to biochemical or submicroscopical alterations of the nervous tissue. It is argued that the possible "special nature" of imprinting may relate more to the form of information storage than to the mode of acquisition of such information.

INTRODUCTION

For the evaluation of the natural history of learning in a particular taxonomic group of animals, it is necessary to consider the specific biological requirements of these species and the adaptive pressures they face. It is also necessary to look at the characteristics of their ontogenetic development which are, in turn, adaptations to such requirements. Thus, to characterize the specific features of bird learning, the special biological requirements of birds and the details of their individual development should be given special attention, especially in comparison with the other class of homeothermic vertebrates, the mammals. This type of characterization will facilitate a biological explanation of some of the peculiarities which are correlated with the occurrence of learning in birds.

The most remarkable characteristic of the ontogenetic development of birds is its rapidity: Birds may reach their adult size and weight within about 1% of their total life expectancy, whereas some mammals need 30% and more, as is the case for most of the higher primates. Such speed in general development has commonly been understood as an adaptation to a quick acquisition of the ideal ratio between body weight and a constant wing surface. This process must necessarily also involve some behavioral characteristics of birds, including the development of learning and memory. It is certainly not by chance, therefore, that the much-discussed phenomenon of imprinting, being a comparatively rapid and early learning process, as compared, for example, to the evolution of prolonged behavioral flexibility in mammals, not only was first discovered but is also especially widespread in avian species (14, 21).

A second characteristic of development which birds share with the majority of poikilothermic vertebrates is their oviparity. Birds lay eggs which usually are incubated by one or both parents. As with the embryos of other egglaying species, the pre-hatching stage of birds seems to be more open to various influences from the environment than are comparable prenatal developmental stages in viviparous animals, particularly in mammals. For example, intensive acoustic contact has been described between embryos as well as between the mother or both parents and the developing embryo (15). (In most birds the father participates in parental care to a larger extent than in most species of mammals (19)). Birds, therefore, represent ideal objects for the study of such phenomena as prenatal learning or the interaction between genetic and environmental influences in general ((24); and see Marler, this volume).

Examples of behavioral areas in which the environmental demands on bird learning, as compared to learning in other animals, are very high and in which the above-mentioned interplay is particularly apparent include - to select only a few - orientation, food storage, song development, and kin recognition. Of these areas, acoustic communication and song learning have been studied most extensively, and the large amount of data available has provided an essential contribution to the general elucidation of the nature of bird learning. Consequently, they will be treated separately (Marler, this volume), whereas the other areas will be discussed in this paper, but in a less extensive way.

One interesting general difference, however, should be mentioned here. This is the difference between birds and mammals with regard to the role played by learning during the development of acoustic communication. In birds, particularly in the oscines, many characteristics of their vocalizations, as expressed not only in the song but also in several types of calls, need to be acquired entirely or at least in part. Within the birds, there even seems to be a phylogenetic trend from a rather definite to a very weak genetic determination in the sense of "closed" and "open" programs, as discussed by Mayr (23). In mammals, by contrast, even in the most advanced forms (e.g., nonhuman primates), an amazingly large number of purely innate vocalizations has been found. In contrast to birds, deprivation experiments with mammals do not result in permanent alterations in the development of the physical structures of vocalizations, and calls of juveniles have been found to be identical or similar to those used by the adults. In mammals it seems to be the social context in which to use a vocalization rather than its acoustic features where learning is more heavily involved (7, 9).

EXAMPLES OF BIRD LEARNING

In principle, learning in birds does not differ from other species, and the main categories of learning which are known in other vertebrates have also been described in birds ((7); and see Lea, this volume). Minor differences, however, do occur. They pertain mainly to the temporal course of learning, its rather pronounced phase-specificity in some cases, and to the special learning and memory capacities correlated with specific requirements of avian biology, e.g., in the areas of orientation and food storage.

Learning and Orientation

One of the areas in which environmental demands are particularly strong in birds is orientation. Migratory birds are able to find their way between their breeding and their wintering quarters, which are often thousands of miles apart. Furthermore, the journey often leads across unfamiliar country or across long distances over the ocean.

In long-distance orientation even small angular errors may mean that the destination is missed, which - especially in species or populations with small breeding and/or wintering ranges - will have lethal consequences. It is to be expected, therefore, that selection pressure has favored the evolution of very accurate mechanisms of orientation in birds. Such mechanisms have indeed been found, and the information available indicates that there are very strict genetic constraints on learning, resulting in a tight interplay between inherited and experiential factors. Such interplay, which probably guarantees an even higher degree of accuracy than a totally preprogrammed system, is found mainly in the temporal course of the ontogeny of orientational mechanisms as well as in the development of route knowledge.

Two important mechanisms of long-distance migration for diurnal and nocturnal migrants, respectively, are the sun and the star compass. Experimental studies, mainly in planetariums, have shown that long-distance migrants must learn to read the star compass. Such learning, however, is possible only during a very brief period of time between shortly after fledging and the start of their first autumnal migration. If a bird is exposed to the natural sky prior to the time of departure, it will orient properly, but if such exposure takes place at a later time, it will never learn to use the star compass even after repeated exposure (17).

Recent studies indicate that for the sun compass of a diurnal bird, the pigeon, a similar temporal restriction in learning seems to exist, and that during the most sensitive age, a very brief exposure to the sun is sufficient to achieve normal sun compass behavior. This is a remarkable parallel to the phase specificity in filial or sexual imprinting (see below) (31).

A second example of the interplay between genetic and environmental factors comes from the development of route learning. Garden warblers were hand-reared under constant laboratory conditions. When they were tested in a circular test cage during their first fall migratory period, they showed a southwest preference in their migratory restlessness during August/September and a shift to a southeasterly preference late in September or early in October. This corresponds exactly with the time at which, under natural conditions, they arrive over southern Spain or northern Africa. In order to remain over continental Africa, they need to make the same change in flight direction. Obviously, these birds have a genetically determined program which includes the direction and the distance to be flown towards the relevant destination. Such an internal program may be the primary guide of the migratory bird for its first journey, during which it collects additional environmental information for all subsequent migratory flights. Such information will help, for example, in returning to the same wintering grounds and stopping at the same resting places en route, which are already familiar to the bird. Site tenacity for both winter quarters and resting places has been reported in several species of birds (10).

Food Storage and Foraging

Like orientation in migratory birds, the storage of food may place specific demands on cognitive abilities. Food storage has evolved in several species of woodpeckers and tits and particularly in corvids which, like the nutcrackers, live in areas with very severe winters and thus must survive through periods of unsufficient food supplies (29). In contrast to mammals (rodents) which are able to construct subterrestrial burrows and to store their food there, birds must cache food somewhere else, e.g., on or in the ground, in holes in or underneath the bark of trees. As a consequence, they must rely on very durable vegetable matter (seeds) and they need to hide them in small quantities or even in single pieces in order to prevent excessive loss of food if a hiding place is discovered by another individual. This, in turn, requires a large number of seed caches and thus a great deal of time and energy. Moreover, it presupposes high memory skills for the recovery of cache sites, especially if the ground is covered with snow, concealing most of the cues available when the food was stored in autumn. Nevertheless, for several species of corvids, stored seeds are the principle and sometimes even the only source of food during harsh winters.

Work on seed-caching behavior of nutcrackers has indicated that these birds use a series of clues to find unexposed caches. The main strategy seems to be a combination of orientational cues from the environment with information stored in memory. Such information may be stored in a hierarchical sequence, which comprises the general area, a particular patch within the area, and the specific cache within the patch. An organized memory pattern for recovering caches has a certain similarity to the "template" laid down during the acquisition of song in passerine birds ((1), and Marler, this volume).

Taken together, the performance of seed-caching birds has been found to be in accordance with the predictions of optimal foraging theory and indicates - together with a highly developed capacity for recognizing spatial relationships - prominent skills of individual memory with regard to both the number of details to be remembered and the duration of memory storage which may extend to up to eleven months. This is a similar duration as was found in some species of birds for songs even acquired in adult life. The further study of these birds will certainly help to elucidate some important general points of avian learning and memory.

The example described here refers only to a rather small number of birds. However, fairly pronounced learning and memory capacities in the area of food collection have also been described for many other species. If their search behavior directs them towards a well camouflaged or partially hidden but still abundant food source, they modify their search strategies accordingly by forming a "search image," which involves selective attention to the stimuli from this particular kind of food and also helps them to "overlook" other items. They also quickly learn to avoid noxious food and may retain the subsequent avoidance behavior for long periods (6).

IMPRINTING

In discussions of learning in birds, the term imprinting immediately comes to mind. For reasons indicated in the introduction, imprinting is particularly widespread in birds. This very accelerated and early learning process may have a selective advantage for organisms characterized by rapid ontogenetic development and by early dispersal of the young.

The "Classical" Cases

The two most intensively studied types of imprinting are filial imprinting, which determines a preferred target for the following response of young precocial birds, and sexual imprinting, which determines subsequent mate preferences in both precocial and altricial birds. These studies have revealed many details about the characteristics of early learning, extending the original, more narrow concept of imprinting proposed by Konrad Lorenz to encompass the general attributes of early learning in birds.

The reason for the more restricted nature of the original concept is quite obvious. Filial and sexual imprinting offer two particularly drastic examples of what has been described, relative to several less important criteria, as the two most prominent characteristics of imprinting in general. These are the existence of rather distinct sensitive phases, especially in filial imprinting, and a comparatively high degree of stability, particularly with regard to sexual preferences. Even these characteristics are now known to be less prominent than described initially. In filial imprinting, sensitivity for the establishment of preferences may last for days instead of hours, and in sexual imprinting, changes in preferences through subsequent experience may occur in adolescent and even - although only in a transient form - in adult individuals. Consequently, the previously used term, irreversibility, has been largely abandoned and replaced by more neutral terms such as permanence or stability (2, 14).

Such findings have led to much controversial discussion about imprinting, whether it is a special form of learning which is more or less different from other learning processes, or whether it is "only" a form of conditioning in a wider sense. Recent views suggest that it is neither. (This question is also addressed by Bateson, this volume.)

Imprinting-like Processes

Based on the results and conclusions of early research on imprinting, observational and experimental studies have indicated that, in addition to mate preferences and to preferences in connection with the following reaction, there are other developmental processes which take place during an early and limited stage during ontogenetic development and which lead to a comparatively stable result. Such processes include the establishment of preferences for habitat, locality, and food, the development of host preferences in parasitic species, and, in numerous species of passerine birds, the acquisition of a song-template (13).

Such findings have extended the original concept of imprinting, which may now be defined as an "early learning process with a rather stable result."

PHASE-SPECIFICITY OF LEARNING - THE ULTIMATE ASPECTS

The widespread occurrence of age-specific learning does provoke a consideration of the possible biological functions of privileged periods of learning in birds. A comparison of the data on sensitive phases in different species of birds, as well as in different functional systems within a species, reveals that the time course of sensitive phases seems to be highly adaptive. It is geared to both the particular time during which the opportunities for acquiring relevant information are favorable and to the specific age-requirements at which such information has to be available for the first time. Young precocial birds have to be able to recognize their mother or their parents as early as a few hours after hatching in order to be able to selectively approach and follow her or them. For altricial birds more time is available. Even for these species, however, a concentration of learning to early stages of development, before they have reached nutritional independence, certainly will be selected for because the young animal's opportunities to acquire knowledge in social and nonsocial domains necessarily will be much greater early in life than after it has left the family and the native breeding grounds. In addition, subsequent information might not only be more difficult to obtain, but it might also be "wrong," for example, if the bird is outside the reproductive season and has to live in suboptimal habitats or if, within the winter flocks, it comes into close contact with members of other species, subspecies, or populations. The strict restraints on plasticity may be an adaptation, therefore, to the rapid speed of development and to the high degree of mobility which, in turn, requires rapid learning during brief privileged periods as well as protection of the results of early learning against some of the subsequent environmental influences (14).

Support for this view comes from the fact that in the species studied so far, the termination of sensitive phases seems to be geared very closely to the age at which the young bird leaves its parents or their breeding grounds. The zebra finch has a very brief sensitive phase for sexual imprinting. In the vast majority of individuals, this phase ends at the very latest at 40 days of age, and the young stay close to the parents only for about five weeks. In the greylag goose, in which the parent-offspring bond lasts for about ten months, the sensitive phase lasts for at least 150 days. Sensitivity to star patterns in indigo buntings (see above) terminates exactly at the time when the young depart from the natal area, and the collared flycatcher likewise needs a short period of exposure just before departure for autumn migration to be able to return to the breeding grounds during the second year (12).

There is also evidence from song learning in birds such that much variety has been found in the occurrence and duration of sensitive phases. These phases range from a learning period of only a few weeks' duration, with subsequent stability of the song repertoire, to an ability to adopt new elements and phrases in later years. Brief sensitive phases have been found, for example, in species which disperse and form flocks very early and where the danger of "misimprinting" is high. An extension of the learning period into the male's first spring occurs in territorial species which disperse but soon settle in a breeding area to which they return in later years. Their extended sensitivity provides the opportunity to adopt new elements from territorial neighbors. This may help to increase the overall song repertoire and to match the neighbor's song with which the yearling interacts socially (18, 25, 33).

A particularly impressive example of the close adaptation of sensitive phases to specific ecological conditions comes from Nicolai's work on the parasitic African viduine finches. In the village indigobird the sensitive phase for learning the acquired parts of the song phrase, which imitate the begging calls of the nestlings of the host species, lasts from the second to the seventh month of life. Such timing enables young birds born early in the season to learn their song from the young of another pair of hosts shortly after fledging. Likewise, it provides young males born late during the brief breeding season with the opportunity to acquire their song from the offspring of the first broods of the following season, which starts about six months after the end of the first (Nicolai, in preparation).

These examples indicate the existence of strong selection pressures which favor great sensitivity to certain environmental stimuli and maximal learning capacities during early stages of development. Furthermore, the available evidence suggests that different ultimate factors have not produced a uniform sensitive phase phenomenon, but rather have led to a variety of specific adaptations.

Phase-specific learning with subsequent stability, so widespread in birds, presents a combination of attributes: It offers the benefits of early ontogenetic plasticity and is thus quickly adaptable to any changes in the social or nonsocial environment of the organism; it may ensure the availability of relevant information at or before the time of its first application, and it has stable and long-lasting effects for any one individual. From the evolutionary point of view, these avian learning processes may have two possible consequences: They may lead to the formation of distinct habitat and other ecological preferences and - once natural selection has led to the evolution of slightly different gene pools adapting groups of individuals to local conditions - they are able to preserve such gene pools by means of selective mating. The result will be a division of populations into a continuous or mosaic system of subpopulations with habitat-linked differences in various characters, as it has been described for numerous species of birds (13).

MECHANISMS OF LEARNING - THE PROXIMATE SIDE

Early learning in birds has also been used as a tool to study the mechanisms which are involved in the regulation of learning processes. Although the issue is far from being resolved, the available information suggests that the two regulatory systems, the hormonal and the central nervous system, may have specific functions, e.g., in the regulation of sensitive phases. For filial imprinting it has been suggested that an increase in the corticosterone level in the blood is one of the factors which restricts the sensitive phase for filial imprinting in ducklings (20). Song learning is likewise under hormonal control. Since Nottebohm provided the first evidence that rising testosterone levels may affect song development in the chaffinch, similar correlations have been found for other species. In the zebra finch the first occurrence of song is correlated with a peak in testosterone production. In male canaries a peak of estradiol corresponds to the onset of full song. Such first results suggest that hormones are involved in the development of song but that the effects may vary from species to species (27).

For sexual imprinting, preliminary evidence suggests that in male zebra finches, androgens have a positive, whereas estrogens have a negative, effect on the establishment of sexual preferences during the sensitive phase. Testosterone and 17 β-estradiol treatment of young males between days 8 and 35, i.e., during the sensitive phase for sexual imprinting, resulted in stronger or weaker preferences for the species of the (foster) parents, respectively, as compared to untreated birds. A quantitative study of developmental changes in testosterone levels in the blood plasma using radioimmunoassay also points to a temporal correlation between peaks in sensitivity to social stimuli and maxima in the production of testosterone (26, 28). In Japanese quails, castration of young males had no effect on sexual imprinting. Castrated males which were treated with testosterone proprionate in adulthood did not differ in their preferences for the imprinted object from control animals (11). Future studies should elucidate whether such differences represent true species differences or whether there are differences between hormonal influences on the sexual imprinting mechanisms and on the executive system underlying sexual behavior.

Evidence for possible regulatory functions of the central nervous system on imprinting comes from both biochemical and neuroanatomical work. The detailed studies by Bateson and Horn (3) on filial imprinting in domestic chicks are reported by Bateson (this volume).

Recent neuroanatomical studies indicate that the greater readiness to learn early in life may be correlated with the formation of new synapses and dendritic branching, with the rearrangement of synapses optimizing the probability of contact for some neurons, and/or with the disconnection of neurons and their physiological death.

In a comparative study on mallards, silky chicks, and zebra finches, developmental degeneration processes in the forebrain, especially in the laminae, have been followed. Using a new technique of impregnating the degenerating axon terminals, it was shown that in all three species degeneration occurs shortly before and in temporal correlation with the sensitive phase for filial and sexual imprinting, respectively. In the two precocial species the maximum of degeneration within the tectum opticum, a primary visual center of the mesencephalon, occurs about five days before the sensitive phase for filial imprinting, i.e., prenatally. In the altricial zebra finch, in contrast, it occurs postnatally, about ten days before the onset of the sensitive phase for sexual imprinting. The longer time lag in the latter case probably reflects a slower speed of maturation in the CNS of an altricial bird. A similar sequence in time occurs in the degeneration processes within the higher-order visual systems of the telencephalon, which only shortly precedes the maximum of sensitivity to imprinting stimuli in all three species (30). Electron microscopic studies have shown that degeneration in different regions of the zebra finch brain affect already established synaptic contacts. By this process, the ability of the innervating neuron to innervate other postsynaptic sites subsequently may be restored. By this process a kind of secondary neuroplasticity is produced, which may be a precondition for rewiring and for a new structural stabilization of neuronal connections. This rewiring process may contribute to the determination of the conditions under which imprinting takes place (5, 32).

Additional evidence for a link between imprinting and the occurrence of phase specificity in neuroanatomical development - although not collected in a bird species - comes from studies on the visual cortex of adult cats. Binocular neurons in area 17 can be driven by visual stimulation of the left as well as of the right eye. If the eye of a kitten is artificially closed during a brief period early in life, most neurons lose their binocular characteristics and can be driven only by the non-deprived eye. Similar monocular deprivation in adult cats, in contrast, does not alter the distribution of ocular dominance, even if it is maintained over a much longer period of time.

This is an almost perfect parallel to at least sexual imprinting in birds, as it not only includes the occurrence of a sensitive phase but also a high degree of stability after its end. Such comparisons between sensitive phases for imprinting and the development of certain parts of the CNS can even be carried one step further. Comparing the time course of the efficiency of external stimulation on sexual imprinting in zebra finches, sexual imprinting in Japanese quail, and the development of ocular dominance in the cat, Bischof (4) developed a model demonstrating remarkable similarities in a number of details: Given standardization of developmental rates, if one superimposes the birth dates as well as the age at which the animal reaches sexual maturity, it becomes evident that the ascent, peak, and decline of the curves for all three processes match very closely. In addition, the storage of information in all three cases seems to be comparably stable: Environmental influences after the end of the sensitive phase, even if they last for a longer period of time than the original stimulation, are at best able only to augment the previously stored information, and they cannot alter or eliminate the influence of the primary stimulation.

The elucidation of possible correlations between imprinting and neuroanatomical development may also serve to focus interest on the possible "special nature" of imprinting as compared to other learning processes. The available evidence points to the following possibility: Early sensory stimulation may lead to morphological alterations in particular brain areas, but after the end of the sensitive phase, morphological plasticity is reduced and environmental stimulation only leads to biochemical and perhaps submicroscopical alterations of nervous tissue. As morphological alterations might be more stable than biochemical ones, this may be the reason for the remarkable permanence of the effects of sensory stimulation during the sensitive phase.

If this is the case, imprinting - as already inferred from behavioral observations - cannot be regarded as a separate learning process. Its main characteristic, the impressive permanence of learning process outcomes, may be seen rather as the result of increased neuronal plasticity during the sensitive phase and of a rapid decrease in plasticity after its end. With this hypothesis, one could explain the great diversity of imprinting and "imprinting-like" processes described in the literature which, at first sight, seem to comprise very different forms of learning (14). Consequently, it could be the type of information storage more than the kind of acquisition of such information which is different in imprinting and in "other" learning processes (4).

LEARNING PROGRAMS

Learning behavior in birds has been shown - as already mentioned for the development of star compass orientation - to be particularly well suited for elucidating the interaction between genetic programs and environmental influences during behavioral development. The most frequently cited examples refer to what has commonly been called learning dispositions. A prominent case is the individual recognition of the offspring by their parents and the stage of the reproductive cycle during which it occurs (Gould and Marler, this volume).

Another example of species-specific predispositions and of the interaction between genetic and environmental factors is again provided by the phenomenon of imprinting, both with regard to the regulation of sensitive phases as well as to the nature of the object of imprinting. In view of the possible correlations between sensitive phases and neuroanatomical development, it seems likely that the critical mechanism underlying the temporal limits of plasticity might consist of genetically determined programs for the time course of neuronal plasticity. These programs could determine at what age and for how long morphological changes are possible and thus put constraints on the age period during which information can be stored with a high degree of permanence.

The great amount of individual variation, however, which has been found with regard to the termination of sensitive phases, indicates that such programs are rather flexible and only determine the limits within which imprinting can appear, whereas the nature and time of occurrence of relevant stimuli is decisive for the exact age at which a particular preference is in fact established (14). These considerations are in accordance with the notion that imprinting is a preemptive self-terminating process (2). It is a challenging thought that the occurrence of adequate stimuli as compared to less acceptable models might accelerate the above-mentioned neuroanatomical processes within the genetically determined limits. This would provide another, although still much less well analyzed, example of neuroselectional learning in birds.

Evidence for different suitability of environmental stimuli for the establishment of preferences is indeed available: In both filial and sexual imprinting, preferences are developed more easily for their own than for another species. If no conspecifics are available, preferences for similar species are established more readily than for dissimilar ones, and the young of precocial birds imprint on living animals more easily than on models moving in a circular apparatus. Apparently, there are some genetically determined biases which facilitate the establishment of preferences for an individual's own species and make imprinting on another species more difficult (12).

It seems to be possible that the selectivity of imprinting reflects its evolution from a stage where parental or species recognition was entirely independent of experience, as still - although probably as a secondary development - must essentially be the case in brood parasites. The general occurrence of very intensive parental care in most modern birds, which has led to the reliable availability of the parent(s) at the time of hatching, constitutes a relaxation of selection allowing the preprogrammed portions of species recognition to decrease and to be supplemented by learning to participate in the many advantages of "open programs" (23). The integration of both elements may lead to a "learned identification program" similar to the "learned motor program" which characterizes song development in birds (Marler, this volume). The nature of such innate preferences, however, has not yet been studied in detail. Their biological significance certainly is to provide for some additional security, so that under natural conditions the young animal is always imprinted on its own species. This is particularly important in situations where several closely related species occur in the same area (2, 12).

The innate learning cues are supplemented by a second set of genetical constraints on learning, influencing the onset, time-course, and termination of sensitive phases, which, as described above, provide another mechanism for selecting the appropriate stimuli for establishing preferences. As in song learning in birds, therefore, the behavioral plasticity of "open programs" for the establishment of stimulus preferences is guided by an important predetermined frame.

The strength of such constraints is different in different species, and, in turn, this may again be an adaptation to the specific needs of a species. An example is offered by two closely related species of ducks, the redhead and the canvasback. Visual imprinting occurs in both species. But only in canvasbacks are the responses selectively enhanced by calls presented in the absence of the visual stimulus. Such differences in strategies can easily be understood. In the nonparasitic canvasback, strong responses to maternal calls when the mother is heard but not seen could facilitate cohesion of the family unit. By contrast, in the semiparasitic redhead, a reduced auditory responsiveness to calls of the foster parents could help to create the preconditions for the necessary subsequent separation of the young from members of the foster species (22).

The most elegant studies on the interactions between genetic and environmental factors have been done with embryos. Young ducklings hatched in isolation in incubators are able to respond preferentially to the maternal call of their own species as compared to the call of another species, even if they have never heard the species-specific call before. This finding strongly suggests a certain genetic background for such selectivity. However, as the developing embryo is, of course, able to listen to its own vocalizations which begin several days before hatching, some exogenous influence via feedback through the auditory system cannot be excluded. To examine such a possibility, the preferences of ducklings were tested which were mechanically prevented from vocalizing prior to the onset of their own vocalizations. Even these ducklings showed a high degree of preference for the species-typical maternal call, demonstrating even more clearly the high degree of innate predispositions for selective responses. The devocalized-isolated ducklings, however, presumably due to an overlap in certain acoustic features, failed in one of the discrimination tests, whereas the sham-operated controls did not. It can be concluded that selectivity in the responses to auditory stimulation is not as perfect if no auditory experience whatsoever is available, and that even such strongly predetermined responses are open to some, although minor, enhancement through sensory stimulation. Such stimulation may be remarkably brief: The deficit in discrimination abilities did not appear if the devocalization procedure was delayed such that the embryos had about 24 hours of exposure to their own vocalizations before they were muted (8). Subsequent studies have provided additional evidence for the fact that the mallard maternal call is uniquely attractive to naive mallard and Peking ducklings and that it obviously plays the role of an "innate perceptory pattern" or "releaser" in the Lorenzian sense (16). Results like these indicate how many fruitful insights a detailed study of learning programs is able to provide. They contradict the occasionally expressed opinion that reference to the genetic background hampers rather than promotes the elucidation of developmental processes.

The main purpose of this paper was to indicate where learning in birds is especially advanced and how far it differs from learning in other species. I have also tried to indicate areas, such as the possible structural background of early learning and the interplay between genetic constraints and environmental influences, where studies on avian learning may contribute to a general elucidation of learning phenomena.

Acknowledgements. I would like to thank R.P. Balda, P. Bateson, H.-J. Bischof, S.E.G. Lea, E. Pröve, J.P. Rauschecker, J.E.R. Staddon, and H.S. Terrace for their comments on an earlier version of the paper and P. Marler for his help with the final manuscript.

REFERENCES

(1) Balda, R.P. 1980. Recovery of cached seeds by a captive Nucifraga caryocatactes. Z. Tierpsych. 52: 331-346.

(2) Bateson, P. 1979. How do sensitive periods arise and what are they for? Anim. Behav. 27: 470-486.

(3) Bateson, P.P.G.; Horn, G.; and Rose, S.P.R. 1975. Imprinting: Correlations between behavior and incorporation of (14 C) uracil into chick brain. Brain Res. 84: 207-220.

(4) Bischof, H.-J. 1983. Imprinting and cortical plasticity: A comparative review. Neurosci. Biobehav. Rev. 7: 213-225.

(5) Bischof, H.-J., and Herrmann, K. 1984. Ontogenetic development of sensory and song control areas in the zebra finch brain. Behav. Brain Res., in press.

(6) Curio, E. 1976. The Ethology of Predation. Berlin, Heidelberg, New York: Springer.

(7) Delius, J.D. 1983. Learning. In Physiology and Behaviour of the Pigeon, ed. M. Abs, pp. 327-355. London: Academic Press.

(8) Gottlieb, G. 1975. Development of species identification in ducklings. I. Nature of perceptual deficit caused by embryonic auditory development. J. Comp. Phys. Psych. 89: 387-399.

(9) Gould, E. 1983. Mechanisms of mammalian auditory communication. In Advances in the Study of Mammalian Behavior, eds. J. Eisenberg and D. Kennan, pp. 265-342. Spec. Publ. American Society of Mammologists.

(10) Gwinner, E., and Wiltschko, W. 1980. Endogenously controlled changes in migratory direction in the garden warbler, Sylvia borin. J. Comp. Physiol. 125: 267-273.

(11) Hutchison, R.E., and Bateson, P. 1982. Sexual imprinting in male Japanese quail: The effects of castration at hatching. Dev. Psychobiol. 15: 471-477.

(12) Immelmann, K. 1972. Sexual and other long-term aspects of imprinting in birds and other species. Adv. Study Behav. 4: 147-174.

(13) Immelmann, K. 1975. Ecological significance of imprinting and early learning. Ann. Rev. Ecol. Syst. 6: 15-37.

(14) Immelmann, K., and Suomi, S.J. 1981. Sensitive phases in development. In Behavioral Development, eds. K. Immelmann, G.W. Barlow, L. Petrinovich, and M. Main, pp. 395-431. New York: Cambridge University Press.

(15) Impekoven, M. 1976. Prenatal parent-young interactions in birds and their long-term effects. Adv. Study Behav. 7: 201-253.

(16) Johnston, T.D., and Gottlieb, G. 1981. Visual preference of imprinted ducklings are altered by the maternal call. J. Comp. Phys. Psychol. 95: 663-675.

(17) Keeton, W.T. 1980. Avian orientation and navigation: New developments in an old mystery. In Acta XVII Congressus Internationalis Ornithologicus, Berlin 1978, ed. R. Nöhring, pp. 137-157. Berlin: Deutsche Ornithologen-Gesellschaft.

(18) Kroodsma, D.E. 1978. Aspects of learning in the ontogeny of bird song: Where, from whom, when, how many, which, and how accurately. In The Development of Behavior: Comparative and Evolutionary Aspects, eds. G.M. Burghardt and M. Bekoff, pp. 215-230. New York: Garland STPM Press.

(19) Lack, D. 1968. Ecological Adaptations for Breeding in Birds. London: Methuen.

(20) Landsberg, J.W. 1980. Hormones and filial imprinting. In Acta XVII Congressus Internationalis Ornithologicus, Berlin 1978, ed. R. Nöhring, pp. 837-841. Berlin: Deutsche Ornithologen-Gesellschaft.

(21) Mason, W. 1979. Ontogeny of social behavior. In Social Behavior and Communication, eds. G.J. Vandenberg and P. Marler, pp. 1-28. New York: Plenum.

(22) Mattson, M.E., and Evans, R.M. 1974. Visual imprinting and auditory discrimination in young of the canvasback and semi-parasitic redhead. Can. J. Zool. 52: 421-427.

(23) Mayr, E. 1974. Behavior programs and evolutionary strategies. Am. Scient. 62: 650-659.

(24) Oppenheim, R.W. 1974. The ontogeny of behavior in the chick embryo. Adv. Study Behav. 5: 133-172.

(25) Payne, R.B. 1981. Song learning and social interaction in Indigo Buntings. Anim. Behav. 29: 688-697.

(26) Pröve, E. 1983. Der Einfluß von Steroiden auf die sexuelle Prägung männlicher Zebrafinken. Verh. Dtsch. Zool. Ges. Stuttgart: Fischer.

(27) Pröve, E. 1983. Hormonal correlates of behavioural development in male Zebra Finches. In Hormones and Behaviour in Higher Vertebrates, eds. J. Balthazart, E. Pröve, and R. Gilles, pp. 368-374. Berlin, Heidelberg, New York: Springer.

(28) Pröve, E. 1984. Physiological basis of sexual imprinting in male Zebra Finches. In Environment and Hormones, eds. S. Ishii, B.K. Follett, and A. Chandola. Tokyo: Japan Scientific Societies Press, in press.

(29) Roberts, R.C. 1979. The evolution of avian food-storing behavior. Am. Nat. 114: 418-438.

(30) Teuchert, G.; Wolff, J.R.; and Immelmann, K. 1982. Physiologische Degeneration in der Ontogenie des ZNS von Vögeln - eine Einflußnahme auf die sensible Phase für Prägung? Verh. Dtsch. Zool. Ges. Stuttgart: Fischer.

(31) Wiltschko, R. 1980. Die Sonnenorientierung der Vögel. J. Orn. 121: 121-143.

(32) Wolff, J.R. 1981. Some morphogenetic aspects of the development of the central nervous system. In Behavioral Development, eds. K. Immelmann, G.W. Barlow, M. Main, and L. Petrinovich, pp. 164-190. New York: Cambridge University Press.

(33) Yasukawa, K.; Blank, J.L.; and Patterson, C.B. 1980. Song repertoires and sexual selection in the Redwinged Blackbird. Behav. Ecol. Sociobiol. 7: 233-238.

The Biology of Learning, eds. P. Marler and H.S. Terrace, pp. 289-309. Dahlem Konferenzen 1984. Berlin, Heidelberg, New York, Tokyo: Springer-Verlag.

Song Learning: Innate Species Differences in the Learning Process

P. Marler
Rockefeller University Field Research Center
Millbrook, NY 12545, USA

Abstract. Normal song development results from the interaction of auditory information that is learned and stored, with central motor programs and auditory templates that are innately different from species to species. Learning is selective and based on a process of stimulus-triggered cue storage. A new model of the auditory storage process is presented. Auditory templates are postulated for detecting species-specific vocal signals, storing them, and using them specifically to guide vocal development by auditory feedback. The templates are of two kinds, active and latent. Lacking species-specific auditory stimulation, some templates remain latent and uninvolved in song development, hence the abnormality of song in birds reared in isolation. When activated, the templates provide the basis for memorizing learned songs and converting them into motor patterns by auditory feedback.

INTRODUCTION

There are about 9000 species of birds and most have something equivalent to a song. In more than half the song probably develops normally in isolation. By current estimates, about 4000 bird species (mainly oscine songbirds, also parrots and hummingbirds) have to learn to sing in the sense that birds reared in isolation will develop abnormally (20). The song is typically a male activity serving the functions of territorial defense and attraction and stimulation of females. It is usually more complex than the set of up to twenty or so simple, often monosyllabic calls that make up a typical avian vocal repertoire, most of which develop normally in isolation, often without the need of auditory feedback. The functions songs serve require that they be species-specific, and most field naturalists rely on song for identification. Yet song development is plastic, local dialects are common, and some degree of song individuality is virtually ubiquitous.

In the evolution of birdsong there is tension between selection pressures to develop along species-specific lines and pressures to systematically or opportunistically develop novel behaviors, adding more options for invading new communities and sound environments, and thus exploiting new resources. It is this tension that makes song development interesting for epigenetic analysis. A case is presented in this paper that the balance is struck by using song components that display innate species differences, with learning controlling which subset an individual will use and, within certain limits, how they are put together to make a song. The interpretation of song learning presented here derives from the auditory template theory of song learning of Konishi (17) and Marler (23) and has features in common with what Changeux (8) and Edelman (11) characterize as learning based on principles of neuroselection or neuronal group selection.

STRUCTURE OF A "TYPICAL" SONG

A typical song is a useful fiction. It might be two seconds in duration, consisting of fifty or so discrete vocal gestures or notes (Fig. 1). High rates of note delivery, up to five times faster than syllable production in normal human speech, are commonplace. Notes are commonly 10 to 100 msec in duration. Groups of notes form syllables 100-200 msecs in duration assembled in turn into phrases, then grouped to form a song. In some birdsongs every note is different, but more commonly, as in speech, notes and syllables are reused to produce trills and to create repertoires of different patterns, with rules for phonological syntax that vary from species to species. There are species-specific repertoire sizes, from one to a hundred or more discrete song types, with specific modes of progression that take tens of minutes to cycle through. Thus, there are several levels of organization in which genes and experience can be seen to interact during song development. Descriptive analyses of natural song reveal some characteristics that are species-universal and others that are species-variable, our first clue that different components and features have different epigenetic histories. Songbirds have evolved epigenetic programs for the embedding together of stereotyped and variable features in an orderly fashion as attributes of a single basic motor coordination.

CENTRAL MOTOR PROGRAMS

Normal oscine song development depends on auditory stimulation from the environment. Internalized representations of this stimulation interact with information provided by the brain from two sources, auditory templates and central motor programs. Consider motor programs first. Some aspects of song structure develop even in birds that are deaf, with the auditory feedback loop severed. Early deafening results in grossly abnormal development of song. Despite the degradation and distortion of the fine structure of song in early deafened individuals (17), Sherman and I (28) have recently shown that significant species-specific differences in duration, frequency characteristics, segmental structure, and repertoire size still emerge. These species contrasts are maintained in the face of drastic changes in the absolute values of many measures. Yet, although the contribution of central motor programs displaying innate species differences is fundamental, they generate only simple versions of typical singing behavior (Fig. 1B).

AUDITORY TEMPLATES FOR SYNTAX

The contribution of auditory mechanisms to innate species differences in song can be assessed by comparing development in birds reared in isolation with hearing intact with that in birds with the auditory feedback loop severed by deafening prior to singing (Fig. 1). Numerous species-specific song differences that fail to develop in deaf birds, including details of syntactical organization, emerge in the intact birds (Marler and Sherman, 1982). These innate species differences are attributable to central auditory templates guiding vocal development by feedback control. They permit development of a degree of species-specific song syntax, irrespective of whether or not there is access to environmental stimuli embodying such syntactical patterning.

The most striking abnormalities in the songs of birds reared in isolation are found in the fine structure of notes and syllables. In sparrows, environmental factors exert their maximum influence at this level of song structure. In these birds most aspects of normal note structure cannot be realized without environmental stimulation by conspecific song. Despite this dependence on auditory stimulation, note structure is often species-specific, a crucial point for the theory of song learning to be developed here.

IMITATION

The impact of experience on song development is most obvious when imitation occurs. In the great majority of natural cases, imitation is

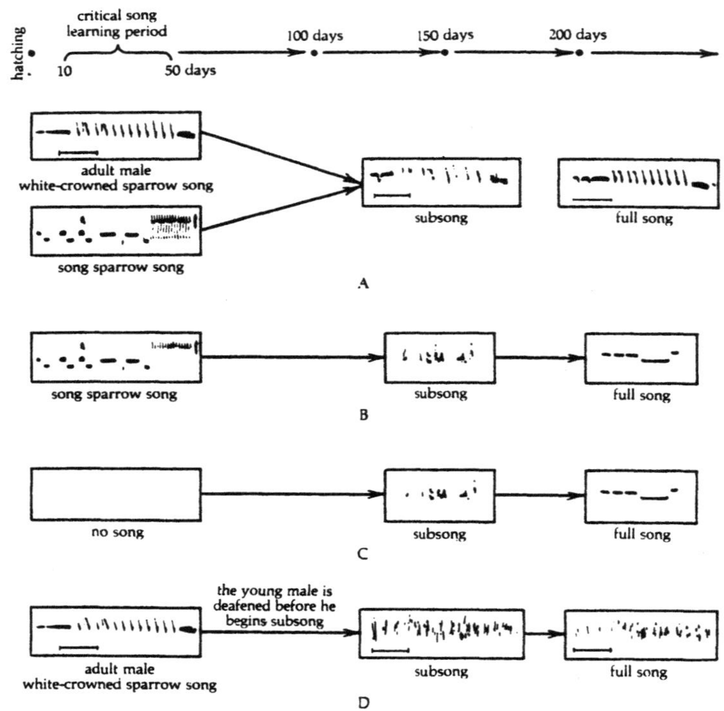

FIG. 1A - A diagram of typical patterns of song development in the white-crowned sparrow. White-crowned sparrows learn their species' song during a sensitive phase from 10 to 50 days of age. At about 150 days they begin to vocalize and practice making song syllables. By 200 days they have developed a stable song which matches parts of the song heard during the sensitive phase. The song learning is selective, so that if offered a choice, birds will prefer to learn their own species' song (A), and if offered the wrong song (B) or no song at all (C), the bird will learn nothing and sing only a simple pattern. If deafened before he begins to practice (D), the bird will never sing anything melodic, while if deafened after the song "crystallizes," subsequent singing is unimpaired (from Gould (11)).

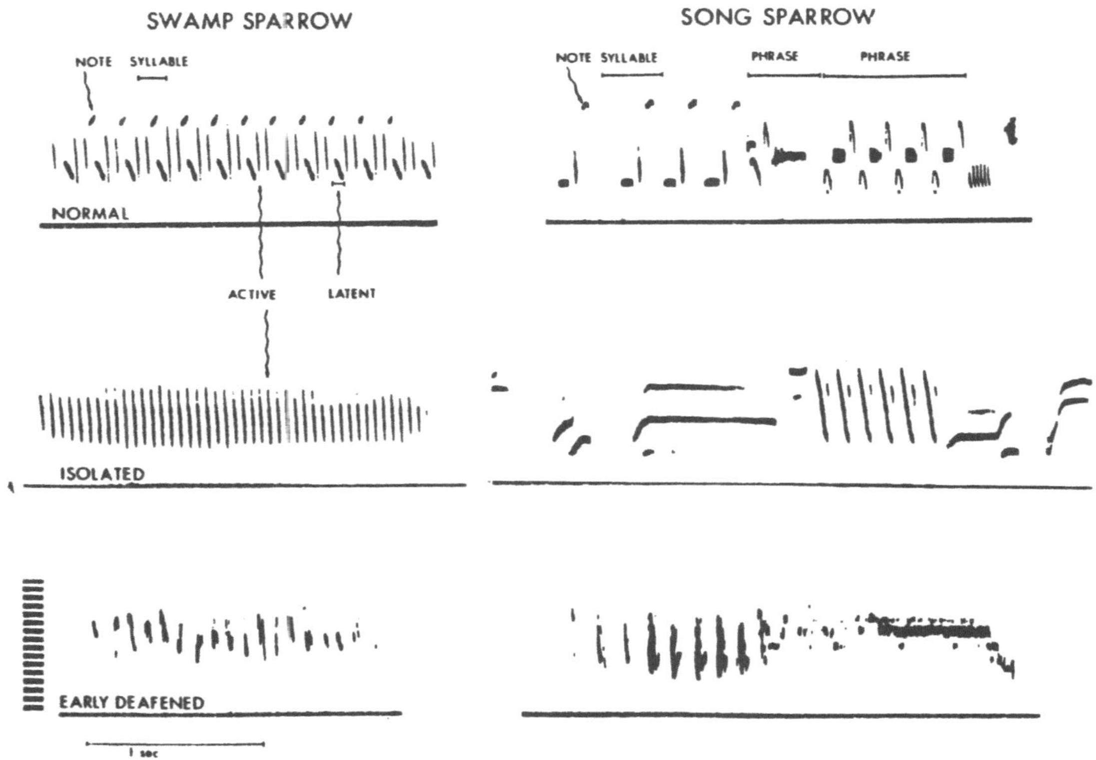

FIG. 1B - Normal, isolate, and deaf songs of swamp and song sparrows. In the swamp sparrow the one normal note type that develops in isolates, encoded by "active" templates, is indicated. Innate species differences emerge in isolated males, and some, though fewer, in deaf males.

restricted to members of the same species. Interspecific mimics, which make up about fifteen percent of oscine songbirds (2), do not sacrifice species-specificity, but achieve it by rearranging material derived from other species in their own fashion and responding themselves to the species-specific composition rules (6). The other eighty-five percent of songbirds learn to sing but typically do not mimic other species. They must, therefore, possess an ability to reject heterospecific songs as models for learning. The basis of this rejection is the subject of much current research. Strict auditory selectivity has been demonstrated in some species. In others it is subordinated to learning guided by social factors, as in the zebra finch (15). Although this is a matter of some contention, zebra finches may also have sensory mechanisms favoring conspecific songs as learning models (22, 42).

INNATE SPECIES DIFFERENCES IN AUDITORY SELECTIVITY

Auditory selectivity that varies innately from species to species is displayed by certain songbirds as a robust phenomenon, leading them to favor sounds of their own kind when given a choice between learning conspecific or heterospecific songs in infancy. Selective responsiveness is focussed on two levels of stimulus organization, the fine acoustic structure of notes and the patterning of the whole song (18, 26). Responsiveness to such stimuli is employed in infancy during the sensory phase of song learning to sift the barrage of acoustic stimuli impinging on the young bird's ear, priming it to be selectively attentive to conspecific songs (Fig. 1).

The selectivity is attributable not to passive sensory filtration but to active central processes. Experiments with synthetic songs in which conspecific and heterospecific song materials are intermingled reveal a facility to learn and reproduce heterospecific sounds, even though the same sounds are rejected when presented alone. The presence of conspecific stimulus triggers to which even isolation-reared birds are responsive renders entire songs acceptable as models. Such stimulus triggering results from the presence of particular notes or syllables, as in swamp sparrows (26) and white-crowned sparrows (Konishi, personal communication), or certain syntactical features, as in the song sparrow (26). The capacity to reproduce alien sounds, along with the abilities for heterospecific song reproduction of habitual "mockingbird" mimics, and rare cases of heterospecific singing in other species, demonstrate that the vocal tracts of songbirds are versatile instruments, capable of producing a wider range of sounds than those typical of the natural song of any one species. Thus, apart from general frequency characteristics, species-specific attributes of song probably derive not from differences in design of the vocal apparatus but from species differences in central instructions.

AUDITORY TEMPLATES AS ACTIVE STIMULUS FILTERS

As originally conceptualized, auditory templates are primarily concerned with the task of guiding the sensorimotor phase of song development by auditory feedback. However, the specific auditory properties that they possess may also be applied earlier, in the sensory phase of song learning, to the task of filtering incoming auditory stimuli and focussing the attention of the young songbird on conspecific songs. Although there is no necessary reason that the same mechanism has to serve both functions, this is our current working hypothesis. The song learning process will impose changes on some properties (especially of an organizational nature) of the auditory templates, but other properties, especially perhaps those of an elemental nature, appear to remain unchanged from the sensory phase through to the sensorimotor phase of song development. Evidence on this point will be presented below.

According to this interpretation, species-specific auditory templates are involved in the earliest stages of song learning. Their role at this stage is to identify appropriate models for learning and to initiate the process of committing them to memory. As already mentioned, this is an active process and not just a matter of passive sensory filtration. At a later stage of song development, after changes have been imposed by the learning experience, auditory templates become involved in song production.

SENSORIMOTOR INVENTION AND IMPROVISATION

Once a song is accepted as a model for learning in the sensory phase of song development, it is committed to memory. Imitations are produced later, in the sensorimotor phase. Production is guided by auditory feedback (17), progressing from an amorphous subsong, through plastic song, and crystallizing into full song. The interval over which the information is stored without overt rehearsal ranges from seconds to months. Often learned models are not reproduced in their entirety, although this can occur. More typically, models are broken down into phrases, syllables, and even notes. This systematic parsing of learned songs may be a motor phenomenon, or a reflection of processes occurring in memory consolidation (Squire, this volume), a possibility that evidence of separate perceptual processing of song components renders worthy of consideration. The motor behavior to reproduce models or parts of them is then perfected by trial and error (Gould and Marler, this volume) in the transition from plastic song to full song, often resulting in recombinations into novel sequences (24). Once the final order has crystallized, singing behavior becomes fixed as a learned motor program. In some cases auditory feedback is no longer even needed (17), though in others it is necessary. At intermediate stages of development learned models may also be subjected to gradual transformation by improvisation, resulting in deviations from strict imitation. There are also outright inventions. Thus several processes serve to promote individuality in song structure.

DEVELOPMENTAL PARTITIONING OF SONG SEGMENTS

There is growing evidence that, in songs with a multipartite structure, separate segments have a different developmental basis. Thus, precisely where individuality is to be found in the song varies from species to species. Not uncommonly, particular predesignated components or segments in a song are slavishly imitated, while others are subject to individual invention. Features characterizing a local dialect may be concentrated in certain song segments while, within the dialect, the structure of other song segments varies individually. Members of a species tend to be consistent in the song features or segments to which imitations, improvisations, and inventions are assigned, and where innate species differences are located.

There are indications that perceptual processing can be partitioned. An apparent case of alerted detection occurs in rufous-sided towhees whose songs consist of a simple, unspecific tone followed by a more complex trill. Towhees seem to use the tone as an alerting signal and the trill for species identification (37). When cowbirds and red-winged blackbirds are required to classify conspecific songs into categories in the laboratory, they base their classifications on different song segments than when they perform an equivalent task with songs of the other species (39). Tests with isolated song segments show that they do equally well with introductory sections of their own and the other species. On the terminal portions, however, which are the most highly species-specific, each does better with conspecific than with heterospecific songs (10). The existence of species-specific auditory templates for song syntax should prepare us not only for separate processing of song segments via auditory feedback in the development of song production, but also for the possibility of differential processing of segments of incoming conspecific song stimuli independently of song development, as Sinnot's data suggest (39).

INNATELY-TRIGGERED CUE STORAGE

In understanding the responses of a young bird to the song as a complex stimulus array to be learned, it is important to distinguish between two factors that influence the acquisition process: a) There are acoustic components, characterized by details of frequency modulation, bandwidth, syntax, etc., that trigger acceptance of a whole song as a model. b) Other song components or features, viewed as cues to be learned, may be endowed with varying degrees of salience (25, 35). While the potential to learn and to reproduce many sounds clearly exists, there appear to be innate species differences in the qualities that specify triggering stimuli and establish the degree to which qualitatively different cue stimuli are attended to. Given a choice, some take precedence and are learned more easily and more faithfully than others, presumably because they mesh with auditory template specifications.

There appears to be an element of selective attention in the process of song memorization. A complex sound is heard, stored in short-term memory, and then forgotten unless a triggering component from the song or some other source (e.g., visual stimuli from the singer) says "store." The "store" command may, in fact, originate with activation of the auditory templates. If triggering stimuli occur, both they and their associated components are likely to be remembered, especially those with high learning-cue salience. Temporal association between the experience of triggering and of cueing stimuli evidently suffices for learning to occur. Auditory stimuli are thus committed to memory, probably with various transformations, either during memorization, or in subsequent memory consolidation.

Imitations are subsequently produced under the guidance of auditory feedback, and appropriate components are fitted into niches in the motor pattern. Contributing to production of the motor pattern, also under auditory feedback control, are the auditory syntax templates - complementing, modifying, and detailing the orchestration by central motor programs of the song as a whole.

IMITATION: ARE THERE MINIMAL STRUCTURAL SONG UNITS?

In common parlance, imitation involves production of new motor patterns under the influence of experience. In fact imitations are rarely completely novel. If we imitate someone else's speech, even in a foreign language, the novelty of the undertaking is limited by comparison with, say, a person imitating a birdsong. The multitude of universal features held in common by all speech sounds relegates the undertaking to one of modifying the nature of sounds we already make and rearranging them in novel sequences.

In the same spirit, a reductionistic analysis of birdsongs down to minimal components reveals instances where, although learned, the song is created entirely or in part by permutations and combinations of a species-universal set of basic vocal gestures. The basic "note" repertoire may be large or surprisingly limited. In the most extreme case yet described, that of the swamp sparrow, all known songs can be broken down into six basic note categories, each with a limited degree of within-category variation (27). The process of song imitation as it occurs in nature consists of the selection of a particular sequence of note types and subtypes to match the model. A given population restricts itself to a subset of the large number of possible combinations, thus constituting a local dialect. By abstracting the same set of combinatorial rules and using them freely, rather than slavishly reproducing all details of a model, a degree of individuality can still be achieved.

In the case just mentioned, that of the swamp sparrow, the feature most obviously universal in all species members is the set of minimal acoustic units or notes from which the songs are constructed. In other birds, species universality also embraces larger units, such as clusters of notes or syllables. This is the case in the indigo bunting which possesses a universal set of about one hundred note clusters, or "syllables," which suffices to characterize all known populations (43). Here learning determines the subset of syllables an individual selects from this universal array, the number of repetitions, and the particular order in which they are sequenced in a song (36).

Thus a close inspection of the structure of "imitated" songs reveals that there are fewer degrees of freedom than meet the eye. Presumably, species differences in the universally shared attributes of imitated songs are innate. Yet some of them fail to appear in birds reared in social isolation. It should be recalled that the note structure of the songs of isolation-reared sparrows is highly abnormal. We are obviously dealing with something more subtle than the purely endogenous guidance of development that used to be thought of as characterizing so-called instinctive behavior.

LATENT AUDITORY TEMPLATES: A NEW INTERPRETATION OF THE LEARNING PROCESS

Certain salient facts have to be accommodated by any theory of song learning. As one side of the coin, vocal production is modifiable as a consequence of auditory experience, often resulting in imitation. On the other hand, we have to account for the selectivity of learning and the presence of many species-universal and species-specific traits in song. Most important of all is the fact that some of these traits that are universally present in normal song fail to emerge in birds reared in isolation. Thus songs of isolation-reared sparrows display a number of abnormalities, especially in the timing and acoustic morphology of notes and syllables. Evidently some species-universal song traits are dependent upon auditory stimulation for their development. Without such stimulation they remain latent. When these latent features emerge, they often play a dominant role in the formation of local dialects.

In addition to motor programs, each songbird species possesses auditory templates for detecting species-specific notes and syllables and syntax. These auditory templates also have sensorimotor capacities in the sense that they are used not only for detecting song stimuli, but also for guiding subsequent song development. Although birds obviously acquire responsiveness to many environmental sounds other than conspecific song during infancy and store representations of them, these do not normally have the same potent influence on motor development that representations of conspecific song stimuli exert.

The deficiencies of the songs of isolation-reared birds suggest that these auditory templates are of two kinds. Type 1 (ACTIVE) templates require no specific auditory stimulation in order to begin influencing song development. Auditory syntax templates in song and swamp sparrows are evidently of the "active" type, as are those for certain note types which also emerge normally in isolation. Type 2 (or LATENT) templates require triggering by matched auditory stimulation from the environment for activation. I propose that when they are subjected to patterned stimulation and thus to patterned facilitation, the two types of template become involved not only in the commitment of the pattern to memory, but also in memory consolidation. We know that both have the further attribute that they are consulted in the guidance of sound production by auditory feedback. Even without song stimulation, the subset of acoustic features under the control of "active" templates emerges normally, as manifest in the songs of birds reared in isolation with hearing intact. But "latent" templates remain unactivated in this case, hence the deficiencies of the isolate song. These deficiencies will be especially manifest in the components that define the local song dialects of that species in nature.

Suppose now that a young bird experiences normal song components during the appropriate sensitive phase (16). In this event, both active and latent templates would be stimulated in particular constellations that capture the essential features of the song, corresponding to Konishi's "learned auditory template" (17). Information about these constellations would be stored, perhaps broken down into syntactical units and later converted into matching sounds, then fitted into appropriate niches in the central motor program. There are several levels of potential plasticity, although only within certain limits. Auditory templates for notes and syntax specify arrays of alternatives rather than single stereotyped patterns. The set of possible specifications allows sufficient latitude to accommodate the syntax and note typology of different models, as long as they lie within the song space of the species. Insofar as a complete, conspecific song will engage genetically preordained mechanisms that are completely attuned to it, from the minimal vocal gestures to the entire syntactical pattern, so the song will be easily and efficiently committed to memory and made available for consultation when the process of motor development begins at some later time.

If normal song stimulation is withheld, the bird will then fill out the central motor program with what is provided by "active" templates for notes, syllables, and syntax. A degree of species-specific structure will thus emerge, but those song features mediated by latent templates will be lacking. Latent templates that are not matched by experience will remain inactive, falling into disuse or relinquishing neural space to the competing services of other, more generalized kinds of perceptual processing. A chronically deaf bird, unable to exploit the potential of either active or latent templates, will fall back on the dictates of the central motor program, to be filled in with the highly variable and imprecise notes and syllables that are typical of deaf birds.

In some species the commitment to sensorimotor processing of a particular set of songs appears to be more or less irreversible, culminating in a learned motor program that persists for life, no longer even requiring auditory feedback, at least for short-term maintenance (17). Some songbirds must learn to sing in a short sensitive phase in late infancy or forego the opportunity altogether. In other species motor plasticity is renewed in successive reproductive cycles, and cases may exist in which potential plasticity is virtually continuous. The time course of the plasticity will be a function of the degree to which learning is endogenously cued to conform to a system of sensitive phases, which may be short or long, one or many (20). The annual cycle of dendritic growth and retraction described by Nottebohm (30) in certain of the brain nuclei controlling song behavior in the canary, a species in which annual song change is possible, may be engaged in the process of reinstating templates for the annual supplementation of the song repertoire.

LEARNING BY INSTRUCTION?

If we ask whether birds are capable of vocal learning by instruction, the fact that even birds that favor conspecific song can be persuaded to imitate songs of other species suggests an affirmative answer. So does the occurrence of habitual heterospecific mimics. The ornithological literature also provides sporadic illustrations of individual birds occasionally incorporating elements or even entire songs of other species into their own. I believe, however, that we can press the concept of innately specified auditory templates further. We can assume that the brains of birds are pre-armed with auditory mechanisms for guiding vocal development that are innately responsive not to conspecific song, but to heterospecific and environmental sounds. As I visualize the patterns of responsiveness of such mechanisms, they would be specific, but at a more basic level of acoustic structure than those considered earlier, and capable of breaking down virtually any complex sound into elementary components.

So we are now postulating a third set of GENERALIZED auditory templates, potentially capable of guiding vocal imitation, with broader capabilities than the two kinds of highly specialized templates already postulated. These generalized systems may in fact share, or be indistinguishable from, neuronal mechanisms that control auditory responsiveness to environmental sounds in general (19). They may also contribute to the more specialized processes of conspecific song detection. The auditory forebrain projection areas of birds may prove to contain both specialized and unspecialized neurons, much as Suga (41) has described in bats, but with specialization for song detection rather than echolocation (Konishi, this volume).

The process of imitation will presumably proceed differently if mediated by generalized auditory templates. In species such as sparrows potentially capable of bringing both specialized and generalized mechanisms to bear on song development, one might predict that the processing of heterospecific sounds, being generalized, would be slower and imitation for a given degree of exposure less precise. Also, it may be predicted that stored representations of complex sounds mediated by generalized mechanisms would be subject to the likelihood of preemption by conspecific sounds. The chances of such preemption would presumably be greatest during the sensitive phase, although we need to know whether this is a matter of memory retention or of specific involvement with sensorimotor aspects of song development. In addition to the use of song performance as an index of what is learned in the sensory phase, there is an urgent need to bring sensitive psychophysical techniques to bear on the measurement of rates of perceptual processing, memorization, and forgetting of song stimuli (10, 38, 39).

In view of Nottebohm's demonstrations of the extent to which song processing in the oscine brain is conducted in highly localized centers ((31), and Konishi, this volume), neuroanatomically distinct brain areas may prove to house generalized and specialized auditory templates or, alternatively, they may be controlled by a single system. Given the clarity with which song control nuclei can be delineated in the bird brain, comparison of the volumes of tissue committed to specialized and generalized auditory processing might be possible in the brains of, say, a swamp sparrow and a mockingbird. The latter presumably has a heavy assignment to generalized processing. In a sparrow brain one would expect specialized processing of conspecific song stimuli to be emphasized. Correlations have already been established between variations in song repertoire size and the volume of song control brain nuclei ((7, 32), and Kroodsma et al., this volume), although their interpretation is still a matter of some controversy (Konishi, this volume).

REINFORCEMENT AND ASSOCIATIVE PREDISPOSITIONS

With regard to the contingencies that must be satisfied for a bird to learn a song pattern, no extrinsic reinforcement seems to be necessary, although there are cases in which food or social reinforcement augments song learning. It suffices for song experience to occur during a sensitive phase, and for the song or other associated stimuli to contain innately specified triggering stimuli. There is evidence that auditory experience of conspecific song is itself reinforcing in the sense of sustaining contingent operant behavior (40); activation of auditory song templates may suffice. In some species there are indications that contingent social stimulation may be both necessary and sufficient for vocal learning (33). Quantitative evidence on this issue is urgently needed. The possibility should be explored that innately specified responsiveness to visual stimuli that trigger learning could be involved in imprinting, such as those generated by plumage patterns and displays. Attempts to gain control over the morphology of vocal behavior by classical extrinsic reinforcers, as contrasted with control of the frequency of production of existing vocalizations, have been largely ineffectual in the past. This may be attributable more to the difficulty of making the nuances of the acoustic structure of vocalizations contingency-dependent rather than to the refractoriness of vocal behavior to such influences (23). It seems unlikely that vocal behavior is completely resistant to shaping by classical reinforcement, even though song is obviously more sensitive to shaping by auditory stimulation. We can conclude that strong vocal learning effects are obtainable in songbirds without extrinsic reinforcement other than that imparted by triggering stimuli present in the stimulus complex to be learned, perhaps as a function of activation of auditory templates for song.

PARALLELS WITH OTHER SYSTEMS: MOTOR DEVELOPMENT

There is a basis for extending certain aspects of this interpretation of song learning to other motor systems (Gould and Marler, this volume). Studies of the motor coordinations used by squirrels to open nuts suggest that the basic motor elements are innately specified, but the squirrel must learn to achieve the proper order, timing, and orientation of the movements (12). The grossly abnormal copulatory behavior of rhesus monkeys reared in social isolation seems to result not from a failure to develop the basic motor components, but rather from the failure to assemble them into a sequence that is properly coordinated with the behavior of the female (29). The roles of sensory templates and central motor programs have not been explored in either instance, however. The skillful use of blended facial expressions by our own species may be a case where experience is crucial not for the development of the minimal muscular coordinations that constitute the building bricks of expressions, but rather for assemblage of the larger-scale organization, and something like sensory templates may be involved.

The early development of human speech sound production is especially amenable to this kind of interpretation. It is generally agreed that speech perception precedes production. There is ample evidence of the universality of many of the basic vocal gestures from which speech behavior is constructed, although they may be lost through disuse. Experience is crucial both in the selection of within-category variations and in the assemblage of phonemic components into morphemes, words, and sentences with their underlying intonations. There are also parallels between speech and birdsong on the sensory side. Infants are capable of some basic processing of speech sounds at the phonemic level, even though here again the patterns of innately specified responsiveness are more properly thought of as establishing points of embarkation for processes of learning, rather than as design features for automata.

PARALLELS BETWEEN SONG LEARNING, IMPRINTING, AND VISUAL DEVELOPMENT

There are numerous points of correspondence between the sensory phase of song development and imprinting (Bateson and Immelmann, both this volume). In both cases early experience has profound consequences for subsequent behavior. There is often strong phase specificity and subsequent resistance to further change. Another shared feature is the "preemptive self-terminating" nature of the sensitive phase (Immelmann, this volume). In imprinting and in song learning the withholding of specific stimuli serves to extend the sensitive phase for learning (3, 21).

As a result of genetic predispositions, certain stimuli or stimulus complexes are favored in imprinting and song learning. In discussions of imprinting, stress is often placed on the conspicuousness of favored imprinting objects, but there are indications that species-specific stimulus features, either visual or auditory, are also involved (Bateson and Immelmann, both this volume). Bischof (4) makes a distinction between species-specific visual mechanisms controlling responsiveness of zebra finches during imprinting that is reminiscent of that made here between "active" and "latent" auditory templates. We need to consider the possibility that stimulus-triggered cue storage plays a role in imprinting, as it does in song learning, with temporal and/or spatial proximity of other stimuli to the triggering events playing a crucial role in the acquisition process. There are similarities in the extension of effects from the particular stimulus complex learned to a larger class of which it is a member. Thus, processes of imprinting and song learning have many features in common.

There are parallels between imprinting (4, 5), song development, and the patterns of interaction between the organism and its environment in the development of the mammalian visual system ((44), and Singer, this volume). All exhibit phases of sensitivity and refractoriness to specific kinds of environmental stimulation. Sensitive phases of visual development again display some "self-terminating" characteristics (9, 34). In each case sensory mechanisms display innately preordained responsiveness to specific stimulus properties. In both visual development and song ontogeny the continued involvement in behavioral control of mechanisms with innately specified responsive properties is contingent upon exposure during a sensitive phase to stimulation that matches specific requirements of those mechanisms. Events in visual development at the cellular level can plausibly be related in turn to more typical learning phenomena by way of Hebb's classic synaptic theory of learning (4, 14, 34). Studies of song learning, imprinting, and visual development thus provide a germinal point of contact between ethology, psychology, and developmental biology and a potential source of insight into how learning theories and studies of neural and behavioral development can be reconciled.

In a recent review of the role of the environment in the development of sensory systems, Atkinson, et al. (1) set aside the classical form of the nativist/empiricist controversy, with its focus on the dichotomy of Nature or Nurture, as obviously outdated. Instead they characterize the basic underlying process as one of interaction between innate, genetically formed neural mechanisms and sensory experience. Having stressed the inevitable involvement of both Nature and Nurture, they offer the caveat that it can still be fruitful to ask the "either-or" question when addressing specific issues such as color blindness (almost entirely Nature) and stereo-blindness (probably mostly Nurture). They conclude, however, that "the greatest interest attaches to the cases where no such simple answer is possible, because these cases demonstrate the intimate relations between the two." Imprinting and song learning are instances offering perhaps our best current prospect for understanding the kinds of epigenetic interactions that underlie behavioral development in general and learning in particular.

Acknowledgements. Criticisms of P. Bateson, H.-J. Bischof, L. Finkel, J. Gould, C. Hopkins, K. Immelmann, J. Rauschecker, and H. Terrace are gratefully acknowledged. Research was supported by grant number NIMH 14651.

REFERENCES

(1) Atkinson, J.; Barlow, H.B.; and Braddick, O.J. 1982. The development of sensory systems and their modification by experience. In The Senses, eds. H.B. Barlow and J.D. Mollon, pp. 448-469. Oxford: Cambridge University Press.

(2) Baylis, J.R. 1982. Avian vocal mimicry: Its function and evolution. In Acoustic Communication in Birds, eds. D.E. Kroodsma and E.H. Miller, vol. 2, pp. 51-83. New York: Academic Press.

(3) Bateson, P.P.G. 1979. How do sensitive periods arise and what are they for? Anim. Behav. 27: 470-486.

(4) Bischof, H.-J. 1983. A model of imprinting evolved from neurophysiological concepts. Z. Tierpsych. 51: 126-139.

(5) Bischof, H.-J. 1983. Imprinting and cortical plasticity: A comparative review. Neurosci. B. 7: 213-225.

(6) Boughey, M.J., and Thompson, N.S. 1976. Species specificity and individual variation in the songs of the brown thrasher (Toxostoma rufum) and catbird (Dumetella carolinensis). Behaviour 57: 64-90.

(7) Canady, R.; Kroodsma, D.; and Nottebohm, F. 1981. Significant differences in volume of song control nuclei are associated with variance in song repertoire in a free ranging songbird. Soc. Neurosci. Abstr. 7: 845.

(8) Changeux, J.-P. 1983. L'Homme Neuronal. Paris: Librairie Arthème Fayard.

(9) Cynader, M. 1983. Prolonged sensitivity to monocular deprivation in dark-reared cats: effects of age and visual exposure. Devel. Brain Res. 8: 155-164.

(10) Dooling, R.J. 1982. Auditory perception in birds. In Acoustic Communication in Birds, eds. D.E. Kroodsma and E.H. Miller, vol. 1, pp. 95-130. New York: Academic Press.

(11) Edelman, G., and Mountcastle, V. 1978. The Mindful Brain. Cortical Organization and the Group-Selective Theory of Higher Brain Function. Cambridge, MA: MIT Press.

(12) Eibl-Eibesfeldt, I. 1961. The interactions of unlearned behavior patterns and learning in mammals. In Brain Mechanisms and Learning, ed. J.F. Delafresnaye, pp. 53-73. Oxford: Blackwell.

(13) Gould, J.L. 1982. Ethology: The Mechanisms and Evolution of Behavior. New York: W.W. Norton & Co.

(14) Hebb, D.O. 1949. The Organization of Behaviour. New York: John Wiley Sons.

(15) Immelmann, K. 1969. Song development in the zebra finch and other estrildid finches. In Bird Vocalizations, ed. R.A. Hinde, pp. 61-74. Cambridge and London: Cambridge University Press.

(16) Immelmann, K., and Suomi, S.J. 1981. Sensitive phases in development. In Behavioral Development, eds. K. Immelmann, G.W. Barlow, L. Petrinovich, and M. Main, pp. 395-431. Cambridge: Cambridge University Press.

(17) Konishi, M. 1965. The role of auditory feedback in the control of vocalization in the white-crowned sparrow. Z. Tierpsych. 22: 770-783.

(18) Konishi, M. 1978. Auditory environment and vocal development in birds. In Perception and Experience, eds. R.D. Walk and H.L. Pick, Jr., pp. 105-118. New York: Plenum Press.

(19) Konishi, M. 1978. Ethological aspects of auditory pattern recognition. In Handbook of Sensory Physiology: Perception, eds. R. Held, H.W. Leibowitz, and H.L. Teuber, vol. 8, pp. 289-309. Berlin: Springer-Verlag.

(20) Kroodsma, D.E. 1982. Learning and the ontogeny of sound signals in birds. In Acoustic Communication in Birds, eds. D.E. Kroodsma and E.H. Miller, vol. 2, pp. 1-23. New York: Academic Press.

(21) Kroodsma, D.E., and Pickert, R. 1980. Environmentally dependent sensitive periods for avian vocal learning. Nature 288: 477-479.

(22) Kruijt, J.P.; Ten Cate, C.J.; and Meeuwissen, G.B. 1983. The influence of siblings on the development of sexual preferences of male zebra finches. Devel. Psych. 16: 233-239.

(23) Marler, P. 1970. A comparative approach to vocal learning: Song development in white-crowned sparrows. J. Comp. Physiol. Psychol. 71(2): 1-25.

(24) Marler, P. 1981. Birdsong: The acquisition of a learned motor skill. TINS 4: 88-94.

(25) Marler, P.; Dooling, R.; and Zoloth, S. 1980. Comparative perspectives on ethology and behavioral development. In The Comparative Method in Psychology, ed. M. Bornstein, pp. 189-230. Hillsdale: Erlbaum.

(26) Marler, P., and Peters, S. 1980. Birdsong and speech: Evidence for special processing. In Perspectives on the Study of Speech, eds. P. Eimas and J. Miller, pp. 75-112. Hillsdale: Erlbaum.

(27) Marler, P., and Pickert, R. 1984. Species-universal microstructure in the learned song of the swamp sparrow (Melospiza georgiana). Anim. Behav. 32: 673-689.

(28) Marler, P., and Sherman, V. 1982. Song structure without auditory feedback: Emendations of the auditory template hypothesis. J. Neurosci. 3: 517-531.

(29) Mason, W.A. 1968. Early social deprivation in the non-human primates: Implications for human behavior. In Environmental Influences, ed. C. Glass, pp. 70-101. New York: Rockefeller University and Russell Sage Foundation.

(30) Nottebohm, F. 1981. A brain for all seasons: Cyclical anatomical changes in song control nuclei of the canary brain. Science 214: 1368-1370.

(31) Nottebohm, F. 1981. Brain pathways for vocal learning in birds: A review of the first 10 years. In Progress in Psychobiology and Physiological Psychology, eds. J.M.S. Sprague and A.N.E. Epstein, vol. 9, pp. 85-124. New York: Academic Press.

(32) Nottebohm, F.; Kasparian, S.; and Pandazis, C. 1981. Brain space for a learned task. Brain Res. 213: 99-109.

(33) Payne, R.B. 1981. Song learning and social interaction in indigo buntings. Anim. Behav. 29: 688-697.

(34) Rauschecker, J.P., and Singer, W. 1981. The effects of early visual experience on the cat's visual cortex and their possible explanation by Hebb synapses. J. Physiol. (Lond.) 310: 215-239.

(35) Rescorla, R.A., and Wagner, A.R. 1972. A theory of Pavlovian conditioning: Variations in the effectiveness of reinforcement and nonreinforcement. In Classical Conditioning II: Current Research and Theory, eds. A.H. Black and W. Prokasy, pp. 64-99. New York: Appleton-Century-Crofts.

(36) Rice, J.O., and Thompson, W.L. 1968. Song development in the indigo bunting. Anim. Behav. 16: 462-469.

(37) Richards, D. 1981. Alerting and message components in songs of rufous-sided towhees. Behaviour 76: 223-249.

(38) Sinnott, J.M. 1980. Species-specific coding in bird song. J. Acoust. Soc. Am. 68: 494-497.

(39) Sinnott, J.M. 1984. Modes of perceiving and processing information in birdsong. In Categorical Perception, ed. S. Harnad. New York: Cambridge University Press, in press.

(40) Stevenson, J.B. 1969. Song as a reinforcer. In Bird Vocalizations, ed. R.A. Hinde, pp. 49-60. Cambridge, England: Cambridge University Press.

(41) Suga, N. 1982. Functional organization of the auditory cortex: Representation beyond tonotopy in the bat. In Cortical Sensory Organization, ed. C.N. Woolsey, vol. 3, pp. 157-218. Clifton, NJ: Humana Press.

(42) Ten Cate, C. 1982. Behavioural differences between zebrafinch and bengalese finch (foster) parents raising zebrafinch offspring. Behaviour 81: 152-172.

(43) Thompson, W.L. 1970. Song variation in a population of indigo buntings. Auk 87: 58-71.

(44) Wiesel, T.N. 1982. Postnatal development of the visual cortex and the influence of environment. Nature 299: 583-591.

The Biology of Learning, eds. P. Marler and H.S. Terrace, pp. 311-324. Dahlem Konferenzen 1984. Berlin, Heidelberg, New York, Tokyo: Springer-Verlag.

A Logical Basis for Single-neuron Study of Learning in Complex Neural Systems

M. Konishi
Div. of Biology 216-76, California Institute of Technology
Pasadena, CA 91125, USA

Abstract. An understanding of learning in complex neural systems requires the knowledge of neuronal connections and signals which together code for stimuli, responses, and their relationships. A higher-order neuron forms a nodal point of convergence of afferent and efferent channels and owes its stimulus selectivity both to its intrinsic properties and to the connections it makes with other neurons. Such a neuron, under certain conditions, shows not only how the brain encodes complex stimuli but also how experience determines or modifies its stimulus selectivity. One of the brain areas for the control of birdsong contains neurons selective for the individual bird's own song. These neurons seem to acquire their stimulus selectivity during song development. This article also discusses recent theories about the neural substrates of song learning.

INTRODUCTION

The discussion of learning cannot be divorced from that of neural codes, for these are the language for all transactions in the nervous system including learning. What is encoded in learning has been an important question in learning theory (Jenkins, this volume). The brain's coding scheme is based both on the anatomical connections and on the methods of signalling between its neurons. In this scheme the establishment of a new relationship between selected variables, whether they are stimuli or responses, involves changes in neuronal connections by modification of synaptic efficacy. Although such changes explain learning in simple neural circuits (Quinn, this volume), learning by modification of complex neuronal connections is little understood. The study of neural signals is one way to identify the nature and site of synaptic changes in learning. Nerve cells communicate using many forms of signal: discrete and graded potentials, chemical transmitters, and hormones originate in one group of neurons and modulate the response of other groups of neurons. Action potentials mediate fast, sensory and motor responses because they are a rapid carrier of signals between distant sensory receptors, neurons, and muscles. How information is encoded with action potentials is a central issue in integrative neurophysiology. Recording action potentials from a single neuron at a time is the method of single-neuron or single-unit electrophysiology. In this article I shall discuss how this method can contribute to the study of learning in complex neural systems.

HIGHER-ORDER NEURONS AND NEURAL CODES

Single-unit neurophysiologists believe that recording action potentials from a single neuron at a time should enable them to decipher the brain's codes. They suppose that the response properties of a single neuron contains information about neural coding, although they do not deny that multi-neuron coding exists. The single-unit approach, particularly with intracellular recording, has been a powerful tool for identifying the neuronal connectivity and interaction that underlie fixed action patterns and modifiable reflexes (Quinn, this volume). The history of extracellular neurophysiology is equally illustrious. Recent advances in the study of the visual cortex show how this method can lead the search for the anatomical and physiological bases for information processing in complex neural systems (41).

The antithesis of the single-unit approach asserts that the response of a single neuron does not reveal anything about neural coding, because it is the distributed properties of a neuronal population that encode information. Multi-neuron coding is attractive primarily because mathematical or logical methods to describe and predict the behavior of a population of simple elements are available. Also, when the methods of recording from many neurons simultaneously in separate channels become practical, theoretical predictions will be necessary both for designing experiments and for testing hypotheses. However, this approach faces all of the difficulties inherent in dealing with complex neural systems. The brain is obviously not an aggregate of identical or simple neurons. It is a system of many different classes of neurons forming complex networks. This heterogeneity and complexity are difficult properties for theoretical treatment.

Whichever the approach, one faces another fundamental problem which is bound to affect research strategy. A search for neural codes can start from many different levels within a given neural system. Starting with primary sensory neurons and going "upstream" to higher-order neurons would seem logical, because higher-order neurons can use only what primary sensory neurons can encode. This approach assumes that the response properties of primary sensory neurons can predict those of higher-order neurons. However, the "upstream" predictability is usually low, because there is no logical method to predict what neural activity codes for what stimulus and how the codes are transformed from one level of neural organization to another. For example, suppose neural responses R_1 and R_2 code for, respectively, stimuli S_1 and S_2 at one level. The code for $S_1 + S_2$ at the next level should be $R_1 + R_2$, if the system is linear. This simple relationship seldom prevails in the nervous system (2).

Why cannot higher-order neurons be the starting point? This approach does not seem feasible because the stimulus selectivity of a higher-order neuron tends to be refractory to logical analysis. On this issue Marr (19) wrote, "Suppose, for example, that one actually found the apocryphal grandmother cell (a cell that fires only when one's grandmother comes into view). Would that really tell us anything much at all? It would tell us that it existed - Gross's hand-detectors tell us almost that - but not why or even how such a thing may be constructed from the outputs of previously discovered cells. Do the single-unit recordings - the simple and complex cells - tell us much about how to detect edges or why one would want to, except in a rather general way through arguments based on economy and redundancy? If we really knew the answers, for example, we should be able to program them on a computer. But finding a hand-detector certainly did not allow us to program one."

A higher-order neuron is not an isolated single channel, but rather a nodal point of afferent and efferent connections. The stimulus selectivity of a higher-order neuron is either relayed to it from its afferent channels or "emerges" from it as a result of integration of the inputs. The selectivity represents either a total sum or a subset of the integrative actions that occur in the neuronal networks converging on it. These integrative actions are the processes by which the excitation of the neuron generates the percept of the stimulus (1). The crucial test of this proposition must include behavioral observations. A recent study on the superior temporal sulcus of the monkey presents an excellent example to illustrate the above point. Most neurons in this area are selective for the direction and speed of stimulus motion. A focal lesion of this area abolishes the pursuit movements of the eye that normally occur in response to a moving stimulus (23). Also, a combination of behavioral, electrophysiological, and anatomical studies designed to understand the stimulus selectivity of the owl's space-specific neurons (which respond only to sounds emanating from a small restricted area in space) has demonstrated the power of this approach (22, 38, 39). In general, the effective use of higher-order neurons for the study of neural coding requires the following conditions: a) They are selective for a complex, yet definable stimulus. b) The mechanisms of their stimulus selectivity can be determined. c) The same stimulus selectivity is subject to behavioral analysis. d) Lower-order neurons in the pathway leading to the higher-order neurons can be identified.

These conditions also allow higher-order neurons to be used for studying the mechanisms of learning in complex neural systems. How a neuron's stimulus selectivity develops is the question that links the discussion of neural codes to that of learning. The development of a neuron's stimulus selectivity may be sensitive to a specific stimulus. If the mechanism of stimulus selectivity is known, then the sites and nature of plasticity underlying its establishment or modification can be determined. For example, plugging one ear of a young barn owl causes a permanent shift in the receptive field locations of space-specific neurons (10). Similarly, binocular and orientation selective neurons in the cat's visual cortex are sensitive to the nature of early visual experience. The changes in the stimulus selectivity of these neurons are perhaps causally related to the behavioral effects of ear plugging in the owl and eye closure or stripe-rearing in the kitten (11, 37).

THEORY OF SONG LEARNING

I propose to use the above idea in the study of song learning. I shall first summarize the attributes of song learning. Song learning consists of sensory and sensorimotor phases (Marler, this volume): young birds memorize a song during an impressionable period and vocally reproduce it later. The interval between the two phases can be as long as several months. In some species they overlap with each other; birds continue to learn new sounds during the sensorimotor phase. This phase is so called because hearing of a bird's own voice is essential for the vocal reproduction of the memorized song (14). A song will also develop without a tutor model, as in acoustic isolation. Isolate songs usually lack some of the characteristics of the wild-type song. The songs of deaf birds typically contain significantly fewer normal features than the isolate songs. These facts indicate that control of voice by auditory feedback is essential for song development with or without a tutor model. Birds use auditory feedback to prevent their songs from deviating from the intended pattern, whether this is innate or acquired. Because this process resembles trial and error, song development always contains elements of learning. This reasoning justifies such a statement as "a song is a learned motor skill," even when no copying from other individuals is involved (17, 27). However, in species such as song sparrows and canaries, certain aspects of the song temporal pattern develop normally in deaf individuals, although the component sounds are abnormal (4, 18). These aspects of song are, therefore, not learned unless nonauditory feedback steers their development. If nonauditory feedback is used, then one has to assume the existence of an internal standard by which the bird corrects motor errors. If no sensory feedback is involved, then these attributes of song must be due to central motor programs. It is important to recognize the contributions of these programs to song learning, although there are large species differences in the nature and amount of central contributions.

THE VOCAL CONTROL SYSTEM

The motor pathway that controls the vocal organ, the syrinx, is a major component of the vocal control system. This system consists of five nuclei in the forebrain, one in the thalamus, one in the midbrain and one in the hindbrain (Fig. 1) (5, 30, 34). In songbirds the medullary nucleus, the syringeotrachealis division of the hypoglossal nucleus, innervates exclusively the ipsilateral syringeal musculature, which modulates sound presumably by regulating the tension of the internal tympaniform membrane on that side. The highly stereotyped song of an adult bird must be due to a fixed pattern of neuromuscular coordination. A recent study shows that the basic neural signals for this coordination occur earliest in NIF and descend to the syrinx via HVc, RA, DM, and nXIIts, in that order (see Fig. 1 for abbreviations) (20). These signals in NIF do not initiate a stored program, rather they are the precursors of the eventual motor discharges that control the patterned contractions of the syringeal musculature. In species such as the white-crowned sparrow, young birds copy the pattern of amplitude and frequency modulation found in the adult song (16), indicating that the neural signals for them are established by learning. The vocal control system shows remarkable versatility in some species: for example, a brown thrasher can produce over 2000 different syllable types in its song, which means over 2000 different motor output patterns (17)! In addition to the motor pathway, the system receives auditory inputs at two of the forebrain nuclei, HVc and RA, and HVc auditory neurons project to X (7-9). These auditory connections may be involved in the control of vocal output by hearing.

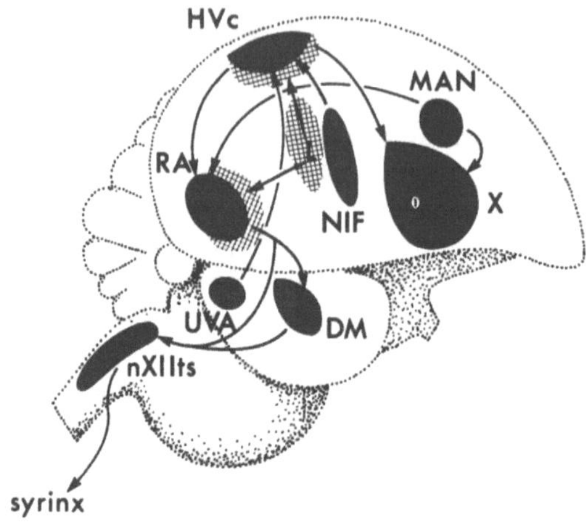

FIG. 1 - Schematic diagram of the vocal control system in songbirds. Arrows indicate anterograde connections between nuclei. Neural signals for song originate in NIF (Nucleus interfacialis) and descend the pathway to the vocal organ, syrinx, via HVc (Nucleus hyperstriatum ventrale, pars caudale), RA (Nucleus robustus archistriatalis), DM (dorsomedial nucleus of nucleus intercollicularis), and nXIIts (Nucleus hypoglossus, pars tracheosyringealis), in that order. Other nuclei (MAN = magnocellular nucleus of the anterior neostriatum, X = Area X, and UVA = Nucleus uva) are inactive during vocalization. Hatched areas indicate known projection zones of the telencephalic auditory area (Field L).

NEURAL THEORY OF SONG LEARNING

Cross-taxa comparisons, brain lesions, and neuroendocrinological studies have recently uncovered several new facts that are thought to link the vocal control system and song learning. I shall review them below.

Cross-Taxa Comparisons

The most direct correlation between brain anatomy and song learning was found in cross-taxa comparisons. The forebrain nuclei HVc, RA, and X occur only in birds capable of vocal learning such as oscine songbirds and parakeets; they are apparently absent in birds without that capability such as suboscine and non-passerine birds (24, 35). Brain structure homologous to these nuclei has not been identified in non-passerine birds except the parakeet. Oscine songbirds which include such songsters as mockingbirds, canaries, and thrushes generally develop abnormal songs when raised in acoustic isolation or deafened as young, whereas suboscines and non-passerine birds such as domestic chickens and ring doves develop normal vocalizations in acoustic isolation (12, 13, 32). If the auditory connections in the forebrain nuclei underlie the control of vocal development by auditory feedback, the correlation may reflect the presence or absence of that control mechanism. This explanation is consistent with the results of deafening on vocal development in oscines and non-passerines. No information is available about the effects of deafening in suboscines.

Brain Anatomy and Song Learning

Comparisons of different individuals or populations of the same species has uncovered another possible correlation between brain anatomy and song learning (Kroodsma, this volume). Canaries with larger numbers of syllables tend to have larger HVc's and RA's than those with smaller numbers of syllables, although it is statistically a weak correlation (29). A song consists of discrete sound components called syllables (see Marler, this volume, for terminology of birdsong). Canaries produce a long, continuous series of syllables in their songs, and the number of different syllables used by an individual can vary between about twenty and fifty. Young canaries develop syllables both by imitating their fathers and other adults and by improvising (42). In contrast with many species that do not change song after the first singing season, canaries "change their song repertoires each year by adding, dropping or modifying components" ((27), see also (33)). Canaries deafened after their first singing season fail to maintain or increase their syllable repertoires in subsequent singing seasons (34). Therefore, the process by which new and modified syllables are created is regarded as learning, and the number of syllables can serve as the index of song learning (24). Why does the size correlation exist? The more learning, the more brain space? If the number of different syllables is correlated with any other variables, then it will be necessary to determine which variable is causally related with the size of HVc and RA. For example, a reasonable hypothesis is that the direct correlation is between the total amount of sound energy and the nuclear size. If this is true, then the size correlation would have nothing directly to do with learning. Although canaries deafened after their first singing season produce abnormal songs, the total amount of sound in them is measurable. It would be of interest to compare the HVc's and RA's of deaf canaries whose songs contain different amounts of sound energy. If there is a size correlation, then this is due to different amounts of use instead of learning, because deaf birds cannot learn song.

The plasticity of canary song predicts the existence of modifiable neural substrates. Testosterone, which stimulates singing in young birds, females, and castrated males presumably acts directly on the vocal control system by inducing growth in the somata and dendrites of some of its neurons (3, 25). In the male canary the size of HVc and RA fluctuates seasonally in parallel with the seasonal variation in circulating testosterone: they are larger in the spring singing season than in the fall and winter (26). The growth and shrinkage of dendrites presumably contribute to this fluctuation. An increase in synaptic sites probably accompanies growth in dendrites. The view that learning necessarily involves the formation of new synapses is not generally accepted (6), although there is physiological evidence such as the new synapses formed in the cat's red nucleus during readjustment of forelimb coordination following cross-innervation of flexor and extensor muscles (40).

The hypothesis that the seasonal fluctuation of HVc and RA volume indicates learning and forgetting of song requires careful scrutiny (26). There is a real possibility that HVc and RA undergo seasonal volume changes in deaf canaries, because their singing fluctuates seasonally as in intact birds. If it is true, then the significance of this phenomenon for song learning is doubtful. The normal temporal pattern of song that develops in deaf canaries is likely to be due to central pattern generator circuits, as mentioned before. Because it develops by the first singing season without auditory feedback, it would not be surprising if some of its generator circuits are annually "reformed" even in deaf birds. Reorganization of central pattern generators is an unorthodox idea, but there is no reason to exclude such a possibility. It will be important, therefore, to distinguish the anatomical and physiological changes associated with feedback-controlled modification of song from those related to central reorganization and use-disuse phenomena mentioned earlier.

Lateralization of Song Control

Birdsongs share some attributes with human speech development. One of them is thought to be hemispheric lateralization. In the canary, a lesion of the left hypoglossal nerve or left HVc causes greater loss or more severe deterioration of syllables than a lesion on the right side. These observations led to the hypotheses that each hemisphere independently controls the ipsilateral half of the syrinx and that the left hemisphere controls more syllables than the right one (31, 34). However, the evidence does not unequivocally prove that the site of lateralization is in the forebrain hemisphere. Because the vocal system controls predominantly the ipsilateral syringeal musculature, a lesion of any part of the descending pathway disables the musculature of that side. If the left half of the syrinx plays a more dominant role in song control than does the right half, then neither cutting the hypoglossal nerve nor lesioning the HVc discriminates between hemispheric and peripheral lateralization, because either operation exposes the syringeal asymmetry, which can mask any hemispheric differences (20). Such a peripheral asymmetry does exist in the canary's syrinx; the left syringeal musculature is clearly more voluminous than the right one. Consistent with this asymmetry, the left bronchus working alone can produce more of the normal syllables than the right one working alone, although the right one contributes to all syllables (20). Some HVc-lesioned canaries showed fewer changes in their songs than hypoglossus-lesioned birds (24, 31, 34). This difference may be due to the fact that cutting a peripheral nerve can be done more cleanly than lesioning a brain nucleus. Furthermore, if the size of HVc can serve as the index of song learning, one would expect left-right anatomical differences in the vocal control nuclei. No such differences are present in any of the vocal control nuclei except the hypoglossal nucleus, where the left side is said to be larger than the right by 6% (28). Recordings from the HVc of the singing canary show that both sides are active during all syllables. This observation is difficult to explain, if HVc is the site of lateralization (20).

How the muscular asymmetry develops in the ontogeny of an individual is an unanswered question; some unknown central asymmetry may drive the two sides differently, or it may develop in response to differential loading in the periphery. In the adult canary a lesion of the left hypoglossus results in a shift of song control to the right side. However, this shift parallels that of the muscular asymmetry to the right side. Why the lateralized control of song evolved is another unanswered question. There does not seem to be any correlation between the ability to learn song and the degree of lateralization. Deaf canaries, which are not supposed to "learn" song, are left-dominant (31). An asymmetry is hardly observable in the song of the zebra finch, which learns song well (36). Thus, the analogy of song lateralization to human speech control should be viewed with caution.

A SINGLE-NEURON APPROACH TO SONG LEARNING

Although some of the correlations mentioned above may link brain anatomy to song learning, the ultimate proof for it requires an understanding of the neural codes involved: how memorized songs, auditory feedback information, and song motor programs are encoded is an important question that calls for physiological analyses at the cellular level. I shall summarize below the results of preliminary studies to address these issues. As mentioned before, HVc contains both motor and auditory neurons: the two classes of neurons interact such that auditory neurons are inhibited while motor neurons fire during song (21). Some HVc auditory and motor neurons show properties suggestive of the effects of song learning. Single-unit recording from the HVc of the singing mockingbird uncovered neurons which fire only before and during the delivery of a specific learned sound pattern in the bird's song (motor-specific neurons) (20). In the white-crowned sparrow some HVc auditory neurons respond selectively to the bird's own song (song-specific neurons) (15). All song-specific neurons that were thoroughly studied require two consecutive sounds (phrases) of the song, each of which alone elicits little or no response. They detect a particular sequence of acoustic cues such as frequency in the two phases. A bird's song-specific neurons may respond to the songs of some other individuals from the same dialect area and typically fail to respond to the songs of birds from other dialect areas. This selectivity is predictable when the relevant acoustic features in the bird's own song are compared with the corresponding features in the other songs. Song-specific neurons in birds singing abnormal songs due to early isolation respond selectively to these abnormal songs and fail to respond to wild-type songs of the species. When a bird's own song differs from its early tutor song, its song-specific neurons are selective not for the tutor song but for the song sung by the bird.

Reference to the memorized song (acquired template) is essential for its vocal reproduction: the vocal output is varied until it matches the template (Marler, this volume). Whether song-specific neurons are involved in some aspect of the matching processes is not known, although their specificities are appropriate for that which is required of the template. Because the template is established as a result of early auditory exposure, its neural representation should be present before the onset of singing. So far, song-specific neurons have not been found in pre-singing birds (Margoliash, unpublished results). Our current hypothesis about the origin of song-specific neurons is as follows: song-specific neurons acquire their specificity during song development; they become matched to the final form of song as this is gradually shaped. A similar argument would apply to the motor-specific neurons mentioned earlier; their specificity may also develop during the sensorimotor phase of song learning.

CONCLUDING REMARKS

Song-specific neurons satisfy some of the conditions for use in the study of neural coding. They are selective for a particular song. This complex stimulus can by physically defined and the neurons' selectivity can be understood in terms of the acoustic cues contained in the song. Although lower-order neurons leading to song-specific neurons are not yet known, presumably they will be found. In the white-crowned sparrow, song selectivity needs further behavioral analysis; acoustic cues for discrimination between different dialects need to be determined so that they may be compared with the cues appropriate for song-specific neurons. The most encouraging aspect in the study of song-specific neurons is the possibility that sensorimotor experience during song crystallization modifies their stimulus selectivity. This phenomenon reminds us of the modifiability of cortical neurons by early exposure to patterned stimuli. In conclusion, I advocate that analysis of the properties of a higher-order neuron can show not only how the brain encodes a complex stimulus, but also how its stimulus selectivity is determined or modified by learning.

Acknowledgements. I thank P. Bateson, D. Margoliash, P. Marler, P. Patterson, J. Rauschecker, and W.E. Sullivan for reviewing and correcting the manuscript at various stages of its preparation. This work was supported by the Bing Chair of Behavioral Biology, Caltech.

REFERENCES

(1) Barlow, B. 1972. Single units and sensation: A neuron doctrine for perceptual psychology? Perception 1: 371-394.

(2) Capranica, R.R. 1972. Why auditory neurophysiologists should be more interested in animal sound communication. Physiologist 15: 55-66.

(3) DeVoogd, T.J., and Nottebohm, F. 1981. Sex differences in dendritic morphology of a song control nucleus in the canary: A quantitative Golgi study. J. comp. Neurol. 196: 309-316.

(4) Guettinger, H.R. 1981. Self differentiation of song organization rules by deaf canaries. Z. Tierpsych. 56: 323-340.

(5) Gurney, M.E. 1981. Hormonal control of cell form and number in the zebra finch song system. J. Neurosci. 1: 658-673.

(6) Kandel, E.R. 1979. Cellular insight into behavior and learning. Harvey Lect. Ser. 73: 19-92.

(7) Katz, L.C. 1982. The avian motor system for song has multiple sites and types of auditory input. Soc. Neurosci. Abstr. 8: 1021.

(8) Katz, L.C., and Gurney, M.E. 1981. Auditory responses in the zebra finch's motor system for song control. Brain Res. 221: 192-197.

(9) Kelly, D.B., and Nottebohm, F. 1979. Projections of a telencephalic auditory nucleus -field L- in the canary. J. comp. Neurol. 183: 455-470.

(10) Knudsen, E.I. 1983. Early auditory experience aligns the auditory map of space in the optic tectum of the barn owl. Science 222: 939-942.

(11) Knudsen, E.I.; Knudsen, P.F.; and Esterly, S.D. 1982. Early auditory experience modifies sound localization in barn owls. Nature 295: 238-240.

(12) Konishi, M. 1963. The role of auditory feedback in the vocal behavior of the domestic fowl. Z. Tierpsych. 20: 349-367.

(13) Konishi, M. 1965. The role of auditory feedback in the control of song in the white-crowned sparrow. Z. Tierpsych. 22: 770-783.

(14) Kroodsma, D.E. 1984. Songs of the Alder Flycatcher (Empidonax alnorum) and Willow Flycatcher (Empidonax traillii) are innate. Auk, in press.

(15) Margoliash, D. 1983. Acoustic parameters underlying the responses of song-specific neurons in the white-crowned sparrow. J. Neurosci. 3: 1039-1057.

(16) Marler, P. 1970. A comparative approach to vocal learning: Song development in white-crowned sparrows. J. comp. Physiol. Psych. 76(2): 1-25.

(17) Marler, P. 1981. Birdsong: The acquisition of a learned motor skill. Trends Neurosci. 4: 88-94.

(18) Marler, P., and Sherman, V. 1982. Song structure without auditory feedback: Emendations of the auditory template hypothesis. J. Neurosci. 3: 517-531.

(19) Marr, D. 1982. Vision. San Francisco: Freeman.

(20) McCasland, J.S. 1983. Neuronal Control of Bird Song Production. Ph. D. Thesis, California Institute of Technology.

(21) McCasland, J.S., and Konishi, M. 1981. Interaction between auditory and motor activities in an avian song control nucleus. Proc. Natl. Acad. Sci. USA 78: 7815-7819.

(22) Moiseff, A., and Konishi, M. 1983. Binaural characteristics of units in the owl's brainstem auditory pathway: Precursors of restricted spatial receptive fields. J. Neurosci. 3: 2553-2562.

(23) Newsome, W.T.; Wurtz, R.H.; Durstler, M.R.; and Mikami, A. 1983. Deficits in pursuit eye movements after chemical lesions of motion-related visual areas in the superior temporal sulcus of the macaque monkey. Soc. Neurosci. Abstr. 9(46.11): 154.

(24) Nottebohm, F. 1980. Brain pathways for vocal learning in birds: A review of the first 10 years. Prog. Psychobiol. Physiol. Psychol. 9: 85-124.

(25) Nottebohm, F. 1980. Testosterone triggers growth of brain vocal control nuclei in adult female canaries. Brain Res. 189: 429-436.

(26) Nottebohm, F. 1981. A brain for all seasons: Cyclical anatomical changes in song control nuclei in the canary brain. Science 214: 1368-1370.

(27) Nottebohm, F. 1981. Laterality, seasons and space govern the learning of a motor skill. Trends Neurosci. 4(5): 104-106.

(28) Nottebohm, F., and Arnold, A.P. 1976. Sexual dimorphism in vocal control areas of the song bird brain. Science 194: 211-213.

(29) Nottebohm, F.; Kasparian, S.; and Pandazis, C. 1981. Brain space for a learned task. Brain Res. 213: 99-109.

(30) Nottebohm, F.; Kelly, D.B.; and Paton, J.A. 1982. Connections of vocal control nuclei in the canary telencephalon. J. comp. Neurol. 207: 344-357.

(31) Nottebohm, F.; Manning, E.; and Nottebohm, M.E. 1976. Left hypoglossal dominance in the control of canary and white-crowned sparrow song. J. comp. Physiol. 108: 171-192.

(32) Nottebohm, F., and Nottebohm, M.E. 1971. Vocalizations and breeding behavior of surgically deafened Ring Doves (Streptopelia risoria). Anim. Behav. 19: 313-327.

(33) Nottebohm, F., and Nottebohm, M.E. 1978. Relationship between song repertoire and age in the canary, Serinus canarius. Z. Tierpsych. 46: 298-305.

(34) Nottebohm, F.; Stokes, T.M.; and Leonard, C.M. 1976. Central control of song in the canary, Serinus canarius. J. comp. Neurol. 165: 457-486.

(35) Paton, J.A.; Manogue, K.R.; and Nottebohm, F. 1981. Bilateral organization of the vocal control pathway in the budgerigar, Melopsittacus undulatus. J. Neurosci. 1: 1276-1288.

(36) Price, P.H. 1977. Determinants of Acoustical Structure in Zebra Finch Song. Ph. D. Thesis, University of Pennsylvania.

(37) Rauschecker, J.P. 1984. Neuronal mechanisms of developmental plasticity in the cat's visual system. In Human Neurobiology, ed. A. Fiorentini, vol. 3. Berlin, Heidelberg: Springer-Verlag, in press.

(38) Sullivan, W.E., and Konishi, M. 1984. Segregation of stimulus phase and intensity coding in the cochlear nucleus of the barn owl. J. Neurosci., in press.

(39) Takahashi, T.T.; Moiseff, A.; and Konishi, M. 1984. Time and intensity cues are separately processed in the owl's auditory brainstem. J. Neurosci., in press.

(40) Tsukahara, N. 1981. Synaptic plasticity in the mammalian central nervous system. Ann. Rev. Neurosci. 4: 351-379.

(41) Van Essen, D.C. 1979. Visual areas of the mammalian cerebral cortex. Ann. Rev. Neurosci. 2: 227-263.

(42) Waser, M.S., and Marler, P. 1977. Song learning in canaries. J. comp. Physiol. Psychol. 91: 1-7.

The Biology of Learning, eds. P. Marler and H.S. Terrace, pp. 325-339. Dahlem Konferenzen 1984. Berlin, Heidelberg, New York, Tokyo: Springer-Verlag.

The Neural Basis of Imprinting

P.P.G. Bateson
Sub-Dept. of Animal Behavior, University of Cambridge
Madingley, Cambridge CB3 8AA, England

Abstract. A site in the intermediate part of the medial hyperstriatum ventrale (IMHV) is strongly implicated in filial imprinting in domestic chicks. Two distinct storage systems may be involved. The left IMHV is almost certainly a site for coded information stored during imprinting. The right IMHV is necessary for storage somewhere else. Bilaterally lesioned birds are incapable of imprinting. When lesions are placed in IMHV after imprinting, chicks fail to recognize the object with which they have been imprinted. However, such birds have no difficulties in learning visual discrimination tasks for heat rewards and will readily work to present themselves with just those moving objects that are effective in imprinting normal birds. The neuroanatomical specializations required for imprinting probably evolved so that kin-recognition, once developed, is protected from change but other forms of behavior can be altered and elaborated after imprinting is complete.

INTRODUCTION

When recently hatched birds such as ducklings or domestic chicks are hand-reared for a few days, they strongly prefer the company of their human keeper to that of their own species. The remarkable process, which can so dramatically influence the development of early social relationships, is called "filial imprinting." A bird's experience later in life can also strikingly influence its sexual preferences. When exposed to another species at a certain stage in development, it may subsequently prefer to mate with that species. The process influencing mate choice is known as "sexual imprinting" (reviews in (3, 22, 29, 41)).

For many years imprinting was thought to enable an animal to recognize its species. However, filial imprinting may serve a more subtle but equally important function, enabling the young to identify one or both of their parents as individuals (4). The rapid attachment to the first conspicuous object encountered after hatching is characteristic of precocial species. In birds like the swallow, which are hatched naked and helpless, learning occurs later in development and the young only respond selectively to their parents when they have left the nest about two weeks after hatching (14). Sexual imprinting probably occurs later in development because of the selection pressure on birds to recognize the characteristics of their siblings' adult appearance when they choose a mate (6). The ideal mate must be a bit different but not too different from close relatives to avoid the dangers of inbreeding and, at the other extreme, the dangers of mating with a genetically incompatible partner such as a member of another species. Japanese quail are so careful in their choice of a sexual partner that, having been reared with their siblings, they then choose first cousins as mates (5). Mice behave in a similar fashion (19). Indeed, the processes similar to those first described in birds are likely to be equally important in mammals, and probably other groups as well. The biological need to recognize close kin, both early in life and when adult, is a general one (17, 24).

In this paper I shall solely be concerned with the neural basis of filial imprinting in domestic chicks and drop the "filial" henceforward. The advantages of such imprinting for neurobiology are as follows: a) The chick can be hatched and reared in darkness until exposed to the imprinting object, so the training procedure provides the animal with its first visual experience. b) The learning process proceeds spontaneously and requires no additional incentives or rewards, given an appropriate stimulus. c) After imprinting the bird continues to respond vigorously to the familiar object. Therefore, some of the general behavioral changes, such as increased attentiveness and motor activity which are frequently confounded with learning, can be dissociated from the process involved in acquisition.

Work such as I shall describe here could not be done without the involvement of many people with expertise in very different fields of biology, and it is worth summarizing both the history and the extent of this collaboration in the imprinting project. When the work started in the late 1960s, the collaboration was primarily between Gabriel Horn (a physiologist), Steven Rose (a biochemist), and myself (an ethologist)

(7). In the mid-70s as the work became more anatomical and physiological in character, Rose moved on to study the biochemistry of learning in other preparations (38). At this stage, Brian McCabe (27, 34, 35) became actively involved in the autoradiography, stimulation, and lesion work, and a little later Philip Bradley (13) collaborated on the anatomical studies. In recent years, as the work has radiated into many areas of neurobiology under Horn's (25) leadership, my own role has dwindled. However, I still maintain close contact with Horn's group and he and I collaborate closely on behaviorally oriented aspects of the work.

REFINING THE PROBLEM

At the outset the broad questions were these. Could neural events involved in imprinting be localized? If so, are these events necessary for and exclusively related to imprinting? Biochemical dependent variables were chosen because these measures were likely to be related to protein synthesis. The presumption was that if storage of information in the brain is associated with alterations in synaptic connections as has often been suggested (16, 20), then such changes should be preceded by protein synthesis. It soon became apparent that imprinting enhances biochemical activity in the roof of the forebrain (7).

Many events in the brain could be responsible for the enhanced biochemical activity. For example, the imprinted and non-imprinted chicks almost certainly differed in amount of motor activity during the period of training. Their levels of stress may have been different. The imprinted birds almost certainly received more sensory stimulation than the unimprinted birds and, at a more subtle level, their attention may have been focused more strongly. In order to reduce such ambiguity a series of experiments were done with a view to separating these various consequences of the training procedure.

The first procedure took advantage of the fact that the great majority of incoming fibers in the avian optic tract originate in the contralateral retina. Visual input can be restricted to one hemisphere by vertically dividing the supra-optic commissure and allowing the chicks to view the imprinting stimulus with one eye only; the other eye is covered with a patch. The biochemical activity (incorporation of radioactive uracil into macromolecules) was some 15% higher on the trained side of the forebrain roof than on the untrained side. No biochemical differences between trained and untrained sides were observed in other regions of the brain (28). The "split brain" technique eliminates some explanations which propose that both sides of the brain are affected equally by training. However, it does not exclude the possibility that the enhanced biochemical activity was due to greater visual stimulation of the trained side. Therefore, other procedures were needed.

One experiment exploited the natural variability of the chicks (8). All the chicks were treated identically. Their behavior was measured while they were being trained and their preferences for the familiar object were tested afterwards. Then the relationships between the behavioral measures and biochemical activity in different parts of the brain was examined. Only one behavioral measure was positively correlated with activity in the roof of the anterior forebrain, namely, how much the chicks preferred the familiar object to the novel object in the choice tests. This index of learning was not correlated with biochemical activity in any other region of the brain and, equally important, was poorly correlated with other behavioral measures such as the birds' activity and responsiveness. The natural variability of the birds <u>had</u>, therefore, dissociated many of the variables such as motor activity which are normally confounded with changes during learning. Furthermore, the hypothesis that biochemical activity in the brain was induced by short-term sensory stimulation had become much less plausible.

In a third experiment the amount of training that the chicks received on the first day after hatching was varied so that some of them were undertrained and some of them were overtrained (10). On the second day, all the birds were trained for the <u>same</u> amount of time. If biochemical activity in the forebrain roof is specifically related to learning, birds that had previously been overtrained and had learned more of the characteristics of the stimulus object should show less biochemical activity in the forebrain roof than undertrained birds on the second day after hatching. This hypothesis assumes that the extent to which further training takes place diminishes as the length of training increases. We found that as the length of exposure on the first day <u>increased</u>, so biochemical activity in the forebrain roof <u>decreased</u> on the second day. No such relationship was found in any of the other brain regions sampled. Possibly biochemical activity in the forebrain roof is generated by attentiveness on the part of the young birds as they learn. This would be plausible if the chicks trained for longer periods on the first day after hatching were less responsive to the familiar stimulus than the other chicks on the second day. If anything, though, they approached more vigorously than the other chicks. Therefore, the lower rate of biochemical activity in their forebrains is not readily explained in terms of reduced vigilance.

None of the three experiments described above ruled out all the alternatives to the hypothesis that increased biochemical activity in the forebrain roof is necessary for, and exclusively related to, the storage of information. However, each experiment ruled out a different subset of possibilities. The procedure is akin to triangulation since the combination of experiments pinpoints the most likely common explanation, namely, that the enhanced biochemical activity in the forebrain roof is due to rather specific changes and is intimately associated with information storage. The range of possibilities had been sufficiently reduced to justify the next step in the program which was more refined anatomical localization using the technique of autoradiography.

The autoradiographic experiment was complex, relying on a comparison of overtrained and undertrained birds, and it was time consuming, but the upshot was straightforward (27). The hot spot proved to be a part of the intermediate region of the medial hyperstriatum ventrale (IMHV). Little had been known about this region beforehand, and it lay outside the well described visual projection areas. However, as might be expected, it subsequently proved to be connected to these projection areas (11). IMHV is not a clearly defined nucleus like the hyperstriatum ventrale pars caudale (HVc) which has been implicated in the song learning of birds (36). However, it is easily localized and readily destroyed selectively by thermocoagulation.

LESION AND STIMULATION STUDIES

Initially, two types of lesion experiments were carried out. In the first, bilateral lesions were placed in IMHV on the first day after hatching and lesioned and non-lesioned animals were trained for 150 min on the second day after hatching (35). Finally, their preferences were tested. The non-lesioned animals had developed a strong preference for the stimulus with which they had been trained, whereas the IMHV-lesioned animals responded at the chance level, having no stronger preference for the training object than for a novel object.

The second lesion experiment examined the effects of bilateral IMHV lesions on retention after imprinting (33). This experiment was more ambitious in scale and contained groups that were lesioned in other parts of the brain. In one group the chicks were lesioned in a visual projection area, the hyperstriatum accessorium. Another group had bilateral lesions placed in the lateral cerebral area, because this area had been implicated in imprinting by previous investigators (39, 40). Sham-operated chicks showed strong retention when they were given a choice between the familiar and the unfamiliar stimulus. Furthermore, no significant difference was found between this control group and chicks that had lesions in the hyperstriatum accessorium and those that had lesions in the lateral cerebral area. However, chicks with lesions in IMHV had a significantly weaker preference for the training stimulus than the control group. This did not derive from any absence of motivation on the part of the IMHV-lesioned birds since their total approach activity during the preference test was not significantly different from that of the control group. In many respects the lesioned birds behaved like brain-damaged humans who have specific disabilities for recognizing faces (prosopagnosia).

The clinical impression of the IMHV-lesioned birds was that they were entirely normal with respect to other aspects of their behavior. Nevertheless, the second lesion experiment incorporated a number of formal checks which confirmed that these birds did not differ from the other three groups in the accuracy and latency with which they pecked a rocking bead, or the accuracy with which they pecked at seeds. Most strikingly, when they were trained in a visual discrimination in which the reward was heat, the IMHV-lesioned birds showed no deficit in learning ability. This was an important discovery which I shall return to later.

While the trivial alternative explanations for the results of the lesion studies can be dismissed, the data can be interpreted in at least the following three ways: a) Defective recall. IMHV is required for reading out information stored in some other part of the brain. b) Access to store impaired. During imprinting and subsequent recognitions tests, information influencing filial preferences passes through IMHV without being stored there. c) Store damaged. Information having a direct influence on filial preferences is stored in IMHV. The first two explanations seem implausible when the results of the lesion experiment are taken in conjunction with the results of the previous biochemical work, particularly those experiments involving comparisons between undertrained and overtrained birds (10, 27). The biochemical changes are much more likely to be associated with acquisition than recall. Possibly, in the early stages of imprinting, the access point to a store might show heightened biochemical activity because of growth or restructuring occurring there. However, no other part of the brain likely to be involved in imprinting, such as the visual projection or motor output areas, behaved in similar fashion to IMHV. Therefore, the access hypothesis rests on the somewhat implausible assumption that only the access to the store grows in response to use. On the basis of the present evidence it would seem that IMHV is critical for the storage of information that has a direct influence on filial preferences.

Our lesion work did not strongly implicate the lateral cerebral area and so seemingly contradicts the finding of Salzen and his colleagues (39, 40). However, the preference of the birds with lateral cerebral lesions were midway between those of IMHV-lesioned birds and sham-operated controls in our experiment, not being significantly different from either. Furthermore, the imprinting and test procedure used by Salzen et al. spanned many more days than ours, and so the lateral cerebral area may be involved in the performance of well imprinted chicks and is not required in the early stages of imprinting. This view is supported by a 2-deoxyglucose study after imprinting which gave as hot spots both IMHV and the lateral cerebral area (31). Furthermore, anatomical studies have revealed connections running from IMHV to the lateral cerebral area (12).

The lesion studies have thrown up an unexpected result. In recent years half the chicks have been trained with a stuffed jungle fowl (from which domestic chicks are supposedly descended); during training the stuffed fowl either rocks or rotates around its central vertical axis. The other half of the chicks have been trained with a flashing rotating light which is a very effective imprinting stimulus and has been used in our work for many years. The jungle fowl is much richer in detail although the flashing light is more straightforwardly conspicuous. The speeds of rotation were adjusted so that the birds found the two stimuli equally attractive at the outset of imprinting. Sham-operated birds subsequently prefer the stimulus to which they have been exposed, and their strength of preference for the familiar was not affected by the one with which they had been trained. However, Horn and McCabe (26) have recently shown that the IMHV-lesioned birds differed according to the stimulus with which they had been trained. The birds exposed to the flashing light responded at random when given a successive choice test, whereas those exposed to the jungle fowl showed a strong and significant preference for the fowl. However, the preference was significantly below that of intact birds. Possibly chicks have a predetermined capacity to recognize jungle fowl which is not located in IMHV and whose development is facilitated by visual experience. The implication would be that chicks are equipped with jungle fowl detectors which reside somewhere other than IMHV and that plasticity occurring in IMHV normally serves to fill in details so that the bird can recognize its mother as an individual. However, the jungle fowl detector idea should not be accepted uncritically. Since chicks have not yet been tested with other natural objects such as the stuffed skins of other species, the possibility remains that visual experience may activate in a relatively general way aspects of the visual pathway and thereby makes richer and more complex stimuli (such as a jungle fowl) more attractive. Chicks without the upper forebrain will approach moving objects (18); the response is presumably mediated through the optic tectum and basal forebrain structures. If rich visual experience activates part of the tectal analyzing mechanisms, chicks lacking IMHV which is part of the upper forebrain could be influenced by that experience.

Further evidence for the involvement of IMHV in imprinting has come from studies in which attempts have been made to implant artificially a preference in the birds by direct electrical stimulation of the structure. Chicks' preferences for the frequency of a flashing light can be modified by prior electrical stimulation of the IMHV with a train of shocks at the same frequency as the flashing light. Chicks were stimulated with 1.5 trains of shock per second, or 4.5 trains per second. After stimulation, all chicks were given a choice between two red lights, one flashing at 1.5 per second, the other at 4.5 per second. Those chicks which had received electrical stimulation at the lower frequency preferred the light flashing at 1.5 per second; those chicks which had been stimulated by the higher frequency preferred the light flashing at 4.5 per second. This effect could not be obtained by stimulating two visual projection areas, hyperstriatum accessorium and ectostriatum (34).

As usual, the results can be explained in lots of different ways. However, in view of the converging evidence pointing to IMHV as a storage system, the stimulation work adds a new dimension. It suggests that driving the visual pathways with rhythmical stimulation is insufficient for storage of the pattern. Stimulation of IMHV not only provides patterned input but may also act as an enabling condition so that, in the appropriate excitatory state, neurons can be modified and thereby store in coded form the patterned input.

CEREBRAL ASYMMETRY

A long-term goal of our imprinting research program has been to uncover ways in which the nervous system might code information received from the outside world. As a step in this direction, the ultrastructure of IMHV has been examined after imprinting. In order to eliminate unresponsive animals, all the birds were trained for 20 min and then responsive animals were matched in pairs. Half the birds were then trained for a further 120 min. Subsequently, electron micrographs of the structure of the overtrained birds' IMHV were compared with those from the undertrained birds. The effect of further training was to increase significantly the size of the synaptic apposition zone in the left hemisphere but not in the right (13). The result was encouraging and is being actively pursued. However, the asymmetry was surprising because none of our previous studies had suggested that the two sides of the brain operated in different ways, even though other work has strongly suggested lateralization in the chick's brain (37). The findings of the ultrastructure experiment prompted a series of unilateral lesion studies by Horn and his colleagues.

After imprinting, the IMHV on one side of the brain was removed. When the chicks had recovered they were tested and then the IMHV on the other side was removed. Finally, the birds were tested again (15). After the first lesion the birds showed full retention, irrespective of which side was lesioned first. However, the order did matter after the second lesion. If the right side had been lesioned first followed by the left, then the birds showed total amnesia when given their final test. By contrast, if the left side had been lesioned first, the animals showed full retention in the final test. One interpretation of these results is that, after imprinting, information is permanently stored in the left IMHV, but on the right side information is transferred to some other structure, as yet unidentified. As a consequence, when first the left and then the right sides are destroyed, the information about the imprinting stimulus has had time to be transferred elsewhere.

If this interpretation is correct, we can only speculate about why the two sides of the brain should differ in the way they store information after imprinting. One possibility is that the left side is a rapid action system which feeds into predetermined response patterns such as approaching the familiar and escaping from the novel. Once a recognition system has been acquired and linked to its executive systems, nothing further is required. By contrast, the right side of the brain could be involved in much more elaborate computing. We know, for instance, that the transfer of training can occur after imprinting and that birds will learn discriminations more rapidly if the stimulus familiar from imprinting is used either as the positive stimulus (i.e., associated with food) or the negative stimulus (2). In these cases, the output requirements need to be more flexible and are presumably more complicated. A straightforward prediction, which is being investigated, is that chicks with lesions placed in the right side of the brain immediately after imprinting should not be able to profit from imprinting in transfer of training tasks.

IS IMPRINTING SPECIAL?

The debate about whether imprinting is the same or different from other kinds of learning processes has been extensive (1, 21). One important strand in this debate has been the extent to which imprinting can be likened to classical conditioning (23, 30). It must be appreciated that the procedures used for studying imprinting are not associative. The experimenter merely exposes the bird at the right age to a single conspicuous object and subsequently shows that its preferences have become restricted to that object. Even so, the underlying process might involve associations. If this were the case, initially significant and attractive features of the stimulus object would be linked to initially neutral features in the course of learning. An object that works well for purposes of imprinting can also be used as a UCS or as a reinforcer (23, 30). For instance, a young duckling or chick can be easily trained to press a pedal that switches on a motor bringing an object into motion (9). The processes underlying operant conditioning could be the same as those underlying preference restriction. Alternatively, two separate learning processes could occur at the same time.

The involvement of IMHV in imprinting has provided a way of examining whether or not operant conditioning reinforced by an imprinting stimulus can be dissociated from the process that restricts preferences to a familiar object. We have already seen that chicks with IMHV lesions have no difficulties in learning a visual discrimination when they are rewarded with heat. The question is, though, whether the IMHV-lesioned birds would be able to learn to press a pedal which turns on a flashing rotating light. We now know that they can. The lesioned chicks learn the operant conditioning task as quickly as normal animals but fail to learn the characteristics of the moving object with which they were rewarded (Johnson, unpublished experiment). In this respect, IMHV-lesioned chicks are quite different from the sham-operated animals which do develop a preference for the reinforcer as they learn to work for it.

Konrad Lorenz (32) championed the view that imprinting is a special kind of learning. I, for one, was critical of many of his claims (1). However, his intuition may have been justified at least in part since particular parts of the nervous system do seem to be dedicated to the recognition of close kin.

The use of anatomically discrete subsystems for imprinting probably evolved from the special need to protect from degradation stored information about close kin without losing the benefits of modifying other forms of behavior later in life. Self-termination of imprinting protects the kin-recognition mechanism from subsequent change (4). But neural machinery dedicated to imprinting is required so that more general forms of learning can take place after imprinting is complete. Such design features would explain the descriptive evidence for sensitive periods and the anatomical specialization. They do not imply, though, that the precise way information from the external world is coded in neural tissue is also special.

CONCLUSION

The similarities between the imprinting project and the studies of the song-learning system in birds are interesting. Both have uncovered a remarkable degree of localization of function. Both sets of data face familiar and formidable problems of interpretation, the most challenging being whether or not the localized structures are involved in readout mechanisms rather than in storage. It would be a rash person who attempted to be dogmatic when dealing with the neural basis of an inferred process such as memory. However, in the case of the imprinting work, the pattern of results suggests rather strongly that the left IMHV is a storage system.

Imprinting, like song learning, has many of the characteristics of a special purpose system. Nonetheless, the imprinting mechanism would be of general interest if we could discover the precise ways in which neural plasticity coded information from the external world. An efficient means of storing information, once evolved, was probably co-opted for many different kinds of job in the course of subsequent evolution. Therefore, any knowledge about the principles by which the external world is represented in the nervous system is likely to be general and important. Such principles are likely to emerge from the study of imprinting in the not too distant future.

Acknowledgements. I am deeply indebted to G. Horn for carefully reading a draft of this paper. I should also like to thank M. Johnson, M. Konishi, and S.P.R. Rose for their comments. The work has been supported by grants from the Science and Engineering Research Council.

REFERENCES

(1) Bateson, P.P.G. 1966. The characteristics and context of imprinting. Biol. Rev. 41: 177-220.

(2) Bateson, P.P.G. 1973. Internal influences on early learning in birds. In Constraints on Learning: Limitations and Predispositions, eds. R.A. Hinde and J. Stevenson Hinde, pp. 101-116. London: Academic Press.

(3) Bateson, P.P.G. 1978. Early experience and sexual preferences. In Biological Determinants of Sexual Behaviour, ed. J.B. Hutchison, pp. 29-54. Chichester: Wiley.

(4) Bateson, P[P.G.]. 1979. How do sensitive periods arise and what are they for? Anim. Behav. 27: 470-486.

(5) Bateson, P[P.G.]. 1982. Preferences for cousins in Japanese quail. Nature 295: 236-237.

(6) Bateson, P[P.G.]., ed. 1983. Mate Choice. Cambridge: Cambridge University Press.

(7) Bateson, P.P.G.; Horn, G.; and Rose, S.P.R. 1972. Effects of early experience on regional incorporation of precursors into RNA and protein in the chick brain. Brain Res. 39: 449-465.

(8) Bateson, P.P.G.; Horn, G.; and Rose, S.P.R. 1975. Imprinting: Correlations between behavior and incorporation of (^{14}C) Uracil into chick brain. Brain Res. 84: 207-220.

(9) Bateson, P.P.G., and Reese, E.P. 1969. The reinforcing properties of conspicuous stimuli in the imprinting situation. Anim. Behav. 17: 692-699.

(10) Bateson, P.P.G.; Rose, S.P.R.; and Horn, G. 1973. Imprinting: lasting effects on uracil incorporation into chick brain. Science 181: 576-578.

(11) Bradley, P., and Horn, G. 1978. Afferent connections of hyperstriatum ventrale in the chick brain. J. Physiol. 278: 46P.

(12) Bradley, P., and Horn, G. 1979. Efferent connections of hyperstriatum ventrale in the chick. J. Anat. 128: 414-415.

(13) Bradley, P.; Horn, G.; and Bateson, P. 1981. Imprinting: an electron microscopic study of chick hyperstriatum ventrale. Exp. Brain Res. 41: 115-120.

(14) Burtt, E.H. 1977. Some factors in the timing of parent-chick recognition in swallows. Anim. Behav. 25: 231-239.

(15) Cipolla-Neto, J.; Horn, G.; and McCabe, B.J. 1982. The effects of sequential lesions to the hyperstriatum ventrale on the retention of a preference acquired through imprinting. Exp. Brain Res. 48: 22-27.

(16) Cajal, S.R. 1911. Histologie du System Nerveux de l'Homme et des Vertebrates, vol. 2, pp. 886-890. Paris: Maloire.

(17) Colgan, P. 1983. Comparative Social Recognition. New York: Wiley.

(18) Collias, N.E. 1980. Basal telencephalon suffices for early socialization in chicks. Physiol. Behav. 24: 93-97.

(19) Hayashi, S., and Kimura, T. 1983. Degree of kinship as a factor regulating preferences among conspecifics in mice. Anim. Behav. 31: 81-85.

(20) Hebb, D.O. 1949. The Organisation of Behaviour. New York: Wiley.

(21) Hess, E.H. 1959. Imprinting. Science 130: 133-141.

(22) Hess, E.H. 1973. Imprinting. New York: Van Nostrand Reinhold.

(23) Hoffman, H.S., and Ratner, A.M. 1973. A reinforcement model of imprinting: implications for socialisation in monkeys and men. Psychol. Rev. 80: 527-544.

(24) Holmes, W.G., and Sherman, P.W. 1983. Kin recognition in animals. Am. Sci. 71: 46-55.

(25) Horn, G. 1981. Neural mechanisms of learning: an analysis of imprinting in the domestic chick. Proc. Roy. Soc. Lond. B 213: 101-137.

(26) Horn, G., and McCabe, B.J. 1984. Predispositions and preferences. Effects of imprinting of lesions to the chick brain. Anim. Behav., in press.

(27) Horn, G.; McCabe, B.J.; and Bateson, P.P.G. 1979. An autoradiographic study of the chick brain after imprinting. Brain Res. 168: 361-373.

(28) Horn, G.; Rose, S.P.R.; and Bateson, P.P.G. 1973. Monocular imprinting and regional incorporation of tritiated uracil into the brains of intact and 'split-brain' chicks. Brain Res. 56: 227-237.

(29) Immelmann, K. 1972. Sexual and other long-term aspects of imprinting in birds and other species. Adv. Stud. Behav. 4: 147-174.

(30) James, H. 1959. Flicker: An unconditioned stimulus for imprinting. Can. J. Psychol. 13: 59-67.

(31) Kohsaka, S.-I.; Takamatsu, K.; Aoki, E.; and Tsukada, Y. 1979. Metabolic mapping of chick brain after imprinting using [^{14}C]2-deoxyglucose technique. Brain Res. 172: 539-544.

(32) Lorenz, K. 1935. Der Kumpan in der Umwelt des Vogels. J. Ornithol. 83: 137-213; 289-413.

(33) McCabe, B.J.; Cipolla-Neto, J.; Horn, G.; and Bateson, P. 1982. Amnesic effects of bilateral lesions in the hyperstriatum ventrale of the chick after imprinting. Exp. Brain Res. 48: 13-21.

(34) McCabe, B.J.; Horn, G.; and Bateson, P.P.G. 1979. Effects of rhythmic hyperstriatal stimulation on chicks' preferences for visual flicker. Physiol. Behav. 23: 137-140.

(35) McCabe, B.J.; Horn, G.; and Bateson, P.P.G. 1981. Effects of restricted lesions of the chick forebrain on the acquisition of filial preferences during imprinting. Brain Res. 205: 29-37.

(36) Nottebohm, F. 1980. Brain pathways for vocal learning in birds. A review of the first 10 years. Progr. Psychobiol. Physiol. 9: 85-124.

(37) Rogers, L.J. 1980. Lateralisation in the avian brain. Bird Behav. 2: 1-12.

(38) Rose, S.P.R. 1981. What should a biochemistry of learning and memory be about? Neurosci. 6: 811-821.

(39) Salzen, E.A.; Parker, D.M.; and Williamson, A.J. 1975. A forebrain lesion preventing imprinting in domestic chicks. Exp. Brain Res. 2: 145-157.

(40) Salzen, E.A.; Williamson, A.J.; and Parker, D.M. 1979. The effects of forebrain lesions on innate and imprinted colour, brightness and shape preferences in domestic chicks. Behav. Proc. 4: 295-313.

(41) Sluckin, W. 1972. Imprinting and Early Learning, 2nd ed. London: Methuen.

The Biology of Learning, eds. P. Marler and H.S. Terrace, pp. 341-355. Dahlem Konferenzen 1984. Berlin, Heidelberg, New York, Tokyo: Springer-Verlag.

Signals, Conditioned Directed Movements, and Species-typical Response Predispositions in Nonmammalian Vertebrates

E. Hearst
Dept. of Psychology, Indiana University
Bloomington, IN 47405, USA

Abstract. Stereotyped directed actions often develop to signals of forthcoming positive or negative events, even though the events would occur regardless of these behaviors (Pavlovian conditioning). Such findings have questioned the "arbitrariness" of certain standard operant responses (e.g., the pigeon's keypeck) and demonstrated that Pavlovian procedures affect a much broader range and type of behaviors than previously suspected. This paper describes some signal-directed species-typical movements that appear in birds, fish, and reptiles, and relates their appearance and persistence to accounts of learning and performance based on extensions of Pavlov's notion of stimulus substitution and Skinner's concept of superstitious reinforcement. The inadequacies of these accounts suggest the need for approaches that more explicitly incorporate evolutionary, ethological, and developmental analyses. Experimental study of simple approach-withdrawal behavior, directed by signals of favorable and unfavorable events, may be useful; such sign tracking reactions are widespread and have obvious survival value.

INTRODUCTION

In Pavlovian conditioning a relatively neutral stimulus (the CS) signals some biologically significant event (the US), which occurs regardless of the subject's behavior during the CS. Aside perhaps from response habituation and sensitization produced by the mere repetition of a single stimulus, no form of behavioral modification due to experience has on the surface seemed simpler than that caused by pairings of a CS and US. For many years experimental psychologists devoted attention to the visceral and glandular changes (e.g., cardiac; salivary) or the local skeletal responses (e.g., eye blinks; paw flexions) that were acquired to CSs in restrained subjects. Pavlov's concept of stimulus substitution was regarded as one plausible description of the underlying process: The CS comes to act as a substitute for the US and thereby evokes a response closely resembling the one innately elicited by the US.

A common belief has been that Pavlovian conditioning primarily involves autonomic responses and isolated movements of parts of the body, whereas learning of spatially-directed, integrated ("voluntary") actions of the whole organism requires the application of instrumental (operant) conditioning procedures (7). In the latter case the US or reinforcer would not occur unless the appropriate and complete behavioral act was performed. Frequently, the experimenter had to carefully and systematically shape successive approximations to the final act in order to achieve its occurrence.

The discovery of autoshaping by Brown and Jenkins (2) was significant for many reasons, one being the demonstration that unrestrained organisms placed on a Pavlovian procedure exhibit strong directed movements toward a CS - even though these actions have no effect on whether the US is delivered. Pigeons exposed to pairings of a briefly illuminated key (CS) and presentation of grain (US) first show increases in activity during the CS, then orient toward the lighted key, and eventually begin to approach and peck directly at it. The usual operant shaping process is unnecessary for producing a keypeck. Furthermore, so long as CS-US pairings continue, this approach-peck sequence is maintained in virtually every pigeon.

Another provocative finding was Williams and Williams's (33) observation that pecking at the CS persists at reasonably high levels even when the experimenters penalize such pecking by cancelation ("omission") of the US scheduled on that trial. The bird is obviously better off not pecking than pecking, but its tendency to approach and contact a localized signal of food is so strong that the consequential loss of reinforcement is surprisingly ineffective. Similarly, Jenkins (12) found that birds would approach and peck a lit key located more than 90 cm away from the food dispenser, even though they were rarely able to scurry back in time to secure most of the available grain (food remained accessible for only 4 sec on each trial). Powerful species-typical behavior emerged that violates certain basic principles of learning, such as those related to the immediacy and maximization of reinforcement for a response. Breland

and Breland's (1) examples of the "misbehavior of organisms" illustrate the same theme.

Thus the pigeon's keypeck, a response formerly viewed by many workers as an arbitrary bit of behavior, selected for convenience in studies of operant conditioning, develops in appetitive settings where it is unnecessary and even counterproductive. In fairness to B.F. Skinner, who has been the main target of criticism regarding the supposed arbitrariness of the keypeck, I should note that years ago he characterized the pecking response as a "preformed unit" that may have special strength and coherence as a form of species-specific behavior ((25), p. 93). At any rate, in the past decade a number of us have come to view the autoshaping experiment in a much broader sense - not so much in terms of particular ways in which organisms contact a predictive signal, but as a manifestation of their tendency to direct behavior toward or away from an environmental feature (a sign) because of the relation between that feature and another (the US or reinforcer, in a typical experiment). "Sign tracking" was the label attached to such tendencies (12). Research concerned with relevant phenomena seems to have enhanced the significance and applicability of Pavlovian conditioning. The obvious adaptiveness of approach toward beneficial or safe stimuli and their signals, and of withdrawal from harmful or frustrative stimuli and their signals, appears especially pertinent for the behavior of organisms in natural settings.

In what follows, some examples of directed behaviors with respect to CSs are described for a variety of species, particularly but not exclusively nonmammalian vertebrates. (Extensive reviews and analyses of the literature can be found in (3, 12, 17, 23, 31).) Within the context of these observations, I try to list and examine several kinds of overall interpretations proposed to account for the appearance of certain responses. "Associative principles" do not receive much discussion. Rather, the stress is on how learning derived from exposure to relations between stimuli may be expressed in specific organismic action. Psychologists have not had too much success in predicting what response or patterns of response will occur during Pavlovian conditioning in freely-moving subjects. This failure may be primarily traceable to an unwarranted lack of interest in the natural behaviors of species when confronted with different kinds of biologically significant events, appetitive or aversive (cf. Bolles, this volume). My opinion is that many gaps between learning and performance can only be filled by cooperative work between scientists studying associative processes and those analyzing species-typical behaviors - or by individual scientists doing both jobs simultaneously. In other words, I am not too worried about the "generality of the laws of learning," but fret more about our ignorance concerning the translation of an organism's knowledge about environmental correlations into definite responses - which, after all, are the events modifiable by evolutionary selection.

STIMULUS SUBSTITUTION: CORRESPONDENCE, SURROGATION, AND ANTICIPATION

Complexity and diversity characterize the integrated movements occurring during signals of USs. They are hard to classify in a satisfactory way. This was not a real problem in earlier, traditional work on Pavlovian conditioning because the subject was restrained and the experimenter measured only conditioned and unconditioned responses (CRs and URs) from the same effector systems. In fact, in their definitions of Pavlovian (classical) conditioning, Konorski (15) and Gormezano and Kehoe (5) went beyond the pairing procedure used and included restrictions about CR-UR selection, too (see (9)).

When an animal is free to move about in its environment, the experimenter loses control over some factors rigidly regulated in those traditional studies. However, one gain is that the organism is behaving in a setting more like those encountered outside the laboratory. Stimulus substitution, as evaluated by CR-UR resemblance, then becomes an ambiguous concept for at least three reasons. These possibilities are not mutually exclusive on a given trial or within a given setting. First of all, the subject might perform the same directed response during the CS as it does during the US; for example, in the case of the pigeon exposed to keylight-food pairings, it would move to the place where food is about to appear and begin pecking in that area. I will call this outcome correspondence of CR and UR. On the other hand, the pigeon might treat the CS as if it were the US and approach and peck the key light. The CS becomes an object substitute for the US; let us call this type of result surrogation. Standard autoshaping provides an obvious example. Finally, the most unorthodox possibility is that motivated organisms are prewired to emit certain behavior patterns when in the presence of natural precursors of biologically significant stimuli. "Unconditioned" anticipatory or preparatory behaviors occurring during, say, search for or the solicitation of food - as opposed to behaviors happening after actual contact with food - might transfer to an artificial signal like a key light. Dogs direct begging and hunting behavior at a CS for food. Substitution in this sense presumes preorganized patterns of anticipation or procurement for a particular US, rather than patterns for its consumption. This third view is best described by Jenkins et al. (14).

If all these complications are not enough for the reader, who may (like me) have some difficulty separating the three alternatives, several other points can be mentioned before I provide additional specific examples of what subjects actually do in the presence of localizable CSs. Is it legitimate to consider approaching and eating from a grain dispenser as "URs"? After all, a bird must learn where the food is and when it is available and must go to the dispenser to obtain it; some critics would argue that these behaviors are "instrumentally" conditioned, precluding their straightforward analysis in terms of standard Pavlovian principles. Furthermore, can the behaviors that appear to a CS be maintained by "superstitious" or "parasitic" reinforcement? They are closely followed by the US, even though they are not required for the US. And how do the results relate to well-known categorizations of species-typical behaviors like appetitive-consummatory response or preparatory-consummatory response (15, 31)? Space is not available here to discuss all these issues; some of them receive attention in Hearst (9) and Holland et al. (this volume).

SOME SIGNAL-DIRECTED ACTIONS OF VARIOUS ORGANISMS

A sampling of experimental findings, described more fully elsewhere (12, 31, 32), can help place the above controversial questions or accounts in perspective. Pigeons exposed to visual signals of different USs behave differently toward the signals. Pecks at a food-predictive key light are evenly spaced, brief, and forceful, whereas pecks at a water-predictive key light are irregularly spaced, sustained, and relatively weak. The birds seem to be "eating" the former signal and "drinking" the latter. When the US is access to a female sexual partner, male pigeons soon begin to approach, nod, and bow to the CS; sometimes they coo, strut, circle, and peck the signal, too. Bengal monitor lizards exhibit open-mouthed bites at an illuminated key that is followed shortly afterward by delivery of a defrosted mouse pup. Fish like tilapia and mullets often show gobbling movements towards a lit key paired with presentation of pieces of shrimp, and goldfish have been autoshaped in several investigations to strike a Plexiglas target whose illumination signals food. Archer fish squirt water at a red light that precedes delivery of a fruit fly on the surface of the water. While squirting is not involved in consuming the fly, it is the response performed when the fish spots an insect on branches above water; it causes the insect to topple to the water's surface, where the prey is readily consumed. To mention one example with mammals, rats come to gnaw a lever whose illumination and insertion into the chamber are followed by inevitable delivery of a food pellet. Thus the two most common responses in operant conditioning experiments, keypecks and lever contacts or presses, appear on Pavlovian procedures - where the response is not required for the reinforcer.

In other instances the response to the CS is not so easily characterized as somehow resembling behavior actually evoked by the US. Wasserman (30) found that baby chicks peck a lit key signaling the delivery of heat in a cold chamber. However, their responses to the onset of the heat lamp do not include pecking; instead, cessation of activity, twittering, and wing extension occur. Interestingly, Hogan (13) argued that approaching, pecking, and snuggling are part of the natural heat-seeking behavior of young chicks; they peck their mother's underside, a response that is part of the action pattern stimulating the hen to sit and brood. Woodruff and Starr (35) never allowed their baby chicks to eat or drink naturally, i.e., by approaching and contacting food or water. The chicks were force-fed food, water, or sand during rearing, as well as during later exposure to pairings of a lit key and one of these three substances. They still directed characteristic feeding movements (approach, peck, scratch) or drinking movements (approach, sustained contact, "scooping") towards signals of food or water, respectively. The delivery of sand, which had (like food and water) elicited swallowing during force feeding, did not produce movements toward a signal for it. "Phylogenetically preorganized behavior patterns" were said to be triggered by distal stimuli paired with biologically significant proximal stimulation, even though the USs were not obtained in a natural way. Woodruff and Starr speculated on the relationship between some of their results and Lorenz's notion of the "innate schoolmaster," whereby the instinctive behavior of an organism teaches its offspring or other conspecifics about biologically important events.

As might be expected, the form and probability of signal-directed actions depend on the type of signal, too. For example, Timberlake and Grant (29) delivered food to rats soon after another rat, tethered on a platform, was automatically introduced into the chamber. Instead of gnawing or biting their visitor - as one version of a stimulus substitution account might predict - the subjects began to approach, sniff at, and show social contact (pawing, grooming) toward the conspecific CS. Rashotte (personal communication, 1974) found that when a standard response key signaling food was imbedded in the breast of a stuffed pigeon, species-typical aggressive behavior, as well as pecking responses, were directed at the signal; pecking rates were lower than when the response key was positioned, as is usual, on the wall of the chamber. In the former case, there often appeared to be a conflict between keypecks and conspecific "attacks" (stereotyped bows, vocalizations, wing strikes, beak swipes, pecks at the head or neck, etc.). Pigeons oriented toward and approached both kinds of signal, but the form of the contact response depended on the "social" context.

Observations from other laboratories and my own indicate that auditory signals for food in pigeons are more likely to produce directed movements toward the grain dispenser than toward the source of the sound. This difference from the effects described above may be partially attributable to the localizability of a sound as compared to a small light. Thus when a diffuse light illuminating the chamber acts as a signal, the most common response also seems to be a movement toward the site of food. LoLordo (18) has presented some relevant but more interesting results. When a compound stimulus (simultaneous presentation of a tone and illumination of the whole chamber with red light) signaled food delivery in pigeons, the compound stimulus - as well as the visual element presented alone, on subsequent test trials - produced pecking at or near the location of grain; tests with the tone element yielded very little such behavior. On the other hand, when the same compound stimulus signaled shock, pigeons displayed head bobbing and prancing behavior to it; tests with the separate auditory and visual elements revealed that the tone was much more likely to evoke prancing and bobbing than the light. In appetitive situations, visual stimuli seem to exert primary control, but with noxious USs auditory stimuli may normally overshadow visual ones in the pigeon.

There have been some noteworthy failures to obtain autoshaping in settings where it might be expected. Powell and his colleagues (22) reported that very few crows come to peck a lighted key signaling delivery of live meal worms or canned dog food, although crows perform well on operant conditioning procedures requiring the keypeck response for these reinforcers. Wilson (34) obtained analogous results in other corvids; rooks, jackdaws, European jays, and crows displayed some acquisition of pecking to a lit key signaling food, but all the subjects soon stopped responding. Powell suggested that crows engage in little pecking in their natural environments because they highly favor eating meat and live insect larvae rather than grain. In contrast, pigeons peck constantly as they feed on small bits of grain and seeds. Nevertheless, in Powell's and Wilson's experiments it would be worth knowing whether the corvids continued to approach the signal, and whether any consistent forms of behavior other than pecking persisted during its presentations. Perhaps responses including components of corvids' normal foraging or consummatory patterns occurred.

THE QUESTION OF SUPERSTITION

Because Pavlovian CRs are contiguous with the reinforcer (US), the possibility exists that some if not all examples of signal-directed behavior may fit the type of account offered in Skinner's classic observations about "superstition in the pigeon" (24). He found that pigeons receiving grain every 15 sec regardless of their behavior still showed acquisition of regular but idiosyncratic movements in the situation (the procedure was the same as what Pavlov called temporal or time conditioning). For instance, one pigeon pecked toward the floor between food presentations and another bird hopped from side to side. Skinner contended that mere accidental pairings of any response with grain would be sufficient to strengthen and maintain the behavior, despite its irrelevance for food delivery and despite the intermittency of the response-food conjunctions.

However, the work of Staddon and Simmelhag (26) cast doubt on the generality of Skinner's results and interpretation (as they might be applied specifically to autoshaping, see also (12)). In a systematic replication of Skinner's experiment, Staddon and Simmelhag observed that behaviors (e.g., wing flapping, head bobbing) that occurred often at the time of reinforcement during subjects' early exposure to the procedure disappeared eventually or became confined to periods not long after grain delivery ("interim activities"). By the end of twenty or thirty sessions, most subjects were pecking at the front wall as the scheduled grain arrived ("terminal activities") - even though the pecking response had been relatively rare during the initial sessions. There is no plausible way to explain Staddon and Simmelhag's results by superstitious conditioning. On that basis, asymptotic behaviors should not be so uniform from subject to subject, and the behaviors that initially happened to coincide with grain delivery should predominate from then on. As in autoshaping experiments with pigeons, a particular directed response displaying great intraspecies consistency rather than idiosyncracy eventually emerged in Staddon and Simmelhag's work.

In an extensive series of recent experiments, two of my colleagues at Indiana University, Gary Lucas and William Timberlake, confirmed some of Staddon and Simmelhag's results but not others. They have found that responses initially contiguous with grain delivery are not the behaviors that prevail after a few sessions of conditioning. Thus their findings, like Staddon and Simmelhag's, contradict accounts of behavioral change in terms of superstitious conditioning. However, Lucas and Timberlake observe that pecking is not a very probable response at the time of grain delivery. Instead, most subjects are exhibiting fairly uniform behavior of another kind: head stretching and bobbing near the front wall, with much side-to-side movement. After experimentally rejecting the possibility that these behaviors are part of a general foraging sequence, Lucas and Timberlake now speculate that these patterns resemble species-typical food-begging behavior in pigeons (see also (14)).

Also pertinent to the question of superstitious reinforcement is the mass of research (see (17)) aimed at assessing the relative importance of CS-reinforcer vs. response-reinforcer relations in autoshaping. The omission procedure, described earlier in connection with Williams and Williams's results, pits these potential controlling relations against each other (all USs follow a CS, but USs never occur after the designated response). Not only is pecking at the CS in pigeons and chicks maintained at comparatively high levels even when it prevents the US, but "approach" toward the CS also remains strong if that movement is the one canceling scheduled USs (21). Obviously, superstitious reinforcement (response-reinforcer contiguity) cannot account for the persistence of pecking or approaching here. However, with other species, particularly mammals, the omission procedure is often much more effective in eliminating CS-directed behavior.

Incidentally, the omission technique is useful for analyzing a variety of other classically conditioned movements. When a visual CS precedes presentation of a mirror (US) in Siamese fighting fish, different components of the acquired display pattern during CS are differentially affected by their specific prevention of the US. Fin erection and gill extension continue to occur frequently during the CS, but "frontal approach" ceases (20). Thus it is unlikely that all parts of an organized pattern of movements will show persistence on omission procedures. Such research is related to the question of the "inhibitability" of various responses in different situations and species, and to the whole topic of involuntary behavior (6, 7). Skinner (25) remarked that reflexes were traditionally considered involuntary, not so much because "they could not be willed as that they could not be willed against."

Another way of checking on whether superstitious reinforcement could account for the growth and maintenance of various conditioned directed movements to CSs involves actual prevention of the movements while the subjects initially "observe" pairings of the CS and US. The overall conclusion from this research, which has been conducted mainly with avians, is that such an observation period produces a strong subsequent tendency to approach and contact the CS when the barrier or harness is removed (3, 9, 12, 31). These outcomes bear on the old issue of latent learning, and on general controversies between response-centered and perception-centered theories of learning (6, 8, 9, 12). In any event, the findings provide additional ammunition against the role of superstitious reinforcement in the findings described above.

SIGN TRACKING AND APPROACH-WITHDRAWAL

Currently, advance predictions are difficult to make about the specific detailed movements that organisms will direct at particular signals. I have mentioned some of the factors that influence response selection, but I cannot list any trustworthy general guidelines to follow. More needs to be known about species-typical action patterns released by various CSs, USs, and their natural precursors in different settings, and about how these movements compare to CRs appearing to artificial stimuli in freely-moving organisms exposed to different conditioning procedures (see Hollis, this volume). The concept of stimulus substitution - whether phrased in terms of correspondence, surrogation, anticipation, or any combination of the three - is too vague and can probably be twisted to account for almost any imaginable result (27). It seems certain that signal-directed CRs are multiply determined anyway, and that no simple principle will handle all the findings and different responses that occur. Lorenz and Timberlake have argued (see (7, 27, 28)) that in a Pavlovian conditioning experiment a whole system of functionally related behaviors is being conditioned (e.g., food-soliciting patterns in dogs).

A less complicated but perhaps more feasible experimental strategy would be to focus first on generalized approach and withdrawal reactions, rather than on very specific types of conditioned directed movements. Numerous writers have commented on the survival value, and ubiquity throughout the animal kingdom, of responses that bring organisms closer to stimuli of a positive character and further away from stimuli of a negative character (pain, threat, danger). Animals possess unconditioned response predispositions of these kinds, and it seems reasonable to expect that corresponding forms of directional behavior would transfer to signals of innately positive and negative objects or events.

The available evidence supports this hypothesis and reveals that approach-withdrawal measures are often more sensitive to manipulations of Pavlovian associative factors than are responses like the pigeon's keypeck. Pigeons approach localizable visual stimuli that signal appetitive USs like food, and the strength of their approach behavior depends on how well the signal predicts the US (10). There is an interesting negative counterpart to this kind of finding: Pigeons withdraw from a visual signal that indicates an appetitive US is not coming (6). The greater the frequency of that US in the absence of the signal, the stronger is the tendency to move away or stay away from the signal (11). In these studies the spatial position of the pigeons was monitored on every trial by means of switches beneath the floor of the chamber.

Furthermore, there is evidence (3, 12, 16, 19, 31) that signals of the forthcoming presence or absence of aversive stimuli like painful shock produce directional tendencies opposite to those described for appetitive signals in the last paragraph. Subjects move away from localized cues predicting unavoidable shock and they approach cues that signal safety from shock. Because these results have been obtained in situations in which the US occurs or does not occur regardless of the subject's behavior, the approach-withdrawal reactions have no obvious instrumental effect. Instead, they seem to represent natural response predispositions to signals of forthcoming danger or safety.

Thus my final message is close to one expressed by Dickinson and Mackintosh ((4), p. 591): "It is perhaps of even greater long-term significance [than autoshaping] that general approach and withdrawal behavior may be conditioned by purely classical [Pavlovian] contingencies ... Since it must be to an animal's general advantage to approach (and investigate and manipulate) stimuli associated with appetitive reinforcers and to withdraw from stimuli or places associated with danger, it is now clear that classical conditioning plays an unexpectedly important role in modifying an animal's behavior."

The leap from Pavlovian conditioning in the laboratory to its operation in natural biological settings is not a move that every biologist or psychologist will readily accept as justifiable or valuable. However, now that our interdisciplinary forces are better joined, the possibility can be assessed in a more congenial, mutually respectful, and cooperative atmosphere.

Acknowledgement. This work was supported by National Institute of Mental Health Grant MH 19300. I thank H.M. Jenkins, who greatly influenced my thinking about several topics covered in this paper.

REFERENCES

(1) Breland, K., and Breland, M. 1961. The misbehavior of organisms. Am. Psychol. 16: 681-684.

(2) Brown, P.L., and Jenkins, H.M. 1968. Auto-shaping of the pigeon's key-peck. J. Exp. Anal. Behav. 11: 1-8.

(3) Buzsáki, G. 1982. The "where is it?" reflex: autoshaping the orienting response. J. Exp. Anal. Behav. 37: 461-484.

(4) Dickinson, A., and Mackintosh, N.J. 1978. Classical conditioning in animals. Ann. Rev. Psychol. 29: 587-612.

(5) Gormezano, I., and Kehoe, E.J. 1975. Classical conditioning: some methodological-conceptual issues. In Handbook of Learning and Cognitive Processes, ed. W.K. Estes, vol. 2, pp. 143-179. Hillsdale, NJ: Lawrence Erlbaum Associates.

(6) Hearst, E. 1975. Pavlovian conditioning and directed movements. In The Psychology of Learning and Motivation, ed. G.H. Bower, vol. 9, pp. 215-262. New York: Academic.

(7) Hearst, E. 1975. The classical-instrumental distinction: reflexes, voluntary behavior, and categories of associative learning. In Handbook of Learning and Cognitive Processes, ed. W.K. Estes, vol 2, pp. 181-223. Hillsdale, NJ: Lawrence Erlbaum Associates.

(8) Hearst, E. 1978. Stimulus relationships and feature selection in learning and behavior. In Cognitive Processes in Animal Behavior, eds. S.H. Hulse, H. Fowler, and W.K. Honig, pp. 51-88. Hillsdale, NJ: Lawrence Erlbaum Associates.

(9) Hearst, E. 1979. Classical conditioning as the formation of interstimulus associations: stimulus substitution, parasitic reinforcement, and autoshaping. In Mechanisms of Learning and Motivation, eds. A. Dickinson and R.A. Boakes, pp. 19-52. Hillsdale, NJ: Lawrence Erlbaum Associates.

(10) Hearst, E.; Bottjer, S.W.; and Walker, E. 1980. Conditioned approach-withdrawal behavior and some signal-food relations in pigeons: performance and positive vs. negative "associative strength." Bull. Psychonom. Soc. 16: 183-186.

(11) Hearst, E., and Franklin, S. 1977. Positive and negative relations between a signal and food: approach-withdrawal behavior to the signal. J. Exp. Psychol.: Anim. Behav. Proc. 3: 37-52.

(12) Hearst, E., and Jenkins, H.M. 1974. Sign-Tracking: The Stimulus-Reinforcer Relation and Directed Action. Austin, TX: The Psychonomic Society.

(13) Hogan, J.A. 1974. Responses in Pavlovian conditioning studies. Science 186: 156-157.

(14) Jenkins, H.M.; Barrera, F.J.; Ireland, C.; and Woodside, B. 1978. Signal-centered action patterns of dogs in appetitive classical conditioning. Learn. Motiv. 9: 272-296.

(15) Konorski, J. 1967. Integrative Activity of the Brain. Chicago: University of Chicago Press.

(16) Leclerc, R., and Reberg, D. 1980. Sign-tracking in aversive conditioning. Learn. Motiv. 11: 302-317.

(17) Locurto, C.M.; Terrace, H.S.; and Gibbon, J. 1981. Autoshaping and Conditioning Theory. New York: Academic.

(18) LoLordo, V.M. 1979. Selective associations. In Mechanisms of Learning and Motivation, eds. A. Dickinson and R.A. Boakes, pp. 367-398. Hillsdale, NJ: Lawrence Erlbaum Associates.

(19) Masterson, F.A., and Crawford, M. 1982. The defense motivation system: a theory of avoidance behavior. Behav. Brain Sci. 5: 661-696.

(20) Murray, C.S. 1973. Conditioning Betta splendens. Ph.D. Thesis, University of Pennsylvania, Philadelphia.

(21) Peden, B.; Browne, M.P.; and Hearst, E. 1977. Persistent approaches to a signal for food despite food omission for approaching. J. Exp. Psychol.: Anim. Behav. Proc. 3: 377-399.

(22) Powell, R.W.; Kelly, W.; and Santisteban, D. 1975. Response-independent reinforcement in the crow: failure to obtain autoshaping or positive automaintenance. Bull. Psychonom. Soc. 6: 513-516.

(23) Schwartz, B., and Gamzu, E. 1977. Pavlovian control of operant behavior. In Handbook of Operant Behavior, eds. W.K. Honig and J.E.R. Staddon, pp. 53-97. Englewood Cliffs, NJ: Prentice-Hall.

(24) Skinner, B.F. 1948. "Superstition" in the pigeon. J. Exp. Psychol. 38: 168-172.

(25) Skinner, B.F. 1953. Science and Human Behavior. New York: Macmillan.

(26) Staddon, J.E.R., and Simmelhag, V.L. 1971. The "superstition" experiment: a reexamination of its implications for the principles of adaptive behavior. Psychol. Rev. 78: 3-43.

(27) Timberlake, W. 1983. Rats' responses to a moving object related to food or water: a behavior-systems analysis. Anim. Learn. Behav. 11: 309-320.

(28) Timberlake, W. 1983. The functional organization of appetitive behavior: behavior systems and learning. In Advances in Analysis of Behaviour, eds. M.D. Zeiler and P. Harzem, vol. 3, pp. 177-221. Chichester: Wiley.

(29) Timberlake, W., and Grant, D.L. 1975. Auto-shaping in rats to the presentation of another rat predicting food. Science 190: 690-692.

(30) Wasserman, E.A. 1973. Pavlovian conditioning with heat reinforcement produces stimulus-directed pecking in chicks. Science 181: 875-877.

(31) Wasserman, E.A. 1981. Response evocation in autoshaping: contributions of cognitive and comparative-evolutionary analyses to an understanding of directed action. In Autoshaping and Conditioning Theory, eds. C.M. Locurto, H.S. Terrace, and J. Gibbon, pp. 21-54. New York: Academic.

(32) Waxman, H.M., and McCleave, J.D. 1978. Autoshaping in the archer fish (Toxotes chatareus). Behav. Biol. 22: 541-544.

(33) Williams, D.R., and Williams, H. 1969. Automaintenance in the pigeon: sustained pecking despite contingent nonreinforcement. J. Exp. Anal. Behav. 12: 511-520.

(34) Wilson, B. 1978. Autoshaping in pigeons and corvids. Ph.D. Thesis, University of Sussex, England.

(35) Woodruff, G., and Starr, M.D. 1978. Autoshaping of initial feeding and drinking reactions in newly hatched chicks. Anim. Learn. Behav. 6: 265-272.

The Biology of Learning, eds. P. Marler and H.S. Terrace, pp. 357-371. Dahlem Konferenzen 1984. Berlin, Heidelberg, New York, Tokyo: Springer-Verlag.

Cause and Function of Animal Learning Processes

K. L. Hollis
Dept. of Psychology, Mount Holyoke College
South Hadley, MA 01075, USA

Abstract. The General Process approach to learning has sustained much criticism for its neglect of species-typical variation. Nonetheless, this approach has been fruitful in identifying many commonalities of animal learning. Pavlovian conditioning and habituation are examples of two general processes which are found in all vertebrate, and many invertebrate, species. Similarities in structure, which these learning phenomena illustrate, often suggest commonality of function. In this paper, the General Process approach is expanded to explore functional, as well as causal, questions. A major theme of this paper is that the reconciliation of learning theory and natural behavior depends upon attempts to link causal and functional analyses.

INTRODUCTION

The goal of most psychological studies of animal learning has been to elucidate the general laws, or processes, which govern behavioral change ((36); see also papers by Jenkins and Terrace, both this volume). This *General Process* approach to learning has had some unfortunate consequences, as any number of people have emphasized. Too few species have been investigated in too many arbitrary, unnatural environments (25). Yet, the goal of General Process Theory has been to discover the basic mechanisms, or causes, of learning and not to detail the perturbations which inevitably exist. If these general mechanisms fail to take full account of variations between species, or if they fall short of a complete description of natural behavior, we should not be too surprised. Nor should the lack of absolute generality require us to discard the whole approach and start anew, as has been suggested (23). An alternative view is to regard General Process Theory as a theoretical device: It is no more general, but no less heuristic, than "the" vertebrate eye. Viewed in this way, animal learning theory is the study of what are more accurately called generalizations of the learning process. The vertebrate eye, too, is a generalization. Its description, richly detailed, is a useful guide to a structure and its function; in reality, however, that detailed description fits hardly a single species. The same might be said of our use of such terms as mammals, monogamy, and motor neurons.

As a study of generalizations, the General Process approach has been fruitful in identifying learning phenomena which are indeed common to all vertebrates (29). Pavlovian conditioning and habituation, which are discussed later in this paper, are two examples. Yet, if we are to reconcile these generalizations with the behavior of animals in the wild, then our theories also must account for any departures from the general process. To do so requires an understanding of function. As is true of the vertebrate eye, form follows function. And, variations in a common form are the result of modifications of a common function. And so I suggest that a General Process approach be expanded to explore functional (ultimate), as well as causal (proximate), questions. Why does Pavlovian conditioning, despite the many variations in this general process, cause all animals to respond in anticipation of certain events? Of what practical significance is habituation and dishabituation and is it the same in fish, amphibia, reptiles, and mammals? The present paper is an attempt to explore some of these functional questions in the familiar conditioning phenomena of the psychological laboratory.

PAVLOVIAN CONDITIONING

Signaled presentations of food, rivals, predators, and mates produce anticipatory responses in all vertebrate, and many invertebrate, species ((24, 29, 35); see also papers by Sahley and Quinn, both this volume). This is *Pavlovian conditioning*. In the classic study of this phenomenon, a bell was rung just prior to the injection of meat powder into a dog's mouth. After a few such pairings of bell and meat powder, the bell alone was able to evoke an anticipatory salivary response (30).

But the anticipatory nature of the learned response (the *conditional response*, or *CR*) is not its only characteristic. The CR is often signal-directed, and it consists of elaborate species-typical behavior. A male Betta splendens, for example, will perform an aggressive display in front of a light which, in the past, has signaled the appearance of a rival (41). Similarly, an archer fish, which obtains its food by squirting insects off overhanging plants, will squirt a light which has signaled the appearance of prey items (44). As well, Pavlovian signals are capable of eliciting courtship dances, threat displays, and various antipredator behaviors ((20); see also paper by Hearst, this volume).

Are these signal-directed conditional responses adaptive? What advantage does an animal gain by responding in anticipation of certain events? Although our knowledge of the Pavlovian conditioning process has expanded considerably in recent years, psychological research has concentrated on causal mechanisms. While the biological function of these anticipatory, signal-directed CRs has been recognized theoretically, it has been ignored experimentally. Elsewhere (20) I have suggested that the function of Pavlovian conditioning is to enable the animal to optimize interaction with biologically important events. The performance of a CR allows the animal better to deal with food, rivals, predators, and mates. In the same way that an anticipatory salivary response, the classic CR of Pavlov's experiments, is known to improve digestion (7), an anticipatory freezing response might improve predator avoidance and an anticipatory aggressive display might increase the likelihood of successful resource or mate defense.

Some evidence for this preparatory role of the CR comes from an investigation of the conditional aggressive display in territorial male fish. If Pavlovian conditioning of this response confers an advantage on the territory holder, then a signaled intrusion should enable a male to defend his territory more aggressively than when the intrusion is unsignaled. Several recent experiments in my laboratory tested this hypothesis (21). In one experiment, pairs of male blue gouramies (Trichogaster trichopterus) formed the basis of a comparison between Pavlovian conditioning and a control group treatment. Pavlovian males daily received fifteen paired presentations of a red light (the *conditional stimulus*, or *CS*) and a rival male (the *unconditional stimulus*, or *US*). Control males also were presented with these stimuli; however, fifteen CSs were presented in the morning and fifteen USs were presented, on average, seven hours later (an "explicitly unpaired" procedure). Following twenty-four days of training, pair members confronted one another for the first time in an encounter preceded by the CS. Thus, although both males were equally familiar with its presentation, the CS was predictive of intrusion only for the Pavlovian male. The benefit, if any, to the classically conditioned male of that signaling was assessed in terms of each fish's ability to defend its own territory against invasion. In that experiment, and also in a second experiment with a different control group, classically conditioned males delivered significantly more bites and tailbeatings, and won more of the territorial contests, than their control group opponents. These results suggest that Pavlovian conditioning could play an important role in the natural habitat of these fish: By means of the conditional aggressive response, rivals could be confronted at the territory boundary by an already aggressively displaying owner, strategically ready for battle.

A similar methodology was used to explore the function of the conditional courtship display in this same species (Hollis, unpublished manuscript). During the training phase, males received either Pavlovian conditioning pairings of a red light CS and receptive female US, or explicitly unpaired presentations of CS and US. In the subsequent test encounter a female was preceded by the CS used in training. In that encounter, Pavlovian males greeted the signaled females with significantly more courtship appeasement behavior, and significantly less aggression, than control group males. Here, too, the adaptive value of the CR would seem to lie in its preparatory nature. In this species, the initial response of a territorial male to a female is often very aggressive, posing a great risk to the female (9), and causing her to flee the territory altogether. This response to the female is not uncommon in territorial species (5). Nor is it surprising given the fact that female courtship and territory defense are seasonally simultaneous male behaviors. Normally, the continued presence of a female results in the gradual inhibition of male aggression. However, were a male somehow able to foreshorten this process and begin to court the female sooner, his chances of a successful mating might be improved greatly. Pavlovian conditioning is a mechanism which could effect such a strategy.

Generally, these experiments would suggest that the function of signal-elicited courtship is to provide the classically conditioned male (or female) with a competitive edge or to insure that mating occurs with greater speed and efficiency. Such efficiency would help avoid predation, a major risk imposed on mating behavior (9). Results similar to those with blue gouramies have been obtained in rats and quail and offer additional support for this interpretation. Signaled presentations of a female produced anticipatory secretion of testosterone in male rats (14) and anticipatory courtship behavior in male Japanese quail (13). When these classically conditioned males encountered females, they were able to mount (12) and ejaculate (Zamble and Hadad, unpublished manuscript) sooner than the control animals of those experiments.

Results such as these support the CR's hypothesized preparatory function. A serious objection to this analysis, however, is that the CR is an epiphenomenon: The biological function of Pavlovian conditioning, as yet to be discovered, lies elsewhere, and the CR is merely a by-product of the conditioning process, perhaps an accidental "false start." This account of the CR, and especially the notion that the CR is a false start, seems unlikely. The CR is not always similar to the response (the *unconditional response*, or *UR*) elicited by the US. Even the prototypic salivary CR differs significantly from the UR in its enzymatic content and pH (7). In some cases the CR is in the opposite direction of the UR (37) or it may not involve even the same response system (22, 43). To cite but one example, some animals flee from predators (USs) and the flight response is accompanied by a heart rate increase; however, in the presence of CSs predicting the appearance of those same predators, the heart rate drops precipitously and the animal freezes instead (6, 10). Moreover, the form of the CR varies with the type, the location, the localizability, the timing, and the intensity of the CS (18, 19, 28). In rats, CRs evoked by an auditory CS differ substantially from CRs evoked by visual CSs; also, localizability of the CS is an important factor in determining whether a particular behavior, rearing, occurs (18). Most importantly, however, the ability of a functional interpretation to explain these variations of the anticipatory conditional response, and to predict others (20), suggests that a functional analysis is correctly centered on the CR.

Biological function is also a good heuristic for understanding the often bewildering number of phenomena assumed under the rubric, Pavlovian conditioning. If Pavlovian conditioning is a mechanism with which animals reduce the unpredictability of certain events - in some sense, decipher cause and effect - then it should contain safeguards for effacing spurious signals. *Latent inhibition* and *blocking* accomplish this. The ability of a CS to evoke a conditional response is reduced substantially when that CS is presented by itself on a few occasions, before it is paired with the US (latent inhibition), or if the CS is added later, in combination with another, already established CS (blocking). Another safeguard of the conditioning process is *extinction*, which functions to denigrate those signals whose current reliability is questionable: The conditional response quickly wanes when the CS is no longer followed by the US. Moreover, the ability to learn that some naturally occurring signals reliably foreshadow the appearance of food, rivals, predators, or mates should be no more important than learning that, in the presence of other, equally reliable, signals certain events never occur. Animals do seem capable of learning this and the phenomenon is called *conditioned inhibition*. Within this same functional framework, a capacity for *discrimination* and *generalization* enables an animal to thresh reliable signals from the confusion of irrelevant stimulus events. And, finally, *higher-order conditioning* permits associations to be formed between the signals themselves; thus, logical worlds are built of concatenated stimulus events.

Thus far, I have emphasized how functional analyses might further our understanding of causal mechanisms. However, this theoretical mutualism could work in reverse as well: An appreciation of causal mechanisms can further our understanding of function. To illustrate, suppose that we were to discover that several naturally occurring phenomena, which have been studied independently of one another, were controlled by the same learning mechanism. We would then be able to predict that they shared many common characteristics of that general process. And such commonalities, undoubtedly, would have implications for functional analyses. For example, elsewhere (20) I discuss the possibility that forming a search image and learning to recognize food may be Pavlovian in origin. With these let us now consider yet another example: Animals which resemble inedible or noxious prey, and thereby exploit a would-be predator's learned aversion, are depending upon Pavlovian conditioning for their protection. The deceit is called *Batesian mimicry* (8).

Not all examples of Batesian mimicry need rely upon Pavlovian conditioning. If the predator's aversion is innate, then the mimic achieves its protection by possessing those stimulus features which spontaneously release prey avoidance (2). If, however, the predator must learn to avoid noxious prey, avoidance of the mimic is based upon Pavlovian *stimulus generalization*: Stimuli which closely resemble an effective CS (the noxious model) also are capable of eliciting the CR (avoidance behavior).

Let us imagine, for the moment, that the role of Pavlovian conditioning in this example of avoidance behavior has just been established, and now let us explore the causal and functional predictions which would result from such a discovery. Functionally speaking, it would be adaptive for mimics to be scarce during the hazardous period when the predator is learning about the noxious prey (8). Later, however, when the avoidance is well established, the mimics probably can risk being more numerous. The ratio of models to mimics is also important at the causal level, and conditioning experiments could be used to support these functional predictions. Predation upon models constitutes a Pavlovian conditioning pairing of stimulus cues, the CS, followed by the noxious effects of stings or distastefulness, the US. Predation upon mimics, on the other hand, constitutes CS-alone trials; the CS is present at the time of predation but is not followed by the noxious US. During the acquisition of an association between CS and US, we know that the strength of that association depends upon the ratio of the number of CS pre-exposures to the number of CS-US pairings (28). Once this functional relationship were determined in simple conditioning experiments, we could then make predictions about the relative abundance of mimics to models while the predator is still learning. Likewise, we know that once an association has been formed, the ratio can afford to vary in the direction of greater CS-alone trials (28). Here, too, another conditioning experiment could determine the optimum of mimic abundance.

Some of these functional predictions are born out by bumblebee models, hover-fly mimics, and two of their predators, the red-winged blackbird and the common grackle (11). Hover-flies emerge early in the season and are then more abundant than their bumblebee models. By flying earlier, the edible mimic avoids the most hazardous period of the year when fledgling insectivorous birds are first learning about the noxious model. Naive birds experience their first few encounters with bumblebees later in the season, when the mimics are scarce. Although these same birds will be foraging the next spring amidst a surfeit of edible hover-flies, the lesson in avoidance will have been learned by then, and the mimic will be able to exploit its deceptive resemblance to the bumblebee.

Further experiments with these species could test the more specific predictions regarding relative mimic abundance. In addition, several other interesting predictions are worth mentioning here. We know that the strength of the CR is directly related to US intensity which, in this example, would be the severity of the model-induced malaise. One might predict then that, for a given ratio of models to mimics, the protection afforded mimics varies directly with the noxiousness of the model. In itself, this prediction might seem obvious; however, it suggests an interesting corollary. Might the severity of the model-induced malaise affect the optimum ratio of mimics to models? Functionally speaking, it would make sense that predators would be more wary of some models than others; the cost of making some mistakes might be too great a risk. If this were so, then the mimics of especially nasty prey probably could risk being more numerous than mimics of prey whose avoidance by predators was less strong. To my knowledge, no conditioning experiments have explored the effect of US intensity on the degree to which CS pre-exposures retard learning or the effective ratio of CS-alone to CS-US trials.

Whether Pavlovian conditioning actually contributes in any way to an avoidance of models and subsequent generalization to their mimics remains to be demonstrated experimentally. However, this particular example is especially intriguing. Elsewhere in the literature (38) is a report that red-winged blackbirds, one of the hover-fly predators, direct anticipatory conditional responses, food pecks, at visual cues which have been paired with food. And, much research on search image formation has been carried out in this species (1). Thus the suggestion that the same learning process, Pavlovian conditioning, is responsible for these birds' ability to avoid - as well as to locate - certain food items is an exciting possibility.

HABITUATION

In the preceding section, a few examples were chosen to demonstrate the fruitful synergism which can exist between causal and functional analyses of Pavlovian conditioning. Here, this same approach is extended to another learning phenomenon, *habituation*.

Habituation is the progressive decrease in responsiveness resulting from the repeated presentation of a stimulus (40). From both causal and functional points of view, habituation is an interesting challenge for a General Process approach; it is clearly the most widespread of the learning phenomena presently studied and, as such, any claims of theoretical generality must be considered carefully. According to Hodos ((17), p. 42), for example, "habituation appears to be a universal phenomenon in all organisms, including protozoans." However, all instances of attenuated responsiveness are not examples of habituation; some are merely the result of sensory adaptation or muscular fatigue. True habituation is a phenomenon of neural memory: The response to repeated sensory stimulation is choked off somewhere in the brain - or nerve net - of the animal. That is, habituation is a learning phenomenon. Nonetheless, even if we were to include only those examples where the decrease in responsiveness can be shown to be more centrally organized and is not attributable to receptor or effector fatigue, is it likely that this phenomenon is the same in such diverse groups of animals as anemones and amphibia, mollusks and mammals?

Surprisingly, the answer to this question seems to be, "Yes." Following an encyclopedic review of the habituation literature in fish, amphibia, reptiles, birds, and mammals, Macphail ((29), p. 192) concludes that "there are neither qualitative nor quantitative differences" in habituation within vertebrate species. And, although the range of responses may be greater in vertebrates, researchers of habituation in invertebrates (4, 24) have arrived at a similar conclusion. There is an obvious explanation of the generality of this learning phenomenon. It is, simply, that habituation is programmable with a minimum of neural circuitry (45) and is thus accessible to all metazoans, even those animals whose brains are little more than a collection of a few neurons. Furthermore, habituation serves a very simple but universal purpose, response economy, and is as likely to be advantageous to a snail as it is to a fish, or bird, or even a human. A General Process approach to habituation will, I believe, reveal these common expressions of cause and function.

Several characteristics of habituated responses are found in all vertebrate, and invertebrate, species (15, 29, 39, 40, 45). The rate at which the response decrement proceeds, for example, is known to vary with the response system measured. Orienting responses to a predator habituate very little, and they do so at the same time that overt fear responses are declining markedly (29). The advantage of this response dissociation would seem to be that response economy is not purchased at the cost of vigilance. Furthermore, in the natural world good and bad events do not occur at random. A predator may sustain its attacks for awhile and then give up. Food frequently occurs in batches. Other characteristics of the habituated response suggest that the mechanism is designed to take advantage of this fact. Habituated responses show *spontaneous recovery* following a stimulation-free interval; presentation of a novel stimulus can "disinhibit," or restore, the habituated response abruptly to its initial level (*dishabituation*); habituation is stimulus-specific; and finally, repeated presentations of a stimulus may produce an initial response increment (*sensitization*) before the response begins to habituate (16, 31).

Each of these characteristics of habituation insures that individuals behave economically (16, 45). As a component of the predator defense system, for example, the function of habituation is to recognize false-positives.

A novel stimulus sometimes represents potential danger, and an immediate defensive reaction to such stimuli is advantageous. However, if in the natural world the appearance of a sudden shadow or a rapidly approaching object reoccurs without incident, the event is probably harmless and foraging - or courtship, or nestbuilding - may be safely resumed. Likewise, when repeated attempts to ingest a potential food item, or court a mate, meet with failure, the more economical strategy would be to try elsewhere.

The foregoing is but a brief sketch of the circumstances in which habituation may be biologically advantageous. However, another example, the waning of the initial aggressive response to a conspecific, has received fuller experimental treatment and more clearly illustrates the symbiotic relationship which can exist between causal and functional analyses of behavior.

The basic observation made by ethologists (3, 42) is that aggression between territorial neighbors (*intraspecific aggression*), which is initially both intense and frequent, gradually wanes to a point of sporadic and relatively mild aggressive interactions. Yet, while neighbors spare one another from their territorial aggression, newcomers face its full measure. Laboratory experiments on habituated aggression have reproduced successfully the basic components of the ethologists' observations: The habituated response is stimulus-specific and is of sufficiently long duration to handle the time intervals observed in natural settings (31). On this basis, Peeke and Peeke (31) have proposed that habituation may be responsible for modulating aggression in natural populations of territorial fish. A similar explanation of territorial behavior in birds has been suggested by Petrinovich and Patterson (34).

Peeke and Peeke's analysis of habituation is all the more compelling because it derives from considerations of both cause and function. Current theories of habituation (15, 39) agree that the response decrement reflects the interaction of opposing processes, one decremental, the other incremental (i.e., sensitization). Although Groves and Thompson (15) posit only one source of incremental input, the target stimulus itself, Peeke and Peeke (32) reasoned on functional grounds that parental factors should provide an additional source. Their prediction was derived from functional considerations; because predation upon the young is a major threat in territorial invasion, the aggression shown intruding neighbors, and the rate of subsequent habituation to those neighbors, should reflect parental investment in the spawn (and, one might argue, offspring vulnerability). The results of experiments with convict cichlids support these predictions. The more advanced in the reproductive cycle, the greater the aggression shown an intruder. But, more importantly, a transition from one stage to the next, from eggs to larvae to free-swimming fry, was marked by an abrupt (sensitized) increase in aggressiveness. These results parallel data collected by Petrinovich and Patterson (33) in birds. In those studies, both the pattern of the aggressive response in white-crowned sparrows, and its habituation, was influenced by the breeding condition of the female.

In close parallel, functionally speaking, to the phenomenon of intraspecific aggression is the aggression which occurs between different species. Much less is known about *interspecific aggression*, but the functional requirements of these two naturally occurring phenomena are similar and the learning may be, too.

Many species of territorial fish are capable of recognizing their competitors; fish of other species whose ecological requirements overlap with the territory holder are driven away while noncompetitive species are permitted to remain within the territory (26). Several experiments (27) suggest that individuals learn which species to attack and are capable of making fine discriminations between different species of the same genera. What cues are the fish using when they discriminate between species or generalize from one genus to another? How does an individual's experience with a particular sequence of competitors and noncompetitors influence its choice of the cues upon which to base a discrimination? Might different populations experience different competitors and thus rely upon different cues for attack? Finally, if interspecific aggression is at all similar to intraspecific aggression, discussed earlier, one might expect to see phenomena of habituation: sensitized increases in aggression when noncompetitive intruders first appear, waning of that aggressive response with repeated exposure, and dishabituation when the situation (but not the intruder) suddenly changes in some way. A General Process approach to these phenomena may reveal many more commonalities of cause and function.

CONCLUDING COMMENTS

One theme of this paper is that the reconciliation of learning theory and natural behavior depends upon attempts to link causal and functional analyses. Some examples of naturally occurring behavior whose bases remain unexplored were mentioned earlier. In our present state of understanding, these phenomena (search image formation, food object learning, Batesian mimicry, interspecific aggression) are identified typically as unknown perceptual changes resulting from experience. But, suppose we were to discover, for example, that search image formation and food object learning were Pavlovian in origin, or that both inter- and intraspecific aggression are modulated by habituation. Would we have gained anything more than a familiar term with which to label the phenomenon? I believe we would; we would have, at our disposal, clear, testable hypotheses about mechanism. We would know how this learning might occur and under what conditions it might be lost. In short, we would know something of the necessary and sufficient conditions for this natural behavior to have occurred. Likewise, when an investigation of mechanism pays attention to function, we might expect similar refinements in our understanding of natural behavior.

Acknowledgements. Supported in part by a Faculty Fellowship Grant from Mount Holyoke College. I thank D. Sherry, H.S. Terrace, S.E.G. Lea, S. Shettleworth, J.E.R. Staddon, and K. Immelmann for their comments on the manuscript and R. Sigmundi for much useful discussion.

REFERENCES

(1) Alcock, J. 1973. Cues used in searching for food by red-winged blackbirds (Agelaius phoeniceus). Behav. 46: 174-188.

(2) Alcock, J. 1979. Animal Behavior: An Evolutionary Approach. Sunderland: Sinauer.

(3) Baerends, G.P., and Baerends-van Roon, J.M. 1950. An introduction to the ethology of cichlid fishes. Behav. Suppl. 1: 1-242.

(4) Bailey, C.H., and Chen, M. 1983. Morphological basis of long-term habituation and sensitization in Aplysia. Science 220: 91-93.

(5) Bastock, M. 1967. Courtship: An Ethological Study. Chicago: Aldine.

(6) Blanchard, R.J., and Blanchard, D.C. 1969. Crouching as an index of fear. J. Comp. Physiol. Psychol. 67: 370-375.

(7) Bykov, K.M. 1969. The Cerebral Cortex and the Internal Organs. Moscow: Foreign Languages Publishing House.

(8) Curio, E. 1976. The Ethology of Predation. New York: Springer-Verlag.

(9) Daly, M. 1978. The cost of mating. Am. Nat. 112: 771-774.

(10) deToledo, L., and Black, A.H. 1966. Heart rate: Changes during conditioned suppression in rats. Science 152: 1404-1406.

(11) Evans, D.L., and Waldbauer, G.P. 1982. Behavior of adult and naive birds when presented with a bumblebee and its mimic. Z. Tierpsych. 59: 247-259.

(12) Farris, H.E. 1964. Behavioral development, social organization, and conditioning of courting behavior in the Japanese quail, Coturnix coturnix japonica. Ph.D. Dissertation, University of Michigan, Ann Arbor.

(13) Farris, H.E. 1967. Classical conditioning of courting behavior in the Japanese quail, Coturnix coturnix japonica. J. Exp. Anal. Behav. 10: 213-217.

(14) Graham, J.M., and Desjardins, C. 1980. Classical conditioning: Induction of luteinizing hormone and testosterone secretion in anticipation of sexual activity. Science 210: 1039-1041.

(15) Groves, P.M., and Thompson, R.F. 1970. Habituation: A dual-process theory. Psychol. Rev. 77: 419-450.

(16) Hinde, R.A. 1970. Behavioural habituation. In Short-term Changes in Neural Activity and Behaviour, eds. G. Horn and R.A. Hinde, pp. 3-40. Cambridge: Cambridge University Press.

(17) Hodos, W. 1981. Some perspectives on the evolution of intelligence and the brain. In Animal Mind - Human Mind, ed. D.R. Griffin. Dahlem Konferenzen. Heidelberg: Springer-Verlag.

(18) Holland, P.C. 1977. Conditioned stimulus as a determinant of the form of the Pavlovian conditioned response. J. Exp. Psychol.: Anim. Behav. Proc. 3: 77-104.

(19) Holland, P.C. 1980. CS-US interval as a determinant of the form of Pavlovian appetitive conditioned responses. J. Exp. Psychol.: Anim. Behav. Proc. 6: 155-174.

(20) Hollis, K.L. 1982. Pavlovian conditioning of signal-centered action patterns and autonomic behavior: A biological analysis of function. Adv. Study Behav. 12: 1-64.

(21) Hollis, K.L. 1984. The biological function of Pavlovian conditioning: The best defense is a good offense. J. Exp. Psychol.: Anim. Behav. Proc. 10: 413-442.

(22) Jenkins, H.M.; Barrera, F.J.; Ireland, C.; and Woodside, B. 1978. Signal-centered action patterns of dogs in appetitive classical conditioning. Learn. Motiv. 9: 272-296.

(23) Johnston, T.D: 1982. Selective costs and benefits in the evolution of learning. Adv. Study Behav. 12: 65-106.

(24) Kandel, E. 1979. Behavioral Biology of Aplysia. San Francisco: Freeman.

(25) LoLordo, V.M. 1979. Constraints on learning. In Animal Learning: Survey and Analysis, eds. M.E. Bitterman, V.M. LoLordo, J.B. Overmier, and M.E. Rashotte, pp. 473-504. New York: Plenum.

(26) Losey, G.S. 1982. Intra- and interspecific aggression by the Central American Midas cichlid fish, Cichlasoma citrinellum. Behaviour 79: 39-80.

(27) Losey, G.S. 1982. Ecological cues and experience modify interspecific aggression by the damselfish, Stegastes fasciolatus. Behaviour 81: 14-37.

(28) Mackintosh, N.J. 1974. The Psychology of Animal Learning. New York: Academic Press.

(29) Macphail, E.M. 1982. Brain and Intelligence in Vertebrates. Oxford: Clarendon Press.

(30) Pavlov, I.P. 1927. Conditioned Reflexes. Translated by G.V. Anrep. London: Oxford University Press.

(31) Peeke, H.V.S., and Peeke, S.C. 1973. Habituation in fish with special reference to intraspecific aggressive behavior. In Habituation: Behavioral Studies, eds. H.V.S. Peeke and M.J. Herz, vol. 1, pp. 59-83. New York: Academic Press.

(32) Peeke, H.V.S., and Peeke, S.C. 1982. Parental factors in the sensitization and habituation of territorial aggression in the convict cichlid (Cichlasoma nigrofasciatum). J. Comp. Physiol. Psychol. 96: 955-966.

(33) Petrinovich, L., and Patterson, T.L. 1979. Field studies of habituation: I. The effect of reproductive condition, number of trials, and different delay intervals on the response of the white-crowned sparrow. J. Comp. Physiol. Psychol. 93: 337-350.

(34) Petrinovich, L., and Patterson, T.L. 1981. Field studies of habituation: IV. Sensitization as a function of the distribution and novelty of song playback to white-crowned sparrows. J. Comp. Physiol. Psychol. 95: 805-812.

(35) Sahley, C.; Rudy, J.W.; and Gelperin, A. 1981. An analysis of associative learning in a terrestrial mollusc: I. Higher-order conditioning, blocking and a transient US pre-exposure effect. J. Comp. Physiol. 114: 1-8.

(36) Seligman, M.E.P. 1970. On the generality of the laws of learning. Psychol. Rev. 77: 406-418.

(37) Siegel, S. 1979. The role of conditioning in drug tolerance and addiction. In Psychopathology in Animals: Research and Clinical Implications, ed. J.D. Keehn, pp. 143-168. New York: Academic Press.

(38) Sinnott, J.M.; Sachs, M.B.; and Hienz, R.D. 1980. Aspects of frequency discrimination in passerine birds and pigeons. J. Comp. Physiol. Psychol. 94: 401-415.

(39) Thompson, R.F.; Groves, P.M.; Teyler, T.J.; and Roemer, R.A: 1973. A dual-process theory of habituation: Theory and behavior. In Habituation: Behavioral Studies, eds. H.V.S. Peeke and M.J. Herz, vol. 1, pp. 239-271. New York: Academic Press.

(40) Thompson, R.F., and Spencer, W.A. 1966. Habituation: A model phenomenon for the study of neuronal substrates of behavior. Psychol. Rev. 73: 16-43.

(41) Thompson, T., and Sturm, T. 1965. Classical conditioning of aggressive display in Siamese fighting fish. J. Exp. Anal. Behav. 8: 397-403.

(42) van den Assem, J., and van der Molen, J.N. 1969. Waning of the aggressive response in the three-spined stickleback upon constant exposure to a conspecific: I. A preliminary analysis of the phenomenon. Behaviour 34: 286-324.

(43) Wasserman, E.A. 1973. Pavlovian conditioning with heat reinforcement produces stimulus-directed pecking in chicks. Science 181: 875-877.

(44) Waxman, H.M., and McCleave, J.D. 1978. Auto-shaping in the archer fish (Toxotes chatareus). Behav. Biol. 22: 541-544.

(45) Wells, M. 1968. Lower Animals. New York: McGraw-Hill.

The Biology of Learning, eds. P. Marler and H.S. Terrace, pp. 373-397. Dahlem Konferenzen 1984. Berlin, Heidelberg, New York, Tokyo: Springer-Verlag.

Complex General Process Learning in Nonmammalian Vertebrates

S.E.G. Lea
Dept. of Psychology, University of Exeter
Washington Singer Laboratories, Exeter EX4 4QG, England

Abstract. General process learning reflects advanced cognitive capacities which may be supposed to derive from a single evolutionary history. It is divided into cases where the complexity resides in the stimulus situation, in the response to be made, or in the cognitive processing that is required to get from one to the other. In all these cases there is substantial evidence of capacities for complex learning in a variety of nonmammalian vertebrates, especially birds; but it is not at present possible to make strong links between the kind of learning capacity shown and either taxonomic status or ecological niche.

INTRODUCTION: SCOPE OF THIS PAPER

This paper is about complex learning. I shall take it for granted that habituation, classical conditioning, and operant conditioning are part of a common vertebrate heritage. Similarly, I shall take for granted basic discrimination abilities - I shall not regard as "complex," and so within the scope of this paper, any discrimination based on a single dimension that is known to be within the perceptual capabilities of the species in question. Instead, I am going to try to ask whether there is anything more complex that we can add to this agreed common core of general learning processes.

I want to consider cases of general process learning rather than what I call special skills. "General-process learning theory" was introduced as a derogatory phrase in Seligman's famous paper (77) on what we now call "biological constraints on learning" (cf. Bolles and Revusky, both this volume). Of course, I am not arguing against the existence of such constraints or of species-specific learning capacities. But the major special kinds of learning, at least in birds, are covered elsewhere in this volume (e.g., by Immelmann and Marler, both this volume). Furthermore, I want to assert the importance of more general learning processes.

The simplest kinds of learning, which psychologists predominantly study in the laboratory, solve problems which must turn up without fail in the life of every vertebrate species. The problem of learning which stimuli are a consistent part of the environment (and so can be ignored) is solved by habituation. The problem of learning causality relations between external stimuli is solved by classical conditioning. And the problem of learning the consequences of the animal's own behavior is solved by operant conditioning. These problems are not laboratory abstractions: both classical and instrumental conditioning have been plausibly implicated as partial mechanisms for successful foraging (46, 65). Hollis's paper (this volume) argues these points in greater detail.

Because these problems are essentially common to all species, I believe that the learning capacities that solve them are evolutionarily conservative. Probably all vertebrates possess them by inheritance from pre-vertebrate ancestors. Special learning abilities, such as the capacity for imprinting or for song learning, are usually supposed to be confined to particular taxonomic groups (though this is more often taken for granted than tested). It makes sense to suppose that they are relatively recent developments, and there is every evolutionary reason to expect them to have quite distinct laws. General process learning, by contrast, should be expected to obey the same laws wherever it is found, and that should be as true of complex general process learning as it is of the simpler forms we have discussed so far. The major question posed in this chapter is, simply, whether there are any complex general learning processes that fit this prescription. If there are, we can reasonably conclude that in evolutionary terms they, too, are part of a common vertebrate heritage.

Organization of this Paper

We know so little about the learning capacities of the cold-blooded vertebrates that it would be pointless to try to organize this paper taxonomically. Instead, I have used a loose classification of complex learning tasks, depending on whether the complexity seems to me to reside in the stimuli presented to the animal, the response it has to make, or in the process by which it has to get from one to the other. I have brought in what few data we have on complex learning in fish, amphibians, and reptiles wherever they are relevant to this systematic analysis, with perhaps a little extra emphasis because of their rarity.

COMPLEX STIMULUS LEARNING

By complex stimulus learning, I mean learning in which a) the stimuli to be discriminated cannot be described in terms of any single, simple, perceptual dimension known to be within the capacities of the species concerned; and b) many different positive and negative stimuli are presented, so that recognition cannot be achieved by a simple "template" of the stimuli. Such learning is usually described as involving "higher-order stimuli," or "concepts." I have argued elsewhere that it is better described as involving "concept discrimination" (47). The stimuli involved are selected on the basis of concepts lying within the mind of the experimenter: we should avoid implying that they exist within the subject - that is at best a hypothesis to be tested.

Two kinds of concept discrimination experiment have been performed. In the first and simpler type, the discrimination to be made could in principle be described in terms of a sufficiently elaborate disjunction of conjunctions of simple perceptual dimensions. The classic experiment of this type is Herrnstein and Loveland's demonstration that pigeons could discriminate color slides on the basis of whether or not they included a picture of a person (33). I shall describe these experiments as involving discriminations of "perceptual concepts." In the other type, the stimuli are described in terms of some more abstract logical rule which may be quite simple at its own level but cannot be reduced to simple statements about basic perceptual dimensions. I shall call these "logical concepts." Examples include symmetry and number.

The general technique used in experiments on complex stimulus learning is the same, regardless of which type of concept is involved. Large numbers of stimuli are prepared and divided into groups on the basis of the concept to be discriminated. Most usually, an operant conditioning technique is used: the stimuli are presented one or two at a time, and the subject is rewarded for responses made to stimuli belonging to the positive class. Ideally, each individual stimulus would only be presented once - there is a prevalent suspicion that birds are capable of very accurate learning of large numbers of stimulus patterns, a view that has at least some support in the literature (90, 91). Even if it is necessary to use each stimulus more than once during training, it is usual to reserve a subset of stimuli to use for post-training transfer tests, so as to ensure that the subject has not simply learned how to respond to each individual member of the stimulus classes.

Discrimination of Perceptual Concepts

In all the nonmammalian vertebrate classes, there are strong sociobiological grounds for thinking that many species are capable of some degree of complex stimulus discrimination. Any species that lives in a society with strong pair bonding, extended parental care, or a consistent dominance hierarchy, must be supposed to have some mechanism for individual recognition, and another individual is an obvious instance of a "higher-order perceptual concept": there is unlikely to be any single perceptual feature that will be sufficient for its recognition ((1, 73), and Ryan, unpublished dissertation). In the case of birds these sociobiological deductions can be supported by experimental data.

There can be no doubt that pigeons are capable of perceptual concept discriminations, at least in the visual modality. In addition to people, they have been taught to detect cartoon characters, letters of the alphabet, pigeons, fish, trees, water, and geometrical figures. Most of these experiments have been reviewed by Herrnstein (31) and Herrnstein and De Villiers (32). Ryan ((73), and unpublished dissertation) has shown that chickens as well as pigeons can perform this kind of discrimination, using as stimuli slides of individuals of the same or another species.

Most such discriminations are learned rapidly and without apparent difficulty. However, Lea and Ryan (49) reported that pigeons acquired a discrimination between groups of letters very slowly. Ryan (unpublished dissertation) was unable to train pigeons to discriminate between photographs of individual pigeons, and although they did learn to discriminate between slides of individual chickens, she could not demonstrate transfer to new photographs following training with photographs of individual chickens (even though chickens discriminated the same groups of slides of both chickens and pigeons readily and transferred the discrimination to new instances after training).

In order to be sure that pigeons can respond to multiple dimensions of stimuli, Lea and Harrison (48) used artificial "concepts" in which pigeons had to respond to three distinct features if they were to perform correctly. They found that such discriminations were readily learned. Rodwald (70), using chickens, also showed that multiple features of a stimulus can control behavior. In conventional concept discrimination, using letters (5, 49, 59) or more naturalistic stimuli (34, 50), pigeons' discrimination accuracy for individual members of the stimulus sets depends on identifiable stimulus features.

What is in doubt is whether anything more than multiple feature discrimination is involved in these experiments. Current data give no grounds for postulating that the pigeons actually possess or form concepts similar to those used by the experimenters in selecting the stimuli (47). There is no evidence at all that pigeons do recognize slides as representations of objects, and some evidence that they may not. Pigeons can discriminate different slides taken of the identical scene, a few seconds after each other, and apparently irrelevant details of the background play an important part in some perceptual concept discrimination learning (27). When pigeons discriminate cartoon characters, they maintain their discriminations even when the cartoons are effectively cut up into quite small pieces and reassembled (10). These results show that very local features of the slides may be used in discrimination, and this would be difficult if the slide were being perceived as a representation of an object.

In direct tests of recognition of objects from photographs (8, 51), pigeons trained to discriminate three-dimensional objects did recognize two-dimensional photographs that were carefully positioned so as to give roughly the same retinal projections as the training objects. Budgerigars trained to discriminate photographs of two other budgerigars transferred the discrimination to live birds, but transfer in the reverse direction did not work (89). The whole question of what object perception in a bird might mean is in any case an open one: it is considered briefly at the end of this paper, in the discussion of experiments on object permanence.

One set of experiments on perceptual concepts deserves special mention in this volume, because of its biological relevance. Pietrewicz and Kamil (63) review a series of studies on blue jays in which they trained birds to respond to slides of particular species of moth. The moths were chosen from among the species that form the blue jays' natural prey and were mounted for photography in varying orientations on backgrounds which gave them varying degrees of crypticity. In other respects the procedure was a conventional concept discrimination. By this means, Pietrewicz and Kamil were able to study ecologically relevant aspects of the stimulus complex presented by the moths, without sacrificing experimental control. They were also able to provide formal evidence of search image formation (the tendency for particular classes of prey to be recognized more easily after a number of exposures). Prey species recognition by blue jays has also been demonstrated by making use of the jays' avoidance of the emetic and poisonous Monarch butterfly and its mimics such as the Viceroy (7). Unfortunately, these experiments were poorly controlled against the influence of genetic factors and pre-experimental experience on the jays' rejection of Monarchs.

All these results involve visual concepts. There are one or two experimental reports of complex auditory discrimination in birds. Blackbirds have been trained to discriminate vowel sounds (35), and starlings to discriminate rhythms (38). Of course, the ecological literature on territorial song provides a rich source of additional information. For example, territory-holding white-throated sparrows show little response to tape-recordings of the songs of long-standing neighbors but sing strongly in response to recordings of songs made in other areas (6); this is clear evidence of song discrimination. But song discrimination is intimately linked to the special skill of song learning and is therefore discussed in Marler's paper (this volume) rather than here.

Logical Concepts

As was said above, the distinction between perceptual and logical concepts is not hard and fast, and right on the borderline sits the concept of symmetry. Delius and Habers (14) advanced convincing ecological arguments for believing that many birds should be able to detect symmetry, and they and Delius and Nowak (15) succeeded in training pigeons to discriminate it.

Closely related to the concept of symmetry is the concept of "same." Discrimination of sameness has been studied extensively in connection with the "matching-to-sample" task, or its close relative, "oddity-from-sample." In the form in which it is used with pigeons, this extremely popular experimental procedure involves presenting the pigeon with three pecking keys, each of which can have a stimulus projected onto it. Most usually the keys can be lit up with different colors. In the first phase of a trial, the central key (the "sample key") is lit; subsequently, the other two keys ("choice keys") are also lit, and the subject is rewarded for responding to the choice key which is the same as (in the matching-to-sample task) or different from (in the oddity-from-sample task) the sample key. Both of these tasks are readily learned by pigeons. Performance at an oddity task may start at a higher level, but perfect performance is harder to obtain on the oddity than on the matching task (2). Goldfish can also learn both tasks (25) but do not show the same differences between them as pigeons.

At first glance, it appears that successful performance at matching-to-sample must involve a concept of "same." However, it is possible that pigeons simply learn which choice key to peck, given each of the different three-key stimulus configurations that are possible in the experiment. Support for this interpretation comes from the discovery that so-called "symbolic matching-to-sample" is learned as quickly as direct matching-to-sample (9). In the symbolic matching-to-sample task, each sample-key stimulus indicates that a particular choice-key stimulus will be correct on the current trial, but there is no direct relation between the sample-key and choice-key stimuli; for example, a triangle on the sample key might signal that a peck to a green choice key will be rewarded.

At present, the balance of evidence suggests that pigeons do have a concept of "same." Zentall, Hogan, and Edwards (96) review a series of experiments in which they transferred pigeons from one matching-to-sample task to another, from one oddity-from-sample task to another, or between matching-to-sample and oddity-from-sample. Transfer between tasks of the same type is consistently easier than transfer between types, even when none of the stimuli in use are the same in the two tasks. This suggests that a concept has been learned and can be applied to the new task. However, the area is a hotly contested one, and we should not assume that Zentall and his colleagues will be allowed to have the last word.

There is a second line of research which bears on the concept of "same." In a sense it combines the "perceptual concept" and "logical concept" traditions. In experiments on visual memory in pigeons, Wright et al. (94) used a task in which a slide of a complex picture was presented, followed by a pause and the presentation of a second slide. One of two available responses was then reinforced if the two slides were the same, and the other was reinforced if they were different. Wright et al. then expanded this experiment to become a "serial probe recognition task" in which the initial slide presentation was replaced by a sequence of one to six slides, and the "same" response was reinforced if the final ("probe") slide had appeared anywhere in the initial sequence. Pigeons performed less well at this task than either monkeys or humans, and they relied longer on the simple strategy of learning responses to individual slides, but they eventually showed significant evidence of responding to the "sameness" relationship. Once again, though, we need to note the possibility of artefact: Greene (27) found that pigeons which had apparently learned to detect the repetition of a stimulus were in fact responding to minute differences between supposedly identical photographs.

Another logical concept which has been extensively investigated is that of number. Two forms of number discrimination have been used. In one, stimuli are presented including different numbers of objects; in the other, the subject is required to discriminate the number of times it has itself made a particular response. Since there is little consistency between the two sets of data, the latter kind of experiment, response number discriminations, will be discussed under the heading of complex response learning, where they more properly belong.

The former kind of experiment should probably be referred to as "numerosity" discriminations, since it is very doubtful whether they involve the subjects in counting or anything remotely like it. Most of the experiments are old and are discussed in standard reviews (e.g., (88), pp. 385 ff.; (93)). The few more recent reports (e.g., (80)) do not alter the main conclusions. At least one species of fish (the blenny, (30)) and numerous species of birds (especially corvids) have been demonstrated to make discriminations between stimuli differentiated on the basis of number. But the numbers involved have always been small (usually less than five), the discriminations are often not good, the numbers of subjects used in the experiments are generally low, and the risk that some artefact is involved is always substantial. Wesley (93) even goes so far as to argue that the absence of controls for odor effects and the use of manual procedures which inevitably risk introducing "Clever Hans" effects mean that little reliance can be placed on these results. Certainly, in one study with pigeons in which automatic procedures avoided these risks, Greene (27) could find no reliable evidence of conceptual discrimination even of the numbers one and two.

Finally, there is one report suggesting that birds may have a concept of dimensionality: corvids looked for a solid bait in a three-dimensional figure rather than in a two-dimensional one, where it logically could not be (45).

COMPLEX RESPONSE LEARNING

I include under this heading all learning tasks whose complexity resides chiefly in the response to be learned. They fall into two broad classes. In the first, the actual response to be made is relatively simple - typically, pecking a conventional pigeon key; the complexity lies in some required patterning of responses, either relative to each other or relative to stimuli. Such tasks can be said to involve "higher-order operants": the consequences of a response are not a property of the response as such but depend on its relation to stimuli or other responses.

In the second kind of complex response task, it is the actual response that is complex. The obvious examples are the various demonstrations of tool use by birds. In this context, it is particularly important to recall that I am excluding from this paper those kinds of learning for which particular species appear to be specially "prepared": otherwise, learned song production, navigation, and possibly some aspects of flight would be obvious candidates for inclusion here.

Counting Schedules

The simplest kind of higher-order operant is repetition of a response for a given number of times, which is usually described as counting. Early experiments by Koehler and his collaborators, reviewed by Thorpe (88), suggested that a variety of bird species could reliably repeat responses such as eating a single pea a fixed number of times, even when the rhythm of responding was disrupted so that the duration of the response sequence could not be used as a cue. The numbers involved were typically below seven in these experiments. However, Wesley (93) argues that much of this work is methodologically inadequate in the same ways as the numerosity discrimination experiments described above, and his arguments about odor cues are particularly telling in this case.

More recent experiments, reviewed by Hobson and Newman (36), have avoided such problems by using standard automated apparatus. Two kinds of counting task have been used. In one, pigeons are reinforced for pecking a key only after they have first pecked another key for a fixed number of times. In the second, they are given a modification of the symbolic matching-to-sample task, in which the cue for which of the two "choice keys" is correct on a given trial is the number of pecks the bird has to make on the "sample key" before the choice keys are lit up. The accuracy of "counting" reported in such experiments varies wildly, though in some cases (e.g., (68)) it can be very good. Accuracy of discrimination in the modified matching-to-sample task falls if a "retention interval" is imposed between responding on the sample key and the opportunity to respond on the choice keys (44).

In all "counting" experiments, it is difficult to be sure that the bird is responding to the number of responses it has made rather than to the time it has taken to make them. The recent experiments do not deal with this issue well, and the earlier experiments, which dealt with it rather better, are unsatisfactory on other grounds. Until experiments on this "counting vs. timing" issue have been carried out with birds with something like the sophistication recently shown in experiments in the same area with rats (e.g., (12)), we had better reserve judgement on how well, if at all, birds can count their own responses.

Response Strategies and the Bitterman Hypothesis

From the repetition of single response, we move to consider the emission of patterned sequences of different responses. There are relatively few direct investigations of this kind of learning, but interest in the field is increasing. Pigeons can be trained to peck keys in prescribed sequences, either of position or color, with considerable accuracy (67, 83, 84). Terrace (86) argues that the results demonstrate that pigeons must learn a cognitive representation of the required sequence.

However, response patterning is also highly relevant to two areas of research where we do have a substantial set of investigations of cold-blooded vertebrates, namely, probability learning (discrimination tasks, either spatial or visual, where both cues are sometimes associated with reward but one "pays off" more often than the other) and serial reversal learning (repeated reversals of the significance of the cues in a simple learning task). Both research fields have largely been inspired by Bitterman's (4) hypothesis that there are qualitative differences in the learning capacities of different classes of vertebrates, with fish, reptiles, and birds forming an ascending series but all falling short of the capacities of mammals, specifically rats. Macphail ((53), chs. 3, 5, and 6) provides a careful and detailed review of the controversy that has been engendered by Bitterman's proposal.

Fish of quite a range of species have been tested on both serial reversal learning and probability learning. It is unfortunate that so far all experiments have been done on teleosts (bony fishes), for the elasmobranches (shark-like fishes) have substantially larger brains relative to their body weights. So far as I know, there is only one study on either of these tasks on amphibia, a demonstration of serial reversal learning in newts (18). Among the reptiles, Bitterman and his colleagues have worked extensively with turtles, which are interesting from an evolutionary perspective since they probably diverged from other reptiles at around the same time as the separation of the groups that gave rise to mammals and birds; Macphail also cites scattered data from the two other main reptile groups.

Macphail's survey leaves no room for doubt that in the majority of experiments, fish perform less well than other vertebrates on both these tasks: there are fewer demonstrations with fish than with rats, pigeons, or reptiles of more rapid reversal on successive sessions of a serial reversal task, and less evidence of "maximizing" (selecting the cue more likely to be associated with reward on all trials) in the probability learning task. But the difference is at best quantitative, and some fish under some conditions show "rat-like" performance. Even with his conservative attitude to inter-taxonal differences, though, Macphail concedes that the data are consistent with a reduced capacity, in fish relative to terrestrial vertebrates, for responding to come under the control of the after-effects of previous responses or stimuli. This would imply that fish will have difficulty in using complex response strategies, which is the point of interest for our present discussion.

However, Kamil and Yoerg (42) have argued that the response strategies available to a species may depend more on its ecological niche than on its taxonomic status or general "intelligence." So far as I have been able to discover, no attempt has been made to use this kind of information in analyzing complex learning in either fish or reptiles, yet the opportunity is plainly there. Fish are at least as variable in their feeding strategies as birds or mammals. The kinds of learning that would be adaptive for a hunting carnivore such as a pike seem likely to be very different from those that a filter-feeder such as a pilchard might need. It would be a valuable test of Kamil and Yoerg's argument to try to deploy it to analyze the response strategies of fish.

All experiments with birds seem to agree that they show improvement in performance on successive reversals of the same task and maximize on spatial probability learning tasks rather than showing the inferior strategy of matching response probabilities to reward probabilities.

Matching may occur in visual probability learning, at least in pigeons, depending on the precise experimental conditions used (26) and on the amount of experience the bird has with the situation (79). On serial reversal learning, Mackintosh (52) claimed that pigeons were inferior to either rats or "higher" birds (particularly corvids), and Wells and Lehner (92) claimed that mallards were inferior to pigeons. But Macphail argues that these differences are not sufficiently consistent across variations in task and training conditions to justify any strong comparative conclusion.

Learning Sets

One particular "higher-order operant" task has attracted much attention as an index of cognitive complexity. This is the "learning set" experiment (29), in which repeated discrimination tasks are given, and any increasing tendency of the subject to perform correctly on the second trial of each new task is taken as a sign of learning set formation, indicating that the "rules of the game" have been abstracted. Macphail ((53), ch. 6) mentions a variety of attempts to demonstrate learning sets in pigeons, chickens, and corvids, but by far the most substantial research effort in this area is the work of Kamil and his collaborators on blue jays. On the last hundred tasks of a 700-task series, blue jays were correct on the second trial of about 75% of tasks (40).

The importance of Kamil's work, however, lies not so much in the mere demonstration of learning set formation, as in the way Kamil has linked it to underlying response strategies and to ecological factors. Blue jays show positive transfer from serial reversal learning to learning set formation (41). The only commonality between these tasks is that both can be solved by adopting a "win-stay/lose-shift" response strategy, and Kamil and Yoerg (42) have argued that the blue jay's ecology is just such as to encourage the evolution of a tendency to just such a higher-order response patterning. It should be noted, though, that Staddon and Frank (82) showed substantial transfer between the same two tasks in pigeons, and it is not obvious how that would be predicted from the pigeon's feeding ecology; however, Staddon and Frank's experiments used multiple schedules of reinforcement rather than the more usual discrete-trial simultaneous discrimination learning technique, and this makes it hard to assess their pigeon's success quantitatively or to compare it with the performance of other species in other experiments.

Imitation

A final "higher-order operant" is imitation. Here reinforcement depends not on the response made, but on its relation to the responses of another organism. As Marler's paper (this volume) shows, imitation is heavily involved in song learning by oscine birds, but this is the kind of special learning process we are not considering here. Certain species show a much more flexible capacity for vocal imitation; parrots and starlings and their relatives (especially mynahs) are, of course, famous for it. This, too, may be a species-specific capacity related to the social ecology of these birds - mutual song imitation might, for example, be involved in the maintenance of pair bonds (16) - but it certainly deserves further study. Pepperberg's work (60, 61) on the vocalizations of an African gray parrot, mentioned later in this paper, seems likely to arouse a good deal of further interest.

Birds certainly can show imitation of more "arbitrary" responses. Pigeons (95) and quail (74) have been taught to peck keys by observing conspecifics doing so, and pigeons can match the location (81) and the rate (57) of their pecks to those of a conspecific. Titmice can learn more complex food-finding skills (e.g., uncovering and opening a food hole) by observation (75), and such observational learning is thought to have played a part in the spread of milk-bottle opening among British tits (22).

Such experiments demonstrate that birds' learned behavior can come under the stimulus control of other birds' behavior. As has long been realized, however, this need not imply that the birds perceive the relation between the model's behavior and their own ((24); (88), pp. 133 ff.). Nor does it really show that a higher-order operant has been established: that would require a demonstration that the tendency to imitate any response increased as a result of reinforced imitation, as in the experiments on "imitation set" in primates (e.g., (13)). I have not been able to find any such demonstration in a nonmammal.

If imitation can be thought of as a stimulus control by another animal, it makes sense to mention here that pigeons' behavior can also be brought under the control of images of themselves: in a simulation of recent work by Gallup (23) on primates, pigeons were taught to use a mirror to guide them in removing adhesive spots from their plumage (19).

Learned Skills

There has been relatively little systematic work on motor skill learning in birds. There have been a number of studies on string-pulling performance: early work, mainly on titmice, is reviewed by Thorpe ((88), pp. 375 ff); more recently Ducker and Rensch (17) have worked with budgerigars and corvids. Apart from that, the literature consists largely of isolated reports, such as Skinner's (81) well-known tour de force of shaping in which two pigeons were taught a version of table tennis.

Many experiments are concerned with tool use. The Galapagos woodpecker finch's use of twigs to extract insects can be regarded as a species-specific special skill. True woodpeckers, however, use niches in trees and cages to assist them in cracking nuts and handling insect prey and may even demonstrate these skills to their young (11). Corvids also use tools: Powell and Kelly (64) used operant "shaping" techniques to train crows to operate a pecking key with a cocktail stick when it was out of reach of their beaks, and Reid (66) observed a rook which learned to plug a drainhole so as to provide a pool of water for drinking and bathing within its cage.

A recent report by Epstein and Medalie (20) is particularly interesting, because the task in question was one made famous in Köhler's (43) studies of chimpanzees. Epstein and Medalie trained a pigeon to push a small stool around the floor of a large arena using food reward, and they also trained the bird to peck a key on the arena wall. When the key was moved up the wall out of the bird's reach, the bird apparently spontaneously pushed the stool under the key, got on it and pecked. This result echoes Birch's (3) conclusion that "insight" in chimpanzees may depend on prior training in components of the "insightful" response.

COMPLEX COGNITIVE PROCESSES

Obviously, many of the tasks so far considered make considerable demands on the subjects' cognitive processing powers. The tasks to be considered in this final section of the paper, however, have that as their most obvious source of complexity.

Many of the experiments to be mentioned come from the recent and rapidly growing literature on "animal cognition," which has been collected and reviewed in several recently contributed books (e.g., (39, 87)). The guiding principle of this literature has been the application to animal behavior of concepts from modern cognitive science, developed to handle human cognition and artificial intelligence.

Short-term Memory in Pigeons

In birds, the cognitive process that has been most intensively investigated is memory, especially short-term or working memory. Almost all the studies I shall mention used a variant of the matching-to-sample and symbolic-matching-to-sample tasks introduced earlier. The variation is that the stimulus on the sample key disappears before the choice keys are illuminated; usually a delay interval is imposed between the two. During this delay, the subject must in some way store the information present in the sample key stimulus if a correct response to the choice keys is to be made.

I have already mentioned experiments showing that pigeons have an impressive short-term memory for pictures (94) and for their own recent behavior (44). The most interesting experiments, however, are those which go on from demonstrating the existence of animal memory to investigating its properties. One series of experiments, reviewed by Maki (55), has shown that pigeons can learn to respond appropriately to a cue which informs them that the information currently being stored is no longer correct (the "directed forgetting" task). Also, Maki (54), using a variant of delayed matching-to-sample in which the "samples" were presentations of different kinds of food, has shown that "surprising" samples (food presentations that were not preceded by the usual pre-food cues) were better remembered than expected ones.

Two strands of evidence suggest that what is stored is not a "trace" of the sample stimuli, but something more like a representation of the expected choice events. First, a number of studies (reviewed by Petersen (62)) have shown that memory is very much better if correct responses to different choice stimuli are reinforced with different kinds of food. Second, in the delayed symbolic matching-to-sample task, the rate of increase in errors as a function of the delay interval depends on the confusability of the choice stimuli, not that of the sample stimuli (71).

These are only a few illustrative examples of what is at present a very active research area. Taken as a whole, this literature builds up a convincing case for the usefulness of a concept of working memory in pigeons. But we should hesitate before adopting it as a paradigm for studies of animal cognition. In delayed alternation tests in T-mazes, pigeons show much more durable memory than they do in conventional delayed matching-to-sample tasks (56). Siamese fighting fish also show quite impressive performance in maze tasks, though unlike rats or pigeons they use response strategies such as always entering an arm adjacent to the one last left (72). It seems unlikely that current work on pigeons' short-term memory is revealing the limits of nonmammalian memory.

Long-term Memory
This last conclusion is strongly reinforced when we turn to studies of long-term memory. These include two of the three investigations of learning in amphibians I have been able to include in this paper: Miller and Berk (58) were able to show that Xenopus toads retained an escape habit across the period of metamorphosis from tadpole to adult, and (in a rather less well controlled study) Schwartz and Cogan (76) obtained similar results for salamanders. This is not strictly complex learning, but the retention is impressive enough to justify mention here.

Our most impressive examples of long-term memory come from birds, however. Shettleworth (78) has reviewed data which she and Krebs have collected on marsh tits and Balda and his colleagues have collected on two species of nutcrackers. All these birds cache nuts during the autumn for consumption during the winter, and experiments in the laboratory have shown that they have exquisitely accurate memories for the places where they have made their caches. Perhaps these performances should be considered as special adaptations, but it seems inappropriate to leave them out. In any case, very extensive long-term memory has also been demonstrated in "arbitrary" contexts. Pigeons can learn to respond correctly to at least 320 different slides (unconnected by any concept known to the experimenter) (90), and the correct response to each of 160 slides can be retained for at least two years (91).

Object Permanence
One or two investigators have followed the logic of the "animal cognition" idea in a slightly different direction and have applied to animals ideas derived from the psychology of human cognitive development. For example, Etienne (21) has reviewed experiments in which she attempted to follow the development of "object permanence" in young chicks. She found that even 3-day-old chicks could learn to search for a prey item that had disappeared, and about 50% of them reliably hunted in the right place when the object could disappear in either of two directions. But, unlike Gruber, Girgus, and Bannazizi (28) in their work with kittens, she was unable to show any correlation of such performance with overall cognitive development, or to find any signs that the chicks had a true concept of an object that had temporarily disappeared, as Piaget described for the human infant.

Another indication that the perceptual and cognitive processes by which birds recognize objects may be different from those used by humans comes from an experiment by Hollard and Delius (37). Pigeons were taught a matching-to-sample task using complex visual patterns in which the sample stimuli were sometimes rotated relative to the comparisons. There was no evidence of the correlation between angular disparity and reaction time commonly reported for humans in similar tasks.

Language

In this review of complex learning in nonmammalian vertebrates, we have come across many instances of apparently intelligent performance. The naive "scala naturae" approach to comparative psychology would expect most birds to lie below mammals, well below primates, and vastly below man in their cognitive capacities. With a few exceptions, the material reviewed so far in this chapter tends to support the alternative view, taken by Macphail (53), that the only important cognitive difference within the vertebrates is the human capacity to learn language.

That view, however, prompts the question whether there are any language-like performances in birds that could parallel the well known sign-language and artificial-language studies using apes, reviewed by, for example, Ristau and Robbins (69). At least three research projects have posed this question. Chauvin-Muckensturm (see (11)), basing her strategy on that used by Premack with the chimpanzee Sarah, taught woodpeckers to emit different rhythmic patterns of pecks (on her hand!) in order to "request" different kinds of food. In rather similar style, Pepperberg (60) has trained an African gray parrot to utter around forty different sounds in response to, or to request, distinct objects; the sounds reported are fair imitations of English words. She also claims (61) that the parrot can name either color, shape, or material on demand. (Unfortunately, both Chauvin-Muckensturm and Pepperberg were always present during tests of their birds, so that control against "Clever Hans" effects is weak in both of these projects.) Finally, Terrace (85) has deployed his response-sequence learning experiments, described above, to question the linguistic nature of the sequential performances found in the "ape-language" experiments.

All these projects have produced at least some positive results. That fact is more likely to convince outside observers that chimpanzees really do not learn language, than that birds really do. From the point of view of this paper, however, it suggests that we have by no means explored the limits of the cognitive capacities of nonmammalian vertebrates.

CONCLUSIONS

My first conclusion is really expressed in the previous sentence. I have been fairly conservative in selecting experiments to mention in this paper, passing over a number of studies where low sample sizes or poor experimental procedures leave the interpretation open to doubt. Yet I have been able to muster a fair range of complex learning capacities in nonmammals. There seems to be plenty of scope for more research and little for the bland assumption that only mammals are capable of anything worthy to be described as "intelligence."

Second, though there certainly are some differences in learning capacity between different nonmammalian vertebrates, we do not yet have the evidence to relate many of them in any simple way either to phylogenetic status or to ecological niche. Certainly, birds in general and corvids in particular are mentioned relatively often in connection with the more "intelligent" performances, but it is not clear how far this is due to experimenters' prejudices in selecting subjects for the more elaborate investigations.

Finally, writing this paper has been very much a matter of collecting scattered reports. Few, if any, kinds of complex learning have been investigated in a really broad range of nonmammalian vertebrates. But the cumulative impression from the literature is that the common stock of vertebrate learning capacities extends well beyond habituation, simple conditioning, and simple discrimination learning. An interesting task for future research will be to identify which complex learning processes are truly general.

REFERENCES

(1) Bateson, P.P.G. 1973. Internal influences on early learning in birds. In Constraints on Learning, eds. R.A. Hinde and J. Stevenson-Hinde, pp. 101-116. London: Academic Press.

(2) Berryman, R.; Cumming, W.W.; Cohen, L.R.; and Johnson, D.F. 1965. Acquisition and transfer of simultaneous oddity. Psychol. Rep. 17: 767-775.

(3) Birch, H.G. 1945. The relation of previous experience to insightful problem-solving. J. Comp. Psychol. 38: 367-383.

(4) Bitterman, M.E. 1965. Phyletic differences in learning. Am. Psychol. 20: 396-410.

(5) Blough, D.S. 1982. Pigeon perception of letters of the alphabet. Science 218: 397-398.

(6) Brooks, R.J., and Falls, J.B. 1975. Individual recognition by song in white-throated sparrows. I. Discrimination of songs of neighbors and strangers. Can. J. Zool. 53: 879-888.

(7) Brower, L.P. 1969. Ecological chemistry. Sci. Am. 220(2): 22-29.

(8) Cabe, P.A. 1976. Transfer of discrimination from solid objects to pictures by pigeons. Percept. Psychophys. 19: 545-550.

(9) Carter, D.E., and Eckerman, D.A. 1975. Symbolic matching by pigeons: Rate of learning complex discriminations predicted from simple discriminations. Science 187: 662-664.

(10) Cerella, J. 1980. The pigeon's analysis of pictures. Pattern Recog. 12: 1-6.

(11) Chauvin, R., and Muckensturm-Chauvin, B. 1980. Behavioral Complexities. New York: International Universities Press.

(12) Church, R.M., and Meck, W.H. 1984. The numerical attribute of stimuli. In Animal Cognition, eds. H.L. Roitblat, T.G. Bever, and H.S. Terrace, pp. 445-464. Hillsdale, NJ: Erlbaum.

(13) Darby, C.L., and Riopelle, A.J. 1959. Observational learning in the rhesus monkey. J. Comp. Physiol. Psychol. 52: 94-98.

(14) Delius, J.D., and Habers, G. 1978. Symmetry: can pigeons conceptualize it? Behav. Biol. 22: 336-342.

(15) Delius, J.D., and Nowak, B. 1982. Visual symmetry recognition by pigeons. Psychol. Res. 44: 199-212.

(16) Dobkin, D.S. 1979. Functional and evolutionary relationships of vocal copying phenomena in birds. Z. Tierpsychol. 50: 348-363.

(17) Ducker, G., and Rensch, B. 1977. The solution of patterned string problems by birds. Behaviour 62: 164-173.

(18) Ellins, S.R.; Cramer, R.E.; and Martin, G.C. 1982. Discrimination reversal learning in newts. Anim. Learn. Behav. 10: 301-304.

(19) Epstein, R.; Lanza, R.P.; and Skinner, B.F. 1981. "Self-awareness" in the pigeon. Science 212: 695-696.

(20) Epstein, R., and Medalie, S.D. 1983. The spontaneous use of a tool by a pigeon. Behav. Anal. Lett. 3: 241-247.

(21) Etienne, A.S. 1977. L'étude comparative de la permanence de l'objet chez l'animal. Bull. Psychol. 30: 187-197.

(22) Fisher, J., and Hinde, R.A. 1949. The opening of milk bottles by tits. Br. Birds 42: 347-357.

(23) Gallup, G.G. 1979. Self-awareness in primates. Am. Sci. 67: 417-421.

(24) Gewirtz, J.L. 1971. The role of overt responding and extrinsic reinforcement in "self-" and "vicarious-reinforcement" phenomena and in "observational learning" and imitation. In The Nature of Reinforcement, ed. R. Glaser, pp. 279-309. New York: Academic Press.

(25) Goldman, M., and Shapiro, S. 1979. Matching-to-sample and oddity-from-sample in goldfish. J. Exp. Anal. Behav. 31: 259-266.

(26) Graf, V.; Bullock, D.H.; and Bitterman, M.E. 1964. Further experiments on probability-matching in the pigeon. J. Exp. Anal. Behav. 7: 151-157.

(27) Greene, S.L. 1983. Feature memorization in pigeon concept performance. In Quantitative Analyses of Behavior, eds. M.L. Commons, R.J. Herrnstein, and A.R. Wagner, vol. IV, pp. 209-229. Cambridge, MA: Ballinger, in press.

(28) Gruber, H.E.; Girgus, J.S.; and Banuazizi, A. 1971. The development of object permanence in the cat. Devel. Psychol. 4: 9-15.

(29) Harlow, H.F. 1949. The formation of learning sets. Psychol. Rev. 56: 51-65.

(30) Hennig, M. 1977. 'Zahl'-Vermögen bei Blennius pavo Risso (Blenniidae, Perciformes). Zool. Anz. 199: 1-18.

(31) Herrnstein, R.J. 1984. Objects, categories, and discriminative stimuli. In Animal Cognition, eds. H.L. Roitblat, T.G. Bever, and H.S. Terrace, pp. 233-261. Hillsdale, NJ: Erlbaum.

(32) Herrnstein, R.J., and De Villiers, P.A. 1980. Fish as a natural category for people and pigeons. In The Psychology of Learning and Motivation, ed. G.H. Bower, vol. 14, pp. 59-95. New York: Academic Press.

(33) Herrnstein, R.J., and Loveland, D.H. 1964. Complex visual concept in the pigeon. Science 146: 549-551.

(34) Herrnstein, R.J.; Loveland, D.H.; and Cable, C. 1976. Natural concepts in pigeons. J. Exp. Psychol.: Anim. Behav. Proc. 2: 285-302.

(35) Hienz, R.D.; Sachs, M.B.; and Sinnott, J.M. 1981. Discrimination of steady state vowels by blackbirds and pigeons. J. Acoust. Soc. Am. 70: 699-706.

(36) Hobson, S.L., and Newman, F. 1981. Fixed-ratio-counting schedules: Response and time measures considered. In Quantitative Analyses of Behavior, eds. M.L. Commons and J.A. Nevin, vol. 1, pp. 193-224. Cambridge, MA: Ballinger.

(37) Hollard, V., and Delius, J.D. 1982. Rotational invariance in visual pattern recognition by pigeons and humans. Science 218: 804-806.

(38) Hulse, S.H.; Cynx, J.; and Humpal, J. 1983. Generalization and discrimination of rhythmic structures by birds. Ann. NY Acad. Sci., in press.

(39) Hulse, S.H.; Fowler, H.; and Honig, W.K. 1978. Cognitive Processes in Animal Behavior. Hillsdale, NJ: Erlbaum.

(40) Hunter, M.W., and Kamil, A.C. 1971. Object-discrimination learning set and hypothesis behavior in the northern blue jay (Cyanocitta cristata). Psychonom. Sci. 22: 271-273.

(41) Kamil, A.C.; Jones, T.B.; Pietrewicz, A.T.; and Mauldin, J.E. 1977. Positive transfer from successive reversal training to learning set in blue jays (Cyanocitta cristata). J. Comp. Physiol. Psychol. 91: 79-86.

(42) Kamil, A.C., and Yoerg, S.I. 1982. Learning and foraging behavior. In Perspectives in Ethology, eds. P.P.G. Bateson and P. Klopfer, vol. 5. New York: Plenum.

(43) Köhler, W. 1927. The Mentality of Apes, 2nd ed. London: Routledge and Kegan Paul.

(44) Kramer, S.P. 1982. Memory for recent behavior in the pigeon. J. Exp. Anal. Behav. 38: 71-86.

(45) Krushinsky, L.V.; Zorina, Z.A.; and Dashevsky, B.A. 1979. [The ability of Corvidae birds to operate empirical dimensions of objects.] Zh. Vyssh. Nervn. Deyat. I.P. Pavlova 29: 590-597.

(46) Lea, S.E.G. 1981. Correlation and contiguity in foraging theory. In Advances in Analysis of Behaviour, eds. P. Harzem and M.D. Zeiler, vol. 2, pp. 355-406. Chichester: Wiley.

(47) Lea, S.E.G. 1984. In what sense do pigeons learn concepts? In Animal Cognition, eds. H.L. Roitblat, T.G. Bever, and H.S. Terrace, pp. 263-276. Hillsdale, NJ: Erlbaum.

(48) Lea, S.E.G., and Harrison, S.N. 1978. Discrimination of polymorphous stimulus sets by pigeons. Q.J. Exp. Psychol. 30: 521-537.

(49) Lea, S.E.G., and Ryan, C.M.E. 1983. Feature analysis of pigeons' acquisition of discrimination between letters. In Quantitative Analyses of Behavior, eds. M.L. Commons, R.J. Herrnstein, and A.R. Wagner, vol. IV. Cambridge, MA: Ballinger, in press.

(50) Lubow, R.E. 1974. High-order concept formation in the pigeon. J. Exp. Anal. Behav. 21: 475-483.

(51) Lumsden, E.A. 1977. Generalisation of an operant response to photographs and drawings/silhouettes of a three-dimensional object at various orientations. Bull. Psychonom. Soc. 10: 405-407.

(52) Mackintosh, N.J. 1969. Comparative studies of reversal and probability learning: rats, birds, and fish. In Animal Discrimination Learning, eds. R.M. Gilbert and N.S. Sutherland, pp. 137-162. London: Academic Press.

(53) Macphail, E.M. 1982. Brain and Intelligence in Vertebrates. Oxford: Clarendon.

(54) Maki, W.S. 1979. Pigeons' short-term memories for surprising vs. expected reinforcement and nonreinforcement. Anim. Learn. Behav. 7: 31-37.

(55) Maki, W.S. 1981. Directed forgetting in animals. In Information Processing in Animals: Memory Mechanisms, eds. N.E. Spear and R.R. Miller, pp. 199-225. Hillsdale, NJ: Erlbaum.

(56) Maki, W.S. 1984. Some problems for a theory of working memory. In Animal Cognition, eds H.L. Roitblat, T.G. Bever, and H.S. Terrace. Hillsdale, NJ: Erlbaum.

(57) Millard, W.J. 1979. Stimulus properties of conspecific behavior. J. Exp. Anal. Behav. 32: 283-296.

(58) Miller, R.R., and Berk, A.M. 1977. Retention over metamorphosis in the African claw-toed frog. J. Exp. Psychol.: Anim. Behav. Proc. 3: 343-356.

(59) Morgan, M.J.; Fitch, M.D.; Holman, J.G.; and Lea, S.E.G. 1976. Pigeons learn the concept of an 'A'. Perception 5: 57-66.

(60) Pepperberg, I.M. 1981. Functional vocalizations by an African grey parrot (Psittacus erithacus). Z. Tierpsychol. 55: 139-160.

(61) Pepperberg, I.M. 1983. Cognition in the African grey parrot: Preliminary evidence for auditory/vocal comprehension of the class concept. Anim. Learn. Behav. 11: 179-185.

(62) Petersen, G.B. 1984. How expectancies guide behavior. In Animal Cognition, eds H.L. Roitblat, T.G. Bever, and H.S. Terrace, pp. 135-148. Hillsdale, NJ: Erlbaum.

(63) Pietrewicz, A.T., and Kamil, A.C. 1981. Search images and the detection of cryptic prey: an operant approach. In Foraging Behavior, eds. A.C. Kamil and T.D. Sargent, pp. 311-331. New York: Garland.

(64) Powell, R.W., and Kelly, W. 1975. A method for the objective study of tool-using behavior. J. Exp. Anal. Behav. 24: 249-253.

(65) Rashotte, M.E.; O'Connell, J.M.; and Beidler, D.L. 1982. Associative influence on the foraging behavior of pigeons (Columbia livia). J. Exp. Psychol.: Anim. Behav. Proc. 8: 142-153.

(66) Reid, J.B. 1982. Tool-use by a rook (Corvus frugilegus), and its causation. Anim. Behav. 30: 1212-1216.

(67) Richardson, W.K., and Warzak, W.J. 1981. Stimulus stringing by pigeons. J. Exp. Anal. Behav. 36: 267-276.

(68) Rilling, M. 1967. Number of responses as a stimulus in fixed interval and fixed ratio schedules. J. Comp. Physiol. Psychol. 63: 60-65.

(69) Ristau, C.A., and Robbins, D. 1982. Cognitive aspects of ape language experiments. In Animal Mind - Human Mind, ed. D.R. Griffin, pp. 299-330. Dahlem Konferenzen. Berlin, Heidelberg, New York: Springer.

(70) Rodwald, H.K. 1974. A conjoint measurement analysis of control by dimensions of compound stimuli. Perc. Mot. Sk. 38: 551-556.

(71) Roitblat, H.L. 1980. Codes and coding processes in pigeon short-term memory. Anim. Learn. Behav. 8: 341-351.

(72) Roitblat, H.L.; Tham, W.; and Golub, L. 1982. Performance of Betta splendens in a radial arm maze. Anim. Learn. Behav. 10: 108-114.

(73) Ryan, C.M.E. 1982. Concept formation and individual recognition in the domestic chicken (Gallus gallus). Behav. Anal. Lett. 2: 213-220.

(74) Sanavio, E., and Savardi, U. 1980. Observational learning in Japanese quail. Behav. Proc. 5: 355-361.

(75) Sasvari, L. 1979. Observational learning in great, marsh and blue tits. Anim. Behav. 27: 767-771.

(76) Schwartz, J.M., and Cogan, D.C. 1974. Position discrimination in the salamander, Ambystoma tigrinium. Devel. Psychobiol. 10: 355-358.

(77) Seligman, M.E.P. 1970. On the generality of the laws of learning. Psychol. Rev. 77: 406-418.

(78) Shettleworth, S.J. 1983. Memory in food-hoarding birds. Sci. Am. 248(3): 86-94.

(79) Shimp, C.P. 1966. Probabilistically reinforced choice behavior in pigeons. J. Exp. Anal. Behav. 9: 443-455.

(80) Simons, D. 1976. "Zahl"-Versuche mit Kolkraben anhand der Methodik der Musterwahl - ein Beitrag zum Verstaendnis von Problem-Loesungs-Verhalten bei hoeheren Tieren. Z. Tierpsychol. 41: 1-33.

(81) Skinner, B.F. 1962. Two "synthetic social relations". J. Exp. Anal. Behav. 5: 531-533.

(82) Staddon, J.E.R., and Frank, J. 1974. Mechanisms of discrimination reversal in pigeons. Anim. Behav. 22: 802-828.

(83) Straub, R.O.; Seidenberg, M.S.; Bever, T.G.; and Terrace, H.S. 1979. Serial learning in the pigeon. J. Exp. Anal. Behav. 32: 137-148.

(84) Straub, R.O., and Terrace, H.S. 1981. Generalization of serial learning in the pigeon. Anim. Learn. Behav. 9: 454-468.

(85) Terrace, H.S. 1979. Is problem-solving language? J. Exp. Anal. Behav. 31: 161-175.

(86) Terrace, H.S. 1983. Simultaneous chaining. In Quantitative Analyses of Behavior, eds. M.L. Commons, R.J. Herrnstein, and A.R. Wagner, vol. IV. Cambridge, MA: Ballinger, in press.

(87) Terrace, H.S.; Bever, T.G.; and Roitblat, H.L., eds. 1984. Animal Cognition. Hillsdale, NJ: Erlbaum, in press.

(88) Thorpe, W.H. 1963. Learning and Instinct in Animals, 2nd ed. London: Methuen.

(89) Trillmich, F. 1976. Learning experiments on individual recognition in budgerigars (Melopsittacus undulatus). Z. Tierpsychol. 41: 372-395.

(90) Vaughan, W., and Greene, S.L. 1983. Acquisition of absolute discriminations in pigeons. In Quantitative Analyses of Behavior, eds. M.L. Commons, R.J. Herrnstein, and A.R. Wagner, vol. IV. Cambridge, MA: Ballinger, in press.

(91) Vaughan, W., and Greene, S.L. 1983. Pigeon visual memory capacity. J. Exp. Psychol.: Anim. Behav. Proc., in press.

(92) Wells, M.C., and Lehner, P.N. 1977. Serial reversal learning in the mallard duck (Anas platyrhyncos). Bull. Psychonom. Soc. 10: 235-237.

(93) Wesley, F. 1961. The number concept: a phylogenetic review. Psychol. Bull. 58: 420-428.

(94) Wright, A.A.; Santiago, H.C.; Sands, S.F.; and Urcuioli, P.J. 1984. Monkey and pigeon serial probe recognition performance. In Animal Cognition, eds. H.L. Roitblat, T.G. Bever, and H.S. Terrace, pp. 353-373. Hillsdale, NJ: Erlbaum.

(95) Zentall, T.R., and Hogan, D.E. 1976. Imitation and social facilitation in the pigeon. Anim. Learn. Behav. 4: 427-430.

(96) Zentall, T.R.; Hogan, D.E.; and Edwards, C.A. 1984. Conditional discrimination learning by pigeons. In Animal Cognition, eds. H.L. Roitblat, T.G. Bever, and H.S. Terrace, pp. 389-405. Hillsdale, NJ: Erlbaum.

Standing, left to right:
Juan Delius, Klaus Immelmann, Mark Konishi, Stephen Lea, Eliot Hearst, Peter Marler, John Staddon.

Seated, left to right:
Pat Bateson, Hans-Joachim Bischof, Herb Jenkins, Karen Hollis, Don Kroodsma.

The Biology of Learning, eds. P. Marler and H.S. Terrace, pp. 399-418. Dahlem Konferenzen 1984. Berlin, Heidelberg, New York, Tokyo: Springer-Verlag.

Biology of Learning in Nonmammalian Vertebrates
Group Report

D.E. Kroodsma, Rapporteur
P.P.G. Bateson
H.-J. Bischof
J.D. Delius
E. Hearst
K.L. Hollis
K. Immelmann
H.M. Jenkins
M. Konishi
S.E.G. Lea
P. Marler
J.E.R. Staddon

INTRODUCTION

The study of nonmammalian vertebrates, especially birds, has been a rich source of material not only for the student of animal learning, who is interested primarily in the mechanisms (causation) of behavior in laboratory environments, but also for the ethologist, who is interested in causation, function, evolution, and development of behavior in more natural settings. In this report we consider reconciliation of these two approaches by discussing a) the inevitable problem with "unnatural" categories of learning, b) productive efforts at the interface between traditional psychology and ethology, c) the apparently "special" processes of birdsong learning and imprinting, and d) how little we know about the perception of complex stimulus patterns. Finally, we e) comment on the potential productivity of a new behavioral biology.

CATEGORIES OF LEARNING

The taxonomies of learning in use today are not necessarily helpful in identifying important psychological processes. The diversity of organisms in the world can be classified according to their phylogenetic histories, and a variety of characteristics, including morphological, behavioral, and cellular features, can be used to construct such evolutionary histories (43); however, there are no currently accepted "natural" organizing principles upon which to base a taxonomy of learning (Holland et al., this volume).

A taxonomy based on laboratory operations, such as a) mere repetition of an event, stimulus, or a setting, b) exposure to relations between stimuli, or c) exposure to relations between stimuli and responses, may be convenient for textbook headings, but such a classification is relatively arbitrary and may lead to apparently different phenomena being classified together (e.g., imprinting with habituation). A considerable improvement on a procedurally based system might be to classify learning via a functional approach, i.e., how learned behavior reflects increases in an animal's fitness, or reproductive success. Categories based on behaviors in nature might involve abstractions such as prediction, control, familiarization, imitation, and so forth, or immediate functional contexts such as feeding, detection and avoidance of danger, finding mates, or parental care.

Although taxonomies aid in organizing a diverse world and "natural" taxonomies are useful in making predictions, they can also prejudice the direction of research. Classification of imprinting or song learning as a special process, for example, may have prevented a closer examination of the processes involved and their relation or similarity to psychological studies in the laboratory. Furthermore, taxonomy is constantly in a state of flux. This is true not only for phylogenetic classifications of organisms (49), but also for taxonomies of learning. The old distinction between classical and operant conditioning is no longer considered very useful in the sense of characterizing fundamental categories; most, if not all, situations involve some stimulus, a response, and an unconditioned stimulus (or reinforcer), so separation of the categories is always problematical anyway. Likewise, taxonomies once used by ethologists included taxes and releasers, and while these are still heuristically useful, they should not be considered rigid categories (18).

One additional category of behavior, "unlearned" or innate responses, also needs clarification. The term has a variety of meanings (Bateson, this volume), and conclusions from all relevant experiments tend to rest on plausibility or negative evidence. Ethologists using the comparative approach use the term "innate" to designate behavioral differences which depend on a genetic difference. Although an organism can never be isolated from itself, and animals often bring into play backup mechanisms for developing a behavior, rearing individuals of different populations or closely related species under very similar environments can yield valuable data on "innate behavioral" differences between groups (39).

PRODUCTIVE INTERDISCIPLINARY ENTERPRISES

There are, of course, many natural situations in which birds and other nonmammalian vertebrates show learning. One that has received a great deal of recent attention involves searching for food. MacArthur and Pianka (37) argued that success in foraging is so crucial to an animal's survival that animals should evolve to become optimal foragers within the environment to which they have adapted - i.e., foraging behavior should be perfectly adaptive, or optimal.

There are three ways in which optimal foraging requires animals to learn. First, the forager must learn the locations in which prey are to be found ("patches"). Second, it must learn the densities of prey in each patch. Third, it must constantly update its estimates of prey density, because these are likely to change as a result of its own or other animals' foraging efforts.

Working from the general principle of optimization, behavioral ecologists have established quantitative generalizations to describe foraging behavior. These models can be applied to behavior observed in the wild (12, 17), but they can also be tested in laboratory experiments (28). Such experimental methods correspond to standard procedures used in operant conditioning, in which quantitative generalizations about relations between response output and reinforcement frequency are a contemporary topic of lengthy discussion ((8, 19); see also (51), Ch. 9). The parallels were quickly noticed by a number of authors (e.g. (33, 50)), and a fruitful area of interdisciplinary interaction arose. For such work, ecologists have gained access to the precise techniques and substantial accumulations of data of the operant psychologist; operant psychology has gained both a sense of ecological relevance and an impetus to study new species, new situations, and new problems - notably the problem of schedule transition, taken up in several recent theoretical papers (25, 35). Both fields have gained a more precise understanding of the operation of decision rules in a wider range of situations. Further collaboration at this interface between two traditional fields can only benefit both.

From the psychological point of view, the interdisciplinary study of foraging is part of a general attempt to understand natural behavior in terms of the components of learning that can be studied using conditioning procedures (Hollis, this volume). Animals in the laboratory learn relationships between events during classical or operant conditioning, and animals in nature undoubtedly use what they have learned in order to predict future events. However, conditioning studies restricted to the study of causation are bound to remain incomplete. They must be extended to the study of function. Understanding the relationship between events in nature is of interest, but it is also important to understand how the relative fitness of the organism is affected by these behavior patterns.

BIRDSONG LEARNING AND IMPRINTING: SPECIAL PROCESSES?

One of the crucial questions that arises repeatedly is whether or not birdsong learning and imprinting are really "special processes." Do the generalizations that derive from the study of conditioning in conventional laboratory preparations (e.g., CER conditioning, autoshaping, classical conditioning of reflexes (eye blinks, salivation, etc.)) allow us to understand or make sense of the properties of birdsong learning and imprinting? Addressing this issue requires a full understanding of the attributes of both song learning and imprinting, and hence we first discuss more thoroughly the characteristics of these two phenomena before returning to the question of special processes.

Imprinting

The characteristics of imprinting are discussed in greater detail by Bateson and Immelmann (both this volume). One of its most distinctive features lies in the way it restricts the range of stimuli that elicit behavior after imprinting has taken place. Preferences become restricted to those stimuli that have been experienced in the course of imprinting. Establishment of this restriction ends the developmental phase when the preferences in question are sensitive to influences from the external world. The self-terminating characteristic of imprinting can in certain circumstances be overridden so that further plasticity of preferences occurs later in development (21), but in general, once formed, the preferences are remarkably stable (6). In this respect, then, imprinting seems to demand special-purpose neural machinery even if the process of storage is the same as in other learning processes (but compare Immelmann, this volume).

Other aspects of imprinting invite comparisons with less specialized mechanisms. Some components of stimuli used in experiments on filial imprinting in chicks and ducklings are powerful attractors (e.g., movement), whereas others are initially neutral. What happens when the neutral stimulus is paired with the initially attractive one and then a bird's responses to the unpaired neutral stimulus alone are subsequently measured? The answer is clear. The birds approach and are pacified by the now familiar but initially neutral stimulus (for review see (21)). So far so good, but the following experiment has yet to be done. Using three different neutral stimuli, A preceding the initially attractive stimulus, B occurring at the same time, and C following it, which stimulus does the bird learn most about? If imprinting works in the same way as typically observed in classical conditioning, the predicted levels of post-conditioning response should be highest for A and lowest for C. At least one biologist's hunch, however, is that B would elicit the most social behavior with C coming next and A last, the reasoning being that when a bird's attention has been attracted by a moving object, the bird continues to gaze at the object once it has stopped moving (Bateson, personal communication). If the hunch turns out to be correct, then the underlying process is likely to be different from those indicated by Pavlovian experiments (though see (24) for an example of "backward conditioning"). The difference might be small and could involve small variations in the parameters influencing when a neutral stimulus can be associated with an initially effective one to produce imprinting. Once again, though, the implication is that special neural machinery would have to be dedicated to imprinting and the behavior it affects. The expectation is strikingly confirmed by the neuroanatomical studies of the systems involved in imprinting (Bateson, this volume).

A further complication that generates frequent misunderstanding involves the fact that imprinting is not "unitary" but is probably subject to a variety of neural and hormonal factors, as well as processes of growth and differentiation. The process which has been studied most intensely is filial imprinting, and the majority of references in the literature refer only to the "following reaction." However, just as interesting is the establishment of sexual and other social preferences (i.e., sexual imprinting), and there are other "imprinting-like phenomena" such as the acquisition of habitat, locality, and food preferences, or perhaps the learning of particular star patterns as a target for homeward orientation in migratory birds (Immelmann, this volume). These processes all share to some degree the two main characteristics of imprinting, namely, its phase specificity, and a certain degree of selectivity about the stimuli that are most effective in achieving stable results (23). The learning processes involved in imprinting, however, might be different from one case of imprinting to another. Filial imprinting, in which immediate reward is available, may be more akin to various conditioning phenomena than, for example, learning of a particular star pattern, which may be closer to what has been called perceptual or exposure learning.

It seems possible that the "special nature" of imprinting refers to the way information is protected from change in the central nervous system rather than to the way it is acquired (10, 22). The discovery of temporal correlations between sensitive phases for imprinting and developmental processes (cell death, synaptogenesis) in certain areas of the brain may be a pointer, together with the anatomical specialization for the control of song, for directions which future research about the nature of both processes could profitably take ((52), and Bateson, this volume).

Birdsong Learning

Selected aspects of birdsong learning have been discussed by Marler and Konishi (both in this volume). We here elaborate upon other factors which might facilitate the comparison of song learning with general psychological processes.

Seemingly trivial differences in experimental methods can affect results. As in the imprinting literature (5), experimental treatments and methods are rarely, if ever, replicated by other independent investigators or even the same investigator. Thus, data must be interpreted cautiously, and with careful attention given to experimental design. Several key treatment factors include the age at which birds are taken from the field (as eggs, nestlings, fledglings, juveniles, or adults), the treatments or experiences before and after tutoring, the age during which tutoring occurs in the laboratory, the means of presentation of the tutor sounds (e.g., via loudspeakers or live birds - see below), the amount of exposure to the tutor songs, the photoperiod on which the birds are maintained, and possible interactions among experimental subjects (30). There are many permutations and probable interactions of these different treatments, and lack of standardized procedures has undoubtedly contributed to some of the apparent differences (and perhaps even similarities) among species.

"Song learning" usually means "vocal imitation." A juvenile bird can be exposed to a number of songs during its sensitive phase, usually occurring during the first one to three months of life. After passing through stages of subsong, which involves a rambling series of more or less amorphous sounds, and plastic song, when one can begin to recognize some of the sound patterns that will eventually emerge, the young (usually male) songbird finally develops its adult song when about a year old. Audio spectrograms of the tutor songs and those "performed" by the subject are then compared visually to determine if "song learning" has occurred.

Song "performance" gives an incomplete picture of what is "learned." In most studies, the performance of a song is the only reliable indication that a bird has imitated (i.e., learned) that particular song. This distinction between learning and performance is well-known to psychologists. Learning is the change in "internal state" brought about by some experience. Performance is the change in subsequent behavior shown as a consequence of some earlier experience. The effect of an earlier experience may or may not be revealed by a particular test situation or set of situations. Without a change of state, no subsequent performance difference is possible, but lack of a performance difference says nothing decisive about whether learning (state change) has occurred - it may just be that the appropriate conditions for performance have not been identified.

Birds can and do learn far more than is revealed in their vocal displays. For example, in the laboratory a non-singing female white-crowned sparrow (Zonotrichia leucophrys) will respond preferentially to songs heard during an early sensitive phase (2) indicating that even though the song is not imitated vocally, it can still be recognized, and that necessary aspects of the signal parameters have been "learned." Furthermore, Marler and Peters (41) have demonstrated that "sparrows learn adult song and more from memory," because in the developmental stage of plastic song juvenile males provide evidence that they have learned the characteristics of song types which never appear in the final adult repertoire of song types (Marler, this volume). Finally, as implied in the foregoing example, songs are usually committed to memory (i.e., "learned") well in advance of the time when the individual attempts to match its vocal output with the song pattern stored in memory. Hence, it may be useful to distinguish a sensory and a sensorimotor stage of learning with song development per se consisting of the second phase in which "errors" in vocal output are corrected by auditory feedback (26, 44).

Songbirds preferentially learn songs of their own species. Because the effects of perceptual choices are usually assessed only through motor performance (but see (14)), it is difficult to say whether the perceptual or motor system is the primary determinant of predispositions in song learning.

As in imprinting, the association of a positive (i.e., conspecific sound) and otherwise neutral stimulus (heterospecific song) in time can lead to the learning of both (Marler, this volume). This observation has prompted several interesting observations and proposed experiments. For example, if A is the species song and B, C, and D are alien songs, A is typically preferred over B, C, and D in a free choice situation (playing back these songs in random sequence and with equal frequency). If during the sensitive phase birds are exposed to A first, the period terminates or shortens, but exposure to B, C, and D first does not block the acceptance of A. While it is tempting to try to interpret these findings in terms of overshadowing or blocking, both of these phenomena demand two stimuli, a "neutral" one (to-be CS) and a hedonic one (US), and the opportunity to learn a predictive relationship; it is unclear which vocal stimulus would be the US and which the CS, and what predictive relationship is to be learned. It is also possible that exposure to B, C, and D first might enhance the acquisition of A, but such studies have not been done. Still another interesting problem would be to see, after pairing stimuli A and B, where A is the appropriate positive stimulus (a "trigger" in the sense of Marler (this volume)), whether B could then serve as a "conditional" trigger of learning C or D. Demonstration of such a relationship would indicate that song learning shares some common properties with classical conditioning.

Song learning may not require any external reinforcement, but several studies now indicate that use of live tutors may enhance the learning process (4, 32, 48, 54). These live tutors may influence the size of the vocal repertoire learned, the acceptability of heterospecific songs otherwise rejected, and the duration of the sensitive phase for song learning. The nature of the interactions between the tutor and the subject have not been studied in detail, but these studies may open a new avenue for the introduction of paradigms like those used in the study of associative learning.

Song learning demonstrates a remarkable capacity for memory. In song learning the sensory and sensorimotor stages may be separated by several months during which birds need neither hear nor vocally rehearse the memorized song (42). Furthermore, once a song is crystallized (adult song is said to "crystallize" from the earlier developmental stages), it remains unchanged from one year to the next, even though there are months during the intervening non-breeding season when the song is not rehearsed. Even after they are deafened, some species can reproduce the crystallized song every year (27).

There are striking species, sex, and age differences in song learning. Among species there is considerable diversity in the size of the learned repertoire (e.g., a male white-crowned sparrow develops one song, whereas California marsh wrens (Cistothorus palustris) may develop over two hundred (11, 39)) and in the variety of sounds produced (marsh warblers (Acrocephalus palustris) and northern mockingbirds (Mimus polyglottos) may mimic many species, whereas North American sparrows generally do not mimic alien songs (9, 15, 40)). In most temperate zone species only the male sings, though in tropical environments females frequently sing as well (16). Finally, one of the most characteristic features of song learning is the presence of a sensitive phase, a time usually early in life when the ability to learn songs is maximal (29, 39). Hence, there is considerable diversity in the ability and inclination to learn songs among different species, sexes, and ages, a fact that is challenging learning theorists who search for commonalities.

The learning and recognition of large song repertoires may provide interesting insights into the problems of stimulus generalization and/or categorization. For example, a male marsh wren in eastern Washington sings about 115 different song types, each of which is learned from other males in the population. These song types are performed in such a stereotyped sequence that a male wren, upon hearing one of his song types broadcast from a loudspeaker within his territory, will respond with the next song in the sequence (55). Why certain song types are associated with one another in these sequences (56), or how the birds recognize the different song types (which may number up to two hundred in some western USA populations (11)) is at present unknown.

One of the major limitations in the study of these and other forms of learning is that it is unclear how the outside world is represented in the brain. How many features of a stimulus must an animal attend to in order to classify a given sound, and what must be the nature of the neuroanatomical or biochemical changes that accompany learning of a song, surrogate mother, or any other stimulus? Until major advances are made in understanding the physiological bases of learning, many details here will remain a mystery (Konishi, this volume).

A series of discrete brain nuclei in the bird brain appear correlated with the ability to learn songs. These brain nuclei ((45), and Konishi, this volume) are involved in both the perceptual and motor pathways of song learning. The most significant features of this song control system is its apparent absence in birds incapable of vocal mimicry (e.g., New World flycatchers, representatives of the suboscine suborder of the order Passeriformes) and its presence in birds well-known for their vocal imitation and/or mimicry (e.g., parakeets of the order Psittaciformes and songbirds of the oscine suborder of the order Passeriformes) (1, 45). In addition to this presence/absence correlation, the volume of these song control nuclei in the brains of several songbirds is correlated with the size of the song repertoire ((11, 46), and Konishi, this volume). The hypothesis that the system of discrete brain nuclei has in fact evolved to accommodate vocal learning will have to be addressed through further comparative studies.

Special processes?

Because birdsong learning and imprinting have not been "explained" in terms of generalizations that derive from the study of conditioning in conventional laboratory preparations, it is tempting to conclude that these are indeed special processes. Most students of birdsong, though, are more interested in the functions or ecological relevance of the behavior (3, 9, 16, 31, 38, 42, 47, 53) than in trying to explain or understand the development of the behavior in terms of any "general processes" gleaned from laboratory studies performed by learning theorists. This fundamental difference in research orientation and interest is unlikely to change. The "special process" issue must be addressed in a direct manner, perhaps designing experiments after those outlined above, before the possible roles of conditioning processes in birdsong learning and imprinting will be clear.

In addition to the phase specificity and the preferential response to appropriate stimuli, birdsong learning and imprinting appear to be processed in discrete portions of the brain (see above, and separate contributions by Bateson, Immelmann, and Konishi, all in this volume). It is tempting to conclude that the anatomical specialization in the brain for both imprinting and song learning provides additional support for labelling these two phenomena "special processes." However, other forms of learning may also occur in functionally discrete but anatomically diffuse regions of the brain, and mere packaging of relevant brain nuclei into discrete and identifiable units is no indication that within these areas typical conditioning processes are not involved (but see Bateson, this volume). Moreover, learning mediated by different parts of the brain is nevertheless likely to involve the same essential biochemical and neural processes.

PERCEPTIONS OF STIMULI

We know that young songbirds identify conspecific songs during song learning, and that nidifugous chicks come to identify appropriate social partners during filial and sexual imprinting. But in many cases it is not clear either exactly what the stimulus identified is or how it should be characterized. The appearance of a bird from different viewpoints is very different (5) and the samples of song that a bird hears may be very diverse (30). Sometimes a consistent output is derived from this varying input. The problem is similar to that described by Lea (this volume) as involving "concept discrimination." The experimental work of Herrnstein and Loveland (20) and others therefore becomes relevant to this problem.

The work of Herrnstein and Loveland (20) initiated a provocative program of research on natural concepts in the pigeon. This research is interesting for what it might tell us about animal perception and the role of adaptive predispositions in perceptual categorization. At present, however, it is not clear what the results indicate about the level of perceptual analysis that mediates the pigeon's performance.

The results show that pigeons can learn to respond for food to each of a large set of photographic slides consisting of highly disparate exemplars of a natural category such as tree, person, or water, while responding less strongly to each of an equally large, heterogeneous set of slides which do not exemplify any category. They are slower to discriminate between randomly selected slide sets. Moreover, they can respond correctly to first presentations of new exemplars of a natural category.

One interpretation of these results is that the pigeons perceive the slides much as we would, namely, as depicting objects. The interpretation is implied by the claim that the pigeon's performance in discriminating between the category and noncategory slides, and generalizing to new instances, is mediated by a "natural concept."

There is, however, a rival hypothesis that has yet to be rejected. The performance could be the result of discrimination and generalization based on a lower-level perceptual analysis of the slides as arrays of meaningless, or nonrepresentational stimulus features made up of color, brightness gradients, and shapes. It is conceivable that at the level of nonrepresentational features, the slides depicting a natural category exhibit a matrix of similarities that would account for more rapid learning to discriminate between a category and a noncategory set than between randomly selected sets. Moreover, generalization based on similarity of new exemplars to some subset of the previously trained category exemplars might account for better than chance performance on new exemplars.

Experiments are needed to provide information on the level of perceptual analysis of the stimuli in the natural concept experiments, but it is by no means clear how to construct those experiments. One would like to be able to remove the representational value of the slides without altering the similarity matrix at the level of nonrepresentational stimulus features, but it may not be possible to change one without the other. A more feasible, although perhaps less direct, approach would be to examine transfer between photographic representations and real objects, and vice versa. At present there are rather few experiments of this type. In addition to the light such experiments might throw on the hypothesis of natural concepts, they could provide new information about perceptual processing in birds.

Furthermore, "category discrimination" may be a safer term than "concept discrimination." The bird's mode of responding serves to organize the perceptual world into categories, but it is not clear how those categories are represented structurally. Perhaps only if abstract representational processes were demonstrated should we talk in terms of "concepts," and perhaps only if the concepts are demonstrably acquired should we talk of concept formation. However, these "semantic adjustments" still leave the empiricist groping for techniques. Although we can show by using artificial stimuli that pigeons can use more than one feature in making a category discrimination (36), this does not really solve the problem. With any real set of stimuli, our evidence that there is no single feature that could support the category discrimination is always negative; furthermore, why should we attribute concepts to a bird that uses several features to solve a category discrimination any more than to one that uses only a single feature? And this query still disregards the problems associated with defining what a feature is. What might be a feature to us might not be such to a pigeon, and vice versa.

Another problem lies in precisely what we mean by a natural category or a natural categorization. Is it one that divides the world neatly and easily ("carves nature at the joints"), presumably because the world of natural objects contains few things that fall near the inevitably fuzzy border between membership and nonmembership? Is it characterized by and limited to the set of phenomena dealt with by the organism in a single set of responses, distinct from the response set for other categories? Or is it one which the bird comes "naturally" well equipped to solve, either by possessing unlearned concepts corresponding to the categories concerned (13), or by being especially attentive to the features that are useful in discriminating them? And can these possibilities even be distinguished? There are needs for at least three new directions for research in this area.

First, more work needs to be done on the problem of finding metrics for expressing similarity between percepts of natural objects. However, opinions are divided about whether the outcome of such research would be the discovery of useful metrics or of a proof that they do not and cannot exist.

Second, more needs to be done, both in logical and experimental analysis, to find ways of determining whether animals develop concepts. Three proposals were considered: a) The "Clumping" procedure used by Bateson and Chantrey (7) in imprinting with chicks, in which stimuli that occurred in close contiguity during imprinting became hard to discriminate in a subsequent instrumental task. In unpublished work on category discriminations of letter by pigeons, Lea and Ryan have failed to obtain this result, though their procedures and subjects were so different from Bateson and Chantrey's that the two results do not really conflict. b) A technique involving transfer from a single instance to all members of a category, described by Lea (34); Lea and Ryan's unpublished data suggest that this effect may occur. c) Investigations of whether birds' discriminations of slides of objects involve them in perceiving these as objects in any sense. Although this third question might in the end be unanswerable, more data on problems such as transfer from slide discrimination to object discrimination, and vice versa, would undoubtedly give us a better feel for the interpretation of data in this entire field. In this connection, the use of naturally occurring category discriminations has potential advantages. Psychological experiments on category discrimination involve such extended training that they are only really practicable if they can be automated, and it is very hard to automate the presentation of entire objects. Natural learning must necessarily be complete in a manageable period of time.

Finally, there is a need for comparative studies. Pigeons come in a variety of shapes and sizes, ranging from forest-dwelling species, to those that live in deserts, to large, ground-dwelling flightless species. If trees are in fact a natural category for "the" pigeon, closely related pigeon species living in an array of different habitat types should show corresponding differences in predispositions to recognize trees as a category. As stressed by Delius, however, concept learning experiments are often very fickle, and weak or negative results cannot be safely interpreted. Very small changes in procedures can lead to drastic though often inexplicable improvements. If different pigeon species were not equally tame in a laboratory setting, for example, results might be difficult to interpret.

In summary, "concept formation," like so many other psychological terms (e.g., learning, concept formation, perception, etc.), seems to be defined largely by exclusion. It may mean something like "the ability to recognize objects (in photographs) with an accuracy comparable to that of human observers, with similar error patterns, and in a way that is not reducible to simpler mechanisms (such as single feature extraction) which all would agree are not concept formation." However, while there is agreement that some phenomena are not "concept" formation, it is not entirely clear that we know exactly what concept formation itself is.

TO RECONCILE LEARNING THEORY AND NATURAL BEHAVIOR

Learning theory has produced a body of knowledge gleaned from studies of a few animal species under laboratory conditions. The basic question, then, is whether we can apply generalizations derived from these sources to behavior occurring in the field. We believe that the answer is a qualified "yes."

Operant conditioning has contributed to our knowledge of decision rules made by foraging birds, and elements of classical conditioning can be applied to functional accounts of some species' behaviors. And perhaps conspecific song elements, or "triggers," can be tested to see if they are "conditionable." To the die-hard ethologist, though, some of these questions seem strained, perhaps because they appear aimed at merely labelling a behavioral phenomenon with some (un)familiar terminology from learning theory, or perhaps because they appear an attempt to infuse conditioning approaches with ecologically relevant paradigms.

Furthermore, the ethologist revels in the diversity of natural behaviors, often using them as a comparative tool in an approach to understanding innate contributions to development and the functions and evolution of those behavior patterns. The mere thought of reducing behavior patterns to combinations of blocking or overshadowing effects, for example, in an attempt to understand the underlying causation and to seek commonalities in the diversity seems at first heretical.

To gain greater relevance, the study of mechanisms should be extended to that of function, or how the fitness of animals is influenced (Hollis, this volume). Furthermore, a major need is to see clearly the difference between the procedures used in laboratory studies - operant and classical conditioning - and the processes that allow these procedures to have their effects. It is these processes that will generalize to natural behavior, phenomena such as song learning and imprinting, not simplistic accounts that assume something called "classical conditioning," which merely relabels the results of Pavlovian procedures. Progress is being made, and one hopes that these somewhat skeptical remarks only reflect healthy growing pains of an interdisciplinary approach to behavioral biology.

Acknowledgments. We thank D. Boudreau and G. Drake for their painstaking care in preparing the final copy of this paper.

REFERENCES

(1) Arnold, A.P. 1982. Neural control of passerine song. In Acoustic Communication in Birds, eds. D.E. Kroodsma and E.H. Miller, vol. 1, pp. 75-94. New York: Academic Press.

(2) Baker, M.C.; Spitler-Nabers, K.J.; and Bradley, D.C. 1981. Early experience determines song dialect responsiveness of female sparrows. Science 214: 819-820.

(3) Baptista, L.F., and Morton, M.L. 1982. Song dialects and mate selection in montane white-crowned sparrows. Auk 99: 537-547.

(4) Baptista, L.F., and Petrinovich, L. 1984. Social interaction, sensitive phases, and the song template hypothesis in the white-crowned sparrow. Anim. Behav. 32: 172-181.

(5) Bateson, P. 1979. How do sensitive periods arise and what are they for? Anim. Behav. 27: 470-486.

(6) Bateson, P. 1983. The interpretation of sensitive periods. In The Behavior of Human Infants, eds. A. Oliverio and M. Zappella, pp. 57-70. New York: Plenum.

(7) Bateson, P.P.G., and Chantrey, D.F. 1972. Retardation of discrimination learning in monkeys and chicks previously exposed to both stimuli. Nature 237: 173-174.

(8) Baum, W.M. 1974. On two types of deviation from the matching law: bias and undermatching. J. Exp. An. Behav. 22: 231-242.

(9) Baylis, J.R. 1982. Avian vocal mimicry: its function and evolution. In Acoustic Communication in Birds, eds. D.E. Kroodsma and E.H. Miller, vol. 2, pp. 51-83. New York: Academic Press.

(10) Bischof, H.-J. 1982. Imprinting and cortical plasticity: a comparative review. Neurosci. Biobehav. Rev. 7: 213-225.

(11) Canady, R.A.; Kroodsma, D.E.; and Nottebohm, F. 1984. A relation between culture and brain traits. Proc. Natl. Acad. Sci., in press.

(12) Davies, N.B. 1977. Prey selection and the search strategy of the spotted flycatcher (Muscicapa striata), a field study on optimal foraging. Anim. Behav. 25: 1016-1033.

(13) Delius, J.D., and Nowak, B. 1982. Visual symmetry recognition by pigeons. Psychol. Res. 44: 199-212.

(14) Dooling, R.J., and Searcy, M.H. 1980. Early perceptual selectivity in the swamp sparrow. Dev. Psychobiol. 13: 499-506.

(15) Dowsett-Lemaire, F. 1979. The imitation range of the song of the marsh warbler, Acrocephalus palustris, with special reference to imitations of African birds. Ibis 121: 453-468.

(16) Farabaugh, S.M. 1982. The ecological and social significance of duetting. In Acoustic Communication in Birds, eds. D.E. Kroodsma and E.H. Miller, vol. 2, pp. 85-124. New York: Academic Press.

(17) Goss-Custard, J.D. 1977. Optimal foraging and the size selection of worms by redshank Tringa totanus. Anim. Behav. 25: 10-29.

(18) Gould, J.L. 1982. Ethology. The Mechanisms and Evolution of Behavior. New York: W.W. Norton & Co.

(19) Herrnstein, R.J. 1970. On the law of effect. J. Exp. An. Behav. 13: 243-266.

(20) Herrnstein, R.J., and Loveland, D.H. 1964. Complex visual concepts in the pigeon. Science 146: 549-551.

(21) Hoffman, H.S., and Ratner, A.M. 1973. A reinforcement model of imprinting: implications for socialization in monkeys and men. Psychol. Rev. 80: 527-544.

(22) Immelmann, K. 1975. Ecological significance of imprinting and early learning. Ann. Rev. Ecol. System. 6: 15-37.

(23) Immelmann, K., and Suomi, S.J. 1981. Sensitive phases in development. In Behavioral Development, The Bielefeld Interdisciplinary Project, eds. K. Immelmann, G.W. Barlow, L. Petrinovich, and M. Main, pp. 395-431. New York: Cambridge University Press.

(24) Keith-Lucas, T., and Guttman, N. 1975. Robust-single-trial delayed backward conditioning. J. Comp. Phys. Psychol. 88: 468-476.

(25) Killeen, P.R. 1984. Adaptive clocks. Ann. NY Acad. Sci., in press.

(26) Konishi, M. 1965. The role of auditory feedback in the control of song in the white-crowned sparrow. Z. Tierpsychol. 22: 770-783.

(27) Konishi, M., and Nottebohm, F. 1969. Experimental studies in the ontogeny of avian vocalizations. In Bird Vocalizations. Their Relation to Current Problems in Biology and Psychology, ed. R.A. Hinde, pp. 29-48. New York: Cambridge University Press.

(28) Krebs, J.R.; Erichsen, J.T.; Webber, M.I.; and Charnov, E.L. 1977. Optimal prey selection in the great tit (Parus major). Anim. Behav. 25: 30-38.

(29) Kroodsma, D.E. 1981. Ontogeny of bird song. In Behavioral Development, The Bielefeld Interdisciplinary Project, eds. K. Immelmann, G.W. Barlow, L. Petrinovich, and M. Main, pp. 518-532. New York: Cambridge University Press.

(30) Kroodsma, D.E.; Baker, M.C.; Baptista, L.F.; and Petrinovich, L. 1984. Vocal 'dialects' in Nuttall's white-crowned sparrow. In Current Ornithology, ed. R.F. Johnston, vol. 2. New York: Plenum, in press.

(31) Kroodsma, D.E., and Pickert, R. 1980. Environmentally dependent sensitive periods for avian vocal learning. Nature 288: 477-479.

(32) Kroodsma, D.E., and Pickert, R. 1984. Sensitive phases for song learning: effects of social interaction and individual variation. Anim. Behav. 32: 389-394.

(33) Lea, S.E.G. 1981. Correlation and contiguity in foraging theory. In Advances in Analysis of Behavior. Predictability, Correlation and Contiguity, eds. P. Harzem and M.D. Zeiler, vol. 2, pp. 355-406. Chichester: Wiley & Sons.

(34) Lea, S.E.G. 1984. In what sense do pigeons learn concepts? In Animal Cognition, eds. H.L. Roitblat, T.G. Bever, and H.S. Terrace. Hillsdale, NJ: Erlbaum, in press.

(35) Lea, S.E.G., and Dow, S.M. 1984. The integration of reinforcements over time. Ann. NY Acad. Sci., in press.

(36) Lea, S.E.G., and Harrison, S.N. 1978. Discrimination of polymorphous stimulus sets by pigeons. Q. J. Exp. Psy. 30: 521-537.

(37) MacArthur, R.W., and Pianka, E.R. 1966. On the optimal use of a patchy environment. Am. Natur. 100: 603-609.

(38) Marler, P. 1967. Comparative study of song development in sparrows. In Proceedings of the XIV International Ornithological Congress, ed. D.W. Snow, pp. 231-244. Oxford, Edinburgh: Blackwell Scientific Publishers.

(39) Marler, P. 1970. A comparative approach to vocal learning: song development in white-crowned sparrows. J. Comp. Physiol. Psychol. 71: 1-25.

(40) Marler, P., and Peters, S. 1980. Birdsong and speech: evidence for special processing. In Perspectives on the Study of Speech, eds. P. Eimas and J. Miller, pp. 75-112. Hillsdale, NJ: Erlbaum.

(41) Marler, P., and Peters, S. 1981. Sparrows learn adult song and more from memory. Science 213: 780-782.

(42) Marler, P., and Peters, S. 1982. Subsong and plastic song: their role in the vocal learning process. In Acoustic Communication in Birds, eds. D.E. Kroodsma and E.H. Miller, vol. 2, pp. 25-50. New York: Academic Press.

(43) Moynihan, M. 1959. A revision of the family Laridae (Aves). American Museum Novitates 1928: 1-42.

(44) Nottebohm, F. 1970. Ontogeny of bird song. Science 167: 950-956.

(45) Nottebohm, F. 1980. Brain pathways for vocal learning in birds: a review of the first ten years. Prog. Psychobiol. Physiol. Psychol. 9: 85-124.

(46) Nottebohm, F.; Kasparian, S.; and Pandazis, C. 1981. Brain space for a learned task. Brain Res. 213: 99-109.

(47) Payne, R.B. 1981. Population structure and social behavior: models for testing the ecological significance of song dialects in birds. In Natural Selection and Social Behavior, eds. R.D. Alexander and D.W. Tinkle, pp. 108-120. New York: Chiron.

(48) Payne, R.B. 1981. Song learning and social interaction in indigo buntings. Anim. Behav. 29: 688-697.

(49) Sibley, C.G., and Ahlquist, J.E. 1983. Phylogeny and classification of birds based on the data of DNA-DNA hybridization. In Current Ornithology, ed. R.F. Johnston, vol. 1, pp. 245-292. New York: Plenum Press.

(50) Staddon, J.E.R. 1980. Optimality analyses of operant behavior and their relation to optimal foraging. In Limits to Action: The Allocation of Individual Behavior, ed. J.E.R. Staddon, pp. 101-141. New York: Academic Press.

(51) Staddon, J.E.R. 1983. Adaptive Behavior and Learning. New York: Cambridge University Press.

(52) Teuchert, G.; Wolff, J.R.; and Immelmann, K. 1982. Physiological degeneration in the ontogenie of the CNS of birds. An influence on the sensitive phase of imprinting? Verh. Dtsch. Zool. Ges. 1982: 259.

(53) Todt, D. 1981. On functions of vocal matching: effect of counter-replies on song post choice and singing. Z. Tierpsychol. 57: 73-93.

(54) Todt, D.; Hultsch, H.; and Heike, D. 1979. Conditions affecting song acquisition in nightingales (Luscinia megarhynchos L.). Z. Tierpsychol. 51: 23-35.

(55) Verner, J. 1975. Complex song repertoire of male long-billed marsh wrens in eastern Washington. Living Bird 14: 263-300.

(56) Whitney, C.L. 1981. Patterns of singing in the varied thrush. II. A model of control. Z. Tierpsychol. 57: 141-162.

The Biology of Learning, eds. P. Marler and H.S. Terrace, pp. 419-433. Dahlem Konferenzen 1984. Berlin, Heidelberg, New York, Tokyo: Springer-Verlag.

Natural History and Evolution of Learning in Nonhuman Mammals

S.J. Shettleworth
Dept. of Psychology, University of Toronto
Toronto, Ontario M5S 1A1, Canada

Abstract. Traditional comparative psychological approaches to the evolution of learning are inadequate for dealing with a broadly based natural history of learning. An approach based on and integrated with other areas of behavioral ecology is outlined. Learning is viewed as the set of ways of solving ecological problems requiring that the individual adjust its behavior to the details of its own environment. Among nonhuman mammals there is a great deal of variation in the nature and amount of learning that different species' life-styles require.

INTRODUCTION

Bats and blue whales, sloths and shrews, elephants and moles. Mammals live in the treetops, in the sea, and underground; they can be big or small, long-lived or short-lived, social or solitary. Eisenberg (5) distinguishes sixteen categories of dietary specialization, eight categories of substrate utilization, and eleven types of mating system among mammals, not to mention ten rearing systems, eight foraging systems, and five refuging systems. Is it possible to make any useful generalizations about the evolution and natural history of learning in such a diverse group?

Questions about the evolution and about the natural history of learning are obviously related in the same way as these questions are for other aspects of behavior. That is, knowledge about the distribution of some trait in contemporary animal groups can suggest what selection pressures affect it and thereby lead to inferences about its probable course of evolution. This is especially so when traits are shared by phylogenetically diverse groups that face similar selection pressures.

Even to begin to make useful generalizations about the natural history and evolution of mammalian learning would require a comprehensive survey of what a wide variety of mammals learn and how they learn it. Such a data base does not exist. Natural histories rarely contain the necessary information for such a survey because much of the information required cannot by its nature be obtained from field observations alone. While aspects of social structure or feeding habits can be observed fairly directly in the field, observations of individuals at a fixed moment in time can only suggest inferences about what and how they may have learned. For example, individuals may appear to have learned the characteristics of different members of their social group because they react differentially to them. An individual may appear to have acquired a cognitive map of its home range because it can find its own way to feeding or refuging sites. However, learning is fundamentally a causal or mechanistic notion; it refers to a process underlying changes in behavior of an individual over time. Therefore, conclusions about the nature and extent of learning in natural situations require as a first step observing the same individuals through time. In most cases, individuals must also be tested in controlled situations in which they can be exposed to systematically differing experiences. A good nonmammalian example of why this is necessary can be drawn from the study of song learning in birds. Among other things, field observations on local dialects of species-typical songs suggested that birds learn their songs from hearing conspecifics sing, but the manner in which this learning takes place, its selectivity and sensitive periods, had to be elucidated in the laboratory (Marler, this volume).

In addition to an appropriate data base, any discussion of the evolution and natural history of learning requires two sets of interrelated assumptions. On the one hand are assumptions about how to identify and categorize instances of learning and the supposed underlying learning processes. On the other hand are assumptions about the sorts of phylogenetic and/or ecological variables likely to be correlated with learning ability. In this paper I shall first briefly sketch two traditional approaches to the evolution of learning and then suggest a possibly more satisfactory framework for the analysis of ecological and evolutionary variables in learning. None of these approaches is uniquely suited to the analysis of mammalian learning, but the discussion will focus on mammalian examples.

COMPARATIVE PSYCHOLOGY AND THE EVOLUTION OF LEARNING

The traditional comparative psychological approach to learning is based on the conventional classification of types of learning such as habituation, classical conditioning, instrumental conditioning, and other ("higher") forms such as cognitive mapping and insight learning. This approach seeks to place species in a linear hierarchy according to their abilities in these categories. Although the idea of a linear progression from lower to higher forms, with man at the top, is an ancient one, it is thoroughly discrepant with modern evolutionary theory. Anyone still unconvinced of the truth of this statement need only read the chapter by Hodos (8) in a previous volume in this series. Further, contemporary work on conditioning raises questions as to whether the traditional paradigms really do tap into different kinds of learning (Jenkins, this volume). Nevertheless, a great deal of experimental work has been devoted to comparing the performance of a great many species on conventional laboratory learning tasks. Macphail (12) concludes his recent review of this work with the remarkable statement that, aside from man's linguistic ability, there is no good evidence for either quantitative or qualitative differences in learning among any of the vertebrates. All vertebrates studied appear to learn similarly when exposed to habituation and classical and instrumental conditioning paradigms.

Within its own context, Macphail's hypothesis is eminently reasonable. Since each species' performance depends on numerous aspects of the specific situation in which it is tested, no single set of data can fully represent the species' relative "abilities" on a given task. In principle, a different ranking of species might be obtained using different stimuli, rewards, and/or apparatuses for each one. This sort of consideration leads Macphail to his very conservative hypothesis. He acknowledges, however, that just as there is no good evidence for species differences in learning among vertebrates, so is there no firm evidence that such differences do not exist.

In the context of this volume, it should be borne in mind that all the conventional comparative psychological data come from a narrow range of laboratory situations, many designed around an anthropocentric notion of intelligence. There is no reason to assume that performance in these situations reflects anything about behavior in an animal's natural niche, nor that it reflects the outcomes of selective pressures for the evolution of learning. Even if different species' performance in successive reversal, delayed response tasks, and the like could be ranked unambiguously, it is not clear that the ranking would tell us any more about the evolution of their learning abilities than it does about their natural history.

BRAIN SIZE AND THE EVOLUTION OF LEARNING

Of course, whatever tasks are chosen, no examination of learning in present-day species can reveal directly anything about the evolution of the capacity to learn. An approach which at least promises direct information on evolution is Jerison's (9) examination of endocranial casts of extinct and living species. Species can be compared according to encephalization quotient, or EQ. The EQ is the ratio between the observed brain weight and the brain weight expected on the basis of the body weight for the given species together with the brain/body weight regression for all species in the set being considered. Thus it represents how a given species compares in brain weight to an average of others in its size class.

Given the many debatable assumptions of this approach, it is surprising how orderly are the data it has generated. Some of these are mentioned later. The major limitation of EQ as a measure of learning abilities (as opposed to "intelligence") is that, insofar as complex perceptual or motor abilities affect brain size, they contribute to EQ just as much as, if not more than, learning ability. Indeed, Macphail (12) notes that the general impression that some species are more "intelligent" than others may be based just on different degrees of complexity and adaptability in perceptual and motor skills rather than on learning ability as defined by laboratory tasks.

APPROACHES TO THE NATURAL HISTORY OF LEARNING

As sketched above, two of the major existing frameworks for classifying species on learning ability are not very helpful in understanding the natural history of learning. The traditional comparative psychology of learning does not focus on species-relevant learning tasks, and the study of brain-body weight ratios is probably not predictive of learning abilities per se. The obvious starting point for organizing data about the natural history of learning and asking productive questions about the distribution of learning across different classes is information about naturally occurring instances of learning. Rather than start by asking how various species perform on standardized laboratory tasks, one might start by asking what present-day species learn and how they learn it. Such an approach has been advocated by Johnston (10), who seems to suggest that it can be productive in the absence of any preexisting system for classifying and analyzing examples of learning. While I agree that a departure from existing learning theory may be the best way to seek a general account of learning phenomena (see below), I do not think Johnston's sort of approach is likely to be the most productive. Moreover, there are plenty of questions framed in terms of contemporary learning theory that can begin to be answered by an attempt to survey the nature and extent of learning that a variety of species do in the wild. Of particular interest are a number of questions that have arisen in recent discussions of "biological" factors in learning. Most important are the related questions of whether a species' learning ability is general or consists of clusters of specialized abilities and whether there are qualitatively specialized kinds of learning shown by some species in functionally appropriate circumstances.

Rozin (15) has suggested that learning abilities may first evolve as isolated specializations and later become accessible to a wider range of behavioral systems. Thus, for example, associative learning capabilities may at first operate only on certain ecologically relevant combinations of events, as in poison-avoidance learning or visual discrimination of distasteful prey. Gradually the ability to form associations becomes more general to stimuli or behavioral systems. Gould and Marler (this volume) discuss a similar view in detail. An alternative view is that poison-avoidance learning and other examples of associative learning with specialized functions are manifestations of a generally accessible associative mechanism, although it has to be granted that the general mechanism is more responsive to some combinations of events than to others. As a possible example of such specialization from the laboratory, some species' ability to profit maximally from instrumental learning experiences seems to be limited to certain response-reinforcer combinations. For example, in golden hamsters grooming is not performed as efficiently for food as are other activities that might be more likely to function in natural food-getting. This may result from a motivational constraint, limiting what hamsters can do when they expect food, or it may be evidence of a "constraint on learning." Whatever its nature, monkeys are apparently free of it (Iversen, personal communication).

These examples suggest that the comparative study of learning in the laboratory could be enriched by considering the specialization-generalization issue. It may be that when each species is tested in a situation designed to elicit its best performance, there are no firm quantitative or qualitative differences attributable to species' differences in learning ability. However, asking about the range of situations over which each species exhibits a given kind of learning might produce more interesting answers, especially if the resulting information about learning were tied to information about natural history. An example is the work of Daly, Rauschenberger, and Behrends (3) on food-aversion learning in specialist vs. generalist species of kangaroo rats. These authors tested the hypothesis that a species that feeds on a wide variety of foods should more readily acquire a food aversion than one that does not. Their results were consistent with this hypothesis. However, the species differences found could not definitely be attributed to species differences in learning ability, because the kangaroo rats tested also differed in initial responses to the novel foods used, and perhaps in other ways relevant to their performance in the particular test situation.

Studies like that of Daly et al. or recommendations (4) for more ecologically based studies of associative learning do not address the question of whether paradigms developed more or less a priori and imposed on animals in the laboratory encompass all the learning that animals do in nature. Are there specialized forms of learning, such as birdsong learning (Marler, this volume) or imprinting (Bateson, this volume), that do not fit associative paradigms? Do animals acquire complex cognitive representations of social companions or properties of the inanimate environment in ways that cannot be understood in terms of existing theories of learning (Lea, this volume)? The mammals are a particularly good group in which to seek the answers to these questions since the most reports of complex social cognitions (14), play (6), tool use (1), and observational learning seem to involve mammalian species. All of these could be candidates for specialized forms of learning.

Looking for evidence of qualitatively specialized kinds of learning can easily lead to a catalog of "gee-whiz" stories. These may be very suggestive, but often they are only that. For example, when vervet monkey females hear an infant's screams, they look toward the infant's mother, while she looks toward the source of the screams (2). The direction of the onlookers' gaze reveals their awareness of social relationships among members of their group other than themselves. How this awareness develops is a question that remains to be investigated explicitly. Premack (14) suggests that the vervets' behavior is evidence of a complex social cognition while Cheney and Seyfarth (2), its discoverers, imply that it may be the outcome of a chain of associations, calls with baby, baby with mother. Perhaps "social awareness" develops out of clusters of interrelated associative experiences. Ideally, field observations will uncover natural experiments which will help to define the roles of experience in cases like this one.

Suppose, then, that one were to survey the naturalistic literature for examples of behavior suggesting that learning must have occurred, much as authors have surveyed the literature for examples of play (6) or tool use (1). The preceding discussion suggests two complementary ways of organizing the information that might be obtained. First, one might look for general vs. specialized or situation-specific uses of learning ability. Then it could be asked whether generality vs. specificity of learning ability has any ecological or phylogenetic correlates. For example, perhaps primates tend to have a very general learning ability, while rodent's learning is highly situation-specific. Perhaps animals that inhabit a narrow niche can learn rapidly in only a few, ecologically relevant situations, while generalists can learn about a broader range of different events (see also Gould and Marler, this volume). Alternatively, information about the natural history of learning might be organized according to the type of learning that can be inferred. For example, are there any systematic differences between species that show evidence of cognitive mapping in their spatial behavior as opposed to simple route following (see Menzel, this volume)? Is the capacity for learning by observation evidenced by only certain sorts of species, highly social ones, for example? Are the correlates of "higher" forms of learning phylogenetic or mainly ecological or (most likely) both? These are examples of the sorts of questions that a survey of naturalistic examples of learning could be directed toward answering.

THE BEHAVIORAL ECOLOGY OF LEARNING

In other areas of behavior study, behavioral ecology (or sociobiology) has provided a powerful means of organizing data on a wide range of species and suggesting testable hypotheses about patterns of differences among species. Its success in accounting for the cross-species distribution and possible evolution of numerous features of behavior suggests attempting to formulate a behavioral ecological approach to learning. This would be parallel to and integrated with behavioral ecological accounts of social structure, reproductive strategies, feeding strategies, and the like. I have previously discussed this approach in a different context (17). Here I develop it further and show how it might be applied to the varieties of learning in nonhuman mammals. Similar ideas are implicit in the writings of others (e.g., (11, 13); see also Bateson, this volume).

Behavioral ecology implicitly views animals as having certain universal problems to solve: finding food, finding mates, allocating reproductive effort, avoiding predators. It is generally possible to work out from first principles (formally in many cases) what solutions to these problems ought to have evolved as a function of environmental and phylogenetic constraints. For instance, the distribution of monogamy, polygyny, and polygamy can be broadly accounted for by taking into consideration such factors as whether a male can be expected to increase his reproductive success more by helping the female he has already inseminated or by leaving her to raise their young alone and seeking other mates. These characteristics of a species are, in turn, functions of and interrelated with feeding niche, developmental patterns, and other factors. The proximal mechanisms (e.g., the conditions under which potential mates are attractive, the details of courtship) by which each species attains its typical pattern will vary, but these are of no immediate concern in a strictly functional approach. It may be that variations in social structure, foraging patterns, and the like, across a wide range of species can be better encompassed by analysis at a functional level than by detailed comparisons of proximal mechanisms.

These considerations suggest that a behavioral ecological approach to learning would be radically different from any psychological approach. First, "learning" would be defined not by its causes ("reinforced practice"; "experience of relations between events") but by its function. Just as animals have to find food and mates, so they have to adjust their behavior to local conditions which are unpredictable on the basis of anything but individual experience. Adjusting behavior in this way could be defined as "learning." Instances of "learning" identified functionally would not be completely congruent with instances of "learning" defined in other ways. For example, the former would include acclimatization or other physiological adjustments to local conditions, while apparently excluding cases where species-typical experience is a necessary part of the species-typical developmental program. The latter cases are not predictable on the basis of individual experience alone.

Definitional problems notwithstanding, it is worth exploring the consequences of defining learning as the class of solutions to a certain kind of ecological problem. "Learning" then cuts across other behavioral ecological problems at the same time as being very much interrelated with them. A requirement to adjust to conditions peculiar to the individual can be an integral part of mating systems (for example, when the species-typical system presupposes recognition of individual conspecifics), reproductive strategies (for example, when reproductive rate is slow because the young need time to learn important skills), and so on. Optimal foraging theory has helped to make explicit the many ways in which individual experience may be involved in animals' choices of where to feed and what to feed on. Foraging theory also includes examples of how "learning" as a solution to a class of problems can include a variety of mechanisms ranging from motivational to purely associative, even when the local conditions to be adjusted to are defined precisely. For example, according to optimal foraging theory, a forager's choice of whether or not to eat a given type of food item should be influenced by the availability of more profitable items. A variety of species do select items roughly as foraging theory predicts. The form of the theory as well as the behavioral results suggest that foragers must use some sort of knowledge of local conditions to do this. However, the mechanism by which this knowledge is acquired can be as simple as one in which the threshold for accepting items is temporarily raised each time a profitable item is eaten. The threshold gradually falls as time passes since eating the last item, so that relatively unprofitable items become more and more acceptable. Such a short-term mechanism is apparent in shore crabs' selection among large and small mussels and may be most appropriate for very patchily distributed prey (17). In contrast, in some species and in a stable situation, accepting and rejecting prey may be governed by learned anticipation of future profitable items. This kind of mechanism is evident in pigeons' behavior in reinforcement schedules mimicking foraging problems. In both cases, item choice depends on "learning" since it can only be achieved by the individual forager's adjusting acceptance of a given item type in accord with its experience of local availability of other items. It is an interesting question as to whether phylogenetic constraints or details of the foraging situation such as long-term stability of prey densities might predict which of several possible mechanisms a given species will use.

THE BEHAVIORAL ECOLOGY OF LEARNING IN MAMMALS

The idea that animals face the general problem of adjusting to local conditions, unpredictable except on the basis of individual experience, immediately suggests what kinds of learning capacities different kinds of animals will require, a point further discussed by Gould and Marler (this volume). The most basic problems of individual adjustment might be classified as identifying, locating, and processing local resources. An individual needs to be able to direct feeding, mating, parental care, and predator avoidance to appropriate objects in its environment. Where the appropriate object varies from one individual to another, some sort of learning will be necessary. As an example of the ways in which requirements for learning differ across species, consider one difference between Eastern chipmunks and wildebeests. The chipmunk mother has no need to recognize her precocial young as hers since they remain in her burrow until they are almost old enough to disperse. Appropriate allocation of parental investment is ensured if the mother returns to her own burrow, whose location she has already learned, and nurses whatever babies she finds there. The situation is very different for ungulates whose young can stand and walk about on the plains a few hours after birth. Mother and young must be able to identify each other, and it is in these species that the mammalian analogue to imprinting has been observed.

Locating local resources might be as simple as making and keeping to runways between burrow and feeding areas in a small terrestrial mammal or as complex as knowing the distribution of many resources over a wide area, as may be the case in species such as elephants and baboons. Knowledge about the temporal as well as the spatial distribution of resources may be important, as in the case of the kestrels that synchronize their hunting with the activity cycles of their rodent prey. Obviously the patchiness and predictability of a species' characteristic food will influence the value of this kind of learning. Baleen whales, sieving plankton, have very different learning requirements from nectarivorous bats. Optimal foraging theory can help to make these requirements explicit.

Processing local resources often seems to involve learning how to direct motor patterns in order to produce some sort of reinforcement with the least possible delay. For instance, squirrels must learn to orient nutcracking appropriately to different kinds of nuts. Chimps learn to "fish" for termites with sticks, and they seem to require observation of adults as well as experience of their own trial and error (1).

Having sketched some of the problems that seem to require learning for their solution, we are in a position to consider some dimensions of difference among mammals that might be correlated with characteristics of their learning. Ultimately, consideration of these sorts of differences should help to answer some of the following questions. All of these questions assume that functional and mechanistic considerations can be interwoven in a productive way, that the details of ecological problems requiring "learning," in the sense of adjustments by individuals to local conditions, will to some extent shape learning mechanisms through evolution by defining the optimal outcomes of learning ((16); also see Bateson, this volume).

Why learn in a specific case as opposed to relying on other solutions, including that of not making the adjustment at all?

What to learn? For example, when an individual must learn to recognize individual conspecifics or the approach of a predator, what cues does he use? Are they the optimal ones in the sense of being the most reliable predictors in the situation?

When to learn? Does learning begin immediately on first exposure to a situation? Does general learning about the environment occur during periods of "sampling" or play and get put to use when it is needed? Fagen (7) has suggested that play promotes generalized flexibility in dealing with the environment in addition, perhaps, to developing motor skills and physical fitness.

How to learn? Trial and error, imprinting, observation, association: is a particular process the only one that can solve the problem, or if several might serve the purpose, what determines which one a given species uses?

How fast to learn and how long to remember? When a wildebeest has just given birth, mother and infant should learn to recognize each other as soon as possible. A young chimp, with plenty of food sources and a protective family group, can take months of years to learn to catch termites. A bat should perhaps remember which trees had fruit last night but not last week, or how many hours ago it last sucked nectar from a particular flower.

Mammals differ in a number of ways that are likely to be predictive of the answers to these questions. In discussing these I rely primarily on Eisenberg's (5) comprehensive treatment of trends in mammalian adaptation. It is worth noting at the outset, however, that any species' adaptations are interrelated and mutually constrained. For example, a carnivore needs a larger home range than a granivore of the same size; only largish mammals can live in the sea or afford to be grazers. Thus, a need to solve one class of problems may always be accompanied by the need to solve others that inevitably accompany them. This is no less true of adaptations requiring learning than of any others. Exploitation of many and varied prey, demanding specialized methods of capture, depends on evolution of learning abilities as much as it depends on evolution of motor equipment or sociality.

Figure 1 ((5), Fig. 33) is a convenient summary of some of the important determinants of mammalian adaptations. Habitat type and feeding adaptations together define a number of types of substrate utilization and dietary specialization. Mammals can move about primarily in two dimensions (terrestrial) or three (arboreal, aquatic, airborne), and these differences in substrate complexity seem to be reflected in EQ. Feeding niches range from those requiring pursuit of mobile prey, such as piscivore and carnivore, to sucking nectar, raiding ant colonies, and harvesting fruit and seeds. The feeding niche of large terrestrial carnivores like wolves and lions entails cooperative hunting and a long period of learning by the young. Clearly their requirements for learning are very different from those of a grazer or browser.

The habitat type for a given species might be thought of as including its predators and competitors. Type of antipredator strategy tends to be correlated with EQ. Passive strategies such as rolling into a ball tend to be correlated with lower EQ than do active individualistic strategies such as directed flight and counterattack (5).

Demographic strategy includes longevity and rate of producing young; in short, whether a species is r- or K-selected. Small mammals with relatively short life spans tend to produce large litters of altricial young which spend relatively little time in a learning situation with the family group. Such species tend to have a relatively low EQ and relatively large reliance on preprogrammed patterns of behavior. Larger, longer-lived mammals tend to invest more in each offspring and to have a longer

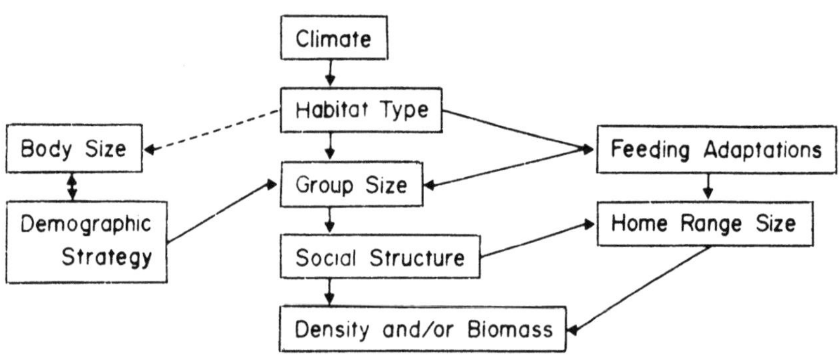

FIG. 1 - Representation of the interrelations among determinants of mammalian adaptations (from (5)).

period of parent-young association during which the young are learning important skills (5). In some animals, such as marmosets, these skills may include learning how to care for the young.

Type of home range and longevity together might be expected to influence what it is worthwhile to learn. For example, it may be worthwhile for a long-lived permanently territorial animal to be sensitive to fairly subtle details of its environment. Similarly, an animal permanently associated with a social group will have more to learn about conspecifics than, say, a mole or a desert rodent living an almost solitary existence.

IMPLICATIONS

The differences between chimpanzees and moles, lions and hedgehogs, are so numerous as to make comparing them seem almost silly. However, the point to be drawn from these comparisons, that different life-styles imply very different requirements for "learning," ought to apply to comparisons of species which differ less extremely along the same dimensions. In principle, one can compare closely related species with different life-styles or unrelated species that have converged to similar life-styles and ask what they learn and how, if at all, their capacities for learning, as revealed in controlled situations, reflect the requirements for learning that they have evolved to meet.

I have sketched some of the ways in which the natural history of learning might be approached and suggested what kind of principles might emerge from such an enterprise. Probably the most productive way to organize a comprehensive natural history of learning in any animal group would be in terms of the ecological problems requiring adjustments to local conditions by the individual. Analysis of learning in terms of such functional problems does not map directly onto the learning theorist's analysis in terms of mechanisms (16). Sometimes different possible mechanisms (association learning, motivational changes, imprinting) might be capable of producing functionally similar outcomes. The comparative study of how similar problems are solved by different species offers some interesting and as yet largely unexplored possibilities. Almost certainly, also, a broad consideration of the mechanisms used by animals to solve problems of adjustment to local conditions will reveal that learning theory up to now has dealt with an artificially restricted set of learning problems.

Acknowledgements. Preparation of this paper was supported by a grant from the Natural Sciences and Engineering Research Council of Canada. I thank R. Sigmundi for comments on an earlier version.

REFERENCES

(1) Beck, B. 1980. Animal Tool Behavior. New York: Garland.

(2) Cheney, D.L., and Seyfarth, R.M. 1980. Vocal recognition in free-ranging vervet monkeys. Anim. Behav. 28: 362-367.

(3) Daly, M.; Rauschenberger, J.; and Behrends, P. 1980. Food-aversion learning in kangaroo rats: a specialist-generalist comparison. Anim. Learn. Behav. 10: 314-320.

(4) Domjan, M., and Galef, B.G. 1983. Biological constraints on instrumental and classical conditioning: retrospect and prospect. Anim. Learn. Behav. 11: 151-161.

(5) Eisenberg, J.F. 1981. The Mammalian Radiations. Chicago: University of Chicago Press.

(6) Fagen, R. 1981. Animal Play Behavior. New York: Oxford University Press.

(7) Fagen, R. 1982. Evolutionary issues in development of behavioral flexibility. Perspect. Ethol. 5: 365-383.

(8) Hodos, W. 1982. Some perspectives on the evolution of intelligence and the brain. In Animal Mind - Human Mind, ed. D.R. Griffin, pp. 33-55. Dahlem Konferenzen. Berlin, Heidelberg, New York: Springer-Verlag.

(9) Jerison, H.J. 1973. Evolution of the Brain and Intelligence. New York: Academic Press.

(10) Johnston, T.D. 1981. Contrasting approaches to a theory of learning. Behav. Brain Sci. 4: 125-139.

(11) Johnston, T.D. 1982. Selective costs and benefits in the evolution of learning. Adv. Stud. Behav. 12: 65-106.

(12) Macphail, E. 1982. Brain and Intelligence in Vertebrates. Oxford: Clarendon Press.

(13) Plotkin, H.C., and Odling-Smee, F.J. 1979. Learning, change, and evolution: an enquiry into the teleonomy of learning. Adv. Stud. Behav. 10: 1-41.

(14) Premack, D. 1983. Animal cognition. Ann. Rev. Psychol. 34: 351-362.

(15) Rozin, P. 1976. The evolution of intelligence and access to the cognitive unconscious. Progr. Psychobiol. Physiol. Psych. 6: 245-280.

(16) Shettleworth, S.J. 1983. Function and mechanism in learning. In Advances in Analysis of Behavior, eds. M.D. Zeiler and P. Harzem, vol. 3, pp. 1-39. Chichester: Wiley.

(17) Shettleworth, S.J. 1984. Learning and behavioral ecology. In Behavioral Ecology, eds. J.R. Krebs and N.B. Davies, pp. 170-194. Oxford: Blackwell Scientific.

The Biology of Learning, eds. P. Marler and H.S. Terrace, pp. 435-446. Dahlem Konferenzen 1984. Berlin, Heidelberg, New York, Tokyo: Springer-Verlag.

Species-typical Response Predispositions

R.C. Bolles
Dept. of Psychology, University of Washington
Seattle, WA 98195, USA

Abstract. The study of stereotyped behaviors, such as grooming in the rat, suggests that these behaviors are coordinated peripherally. The fact that so many of the compensatory and correction mechanisms and other motor adjustments are also peripheral suggests that the entire motor system is under hierarchical control. We see the motor system as controlled by very simple central commands and being largely locally autonomous. We see it as being very elaborate but quite inflexible.

INTRODUCTION

Along with a number of other people, I first became interested in grooming behavior back about 1960. At that time, ethologists were concerned with grooming because it was regarded as evidence of displacement. The occurrence of grooming could be taken as evidence of the "spilling over" of motivational energy whose customary expression had been blocked in some way. I was one of several people who preferred to think of grooming in an alternative framework, namely, as evidence for the stimulus control of behavior. The animal groomed not because it had pent-up energy spilling over into the grooming compartment, but rather because the stimuli that elicit grooming were present. These eliciting stimuli are presumably always present and will always elicit grooming behavior unless the animal has something more urgent to do, that is, unless it is responding to some more salient stimulus. Thus, the idle rat tends to groom in interludes between more important kinds of behavior such as eating and drinking (2).

One of the first things an observer of grooming sees is the regular chaining of different kinds of responses. Typically, rats lick their fur, then scratch themselves, and then fiddle with their feet. The different components usually appear in this order. For example, following an approximately two-second burst of scratching, the rat will almost invariably lick the paw of the foot that has done the scratching. The stimulus control interpretation provides a ready account of this sequencing. We may suppose that after a bout of scratching, there are hairs, skin particles, or other kinds of material attached to the scratching foot. The itch stimulus that initiated the scratch is now gone, and the animal is now confronted with this new stimulus, so it proceeds to clean the foot that did the scratching. Then having cleaned the foot, the animal is ready for the next itch. Sure enough, we are likely to see very soon another two-second burst of scratching, to be followed by another short period of foot licking. The behavior changes as the stimuli change.

In those days, I was content with this simple-minded S-R view of grooming behavior. I could not observe the stimuli that were supposed to be controlling the grooming, but that was of little concern because other investigators (e.g., (8)) had manipulated the controlling stimuli with birds and found that, in fact, the incidence of different forms of grooming could be controlled and that they were appropriate for the stimuli that had been applied, such as sticky material on the bill. And so we had the concept of chains of grooming responses occurring because they were elicited by chains of controlling stimuli, stimuli that arose as a result of the grooming. It was all under deterministic control.

My complacency with the stimulus control approach was short-lived, however. When Woods and I started looking at the behavior of very young rats (4), we discovered a curious thing. Infant rats had grooming sequences that were much like those of adult rats. They too showed the same chaining of response segments, and approximately the same duration of response segments, even in the first days of life. And yet, there was a peculiar difference. Until a rat is about seven days old, when it scratches itself, the scratching foot does not make contact with the body! What this means is that the short burst of scratching is not terminated because the itch goes away, nor is the scratching foot licked because it is the source of new stimuli. The stimulus control concept is clearly not valid. Some quite different conception is needed.

It would appear as though the various components of grooming behavior and their sequences have a ballistic quality. The animal initiates scratching and the whole scratching segment is executed. We take such a thing for granted in terms of the small details of the scratch. That is, one motion follows another at a fixed pace. As long as the rat is scratching, it scratches at the rat's rate of six scratches per second, or whatever. So we do not need a separate initiating command signal for each scratch, we only need to turn on the oscillator. But what we have discovered here is that the whole segment, its duration, and organization, and probably everything about the segment, is pretty much preordained once the command signal is given. Once the command signal initiates the behavior, virtually all of the details of that segment are predictable. Moreover, the subsequent behavior is largely predictable. Having finished a short bout of scratching, the rat will lick its foot. The foot licking is not controlled by the occurrence of a new stimulus, it is more simply just part of the whole program. The whole program is remarkably detailed.

Consider that when a rat begins to scratch, it not only has to turn on the scratch oscillator, it also has to adjust its entire body posture. Before it begins scratching, the rat adopts a characteristic whole body position. It sits on one haunch, it bends its back, it redistributes its weight on the other feet, and it may cock its head to one side. Only at that point, only when the whole body is in place for scratching, does the oscillator fire to produce the reflexive scratching motion. Is this posturing a learned coordination? Do infant rats <u>learn</u> how to position themselves so that the scratching leg can do its work? Do they learn this because if they do not do it right they will fall over and that is aversive? I do not think so; I believe the whole thing is programmed.

I have let the concepts get ahead of the data here because I wanted to indicate where I was headed. There are substantiating data, though. Fentress (5) reported a series of studies that carried the Bolles and Woods observation much further. He used mice that had a scratching limb removed surgically and then recorded the movements of the stump. He found patterns of stump movements that were, from a neuromuscular point of view, indistinguishable from the movements of the limb of animals that were scratching themselves - the same durations, the same sequencing, the same patterning - it was all there. More recently, Provine has studied wing movements in wingless chicks (7). The wings were removed surgically, straws were placed over the stumps, and this preparation was then photographed under stroboscopic light when the chick was dropped (if you drop a chick a few inches, it flaps its wings). If you drop a chick whose wings have been surgically removed, the movement of the stumps is, from a neuromuscular point of view, indistinguishable from the movements of a normal animal. Thus, it is clear that this kind of wing flapping is very much like scratching. The whole coordination is preprogrammed.

Further evidence from another species is the oft-cited evidence that you do not have to teach a child to walk. Indeed, you probably cannot teach a child to walk. All the many hours that so many mothers have devoted to this noble task have probably been wasted. To get a child to walk all you have to do is let it reach the proper maturational age; then you just stand it on its feet. The child is already programmed to walk. And it is, moreover, extraordinarily difficult to alter that programming. Animals that walk have automatic correction systems already built in, so that if the ground is sloping they immediately compensate for the slope. Put weights on their feet and they automatically compensate for that. Change their balance, or their load, or the distribution of body weight, and they compensate automatically for those factors. For example, to maintain her balance a heavily pregnant woman would only need to lean backward from the ankles about one inch. But heavily pregnant women do not do that. If anything, they stand with the pelvis further forward, and tilt backward from the hip several inches. That is uncomfortable and awkward, and it does not seem very natural, but that is the way it has to be done. It is programmed that way.

COMMAND SIGNALS

The fundamental point about command signals is that the more centrally we look at them the simpler they appear. No matter how complex the details of the behavior are, it operates from very simple commands. The corollary is that a quite simple command signal arising centrally can put into action a large, complicated, coordinated program of action. Historically, psychologists have thought of higher brain centers being involved in the coordination of behavior. If coordination was carried out at the cortex, then the assumption was, we had the opportunity to get learning involved, so that motor skills could be acquired through the facilities provided by the cortex. The central computer, as it were, keeps track of everything and tells each of the muscles what to do and when to do it. But the concept I have in mind is much more like that of Gallistel (6), who refers to it as "hierarchical control." The concept is that higher centers issue very simple commands to intermediary centers telling them what to do. They in turn give simple commands to the next lower echelon of control units telling them what to do, and so on down the line. Thus the orders cascade and multiply, so that by the time we come to individual muscles and contractile fibers, there are a lot of different things happening. But everything gets executed in proper sequence.

The proper analogy is the high ranking military officer who wants to move a large number of soldiers over to the other side of the hill. It is not necessary for the general to tell each soldier where to go and how to get there. All he has to do is tell the colonel to get his men over to the other side of the hill. The colonel will tell his majors, the majors will tell their captains, who will tell their lieutenants, and so on. The first thing you know, all the men are on the other side of the hill. The general gives a very simple and very encompassing order which cascades down through the ranks, multiplying as it goes. The way the army works - and the only way it can work - is for each unit to know how to interpret the command signals that come from above and to know how to translate them into the several command signals that must be given to units below. That is the way the army works, and it is also, I propose, the way the motor system works.

We may extend the analogy a little further. Just as one discovers a lot of standard operating procedures in the army, routine ways of executing commands, much the same sort of thing undoubtedly occurs in the nervous system. For example, a muscle is told to contract. That command goes down to a pool of motor neurons with different thresholds. The result is that the muscle starts by contracting minimally, then soon more and more fibers become involved until finally a large number are firing, and all the muscle fibers are contracting to exert the full force of the muscle. What we have is "recruitment," a gradual buildup of contractile power, all in a disciplined and orderly manner. The troops move out on an orderly schedule, rather than every man dashing on his own to the other side of the hill.

Note a corollary of the hierarchical control concept; there is always considerable "local autonomy." That is to say, the general does not tell each lieutenant how or when to move his troops out. That can only be done by the lieutenants. And so the recruitment of contracting muscle fibers cannot be controlled centrally; it has to be done peripherally. And so too, just as when the army moves it must take with it various support functions - communications, the quartermaster, construction people, and so on - so the muscle contraction can only occur in a coordinated context. Other muscles have to be relaxed, body balance has to be maintained, and perhaps the eyes have to be reoriented. It is a very complicated business contracting a muscle. It is like trying to move an army. And it is, I propose, really only possible to do it effectively according to the two principles of hierarchial control and local autonomy. In the army again, it is only the lieutenant who can give the lieutenant's orders and who can properly translate the captain's orders into the next layer of causal determination. That is what I mean by local autonomy. One suspects that if the general were to intervene, trying to substitute for the lieutenant and tell his sergeants what they should do, the general would only mess it up. We know that that kind of thing can happen. The human is uniquely and interestingly capable of intervening in his or her own motor activity. For example, I can decide to do weird and wondrous things. I can upon either self-instruction or someone else's instruction hold my breath, or cross my eyes, or stick my tongue out, or do any combination of those things. Having that kind of response flexibility is part of what it means to be human. But I suspect that, just as the general will typically mess up the lieutenant's business, so my intervention in my motor behavior messes it up, too. How much more effectively can I breathe when I am asleep? How much more effectively does my visual system work when I am scanning the horizon for something? With how very much more facility does my tongue move when I do not attempt to control its motion but simply begin to speak? I suspect that it is far more human to merely act like a human than it is to be able to intervene in one's actions. I suspect that the local autonomy principle is very important.

It is a very great advantage that we have, in effect, all of the minor adjustments and corrections, all of the fine tuning of behavior, done for us by the motor system itself. The person who wants to walk from here to there - that is what has to be kept in mind. The execution of that desire, the carrying out of that command, is done at a local level, peripherally. And it has to be so because walking is such a complex coordination. All of the balancing, all of the compensating, all of the rhythmic coordinations that have to be carried out have to be hardwired because the job cannot be done in any other way.

If you are a human you are going to walk like a human. For example, you are going to swing your arms as you go, because that is the way you are wired. It is not a throwback to earlier evolutionary stages when you were a quadruped, or at least not necessarily so. It is primarily a matter of balance. The last thing in the world you want to think about when you are walking down the street to meet your friend in front of the store is how to keep your balance while you are walking. Fortunately, we do not have to worry about that most of the time. We simply put out the command signal to proceed and away we go. The balancing is taken care of. And all the delicate timing is taken care of. All of the oscillators are coordinated and synchronized for us so that with any luck at all we will get where we are going without having to think about how to get there. And this same basic principle applies to rats, and birds, and everyone else. The human may be able to intervene, literally, to place one foot in front of the other, but most of the time we just walk, just like everyone else.

Notice that the problems of balance, and energetics, and temporal sequencing can vary greatly from species to species. A galloping horse and a galloping dog have to do it differently because their legs are different. And they do it differently because their nervous systems are correspondingly different, and everything has to be timed differently. The size of the animal is enormously important. A large animal has to have protection against falling down, because in falling it could break a leg. A small animal does not have that problem, and so it can have a different kind of coordination. And the speed with which a part can be moved varies enormously with the size of the animal. A fast-running horse completes a gait in about one second, while a small, fast-running dog can complete three gaits in a second. And we may suppose that because there is very little latitude in such matters the whole program can be hardwired.

Consider one advantage of fixed circuitry. If the control system for a particular kind of rhythmic movement, perhaps grooming, or eating, or copulating, or running, were once established and fixed, so that all the correction mechanisms could be built in at the periphery, then the command signal could be very simple indeed. "Let's move forward." "Let's scratch the left side of our neck." Behavior would almost take care of itself, and the mind would be left free to make discriminations, to anticipate the future, and to try to figure out what kinds of contingencies prevail.

MECHANISMS

A fair amount is known about the particular neural mechanisms that are involved in motor coordination. Gallistel's recent book, which seems destined to become a classic, The Organization of Action: A New Synthesis (6), focuses upon three kinds of mechanisms that are known to be important in motor coordination. Gallistel gives extensive quotes from some of the pioneering investigators, such as Sherrington and Von Holtz, and then gives commentary on the mechanisms that they have proposed. He begins with Sherrington and multi-segmental limb reflexes, crossed extension, and the like. He then moves on to the fascinating work that has been done with oscillators, neural networks that produce grooming and locomotion, and other kinds of highly rhythmic behaviors. Finally, he directs attention to the reafference principle. The reafference principle, to oversimplify it, is that when certain responses such as visual orientation movements are made voluntarily, a copy of the expected outcome of the movement is filed away momentarily. Then the actual result of the movement is compared with that copy. If there is a discrepancy, an appropriate correction movement is initiated. Abundant evidence from insects demonstrates the existence of the reafference mechanism, and, indeed, there is good evidence for all three kinds of devices.

These three kinds of mechanisms, the multi-segmental reflexes, the oscillators, and the reafference feedback mechanisms do not, of course, exhaust the variety of neural coordination mechanisms that animals possess. There are simpler things, such as the recruitment mechanisms I have already alluded to, along with stretch reflexes, and a variety of general muscle tonus mechanisms. And then there are some more complicated and more interesting devices, such as feedforward mechanisms. Consider the following. When a man raises his hand in front of him, the body has to tilt back in order to maintain balance. The arm is heavy enough and long enough that some compensation has to be made. So the body tilts back. Typically, however, when the arm movement occurs, the tilting back of the body occurs before the movement of the arm. This may be hard to believe, but try it and you will see that it is true. The intention to raise the arm produces the arm movement, of course, but before it does so it initiates the postural adjustment that has to occur.

The existence of these feedforward mechanisms tells us that the coordination of behavior is really a very complicated business. It involves not only local events such as the scratch reflex controlled by a relatively peripheral oscillator, it also involves whole body postural adjustments that must be controlled much more centrally. Unfortunately, we do not know much about the details of these mechanisms. We do not know how they work, we only know something about the results they produce; they help to create the whole body adjustments that make possible the execution of discrete, peripherally organized motor coordinations.

To the extent that behavior has a fundamentally ballistic quality, to that extent it is not dependent upon its consequences. It is not the sort of thing that has been shaped by the reinforcing effects that it has previously produced in the history of the organism. And so we begin to conceive of a strange new hypothesis. The pieces and parts of behavior that we have always assumed to be learned coordinations may not be anything like that. They may be preformed units of behavior that have been organized over evolutionary time. The behavioral software packages may be useful but they are not very modifiable. The old concept was that all of the details of behavior were modified and polished by feedback. What I am suggesting is that that concept is very much overvalued. The basic model we should be considering as an alternative is the feedforward notion.

RESPONSE FLEXIBILITY

It should be clear from what I have been saying that there can be very little learning of responses per se. There can be very little response flexibility as such. Much of the predictability of behavior, particularly the details of behavior, has to do with physical structures, the size of the animal, and the kind of animal it is. Thus, we need to know if it is a predator or a herbivore, and whether it is large or small. But once we are committed to examining the motor systems of a particular animal, we will find that its behavior, at least in its detail, is controlled by fixed mechanisms. For example, there are various bipeds: a lot of birds, humans, Russian bears, and trained poodles. But each of these bipeds walks in its own way. You can train a poodle to walk, but you cannot train it to walk like a bird. You cannot train a Russian bear to walk like a human. You cannot even train a human to walk like a Russian bear. The details of behavior are controlled by locally autonomous mechanisms that have evolved to take advantage of the size, shape, and kind of animal. Most of an animal's motor coordination come to it in fixed packages across evolutionary time. An animal's coordination comes in software programs, if you will, that tell it how to walk, fly, jump, groom, and so on. And these programs have very little flexibility. They are not subject to modification by their consequences. I suggest that the motor system is quite inflexible. I contend that it is that way because, like the army, that is the only way it can function.

We are now confronted with an important issue. If there is no flexibility in the coordination of behavior, then how is it that we can change the morphology of behavior by applying a reinforcement contingency to a given topography? It is said that we can teach a pigeon to play Ping-Pong. Certainly we can teach a golfer to make a drive of 200 yards, or 175 yards, and to make these different strokes subject to the golfer's own voluntary control. The issue is hard to settle empirically. The late Dalbir Bindra suggested that all learning is really learning about stimuli (1), and that idea intrigues me. I do not think it is quite true, but it is nonetheless an intriguing idea, and probably almost true. What is the golfer learning when he is learning to hit a long drive, or a medium drive? Any golf pro will tell you that he is mainly learning about stimuli. He is learning a) to keep his eye on the ball, b) about the feel of just so much follow through, c) how much the wrist should be bent, and d) most importantly, about the initial tilt of the body. What the golf student is learning, in short, is to make a number of different stimulus discriminations. A reinforcement theorist would no doubt say that if the stroke comes out in a way that is not intended, then the reinforcement contingencies will alter the response. But Bindra pointed out that one could just as well say that if the stimulus consequences are not those that are wanted, then one should seek alternative, antecedent stimulus conditions: a different tilt of the body, a different set of the wrist, or a different way of looking at the ball. Once you have the right stimuli, the response will take care of itself. Once the golfer can find the right antecedent conditions so that the stroke feels right to him, he does not have to worry about the stroke itself. Tut, tut, you say, we know that humans are capable of learning arbitrary motor coordinations; how about the pianist who learns to play a scale by making arbitrary finger movements? Or what about the child learning to speak? Is that not an instance of learning a perfectly arbitrary response specifically because of the reinforcement contingencies that prevail in the community?

A few years ago, I did an experiment which helped me, at least, to understand this question. In this experiment I became a musician. A young girl I know started playing classical guitar. I watched her rapid improvement with amazement; she made it look so easy. I can do that, I said. I went to the music store and bought a guitar and an instruction book. Then I got to work being both experimenter and subject, and I discovered a couple of interesting things. One is that there is probably something like a critical period in learning to play an instrument and that I was well over the critical age. I am intelligent, well motivated and well disciplined. I knew how to go about it and I worked at it very hard, but I made extraordinarily slow progress - nothing like my ten-year-old friend. But with minimum ability and a lot of determination, I persevered, and now I can play a little on the classical guitar. I have discovered that the playing of different pieces is controlled by different kinds of stimuli. With one piece I have to look at the music. Take away the notes and I am helpless, unable to remember a note or do anything with my hands. Nothing happens; the music just stops. But I memorized another little piece; so it is all in my head somewhere. However, in order to play this one I have to look at my hands moving back and forth. I look away and the music, such as it is, stops. Then there are a few pieces where I do not need to look at my hands or at the music. Actually, I cannot tell you what the notes are. I cannot remember what the music itself looks like or what my fingers are supposed to be doing - I just look at them in amazement. My fingers know how to play the piece, but I do not. I have forgotten it, but my hands remember. Is that not response learning? Yes, I think it is genuine response learning. My fingers go back and forth, and all I have to do is interpret the music, putting a little pause here, a little crescendo there. Yes, I think that is response learning. But I think it is rare, and I feel rather dissociated when it happens. And I only get it when I have played a piece a couple of thousand times.

As I have indicated elsewhere (3), there are a number of ways in which stimulus learning, that is, learning about the significance of a new stimuli, can lead to behavior change. Thus, if a rat learns that this tone means shock it will run away from the tone, if it is permitted to do so. We do not have to train the running-away response, we only have to teach the rat that the tone means shock and then this stimulus learning will control the behavior. If those apparatus cues mean food, the rat will soon "learn" to run forward in the apparatus, but again, the content of the learning may be only the changed significance of the apparatus.

People want to believe that they can control their behavior and that their responses are voluntary. No doubt that can happen, but I am not sure it happens very often. I am impressed from my own experience of playing the guitar that my behavior is, after all, not very flexible. I have a great deal more facility in learning the meaning of a new word, or learning to recognize a new piece of music, or learning the name of a lovely woman, or who are the players on a new basketball team. I can still learn a lot about stimuli, but I regret having to say that I am quite unimpressed with my ability to acquire a new motor coordination.

Most of my behavior appears to consist of built-in correction mechanisms, fixed patterns of responding, and reflexive patterns. It appears to me that the notion that responses are readily conditionable is a hoax.

REFERENCES

(1) Bindra, D. 1978. How adaptive behavior is produced: A perceptual-motivational alternative to response reinforcement. Behav. Brain Sci. 1: 41-52.

(2) Bolles, R.C. 1960. Grooming behavior in the rat. Comp. Physiol. Psychol. 53: 306-310.

(3) Bolles, R.C. 1983. The explanation of behavior. Psychol. Rec. 33: 31-48.

(4) Bolles, R.C., and Woods, P.J. 1964. The ontogeny of behaviour in the albino rat. Anim. Behav. 12: 427-441.

(5) Fentress, J.C. 1972. Development and patterning of movement sequences in inbred mice. In The Biology of Behavior, ed. J.A. Kiger, pp. 83-131. Corvallis: University of Oregon Press.

(6) Gallistel, C.R. 1980. The Organization of Action: A New Synthesis. Hillsdale, NJ: Erlbaum.

(7) Provine, R.R. 1979. "Wing-flapping" develops in wingless chicks. Behav. Neur. Biol. 27: 233-237.

(8) Rowell, C.H.F. 1961. Displacement grooming in the chaffinch. Anim. Behav. 9: 38-63.

The Biology of Learning, eds. P. Marler and H.S. Terrace, pp. 447-460. Dahlem Konferenzen 1984. Berlin, Heidelberg, New York, Tokyo: Springer-Verlag.

Associative Predispositions

S. Revusky
Memorial University of Newfoundland
St. John's, Newfoundland A1B 5S7, Canada

Abstract. A failure to understand instances of selective association in the context of a general learning process results in a very superficial analysis.

INTRODUCTION

Learning is a biological process that allows animals to adjust to causal relationships through experience. Built into it are various criteria (or parameters of learning) that make for valid inferences and focus the process on what is biologically important for the animal to learn. Among these criteria are tendencies toward selective association between specific types of events. For example, if events A and B both precede events X and Y, A may become more strongly associated with X than with Y, while the opposite may be true for B. I believe that all learning involves a certain amount of selective association, but the only case about which there is nearly universal agreement is the selective association of tastes with gastrointestinal sickness. The primary focus here will be upon mammals.

If associative learning occurs over long delays without the aid of secondary reinforcement or similarity, selective association must be involved. The animal must have some criterion to associate between two specific events out of a large number of events that occur during the delay. For instance, long-delay taste aversions obtained after a single taste-sickness pairing must be due to selective association. If this were not so, associations of the sickness with the many extraneous events bound to occur over a delay of several hours between the taste and the sickness would drown out the taste-sickness association. Due to the slowness of digestion and absorption, it is critical for the regulation of food intake and the avoidance of poisons that the type of food consumed become associated with its long delayed physiological consequences.

Certain psychologists long denied what I take here to be self-evident: a) that taste aversion learning is due to selective association, and b) that it involves association over a delay. These denials were due to attempts to defend the extreme environmentalistic tradition, the tabula rasa model, in animal learning. Selective association threatened this tradition because it is the equivalent of innate knowledge as to what types of events are likely to be causally related. Hence a burden of proof so heavy as to prevent discovery of all but the most robust effects was placed upon any demonstration of selective association. From a vantage point of evolutionary adaptation, this was absurd since selective association makes learning more adaptive than it otherwise would be.

Selective association also acts to prevent some types of maladaptive learning from occurring. The importance of not learning maladaptively is underestimated, perhaps because professors are accustomed to classrooms, where whatever the instructor says will be correct on an examination prepared by the instructor. But in nature and in human daily life, events can occur in sequence without being causally related. A good learning mechanism is distinguished more by its avoidance of maladaptive conclusions than by the speed with which it reaches conclusions (11). Thus, selective failures to associate dovetail with selective association. For instance, the failure to associate between environmental cues and the physiological consequences of ingestion prevents the animal from avoiding the environment in which it became sick and thus driving itself out of its ecological niche for no good reason.

Although very strong constraints are necessary in the feeding situations, I believe they occur in all learning, albeit to a less conspicuous extent (14). So do Gould and Marler and Terrace (see both papers, this volume). I was surprised that Menzel et al. (this volume) indicated that selective association has not definitively been shown in invertebrates. If a honey bee can store information until an appropriate time to dance, it also ought to be able to store information until other information comes with which the original information can become associated. On this basis, it seems highly probable that selective association will soon be demonstrated in bees. It is even conceivable that, as Rozin has suggested (15), associative learning started out as highly selective and then developed into a more general process.

SELECTIVE ASSOCIATION IS PART OF A GENERAL LEARNING PROCESS

Even when the importance of selective association is freely admitted, ideology tends to distort its analysis. The popular phrase "biological constraints on learning," which also refers to the innate response tendencies discussed in the paper by Hearst (this volume), illustrates this. By suggesting that biological factors are somehow extraneous to the learning process, it allows the traditional tabula rasa description of learning to be accepted as the standard and "constraints" to be a set of biological intrusions upon this correct standard. This prevents us from modifying our view of learning and from understanding "constraints" within the context of a learning process. Thus, instances of selective association are defined as anomalies rather than as intrinsic parts of a biological learning process.

In reality, selective association is a parameter of the learning process that interacts with such traditionally studied parameters as practice, reinforcement, extinction, generalization, and discrimination. Dramatic one-trial learning over delays of hours admittedly is not obtained in traditionally studied situations, but this does not make selective association a basically different parameter of learning. The measured extent of taste aversion learning, like any other type of learning, is due to a trade-off involving associative predispositions and the effects of the traditional parameters of learning. If the traditional parameters are not at an optimal level, one-trial long-delay taste aversion learning does not occur. There are also more specific parametric considerations that lead to the conclusion that associative predispositions do not give taste aversion learning a special status (11). Presumably, as other instances of learning based largely on associative predispositions are investigated, similar conclusions will be reached.

Another way in which selective association is treated as an exception to the rules of learning is in the preconception that it is specific to various genera or species. However, I do not know of any mammal that has been shown not to associate selectively between tastes and sickness, although there almost certainly are differences among mammalian species in how well non-taste cues become associated with sickness. Instances of species or genus differences in selective association that do not involve feeding also have not been demonstrated for mammals. I am sure some such instances will soon be discovered, but they will not justify divorcing selective association from the rest of the learning process.

In presupposing a meaningful general learning process, I am combating a tendency among a few students of behavior to claim that learning is like the more flexible biological functions, such as body size and coloring, in the way it conforms to the requirements of species-typical and genera-typical environments. At the very extreme, the naturalistic attitude can take the form of a claim that learning is "a diverse array of peculiar behavioral adaptations" rather than a basic physiological process (21). When confronted with close similarities in learning among disparate types of animals, I have heard adherents of such views respond that the similarities are analogous, that is, independently evolved in response to similar evolutionary pressures. This is tantamount to a claim that the learning process has evolved and disappeared repeatedly among different vertebrate groups (since if it did not disappear, there would be no need for it to evolve again).

The notion that learning is highly flexible is contradicted by strong (and non-trivial) similarities in learning in very different types of vertebrates (11). Part of the reason why the same basic rules are useful in the vast majority of learning situations is that principles by which causal relationships can be inferred are remarkably uniform. This is why, for instance, human logic is so remarkably transsituational in its effectiveness. Probably another part of the reason is that the nature of the nervous system constrains evolutionary adaptations in learning. It is strange that many who ought to know about evolutionary conservatism are ready to tacitly deny its applicability in the case of learning. Gould and Marler (this volume) actively affirm evolutionary conservatism with regard to behavior when they claim that "for evolution to weed out a behavior with such thoroughness, the behavior would probably have to be actively maladaptive." The same applies a fortiori to a basic process like learning. Only if learning were actively maladaptive would evolution weed it out and then allow it to develop again.

GARCIA'S NEURAL MODEL OF TASTE AVERSION

Both perception and the types of learning traditionally studied are equally dependent upon the nervous system. But in practice, it has been easier to relate perception to the nervous system in a concrete way. This is because the destinations of the specific afferent pathways emanating from particular sense organs are straightforward clues about which central structures are most relevant. A similar general structure is not available for most learning effects because they do not depend on a particular sensory modality. For instance, extinction is much the same regardless of whether an auditory or visual CS is being extinguished. This lack of relevant peripheral information makes it harder to be certain about which central mechanisms are involved. In fact, there may be no specific locus for many learning effects because, as indicated in Menzel et al. (this volume), they occur in animals with relatively few neurons. It is likely that in a complex mammalian brain the structural requirements for learning may be duplicated many times over.

Selective association is unlike most other topics in learning and more like perception in that the specific type of sensory input is bound to be relevant and hence to supply structure for a neuroanatomical analysis. Selective association betwewen events in two modalities implies a propensity for neural connections to develop between the brain centers corresponding to the two modalities. This propensity is likely to be related to the physical proximity of the two centers, and/or the number of neural connections between them. Conversely, if the afferents from two sensory modalities end up in the same brain center, one might expect some selective association between these two modalities. While such considerations are likely to crop up in any neuroanatomical approach to selective association, they have been developed most fully by Garcia. According to him, selective association between tastes and sickness occurs because "gustatory and visceral afferents converge directly to the brain stem, indicating an intimate relationship among tastes, ingestion, and emesis" (17).

Although I cavil at some details of Garcia's analysis (particularly the heavy emphasis on emesis), I think his approach to taste-sickness selective association should be an example for the analysis of other types of selective association. It bridges learning theory and natural behavior by means of genuine physiological considerations, not by means of the hypothetical physiological processes so misused by learning theorists and ethologists. It brings in cogent facts, where these are known, and is suitably vague where facts are not available. This typical biological strategy ought to be more usual in the study of behavior.

Evolutionary Conservatism and Innateness

An important implication of Garcia's position is that the determinant of taste-sickness selective association is the organization of the vertebrate brain, not the organization of the brain of some particular species or genus. Garcia claims that the neural organization responsible for this process is already evident in the brain of the tiger salamander (2). This implies considerable evolutionary conservatism. Also, since taste-sickness selective association depends upon the organization of the vertebrate brain, it must be innate. That is, its specific nature must be predetermined by neuroanatomical and genetic factors. This is contrary to the theory that such selective association is due to early learning of a concept that stimuli other than tastes are not valid predictors of sickness (10). Admittedly, this alternative theory now seems to have few adherents, but it has enough residual credibility that I should explain here why it is false. Taste-sickness selective association is present in very young rats that have never consumed any food other than mother's milk or laboratory chow and were never subjected to severe sickness. This indicates innateness by Bateson's canon of plausibility (see paper entitled Genes, Evolution, and Learning, this volume) since rats do not learn complex arbitrary concepts without extensive and careful training. It is possible that some exposure to feeding contingencies is necessary before there is selective association of tastes with sickness. If so, this is quite typical of genetic encoding, which often relies upon an input that will be present in all normal environments.

Different USs

Some very toxic substances injected into rats after consumption of a flavored substance do not produce pronounced taste aversions. Lett (personal communication) explained this in terms of the suggestion inherent in the Garcia model that toxins that do not affect the emesis center in the brain stem ought not to become easily associated with tastes. However, it was likely that they might become associated with a distinctive place. Her experimental design involved two flavored solutions, two distinct places, and two types of sickness. One type of sickness was always produced by an injection of lithium which is known to produce strong taste aversions presumably because it operates upon the emetic center in the brain stem. In one experiment, the second type of sickness was produced by a gallamine injection. In the other experiment, it was produced by a naloxone injection. These drugs produce relatively weak taste aversions, and Lett supposed that this was because they do not act upon the emesis center. Each rat was subjected to two types of training trials: a) Taste A followed by Place X and an injection of lithium; b) Taste B followed by Place Y and an injection of either gallamine or naloxone. In a two-choice test between Tastes A and B, the rats avoided

Taste A, indicating that lithium produced a stronger taste aversion than either gallamine or naloxone. In a two-choice test between the two places, the rats avoided Place Y, indicating that the gallamine or naloxone was more strongly associated with the place. Thus, the ability of a particular drug to make a particular type of cue aversive is understandable within the framework of Garcia's neural model.

The Garcia approach also suggests questions about the results of lesion studies. I will describe one of a number of examples. The area postrema seems to sense toxins in the circulatory system and to relay the information to the main emetic center. Toxins that stay mainly in the bloodstream might be expected to produce taste aversions only if the area postrema is intact. Toxins that also operate centrally might be able to produce taste aversions regardless of whether the area postrema is intact. The experimental evidence seems to confirm this (2).

Primacy of Taste in Food Aversion Learning

A specific example of evolutionary conservatism in the Garcia model is its insistence that because of the organization of the brain, taste must be the primary modality for association with sickness. A contrary view that supposes more flexibility in the learning system is that the animal will selectively associate with sickness whatever type of cue it tends to use in foraging for food (20). The difference between these views is illustrated by the possible role of olfaction in the feeding of the rat. Garcia's model implies that a rat cannot associate directly between the smell of food or drink and delayed sickness because olfactory afferents do not meet visceral afferents in the brain stem. The assumption of great evolutionary flexibility implies that the rat ought to be able to associate between smells and delayed sickness because of its high olfactory acuity and its use of smell in foraging. I long inclined toward the latter view, but the weight of the evidence is now in favor of Garcia's approach.

Olfaction is important in the feeding behavior of rats, but only within the context of a vertebrate brain in which taste is the primary feeding modality. There is an important difference between olfaction and taste under natural conditions (17). Tastes are, for practical purposes, only experienced in connection with ingestion. Avoidable sickness is very largely a result of ingestion. Thus there can be no doubt that taste-sickness selective association is adaptive even when the sickness occurs hours after the taste. Olfaction, however, has a variety of functions. If animals were to associate directly between smells and delayed sickness, they would associate many smells irrelevant to sickness with sickness. This would interfere with effective avoidance of sickness and also produce avoidance of harmless smells so as to make the animal's usual behavior "neurotic." Such maladaptive results could be avoided if association of a smell with delayed sickness were to occur only when the smell occurs together with a taste. That way, smells unconnected with feeding, and hence probably irrelevant to the occurrence of sickness, will not be associated with sickness and confuse the animal. Simply not associating smells with sickness because taste is a more valid cue is not an optimal strategy, since smell allows avoidance of food from a distance so that the animal does not have to play Russian roulette by taking the risk of tasting poisons.

This analysis describes the learned avoidance of food odors remarkably well. If a rat smells something while drinking tasteless water and then becomes sick, its subsequent aversion to the smell will be relatively weak. (The aversion to the smell is measured in terms of the amount of tasteless water the rat will drink in the presence of the smell.) However, if a taste is added to the water during conditioning, a strong aversion to the smell is obtained. The interaction of a smell with a taste is different in this respect from the interaction of two tastes. If two tasty substances are consumed prior to sickness, the stronger the aversion to one, the weaker the aversion to the other (11). Thus two tastes compete with each other to become associated with sickness, while a taste strengthens the association of a smell with sickness. Garcia and Rusiniak called this strengthening of the smell aversion potentiation and described it as the mouth teaching the nose (5). Odors can become directly associated with sickness under optimal conditions without the mediation of taste. But under difficult conditions, for instance, if the sickness is delayed for several hours after ingestion, the potentiation process is probably necessary. The net result is that aversions to odors are likely to develop only when they are adaptive.

Although I have emphasized evolutionary conservatism in this paper, the process of potentiation is not as conservative as it might conceivably be. If it were maximally conservative, potentiation would operate according to the same stimulus chaining principles that operate in other types of Pavlovian conditioning. If stimulus chaining applied, the taste would become associated with the sickness and the odor would become associated with the taste. The rat would avoid the odor because it reminded him of a taste that became aversive through its association with sickness, but the odor would not become directly associated with sickness. An influential report claimed potentiation was explicable in terms of such traditional stimulus chaining (3), but it was based on flawed experimentation (8). It is now clear that the potentiation effect operates independently of such traditional mechanisms. The presence of a taste at the time of ingestion allows the odor to become directly associated with a sickness, but the odor-sickness association is independent of the odor-taste association. Some evidence comes from demonstrations that a weak taste can potentiate a strong odor. Under such circumstances, the odor aversion, as measured by the amount consumed, can be stronger than the taste aversion (2). This makes it unreasonable to suppose that the odor is aversive because it reminds the rat of the taste. Other evidence is that after the aversion to the taste is extinguished, the aversion to the potentiated odor remains strong (8).

Odor potentiation may not occur in all cases since some have not been able to obtain it. Nor is it certain that potentiation is invariably necessary for the association of odor cues with delayed sickness. Taukulis (19) obtained odor aversion learning in rats that drank unflavored water while being subjected to a jet of air odorized with amyl acetate. Lithium sickness was induced by injection 4 hours later. In theory, this procedure may contain an artifact since some amyl acetate was bound to have gone into solution on the rat's tongue and thus might have produced a taste. However, Taukulis discounted this possibility with the argument that any such taste would have been very weak. I myself have tasted undiluted amyl acetate (but carefully, since it is a poison) and feel it produces a type of irritation but has little taste. Still, in light of the other evidence in favor of the importance of potentiation for long-delay odor-sickness associations, I incline to believe that the Taukulis result is somehow due to potentiation. Perhaps irritation of the tongue that I do not consider a taste somehow still activates centers that normally are stimulated by taste.

The potentiation process is also capable of causing feeding of rats to depend upon the appearance of food (4). In birds, the association of the appearance of drink with delayed sickness is usually due to potentiation (6, 8). Indeed, color potentiation is often easier to obtain with birds than odor potentiation with rats. An early report that color is the primary drinking cue for quail and overshadows taste (20) is almost certainly erroneous (8). However, there are instances of genus-typical selective association among birds in their selection of solid food. Quail, pigeons, and chickens, but not hawks, learn strong color aversions to food without taste potentiation (8). For chickens, a taste imbedded in a solid food will not become aversive as a result of a later sickness US even when the taste is so strong that food intake will be reduced by nearly 50%. Instead, the color cue will be used. However, when the same taste is in a liquid, taste becomes the primary feeding cue for these same chickens (6). Thus the learning of a color aversion without potentiation is a specialized adaptation mainly useful for solid food in certain bird species, but the primary cue for the feeding system even in birds is still taste.

OTHER TYPES OF SELECTIVE ASSOCIATION

Distinction Between Visual and Auditory Stimuli

I believe mammals selectively associate auditory stimuli with dangerous aftereffects. This has not been studied extensively probably because the rat, the most frequent experimental subject for mammalian learning, is a poor choice for studies of visual and auditory perception. However, LoLordo and his co-workers (9) have shown that in pigeons, auditory stimuli tend to be selectively associated with danger (an imminent shock). There are many unrigorous reasons to suppose that something similar is true for mammals. We all know the mild fear produced in us by screechy sounds and it seems likely that such sounds are easily associated with danger. Furthermore, there is laboratory lore' to the same effect. Before LoLordo did his work on pigeons, Jim Miller, a student of audition, commented to me that audition thresholds in cats were obtained by using sound as a signal for shock because cats use sound to escape from danger and vision to obtain food. Of course, my readiness to accept such anecdotal evidence and to extrapolate liberally from it shows that I think of selective association as something to be expected rather than as something to be explained away.

Selective Association of Motor Responses

Another possibility is selective association of responses with specific consequences. There can be no doubt that the control of responses by reward, punishment, and avoidance contingencies depends on the relationship of the response to the contingency: a rat will learn to barpress readily to obtain food but not to avoid shock (1). A golden hamster will not readily learn to groom in order to obtain food although, inexplicably, it can learn to avoid grooming if it is punished with shock (18). However, rigorous proof that these effects are due to selective association would be very difficult to obtain due to likely additional factors. For instance, fear and hunger tend to elicit different natural reactions, and these can prevent other responses from occurring or from becoming more frequent. Then there is the potential role of early learning. The animal might conceivably have learned the characteristic types of outcomes particular types of responses tend to have or tend not to have. For instance, the animal might have learned that food is not acquired by such actions as grooming but only by actions that manipulate the outside world. However, I think it is unreasonable to suppose that selective association is not involved simply because these other factors may also be present. Animals typically have a number of alternative mechanisms that interact to accomplish the same homeostatic end and I do not see why this should not also be true in the case of learning.

Conditioned Antisickness

I will start with a concrete example. A rat is injected with a sedative dose of pentobarbital that functions as a CS in a conditioning paradigm. Some time later, it is injected with a highly toxic dose of lithium that acts as a sickness US. Within four pairings, the CS drug comes to elicit an antisickness response that can attenuate the capacity of the drug US to produce a taste aversion. The evidence for this is obtained by allowing the rat to drink saccharin solution, then injecting it with the pentobarbital CS, and then injecting it with the lithium US. Such a rat will exhibit a weaker saccharin aversion than a control rat subjected to a similar procedure except that the pentobarbital is not injected between the saccharin consumption and the lithium injection. This indicates that the pentobarbital has weakened the capacity of lithium to produce a flavor aversion. Pentobarbital does not have this effect if it has not previously been paired with a lithium US. The primary evidence that this drug-drug conditioning depends upon selective association is that the pentobarbital CS and lithium US can be injected as much as 320 min. apart and still the pentobarbital will become conditioned to yield an antisickness response (12). The effect is not due to a specific pharmacological interaction because it occurs with a variety of drugs other than pentobarbital and lithium, although it is by no means universal (7).

It might seem reasonable to attribute this conditioned capacity of a CS drug to interfere with taste aversion learning to blocking, that is, the CS drug might interfere with the aversion to the saccharin because the rat attributes the sickness US to the CS drug instead. The blocking explanation is wrong, since there is very strong evidence that the CS drug state does not become directly associated with sickness except very weakly and after prolonged training (13). In fact, the propensity to associate internal feeling states with sickness is so low that even the onset of sickness itself is a poor cue for the inhibition of ingestion (16). Further evidence that the CS drug does not become associated directly with the sickness US is its failure to become secondarily aversive as a result of being paired with a sickness US. In fact, exactly the opposite happens (12).

Conditioned antisickness occurs because the drug CS becomes associated with a homeostatic antisickness reaction (UR) instead of with the sickness US itself. The CS drug comes to elicit a conditioned antisickness response CR resembling this homeostatic UR. The selective association of a taste with sickness and of a CS drug state with the homeostatic reaction to the sickness make sense in terms of adaptation (Lett, personal communication). The secondary aversiveness of the taste is adaptive because it makes the animal avoid the substance that contains poison. Secondary aversiveness of an internal stimulus state would not be adaptive because the animal could not directly avoid the drug stimulus state and thus avoid a sickness. Kent Harding and I have unpublished evidence that the mechanism for the conditioned antisickness effect resembles the endorphin effects that are responsible for the capacity of a pain CS to alleviate pain.

Acknowledgements. Supported in part by grants from the Natural Sciences and Engineering Research Council of Canada and the National Cancer Institute of Canada. G.M. Martin, B.T. Lett, M.D. Krank, and V. Grant commented on an earlier version of this article.

REFERENCES

(1) Bolles, R.C. 1970. Species-specific defense reactions and avoidance learning. Psychol. Rev. 77: 32-48.

(2) Coil, J.D., and Garcia, J. 1977. Conditioned taste aversion: The illness US. In Biological Approaches to Learning: Papers presented at the Northeastern Regional Meeting of the Animal Behavior Society, pp. 1-28. St. John's, Newfoundland, Canada.

(3) Durlach, P.J., and Rescorla, R.A. 1980. Potentiation rather than overshadowing in flavor-aversion learning: An analysis in terms of within compound associations. J. Exp. Psychol.: Anim. Behav. Proc. 6: 175-187.

(4) Galef, B.G., Jr., and Osborne, B. 1978. Novel taste facilitation of the association of visual cues with toxicosis in rats. J. Comp. Physiol. Psychol. 92: 907-916.

(5) Garcia, J., and Rusiniak, K.W. 1980. What the nose learns from the mouth. In Chemical Signals: Vertebrates and Aquatic Invertebrates, eds. D. Muller-Schwarze and R.M. Silverstein, pp. 141-156. New York: Plenum Press.

(6) Gillette, K.; Martin, G.M.; and Bellingham, W.P. 1980. Differential use of food and water cues in the formation of conditioned aversions by domestic chicks (Gallus gallus). J. Exp. Psychol.: Anim. Behav. Proc. 6: 99-111.

(7) Lett, B.T. 1983. Pavlovian drug-sickness pairings result in the conditioning of an antisickness response. Behav. Neurosci. 97: 779-784.

(8) Lett, B.T. 1984. Taste potentiation in poison avoidance learning. In Quantitative Analysis of Behavior: Acquisition, eds. M.L. Commons, R.J. Herrnstein, and A.R. Wagner, vol. 3. New York: Ballinger, in press.

(9) LoLordo, V.M. 1979. Selective associations. In Mechanisms of Learning and Motivation: A Memorial Volume to Jerzy Konorski, eds. A. Dickinson and R.A. Boakes, pp. 367-398. Hillsdale, NJ: Lawrence Erlbaum Associates.

(10) Mackintosh, N.J. 1977. Stimulus selection: Learning to ignore stimuli that predict no change in reinforcement. In Constraints on Learning: Limitations and Predispositions, eds. R.A. Hinde and J. Stevenson-Hinde, pp. 75-96. London: Academic Press.

(11) Revusky, S. 1977. Learning as a general process with an emphasis on data from feeding experiments. In Food Aversion Learning, eds. N.W. Milgram, L. Krames, and T.M. Alloway, pp. 1-51. New York: Plenum Press.

(12) Revusky, S., and Coombes, S. 1982. Long-delay associations produced in rats by injecting two drugs in sequence. J. Comp. Physiol. Psychol. 96: 549-556.

(13) Revusky, S.; Coombes, S.; and Pohl, R.W. 1982. Drug states as discriminative stimuli in a flavor-aversion learning experiment. J. Comp. Physiol. Psychol. 96: 200-211.

(14) Revusky, S., and Garcia, J. Learned associations over long delays. In The Psychology of Learning and Motivation: Advances in Theory and Research, ed. G. Bower, vol. 4, pp. 1-83. New York: Academic Press.

(15) Rozin, P. 1976. The evolution of intelligence and access to the cognitive unconscious. In Progress in Psychobiology and Physiological Psychology, eds. J.M. Sprague and A.N. Epstein, vol. 6, pp. 245-280. New York: Academic Press.

(16) Rusiniak, K.W.; Garcia, J.; and Hankins, W.G. 1976. Bait shyness: avoidance of the taste without escape from the illness in rats. J. Comp. Physiol. Psychol. 90: 460-467.

(17) Rusiniak, K.W.; Palmerino, C.P.; Rice, A.G.; Forthman, D.L.; and Garcia, J. 1982. Flavor-illness aversions: Potentiation of odor by taste with toxin but not shock in rats. J. Comp. Physiol. Psychol. 94: 527-539.

(18) Shettleworth, S.J. 1978. Reinforcement and the organization of behavior in golden hamsters: Punishment of three action patterns. Learn. Motiv. 9: 93-123.

(19) Taukulis, H.K. 1974. Odor aversions produced over long CS-US delays. Behav. Biol. 10: 505-510

(20) Wilcoxin, H.C.; Dragoin, W.B.; and Kral, P.A. 1971. Illness induced aversions in rat and quail: Relative salience of visual and gustatory cues. Science 171: 826-828.

(21) Wilson, E.O. 1975. Sociobiology: The New Synthesis. Cambridge, MA: Belknap Press of Harvard University.

The Biology of Learning, eds. P. Marler and H.S. Terrace, pp. 461-477. Dahlem Konferenzen 1984. Berlin, Heidelberg, New York, Tokyo: Springer-Verlag.

Learning to See: Mechanisms in Experience-dependent Development

W. Singer
Max-Planck-Institut für Hirnforschung
6000 Frankfurt/Main 71, F.R. Germany

Abstract. Data are reviewed which suggest the following conclusions:

1) Neuronal activity is an important shaping factor in the self-organization of the developing nervous system.
2) Postnatal signals from sensory surfaces modulate neuronal activity and hence interfere with the self-organizing processes.
3) In the mammalian visual cortex these experience-dependent modifications are restricted to a critical period of postnatal development.
4) The rules which determine the direction of an activity-dependent change of neuronal connectivity resemble those postulated by Hebb for adaptive synaptic connections: Whether a connection is strengthened or weakened depends on the correlation between pre- and postsynaptic activity.
5) For a change to occur it is a prerequisite that the postsynaptic neuron is active. Hence, only sensory patterns capable of activating cortical neurons can induce modifications.
6) In addition to appropriate senory activity, internally generated permissive gating signals are necessary to permit experience-dependent modification. Thus, whether a change can occur in response to sensory stimulation does depend on the central state of the nervous system.
7) Stimulation conditions suitable for inducing long-term modifications are associated with an entry of Ca^{++}-ions into intracellular compartments, suggesting the possibility that Ca^{++}-ions serve as a trigger signal for the processes which cause long-term modifications of excitatory transmission. It is proposed that the experience-dependent

modifications of neuronal interactions have an associative function and serve to assemble neurons according to functional criteria. The resulting selective interactions are thought to be the prerequisite for the development of cooperatively-coupled neuron assemblies.

INTRODUCTION

The neuroreductionistic approach to the problem of genetic influences on learning requires knowledge about two basic properties of the central nervous system: First, we have to have some knowledge about the neuronal substrate of cognitive interactions. Second, we ought to understand the mechanisms through which these interactions induce long lasting changes in the nervous system, and we should be able to relate these changes to modifications of behavior. Clearly, in the mammalian nervous system neither of these prerequisites is fulfilled. None of our present concepts about the algorithms of information processing in nerve nets gets away without numerous unproven assumptions. The situation with regard to the mechanisms of learning processes is somewhat more encouraging. Evidence is accumulating that neuronal activity can, under specific conditions, induce long lasting changes of neuronal properties, both structural and functional, and in some cases we are beginning to see the mechanisms that mediate the change. However, apart from very simple nervous systems such as in Aplysia (Quinn, this volume), the correlation between these neuronal changes and behavioral learning is loose, if present at all. Thus, a neuroreductionistic analysis of the genetic boundaries of learning processes may appear premature.

I propose, therefore, to consider the problem from a different angle and to ask the question to what extent the functions of the nervous system are determined by genetic versus experiential factors. This classical nature-nuture problem has received considerable attention by neurobiologists once it became established that early sensory experience influences profoundly the structural and functional organization of the nervous system. These experience-dependent modifications of the developing CNS have been studied most extensively in the visual system of mammals and amphibia (for review see (24, 28)).

At a formal, descriptive level, adult learning and developmental adaptation have numerous features in common, and as our knowledge about experience-dependent maturation increases, these similarities keep growing. The developing brain very soon starts to interact with its environment and these interactions modify the structural and functional organization of the nervous system. These modifications in turn influence future interactions of the system with the outer world. To an observer this appears as if the earlier interactions had left engrams which contain information about the effect of previous interactions and serve to modify future behavior. The developing system "learns" through interactions with the outside world. I have chosen the word interaction deliberately to emphasize the active part played by the developing system. As will be discussed in more detail below, there is experimental evidence which indicates that sensory signals induce long-term changes only when they are processed by an aroused brain, when they are paid attention to, and when they are used for the control of behavior. The conditions that have to be met in order to bring about experience-dependent modifications of the developing nervous system are thus very similar to those required for adult learning.

In the following I shall illustrate with a few examples the nature of the changes that occur in the visual system of mammals as a function of sensory experience. Subsequently, I shall try to deduce the limits of this malleability from the neuronal mechanism that mediate the change.

EXPERIENCE-DEPENDENT MODIFICATIONS OF THE DEVELOPING VISUAL CORTEX

By the time kittens or monkeys open their eyes, most neurons in the visual cortex respond to stimulation of both eyes. With normal visual experience, but also with complete deprivation of contour vision, this condition is maintained (14, 15, 20). However, when visual signals are available but not identical in the two eyes, either because one eye is occluded (43) or because the images on the two retinae are not in register - as is the case with strabismus (16), cyclotorsion (4), or anisometropia (3) - cortical cells lose their binocular receptive fields. In the first case they stop responding to the deprived eye; in the other cases they segregate into two groups of approximately equal size, one responding exclusively to the ipsilateral and the other exclusively to the contralateral eye. These functional changes in eye preference are associated with distortions of the characteristic columnar organization of the visual cortex. The territories occupied by afferents from the normal eye and by cells responding preferentially to this eye increase at the expense of territories innervated by the deprived eye (18). These effects are obtainable only during a critical period of early development. During this period, but not thereafter, the effects of monocular deprivation can be fully reversed by closing the open eye and at the same time reopening the previously closed eye (44). This indicates that the efficacy of connections can both decrease and increase as a function of retinal stimulation.

Not only the degree of binocularity but also the selectivity of cortical cells for stimulus orientation can be modified by manipulating visual experience. Normally, nearly all cells in the striate cortex of cats respond selectively to contours with a particular orientation. Cells preferring the same orientation are clustered together, forming a system of fairly regularly spaced iso-orientation columns or bands (19). In this columnar system, preferences for all orientations are equally represented. However, when contour vision is prevented by dark rearing or binocular lid suture, only a small fraction of cortical cells develop orientation selectivity, the majority remain or become responsive to contours of all orientations. In addition, the vigor of responses to light decreases and about 30% to 50% of the cells stop responding to retinal stimulation altogether. Throughout the critical period, but not thereafter, these deprivation effects, too, are fully reversible. Eight hours of normal vision suffice to reinstall orientation selectivity (5). When visual experience is available throughout the critical period but restricted to contours of a single orientation, the majority of cortical cells come to prefer this orientation (2, 12, 31). Cortical territories which contain cells preferring the experienced orientation expand at the expense of territories which normally would have been reserved for cells preferring the other orientations (35). All of the experience-dependent modifications of the functional architecture of the visual cortex have distinct behavioral correlates as reflected by the various forms of amblyopia.

NEURONAL MECHANISMS OF EXPERIENCE-DEPENDENT MODIFICATIONS

Activity-dependent growth processes, selective stabilization and repression of newly formed connections (7), and changes of the gain of synaptic connections have been implicated as possible mechanisms of experience-dependent modifications. Which of these processes is prevailing is still unknown. Evidence is available, however, that the activation of the postsynaptic neuron is a necessary prerequisite for any of these long-term modifications (30, 37). If, for example, one eye is occluded and the other exposed to contours of only a single orientation, changes in the efficacy of excitatory transmission occur only for pathways connecting to those postsynaptic cells whose orientation preference corresponds to the orientations seen by the open eye. For these responsive cells the efficacy of afferents from the stimulated eye increases while that of afferents from the deprived eye decreases. The cells with different orientation preferences by contrast, which cannot respond to the retinal signals, maintain their initial eye preference. The results of these and related experiments (31) made it possible to establish the following rules for the modification of excitatory transmission: a) The efficacy of excitatory transmission increases for afferent pathways if they are active in temporal contiguity with the postsynaptic target. b) The gain decreases when the postsynaptic target is active while the presynaptic terminal is silent. c) Irrespective of the amount of activation of presynaptic terminals, differential gain changes do not occur when the postsynaptic cell is inactive. Formally, these rules closely resemble those postulated by Hebb (11) for adaptive neuronal connections and so far have proven sufficient to account for the results of the various deprivation experiments.

STATE-DEPENDENT CENTRAL CONTROL OF PLASTICITY

The modification rules listed above define only the local requirements for experience-dependent modifications. Evidence is available, however, that the Hebbian modifications do not solely depend on the interaction between specific sensory afferents and their cortical target cells but are gated in addition by central control systems.

Neurons of the cat striate cortex remain binocular despite monocular deprivation when cortical norepinephrine (NE) is depleted shortly before the beginning of monocular deprivation (22) and local microperfusion of the cortical tissue with NE reinstalls plasticity (23). This indicates that a certain level of NE in the extracellular space is necessary for the occurrence of changes. However, when NE is chronically depleted, plasticity is not impeded, suggesting that other factors can compensate for the lack of NE (1). Cortical cells also maintain binocular receptive fields when the open eye of monocularly-deprived kittens is surgically rotated within the orbit (39). In this case contour vision per se is unimpaired, but the abnormal eye position and motility lead to massive disturbances of the kittens' visuo-motor coordination. As a consequence, the kittens rely less and less on visual cues and develop a near complete neglect of the visual modality. In this phase, retinal signals no longer modify ocular dominance, and they also fail to support the development of orientation-selective receptive fields.

Another manipulation which prevents retinal signals from inducing cortical modifications is the abolition of proprioceptive feedback signals from the extraocular muscles. When this input is disrupted by severing the ophthalmic branch of the IIIrd cranial nerve bilaterally, retinal signals neither stimulate the development of orientation selectivity (42) nor do they induce changes of ocular dominance (6). As in the kittens with the rotated eye, the kittens with the severed proprioceptive afferents develop a neglect for the visual modality and rely on their other sensory systems for orientation.

These latter results suggest that retinal signals only influence the development of cortical functions when the animal pays attention to these signals and uses them for the control of behavior. This view is supported by two lines of evidence: First, retinal signals never lead to changes of cortical functions when the kittens are paralyzed and/or anaesthetized while exposed to light. Even though the light stimuli undoubtedly drive cortical cells vigorously, they fail to bring about changes of ocular dominance (9, 33) or to develop orientation selectivity (5). Second, the very same retinal signals may induce changes in the visual cortex of one hemisphere but not in the other when the latter is "paying less attention" to the visual signals than the former (34). Small unilateral lesions in the intralaminar nuclear complex of the thalamus produce a sensory hemi-neglect: kittens with such lesions consistently neglect stimuli presented in the hemi-field contralateral to the lesion. If such kittens undergo monocular deprivation, the normal hemisphere develops as with conventional monocular deprivation, i.e., most cortical cells become monocular and develop normal receptive field selectivity. By contrast, in the hemisphere containing the lesion, the majority of the cells remain binocular and unselective (34). Thus, although both hemispheres have received exactly the same signals from the open eye, these signals induce modifications only in the normal hemisphere and remain ineffective in the hemisphere which - because of the lesion - "attends" less to retinal stimulation.

Another interesting abnormality of the hemisphere containing the lesion is the reduced efficacy of reticular activation. In normal animals electrical stimulation of the mesencephalic reticular formation produces a massive facilitation of thalamic and cortical transmission (for review see (32, 33)). In the experimental kittens these effects were greatly attenuated in the hemisphere containing the lesion while they were fully developed in the other. The thalamic lesion had thus affected ascending modulatory systems that enhance thalamic and cortical excitability. This agrees well with the behavioral evidence that the lesion had caused attentional deficits in the form of a sensory hemi-neglect.

The conclusion that ascending modulatory systems have a permissive role in cortical plasticity receives further support from stimulation experiments. By pairing monocular light stimulation with electrical activation of central core structures, it proved to be possible to induce changes of ocular dominance in kittens even if they are anaesthetized and paralyzed. Clear changes in ocular dominance towards the open, stimulated eye became apparent after one night of monocular conditioning (36). Moreover, the gain of excitatory transmission in the pathways from the conditioned eye had increased and cortical cells had become more selective for contrast gradients and stimulus orientation. These results complement the issue of the lesion experiments in suggesting that nonspecific modulatory systems which increase cortical excitability facilitate experience-dependent modifications.

THE GATING MECHANISM

At present it is difficult to decide whether the different possibilities of influencing cortical plasticity reflect the existence of several independent gating systems or whether the various manipulations act through a common final pathway. We obtained evidence that the induction of a change is associated with an entry of Ca^{++}-ions into intracellular compartments (10). The hypothesis that Ca^{++} serves as the permissive "now print" signal which enables Hebbian modifications has two implications that agree well with experimental evidence: Since voltage-dependent Ca^{++}-channels have a higher threshold than the voltage-dependent Na^{+}-channels (25), the generation and propagation of action potentials can be dissociated from the process that mediates long-term changes. Furthermore, since the probability of dendritic Ca^{++}-channel activation depends on the activation state of a large number of converging inputs, many different input systems have the possibility of participating in the generation of the "print" command. Such a heterosynaptic control of plasticity could account for the evidence cited above that interference with quite different projection systems can impede cortical plasticity. The Ca^{++}-hypothesis clearly needs to be substantiated further, but it appears safe to conclude that there is a threshold for the adaptive change which is only reached when the specific pathways and additional converging input systems are in the appropriate activation state. Thus, there is a multifactorial control of the adaptive changes: the direction of the change, increase or decrease of synaptic efficacy, is determined by the contiguity of activity patterns in the pre- and postsynaptic elements that are part of the modifiable transmission chain.

However, whether a change occurs depends, to a crucial extent, on the actual functional state of larger cell assemblies.

CONSTRAINTS OF EXPERIENCE-DEPENDENT MODIFICATIONS

A clear boundary condition for experiential modifications in the visual cortex is set by the duration of the critical period. After the end of this critical period, which in kittens lasts about three months postnatally, visual experience becomes ineffective in altering cortical response properties. This termination of adaptability is relatively independent of the degree of commitment which the neurons have attained previously. With dark rearing, which leaves most of the neurons in an uncommitted state, the plastic phase lasts only a little longer than with normal rearing (8). Thus there is a genetically determined temporal window which limits the time span during which modifications can occur. Some preliminary evidence is available that the duration of such critical periods may differ at different levels of the visual system, but further experimentation is required to clarify this interesting question. The mechanisms which limit the duration of the critical period are unknown.

The identification of genetic factors limiting adaptive changes during the critical period is considerably more difficult. From the modification rules summarized above, it follows that long-term changes of neuronal properties occur only when the cells, whose input connections are to be modified, are active. This implies that only those stimuli are capable of inducing a change that are capable of driving neurons, or in other terms, changes can be initiated only when the pattern of afferent activity matches the response properties (receptive fields) of the cells. The range of effective stimuli is thus determined by the response range of the neurons. The more specific the response properties of the neurons, the narrower the range of stimuli potentially capable of inducing a change. Similarly, the range within which the response properties of a neuron can be modified depends on the neuron's initial selectivity. The orientation preference of cortical neurons can be modified by exposing kittens to a single orientation. However, neurons only develop a preference for this orientation if their initial orientation preference allows them to respond to the stimulation; to be effective a stimulus has to fall within the response range of the neurons whose afferent connections are to be modified (31). Thus, the degree of response selectivity attained prior to experience limits first, the range of stimuli capable of inducing a change and second, the range within which a modification can occur. The degree of pre-experience commitment of cortical cells has been studied extensively in kittens. It differs for the various classes of neurons and changes with age (see, e.g., (20)).

The fact that there is a threshold for the adaptive change which differs from the threshold for action potential generation and which is trespassed only when specific afferents and additional excitatory input systems are active extends the constraints for plasticity far beyond the response properties of a particular cortical neuron. Most of the excitatory synapses on the dendrites of cortical neurons are provided by intrinsic circuits and by cortico-cortical projections. It is to be expected, therefore, that those patterns of afferent activity are most effective in trespassing the plasticity threshold which are capable of driving a large number of interconnected cells both within the cortical area under consideration and within areas that project to this part of the cortex. Thus, the constraints that define the properties of potentially effective stimuli and the boundaries within which a modification can occur are set not only by the receptive field properties of individual neurons but by the resonance properties of - presumably very large - assemblies of interconnected nerve cells. Again, however, this extended matching criterion defines only one necessary condition. A further requirement for the occurrence of a change appears to be that the processing arrays are in a state of high responsiveness. This follows from the experimental evidence that changes occur only when the ascending modulatory systems are active or - in behavioral terms - when the brain is in an aroused state, pays attention to the stimulus, and uses it for the control of behavior. All these conditions increase neuronal excitability and enhance responses to sensory stimulation (13, 32, 33, 40, 45).

The "print" decision is thus likely to depend on a number of local and nonlocal parameters which reflect both the hardware organization of large, interconnected neuron arrays and the actual functional state of the brain. Thus the gating signals which control plasticity appear to emerge from the distributed information processing networks themselves, rather than from a specific superimposed control system. This, in turn, implies that the constraints for experience-dependent modifications are at least in part set by parameters which are themselves subject to continuous modifications. Neuronal assemblies are dynamic entities which permanently change their functional state. Moreover, in a developing system the network properties also change because of the maturation process.

The most serious complication with respect to the identification of genetic constraints arises, however, from the fact that some of the changes of the neuronal networks are in turn caused by experiential factors. In the developing visual system we observe a steady increase in selectivity of neuronal connections and of receptive field properties, but we often ignore to which extent these changes reflect the action of genetic or experiential factors. Since these changes limit the degrees of freedom for further experience-dependent modifications, it soon becomes impossible to distinguish to which extent the constraints for further changes are genetically determined or reflect the past history of epigenetic modifications. For the following reason, it may even turn out that the attempt to make such a distinction is of no heuristic value at all.

We know that from a certain developmental stage onwards neuronal activity becomes an important shaping factor in the self-organization of neuronal networks. It has been shown, for example, that the segregation of the afferents from the two eyes and hence the development of ocular dominance columns is activity-dependent (27, 41). The same is the case for the elimination of excess climbing fibers from Purkinje cells and of supernumerary motor neuron axons from muscle cells (21, 26). Depending on the species, these elimination processes may start prenatally. In monkeys, the segregation of ocular dominance columns is completed at birth and hence occurs definitively without sensory experience (29). We have to assume, therefore, that the electrical activity generated by the brain itself does play an important role in the selection and specification of neuronal connections during development. With regard to these processes, birth introduces only a quantitative but not a qualitative change. After birth, more sense organs become functional and more possibilities for interactions with the environment become available, but the mechanisms of self-organization are likely to remain the same. Thus, there is a closed loop of reciprocally interacting factors: The growing nerve nets generate activity patterns, these modify the structure of the nets, and these structural modifications in turn are likely to modify the pattern of self-generated activity and hence alter the constraints for future modifications. Signals from sensory surfaces are only one additional factor that modulates these internal activity patterns. The attempt to distinguish between genetic and epigenetic influences in such reciprocally coupled processes is thus not very promising from a neuroreductionistic point of view.

THE RELATION BETWEEN DEVELOPMENTAL PLASTICITY AND ADULT LEARNING

In conclusion, I would like to briefly discuss some of the functional implications of the experience-dependent modifications in the visual cortex, since this will illustrate from a more substrate-oriented angle the relation between activity or experience-dependent self-organization and "learning."

Hebbian modifications have an associative function in that they selectively stabilize those connections between neurons that convey correlated activity. This has different consequences at different levels of cortical processing. At the level where afferents from the two eyes converge onto common cortical target cells, such a selection can assure that only those afferents become consolidated which come from corresponding retinal loci in the two eyes. This specification cannot be accomplished without relying on sensory signals, because precise "retinal correspondence" is defined only in a functional domain: Retinal pathways which convey identical activity patterns while the animal is fixating binocularly are defined as corresponding because they encode information from the same point in visual space. However, which pathways actually come from precisely "corresponding" retinal sites cannot be anticipated before eye opening because it depends on parameters such as interocular distance, size of the eye balls, and position of the eyes in the orbit. Thus, the only way to achieve the required precision of binocular convergence is to select the appropriate pathways according to functional criteria. The Hebbian modification algorithm is ideally suited to accomplish this selection since it selectively stabilizes pathways which convey correlated activity.

Likewise, in the domain of orientation selectivity, Hebbian modifications in response to retinal signals can assure that second-order cortical cells receive excitatory input only from those first-order cells that share the same orientation preference. Two considerations suggest that this selection problem is again not a trivial one: Second-order cells with large receptive fields have to receive input from numerous first-order cells that may be distributed over several hypercolumns (17). Because first-order cells which share the same orientation preference are clustered within discrete, regularly spaced columns, this implies discontinous sampling from clusters of first-order cells which may be several millimeters apart. Because of the continuity of contours in the natural environment, this extremely complex specification of connections can again be achieved by selectively consolidating connections which have high probability of being activated simultaneously.

Evidence is further available that such activity-dependent association may occur over rather large cortical distances: When young animals are exposed selectively to regularly spaced contours which share the same orientation, about one third of the cortical neurons develop large unconventional receptive fields with several, widely spaced excitatory regions. The spacing of these excitatory regions corresponds to the angular distance between the contrast borders of the periodic patterns which the animals had experienced previously (38). With normal experience the combinatory complexity of possible contingencies in the spatiotemporal pattern of retinal activity becomes so exceedingly large that it is impossible to predict the resulting pattern of differentially weighted interactions in the cortex. However, it can be extrapolated from the examples cited above that those cells become associated preferentially which have a high probability of responding simultaneously in the presence of particular feature combinations. The establishment of selective interactions between neurons with particular response properties is equivalent to the formation of neuronal representations of particular feature combinations. The effect of selective coupling is to enhance and prolong by reverberation the responses of distinct cell assemblies to particular, frequently occurring patterns. Hebbian modifications at this level of cortical processing could thus be a crucial step toward the formation of cooperative cell assemblies whose coherent and reverberating responses would represent the neuronal code for particular activation patterns at the sensory surfaces. The fact that the processes leading to selective interactions are not solely dependent on the pattern of activity on the sensory surface but are gated by the resonance properties of large cell assemblies provides the developing system with the option to create neuronal assemblies not only as a function of the patterns in afferent sensory pathways, but also as a function of the central state of the system itself.

This implies that the formation of assemblies is a process of active selection whereby the selection criteria emerge from the genetically determined properties of the system, from previous epigenetic modifications, and from the actual dynamic (behavioral) state which the system maintains while it interacts through its sensory surfaces with the "outer" world.

These considerations suggest that the adaptive processes during early ontogeny - at least at this descriptive level - closely resemble what is usually termed associative learning. Conversely, the adaptive processes in the adult brain, commonly described as "learning" processes, might differ only quantitatively from the processes of self-organization occurring during ontogeny. Certainly, the degrees of freedom for a modification decrease with aging: with the end of "critical periods," nerve nets maintain rigidly the once attained properties, neuropile differentiation slows down (reducing the option to select from newly formed connections), neuronal responses become more selective, and hence the range of "matching" stimuli - and only those can leave traces - becomes narrower. At present, however, there is neither a logical nor a factual argument against the view that being is becoming and that "learning" is the continuation of adaptive self-organization. The fact that we are more inclined to accept, for example, imprinting as a "learning" process rather than the activity-dependent formation of ocular dominance, or orientation columns, or the elimination of supernumerary climbing fibers, probably reflects the different levels of description - neuroreductionistic versus behavioral - rather than qualitative differences in mechanisms.

REFERENCES

(1) Bear, M.F.; Paradiso, M.A.; and Daniels, J.D. 1982. Visual cortical plasticity: deficit after acute, but not chronic, noradrenergic denervation with 6-hydroxydopamine. Soc. Neurosci. Abstr. 8: 4.

(2) Blakemore, C., and Cooper, G.F. 1970. Development of the brain depends on the visual environment. Nature 228: 477-478.

(3) Blakemore, C., and Van Sluyters, R.C. 1974. Experimental analysis of amblyopia and strabismus. Brit. J. Ophthal. 58: 176-182.

(4) Blakemore, C.; Van Sluyters, R.C.; Peck, C.K.; and Hein, A. 1975. Development of cat visual cortex following rotation of one eye. Nature 257: 584-586.

(5) Buisseret, P.; Gary-Bobo, E.; and Imbert, M. 1978. Ocular motility and recovery of orientational properties of visual cortical neurones in dark-reared kittens. Nature 272: 816-817.

(6) Buisseret, P., and Singer, W. 1983. Proprioceptive signals from extraocular muscles gate experience dependent modifications of receptive fields in the kitten visual cortex. Exp. Brain Res. 51: 443-450.

(7) Changeux, J.-P., and Danchin, A. 1976. Selective stabilization of developing synapse as a mechanism for the specification of neuronal networks. Nature 264: 705-712.

(8) Cynader, M. 1977. Extension of the critical period in cat visual cortex. Paper presented to the Association for Research in Vision and Ophthalmology, Sarasota, Florida.

(9) Freeman, R.D., and Bonds, A.B. 1979. Cortical plasticity in monocularly deprived immobilized kittens depends on eye movement. Science 206: 1093-1095.

(10) Geiger, H., and Singer, W. 1982. The role of Ca^{++}-ions in developmental plasticity of cat striate cortex. Int. I. Dev. Neurosci. Suppl. R328.

(11) Hebb, D.O. 1949. The Organization of Behaviour. New York: John Wiley and Sons.

(12) Hirsch, H.V.B., and Spinelli, D.N. 1970. Visual experience modifies distribution of horizontally and vertically oriented receptive fields in cats. Science 168: 869-871.

(13) Hubel, D.H. 1960. Single unit activity in lateral geniculate body and optic tract of unrestrained cats. J. Physiol. (Long.) 150: 91-104.

(14) Hubel, D.H., and Wiesel, T.N. 1962. Receptive fields, binocular interaction and functional architecture in the cat's visual cortex. J. Physiol. (Lond.) 160: 106-154.

(15) Hubel, D.H., and Wiesel, T.N. 1963. Receptive field of cells in striate cortex of very young, visually inexperienced kittens. J. Neurophysiol. 26: 994-1002.

(16) Hubel, D.H., and Wiesel, T.N. 1965. Binocular interaction in striate cortex of kittens reared with artificial squint. J. Neurophysiol. 28: 1041-1059.

(17) Hubel, D.H., and Wiesel, T.N. 1974. Uniformity of monkey striate cortex: A parallel relationship between field size, scatter and magnification factor. J. Comp. Neurol. 158: 295-306.

(18) Hubel, D.H.; Wiesel, T.N.; and LeVay, S. 1977. Plasticity of ocular dominance columns in monkey striate cortex. Phil. Trans. Roy. Soc. Lond. B. 278: 377-409.

(19) Hubel, D.H.; Wiesel, T.N.; and Stryker, M.P. 1978. Anatomical demonstration of orientation columns in macaque monkey. J. Comp. Neurol. 177: 361-380.

(20) Imbert, M., and Buisseret, P. 1975. Receptive field characteristics and plastic properties of visual cortical cells in kittens reared with or without visual experience. Exp. Brain Res. 22: 25-36.

(21) Jansen, J.K.S., and Lomo, T. 1981. Development of neuromuscular connections. Trends Neurosci. July: 178-181.

(22) Kasamatsu, T., and Pettigrew, J.D. 1979. Preservation of binocularity after monocular deprivation in the striate cortex of kittens treated with 6-hydroxydopamine. J. Comp. Neurol. 185: 139-162.

(23) Kasamatsu, T.; Pettigrew, J.D.; and Ary, M. 1979. Restoration of visual cortical plasticity by local microperfusion of norepinephrine. J. Comp. Neurol. 185: 163-181.

(24) Keating, M.J. 1976. The formation of visual neuronal connections: An appraisal of the present status of the theory of "neuronal specificity". Stud. Dev. Behav. Nerv. Syst. 3: 59-110.

(25) Llinas, R. 1979. The role of calcium in neuronal function. In The Neurosciences, Fourth Study Program, eds. F.O. Schmitt and F.G. Worden, pp. 555-571. Cambridge, MA: MIT Press.

(26) Mariani, J., and Changeux, J.-P. 1981. Ontogenesis of olivocerebral relationships II. Spontaneous activity of inferior olivery neurons and climbing fiber mediated activity of cerebellar purkinje cells in developing rats. J. Neurosci. 1: 703-709.

(27) Meyer, R.L. 1982. Tetrodotoxin blocks the formation of ocular dominance columns in goldfish. Science 218: 589-591.

(28) Movshon, J.A., and Van Sluyters, R. 1981. Visual neural development. Ann. Rev. Psychol. 32: 477-522.

(29) Rakic, P. 1977. Prenatal development of the visual system in the rhesus monkey. Phil. Trans. Roy. Soc. Lond. B. 278: 245-260.

(30) Rauschecker, J.P., and Singer, W. 1979. Changes in the circuitry of the kitten visual cortex are gated by postsynaptic activity. Nature 280: 58-60.

(31) Rauschecker, J.P., and Singer, W. 1981. The effects of early visual experience on the cat's visual cortex and their possible explanation by Hebb synapses. J. Physiol. (Lond.) 310: 215-239.

(32) Singer, W. 1977. Control of thalamic transmission by corticofugal and ascending reticular pathways in the visual system. Physiol. Rev. 57: 386-420.

(33) Singer, W. 1979. Central-core control of visual cortex functions. In The Neurosciences, Fourth Study Program, eds. F.O. Schmitt and F.G. Worden, pp. 1093-1109. Cambridge, MA: MIT Press.

(34) Singer, W. 1982. Central core control of developmental plasticity in the kitten visual cortex: I. Diencephalic lesions. Exp. Brain Res. 47: 209-222.

(35) Singer, W.; Freeman, B.; and Rauschecker, J. 1981. Restriction of visual experience to a single orientation affects the organization or orientation columns in cat visual vortex: A study with Deoxyglucose. Exp. Brain Res. 41: 199-215.

(36) Singer, W., and Rauschecker, J. 1982. Central core control of developmental plasticity in the kitten visual cortex: II. Electrical activation of mesencephalic and diencephalic projections. Exp. Brain Res. 47: 223-233.

(37) Singer, W.; Rauschecker, J.; and Werth, R. 1977. The effect of monocular exposure to temporal contrasts on ocular dominance in kittens. Brain Res. 134: 568-572.

(38) Singer, W., and Tretter, F. 1976. Unusually large receptive fields in cats with restricted visual experience. Exp. Brain Res. 26: 171-184.

(39) Singer, W.; Tretter, F.; and Yinon, U. 1982. Central gating of developmental plasticity in kitten visual cortex. J. Physiol. 324: 221-237.

(40) Steriade, M. 1981. Mechanisms underlying cortical activation: Neuronal organization and properties of the midbrain reticular core and intralaminar thalamic nuclei. In Brain Mechanisms and Perceptual Awareness, eds. O. Pompeiano and C. Ajmone Marsan, pp. 327-377. New York: Raven Press.

(41) Stryker, M.P. 1981. Late segregation of geniculate afferents to the cat visual cortex after recovery from binocular impulse blockade. Neurosci. Abstr. 7: 842.

(42) Trotter, Y.; Gary-Bobo, E.; and Buisseret, P. 1981. Recovery of orientation selectivity in kitten primary visual cortex is slowed down by bilateral section of ophthalmic trigeminal effects. Dev. Brain Res. 1: 450-454.

(43) Wiesel, T.N., and Hubel, D.H. 1965. Comparison of the effects of unilateral and bilateral eye closure on cortical unit responses in kittens. J. Neurophysiol. 28: 1029-1040.

(44) Wiesel, T.N., and Hubel, D.H. 1965. Extent of recovery from the effects of visual deprivation in kittens. J. Neurophysiol. 28: 1060-1072.

(45) Wurtz, R.H.; Goldberg, M.E.; and Robinson, D.L. 1980. Behavioral modulation of visual response in the monkey: stimulus selection for attention and movement. Progr. Psychobiol. Physiol. Psychol. 9: 43-83.

The Biology of Learning, eds. P. Marler and H.S. Terrace, pp. 479-508. Dahlem Konferenzen 1984. Berlin, Heidelberg, New York, Tokyo: Springer-Verlag.

Physiological and Anatomical Mechanisms: Neural Bases of Learning

R.F. Thompson* and T.W. Berger**
Dept. of Psychology, Stanford University Medical Center
Stanford, CA 94305
**Dept. of Psychology, University of Pittsburgh
Pittsburgh, PA 15260, USA

INTRODUCTION

The most productive laboratory studies concerned with neuronal substrates of learning and memory in the mammalian brain have utilized highly simplified and controlled experimental situations and paradigms, what is often termed a "model system" approach. The basic strategy is to select an organism capable of exhibiting the range of behavioral phenomena one wishes to explain and whose nervous system possesses the properties and/or is understood at a level that makes neurobiological analysis feasible. The trade-off typically encountered is that the more complex the behavior one wishes to explain, the less tractable are the nervous systems of organisms capable of exhibiting such behavior. The chief advantage of the model system approach is that the facts gained from anatomical, physiological, and behavioral investigations for a particular preparation are cumulative and tend to have synergistic effects on theory development and research.

Each approach and model preparation has particular advantages. The value of invertebrate preparations as model systems results from the fact that certain behavioral functions are controlled by ganglia containing relatively small numbers of large, identifiable cells - cells which can be consistently identified across individuals of the species (Quinn and Menzel et al., both this volume). With vertebrate model systems, these goals are considerably more difficult to attain. But if one is to understand vertebrate nervous systems, one must at some point study vertebrates. In addition, if the behavior of interest is complex, it might only be observed in vertebrates. It is clear that higher vertebrates have developed increasing capacities for learning and have made use of these capacities in the development of adaptive behavior. It would seem that the evolution of the mammalian brain has resulted in systems especially well adapted for information processing, learning, and memory.

This model system or analytic approach is held by some ethologically oriented workers to be artificial - a wolf may not learn a discrete leg flexion response in the natural environment. However, the learning paradigms used in the laboratory were selected as simplified examples of natural learning. Learning to obtain food and avoid noxious stimuli are perhaps the principle categories of learned behaviors in the laboratory and in nature. In this paper we focus on two examples or categories of learning in the mammal - classically conditioned autonomic responses and classical conditioning of discrete, adaptive skeletal muscle responses in the attempt to avoid noxious stimuli. We conclude with a general consideration of possible physical/anatomical substrates that may serve to code memory in the mammalian brain.

The nature of the memory trace has proved to be among the most baffling questions in science. The problem of localization has been the greatest barrier in analysis of neuronal substrates of learning and memory in the mammalian brain (52, 68, 109). In order to characterize cellular mechanisms of information storage and retrieval, it is first necessary to identify and localize the brain systems, structures, and regions that are critically involved. In simpler learning paradigms the problem of localization appears finally to be yielding.

For simpler forms of learning, it seems evident that at least some components of the memory circuit must be localized. An animal trained to a particular conditioned stimulus will not respond to a very different conditioned stimulus and must be given additional training. This fact, the existence of a stimulus generalization gradient, argues strongly that sensory-specific information is to some degree preserved in the elements that develop the plasticity coding the learned response. A well trained animal exhibits a particular learned response complex. Activation of motor neurons can be highly selective and specific. This implies that the "motor program" aspect of the memory circuit must itself have specificity. Both the sensory-specific and motor-specific aspects of learning imply localization of the memory circuit.

AUTONOMIC CONDITIONING

A relatively consistent picture is emerging from studies on cardiovascular conditioning in three different species: baboon, rabbit, and pigeon. The paradigm is classical conditioning with a several-second auditory or visual CS terminating with an electric shock US. The conditioned heart rate response is viewed by most workers as a component or reflection of the more general process of "fear" conditioning, i.e., the conditioned emotional response - CER (e.g., (98)). Smith and associates (101) found that small, discrete bilateral lesions of the perifornical region of the hypothalamus in the baboon abolish the entire learned cardiovascular response complex completely, permanently, and selectively. The lesion has no effect on reflex cardiovascular responses and, most important, has no effect on a behavioral measure of the CER (lever press suppression) or on cardiovascular changes associated with exercise. This hypothalamic lesion effect seems to be on the efferent or motor-specific side of the learned response circuit since the behavioral signs of conditioned fear are still present. In some sense it is surprising that a structure as ancient as the hypothalamus exhibits such a highly selective action for a learned response, as opposed to reflex and general regulation. Pribram et al. (93) found that bilateral ablation of the amygdala also abolishes heart rate conditioning in the monkey.

In a recent series of papers, Kapp and associates have shown that the amygdala plays a critical role in heart rate conditioning in the rabbit (64). Lesions of the central nucleus virtually abolish conditioned heart rate slowing in the rabbit but have no effect on the reflex responses or on the initial orienting response to the CS, which itself is a slowing. Injection of beta-adrenergic blockers in the central nucleus partially attenuates the CR, as does injection of the opiate levorphanol. Furthermore, injection of the opiate antagonist naloxone enhances the CR. A direct pathway exists from the central nucleus of the amygdala to the vagal preganglionic cardioinhibitory neurons in the dorsal motor nucleus of the vagus. Finally, unit recordings in the central nucleus indicate that at least some neurons show increases during the development of conditioned bradycardia and a few show significant correlations with the magnitude of the CR.

Cohen and associates have completed the most extensive and detailed analysis of both the efferent and afferent limits of the learned response circuit for heart rate conditioning in their work on the pigeon (e.g., (19, 20)). Both the vagi and sympathetic nerves participate, but the predominant influence is from the right sympathetic cardiac nerve. The brain stem was mapped with electrical stimulation, together with electrophysiological recording, anatomical and lesion-behavior studies, and the efferent pathway defined. In brief, there is a system from the avian homologue of the amygdala to the hypothalamus, which then projects down to the final common path neurons via a ventral brain stem pathway. The CR is completely prevented by lesions of the amygdala, of the terminal field of the amygdala fibers in the hypothalamus, and of the brain stem course of fibers from this region of the hypothalamus. As in the mammal, these lesion effects are selective in that they do not abolish reflex cardiovascular responses, only the conditioned response. The pathway from amygdala to motor neurons in clearly essential for expression of the conditioned heart rate response in the pigeon, as in the mammal. It is not yet known whether some part of the memory trace circuit may be included in the amygdala.

Kapp et al. (64) interpret the role of the amygdala in aversive conditioning to be motoric, e.g., on the motor-specific side of the learned response circuit, at least insofar as conditioned bradycardia is concerned in the rabbit. The more general role of the amygdala in behavioral aspects of fear conditioning is less than clear. There is no question that it is much involved (for reviews see (40, 63)). In a recent study (70), the impairment of retention of aversive learning (one-trial inhibitory avoidance in rats) produced by amygdala lesions was found to depend strongly on the time of lesion. If the lesion was made ten days after training, it produced no deficit in retention. This result was interpreted to mean that the amygdala plays a modulatory role in aversive learning, e.g., that it acts on a memory trace system established elsewhere in the brain. The amygdala appears also to play a critical role in taste and odor aversion learning. Large lesions of the basolateral amygdala significantly disrupt a prior learned taste aversion in rats (83). Finally, it should be noted that the amygdala is also implicated in appetitive learning. Spiegler and Mishkin (102) showed that lesions of the amygdala markedly impair one-trial learning of object reward associations in the monkey but do not impair one-trial object recognition learning (80) (see below).

"ALPHA" VS. "ADAPTIVE" CONDITIONED RESPONSES

At this point it seems necessary to distinguish two types of conditioned responses, which we will term "alpha" and "adaptive." By definition, an alpha response is a reflex response elicited by the CS before learning. If this alpha response increases in strength as a result of pairing with an aversive US - compared to animals given unpaired presentations of the CS and US - it is a conditioned alpha response. The conditioned change in heart rate discussed above is an alpha response in that CS presentations before training produce the same kind of change in heart rate that then becomes conditioned to the US. Heart rate is, of course, controlled by the autonomic nervous system, and the key brain structures involved in learned changes in heart rate are visceral - hypothalamus and amygdala. Different brain systems underlie learning of discrete behaviors, e.g., skeletal muscle responses, as we will see.

We choose the term "adaptive" to refer to standard learning of discrete behavioral responses because the response is adaptive. In leg-flexion conditioning with a paw shock US, for example, over the entire range of CS-US intervals where learning will occur, the learned response is so timed that leg flexion is maximal at the time of onset of the US. In contrast, it is possible to establish an alpha-conditioned leg-flexion response in a spinalized animal. A weak flexion response elicited by a cutaneous stimulus can be increased in strength by pairing with a strong cutaneous stimulus (e.g., paw shock), relative to unpaired controls (89). But the initial short latency reflex response does not change in latency with conditioning, it merely increases in strength - it is not adaptive and is completed before the US occurs. Interestingly, spinal alpha-conditioning shows a similar CS-US interval effect in that best conditioning occurs over the range from about 200 to 500 msec. of CS onset precedence, but the latency of the alpha CR does not itself change over this interval (88).

CONDITIONING OF DISCRETE BEHAVIORAL RESPONSES

"Alpha" Conditioning

Woody and associates have focussed on brain substrates of conditioned alpha responses in the intact behaving mammal - the short latency, conditioned eyeblink response to click paired with glabellar tap (400 msec. CS-US interval) in the cat (14, 122). This short latency, alpha-conditioned response differs from conventional adaptive eyelid or NM conditioning (see below) in several ways: it is elicited by the click CS at the beginning of training, it has a very short latency (20 msec.), the latency does not shift over training, it is not "adaptive" in the sense that the CR is not present at the time of the UR, it requires a much longer time (600-900 trials) for learning, and the motor cortex is essential for development of the conditioned response.

The development of the conditioned alpha response is selective. If the CS is paired with glabellar tap, the eyelid response becomes conditioned; if paired with stimulation of the nose, the nose twitch develops. Extracellular microstimulation of the motor cortex indicates that threshold activation level of those neurons electing the CR is reduced following conditioning - they are more excitable. This effect is specific, as in eyelid vs. nose twitch training. In a technical tour de force, Brons and Woody (14) recorded intracellularly from 290 neurons in motor cortex in the absence of peripheral stimulation and measured threshold currents in already trained animals. In brief, results indicated that, although a group given US presentations alone showed a decrease in threshold relative to a group given CS trials alone, only groups given prior paired CS-US trials showed long-lasting decreases. This decrease persisted over extinction trials. In other work, the increased excitability of motor cortex neurons was found not to be accompanied by increases in spontaneous discharge activity or detectable changes in resting membrane potential. Woody (123) suggests that these results could be due to long-lasting changes in the postsynaptic neurons, e.g., an increase in dendritic resistance.

Woody's work indicates that the motor cortex appears to be a critical site of neuronal plasticity for the conditioned alpha eyelid response and that this plasticity involves increased excitability of neurons there, probably due to persisting postsynaptic changes in the neurons studied. Most conditioned responses, both classical and instrumental, do not require the motor cortex (see below). It would be most interesting if Woody's results were found to hold for all instances of alpha-conditioning in intact mammals, at least for skeletal muscle responses. If so, it would suggest a general fundamental difference in neuronal substrates between alpha-conditioning and associative conditioning.

Voronin (115) has used simple analogues of classical conditioning in studies of cellular plasticity in the motor cortex (see (111)). Recently he developed a paradigm somewhat analogous to Woody's in that a click was used as the CS (116). The UCS was motor cortex stimulation plus hypothalamic stimulation, and the behavioral CR was EMG activity in contralateral forepaw. Intercellular recordings were made from cortical neurons. In general, cortical neurons showed increased excitability over the course of pairings. Some neurons showed initial responses to the neutral click and virtually all responded to the loud click with the same short latency distribution as to the neutral click after paired training. Voronin notes that these cortical and EMG responses to loud clicks are components of the startle response (probably analogous to an alpha response). The conditioning regime increases the excitability of the startle circuit.

O'Brien and associates have developed an analogue of differential classical conditioning for cells of origin of pyramidal tract fibers in the motor cortex of the paralyzed cat (86, 87). Left and right hindpaw shocks were used as the CS+ and CS and the CS+ were paired with antidromic stimulation of the pyramidal tract as the UCS. Activity of neurons so identified in motor cortex were recorded. A significantly greater increase in response developed for the CS+. When electrical stimulation of VL or VPL regions of the thalamus were used as CS+ and CS-, no differential conditioning developed, an interesting result because the peripheral hindpaw stimuli were activating some portions of VPL and VL. As O'Brien notes, these data cannot be explained by a simple pairing theory of conditioning in which any two pathways to a neuron can be used as the CS and US. Voronin (116) emphasized the same point in noting that paired motor cortex and lateral hypothalamic stimulation was a more effective UCS than motor cortex stimulation alone.

"Adaptive" Conditioning

Adaptive classical conditioning of the eyelid or NM response (42) is widely used for the study of neuronal substrates of associative learning, usually in the rabbit but also in other species. This system has a number of advantages, including the very large behavioral literature on eyelid conditioning in a number of species, particularly humans (30, 53, 108)). We view it as but one example of the general class of learning discrete, adaptive behavioral responses in the attempt to avoid an aversive US. A number of studies report changes in neuronal activity in various brain structures during learning in this paradigm. Here, we focus on the essential learned response circuit for the standard delay-conditioned response. Decorticate and thalamic rabbits can learn the conditioned response, as can decerebrate cats (84, 85).

Several recent studies have reported abolition of the learned eyelid/NM response by selective brain lesions. In particular, lesions ipsilateral to the trained eye in several locations in the neocerebellum - large ablations of the lateral portion of the hemisphere, localized electrolytic lesions of the dentate/interpositus nuclei, and small, discrete lesions of the superior cerebellar peduncle - permanently abolish the CR but have no effect on the UR and do not prevent learning by the contralateral eye (69, 75-77, 107, 109). If the unilateral cerebellar lesion is made before training, the ipsilateral eye cannot learn, but the contralateral eye subsequently learns as though the animal is normal and new to the situation (71). If training is given before unilateral cerebellar lesion, the learned response is abolished in the ipsilateral eye, but the contralateral eye learns rapidly, with significant savings (75, 76).

Neuronal unit recordings from the dentate/interpositus nuclear region show evoked responses to CS and US onsets and, in some cases, the development of a temporal neuronal model (a pattern of increased frequency of unit discharges that models the amplitude-time course of the behavioral response) of the learned response but not the unlearned reflex response. This neuronal temporal model of the learned response appears to develop over the course of training in very close association with the development of the learned behavioral (NM) response and precedes it within a trial (66, 75).

Taken together, these results indicate that the cerebellum is an obligatory part of the learned response circuit for eyelid/NM conditioning. Since decerebrate animals can learn the response, this would seem to localize an essential component of the memory trace to the ipsilateral cerebellum and/or its major afferent/efferent systems. The fact that a neuronal unit "model" of the learned behavioral response develops in the cerebellar deep nuclei would seem to localize the process to cerebellum or its afferents. The possibility that cerebellar lesions produce a modulatory disruption of a memory trace localized elsewhere in the brain seems unlikely. If so, it must be efferent from the cerebellum, since discrete lesions of the superior cerebellar peduncle abolish the behavioral learned response. Yet the neuronal model of the learned response is present within the cerebellum. In this context, there is an earlier Soviet literature indicating that in dogs well trained in leg-flexion conditioning, complete removal of the cerebellum permanently abolishes the ability of the animals to make the learned discrete leg-flexion response, but not to show conditioned generalized motor activity (65).

We have recently found that hindlimb flexion conditioning in the rabbit is abolished by lesions of the interpositus nucleus ipsilateral to the trained limb, but critical locus is more medial than that for the conditioned NM/eyelid response (32). From the results, we conclude that the interpositus nucleus of the cerebellum is probably essential for all discrete conditioned behavioral responses learned with an aversive US and suggest that the memory traces for such learned response may, in fact, be stored there.

The cerebellum has been suggested by several authors as a possible locus for the coding of learned motor responses (1, 34, 58, 74). Cerebellar lesions impair a variety of skilled movements in animals (15, 16). In addition, neuronal recordings from Purkinje cells of the cerebellar cortex have implicated these cells in the plasticity of various responses (33,38).

Ito has developed a most interesting experimental model of induced neuronal plasticity using the vestibuto-ocular reflex (VOR) and applied the general Marr-Albus cerebellar models of motor learning to it (e.g., (58-62)). Plasticity of the VOR was reviewed recently in depth and with differing interpretations by Miles and Lisberger (78) and by Ito (62). Ito has also demonstrated a striking simpler form of conjunctive use-dependent plasticity of Purkinje cells in the flocculus and suggests two possible cellular mechanisms, both of which involve decreased sensitivity of chemical (glutamate?) receptors on Purkinje cells (61, 62).

A major efferent target of the cerebellar hemisphere, via the superior cerebellar peduncle, is the contralateral red nucleus. Smith (100) reported that large unilateral lesions in the red nucleus and vicinity markedly impaired a classically conditioned flexion response of the forelimb contralateral to, but not ipsilateral to, the lesion in cats. Tsukahara (112) developed a simplified preparation based on this paradigm. A stimulating electrode serving as the CS is implanted in the cerebral peduncle and the peduncle is lesioned caudal to the corticorubral fibers. The unconditioned stimulus is shock to the contralateral forepaw. The CS pulse train is adjusted to produce a weak flexion response to the forelimb. Animals learn the leg-flexion response to peduncle stimulation and the excitability of the pathway increases. Tsukahara argues that the excitability increase is at the synaptic junctions of peduncle fibers on red nucleus.

An older literature indicates that carnivores can learn a discrete leg-flexion response following complete decortication (e.g., (13, 92)). Consequently, although the corticorubral tract may normally be involved in leg-flexion conditioning in the intact animal, it is not essential. Cerebellar lesions, on the other hand, do apparently permanently abolish the discrete leg-flexion conditioned response (see above). The effects of cerebellar lesions have not yet been examined in Tsukahara's simplified preparation. In the rabbit NM/eyelid paradigm the magnocellular division of the contralateral red nucleus is also essential for the learned response (50, 72).

In general, higher regions of the brain such as the hippocampus and cerebral cortex are not essential for simpler forms of classical and instrumental conditioning. In spite of this, neuronal activity, e.g., in the hippocampus, becomes massively engaged in those same simple learning tasks such as NM/eyelid conditioning and is predictive of the development of behavioral learning. For example, Berger, Thompson, and associates showed a marked learning-induced increase in the activity of identified pyramidal neurons during eyelid/NM conditioning in the rabbit that occurs very early during the course of conditioning and substantially precedes behavioral learning (5). In addition, just prior to a conditioned NM response, hippocampal pyramidal neurons begin discharging with a distinctive pattern of firing that forms a clear temporal model of the amplitude-time course of the learned behavior (4, 5). Analysis of hippocampal activity during a variety of control conditions has shown that these cellular changes occur only during associative learning and are not due to nonassociative aspects of the conditioning process (5). While the hippocampus is not necessary for simpler forms of classical and instrumental conditioning, it does become essential when greater demands are placed on the memory system, even when using the same conditioning paradigms for both simple and complex learning tasks (3, 95, 109). Several reports have shown that changes in hippocampal cellular activity occur during learning of these more complex tasks for which the hippocampus is necessary (2, 27, 31, 109).

In primates, the hippocampus also plays an essential role in certain forms of memory. This is particularly evident in humans with amnesia due to brain damage (106). Mishkin (79) has recently shown that the hippocampus and amygdala are critical parts of the memory circuit for short-term visual recognition memory in the monkey that also includes the primary and secondary visual areas of the cerebral cortex and area TE in the temporal lobe.

NEUROANATOMICAL PLASTICITY AND ASSOCIATIVE LEARNING

Once a change in cellular activity as a result of learning is localized to an identified cell or synapse, a major question concerns the underlying mechanism of that cellular change. Historically, it has often been suggested that learning results in morphological alterations of the nervous system (e.g., (52)). That is, learning-induced changes in behavior might be mediated by alterations in the structural and connectional organization of the CNS, rather than (or in addition to) changes in the biophysical properties of nerve cells or their biochemical capacities. Several studies have now shown that behavioral learning is, in fact, accompanied by neuroanatomical plasticity in mammalian species. However, these reports are best considered as a subset of a larger number of investigations documenting that neuroanatomical plasticity readily occurs in mammals in response to changes in peripheral stimulation, particularly during development. For example, research on the effects of environmental variables has shown that exposing animals to an "enriched" environment (EC) of a variety of manipulanda (or "toys"), social situations, larger home cages, etc., results in dramatic increases in neuronal dendritic branching patterns, numbers of dendritic spines, postsynaptic density length, synapse shape, and the number of neuroglia when compared to animals exposed to an isolated condition (IC) of no toys, no other animals, and small cages (18, 28, 29, 39, 46, 47, 81, 94, 99).

Alternatively, visual deprivation experiments have demonstrated a reduction in many of the same structural features after limiting sensory stimulation of an organism. For example, changes in neuron soma size, dendritic branching, number of dendritic spines, distribution of dendritic and terminal processes, synapse size, vesicle density, and synaptic grid features of visual neocortical neurons have all been shown to occur in response to differences in visual stimulation (9, 12, 21, 24, 35, 37, 49, 56, 114, 118, 121).

In many cases, environmental enrichment and visual deprivation experiments are not readily interpretable within the framework of associative learning, because behavioral training in the sense of traditionally defined conditioning paradigms (i.e., classical and instrumental conditioning procedures) is not involved. Nevertheless, this literature provides an appropriate and instructive context for considering neuroanatomical plasticity as a potential mechanism for associative learning for the following reasons: a) just as learning is ubiquitous in mammalian species, results from enriched environment and visual deprivation experiments provide a wealth of evidence that structural alterations of mammalian nerve cells routinely occur in response to peripheral stimulation; b) the traditionally defined stimulus conditions that induce behavioral learning (e.g., CS-UCS intervals) can be seen simply as unique combinations of peripheral stimulation; and c) psychological research has shown that associative learning can occur in behavioral situations in which explicit relationships between discrete conditioning stimuli do not exist (i.e., latent learning, see (110)).

Neuroanatomical Plasticity Induced by Differences in Environmental Conditions

While the implications of experientially induced neuroanatomical plasticity for associative learning are intriguing, such phenomena must exhibit several characteristics for it to be seriously considered as a candidate mechanism.

First, neuroanatomical lability cannot be limited to development but must extend into adulthood. Several studies have clearly shown that mature animals are capable of exhibiting structural changes induced by environmental variables (see (94, 113)). Several recent analyses have revealed that enriched environmental conditions can induce increased higher-order dendritic branching even when the EC treatment begins at middle age (43) or old age, i.e., 600 days (22). In contrast, structural changes associated with visual deprivation have been thought not to meet this first criterion of generalization to the adult organism. Various studies have demonstrated that manipulations of visual stimulation are most effective in inducing structural and functional changes only during a "critical period" approximating the first 6-12 weeks of life and are generally ineffective in the adult (55, 120).

Although the immature organism clearly is more susceptible than adults to environmental influence, several studies have documented experiential modification of the adult visual system as well ((17, 25, 73), and Singer, this volume). In addition, preceding sections of this paper discuss a number of studies demonstrating that behavioral conditioning procedures alter the electrophysiological properties of sensory cortical neurons in adult animals. Together, these data raise the possibility that, even in the adult organism, physiological plasticity is accompanied or mediated by anatomical plasticity. Thus, although the specific procedures and experimental conditions reviewed here may be more effective during development, plasticity of visual or sensory brain systems is not excluded in the adult.

What may distinguish "critical period" neuroanatomical plasticity in the developing animal from learning-induced neuroanatomical plasticity in the adult is that the latter requires for its expression reinforcement of behavior, or temporal contiguity between the UCS and the sensory stimulus used as the CS. In other words, the stimulus prerequisites sufficient to induce neuroanatomical plasticity may simply be more restricted for the adult than for the developing animal. From this perspective, the critical period may reflect whatever processes are responsible for narrowing the prerequisites, rather than reflecting a decreasing capacity of the CNS for structural alteration.

Second, if neuroanatomical plasticity is to be considered as a candidate mechanism, it must have functional consequences. The functional relevance of different environments has been determined primarily by behavioral measures of learning ability. Animals raised in an EC environment are consistently superior to those raised in an IC environment at learning a variety of tasks (6, 7, 23, 36, 44, 48, 57). While these data do not show that structural changes due to environmental differences increase learning ability per se, they are important in demonstrating that environmental manipulations that induce structural changes in neurons also induce behavioral changes.

Results of visual deprivation experiments have been very useful in demonstrating that experientially induced structural changes are paralleled by physiological changes. For example, monocular deprivation results in a widening or "sprouting" or corticopetal terminals in layer IV of visual cortex from the non-deprived eye, and a corresponding narrowing or retracting of corticopetal fibers from the deprived eye (9, 56). Consistent with these neuroanatomical data, physiological analyses have demonstrated that monocular deprivation causes a dramatic decrease in effective activation of visual stimuli presented to the deprived eye (120), so that the majority of cells are driven only monocularly. In addition to binocularity, the orientation specificity of visual cortical neurons is also stimulation-dependent (10, 11, 82, 91). Visual exposure to only certain linear orientations can shape or determine the orientation preference of visual cortical cells (8, 54, 90).

Third, experientially induced changes must be relatively long-lasting. Very few studies have systematically tested the permanence of structural changes due to differences in environmental rearing. While Cummins et al. (26) have shown that EC-IC differences in the structural features of neocortical neurons are maintained after over 500 days of exposure to each condition, these data do not address the question of how long environmentally dependent neuroanatomical plasticities remain after environmental differences are eliminated. The issue of permanence of experientially induced changes in the visual system is complicated by the existence of a critical period during development. Many studies have shown that if visual deprivation conditions are continued past the end of the critical period, the anatomical and physiological consequences are "permanent" (10, 55, 103). But in this context, "permanence" has little significance, because the nervous system may require different stimulus prerequisites for plasticity, or may not have the same capacity for plasticity, when initial conditions are imposed (within the critical period) and when their consequences are tested (past the critical period). To address the issue of structural changes as a mechanism for associative learning, permanence of visual system plasticity must be examined in the adult animal, and such studies have yet to be reported (though see (25)). Though experientially induced anatomical changes should be long-lasting if they are to serve as a mechanism for associative learning, they need not be permanent. The process of forgetting is as robust and well documented behaviorally as the process of learning, so not all experientially induced anatomical alterations should be expected to last indefinitely. In this regard, studies of the effects of visual deprivation are instructive in demonstrating that deprivation-induced anatomical plasticity is also reversible (e.g., (9)).

Finally, neuroanatomical plasticity must be capable of supporting the capacity and variety of mammalian memory. It is clear that environmental studies have catalogued a considerable number of neuroanatomical changes as a result of differential experience. The quantitative and qualitative differences in peripheral stimulation that distinguish different environments are capable of inducing structural changes in many different morphological features of nerve cells, as well as the number of supportive glia. Additionally, visual deprivation studies have documented structural changes at the level of the neuron soma, dendritic branch, dendritic spine, terminal process, terminal vesicle, and synapse. Both lines of investigation, then, have shown that a rich substrate of anatomical characteristics is available for environmentally induced alteration.

Neuroanatomical Plasticity Induced by Associative Learning

Relatively few studies have specifically investigated potential neuroanatomical plasticity coincident with associative learning in adult mammals. Of those that have, many are extensions of previous work by authors on neuroanatomical plasticity induced by environmental manipulation or visual deprivation, and the results of these experiments again support a conception of the adult CNS as remarkably dynamic in terms of its structural organization.

One of the first reports of the effects of learning procedures on morphological features of mature cortical neurons was conducted as part of an EC-IC experiment (26). One group of adult animals was given only training on a maze task after being exposed to IC conditions. Measures of forebrain weight revealed significant increases in material from trained animals compared to other animals maintained in IC throughout the training period. A later study by Greenough et al. (45) revealed that at least one factor contributing to the training-induced weight change is increased branching of higher-order processes on apical dendrites of pyramidal neurons. More recently, Larson and Greenough (67) trained rats in a forepaw motor task (handedness reversal) and examined dendritic branching patterns of layer V pyramidal neurons in the forelimb region of motor cortex. Distal portions of apical dendrites from cells of cortical areas projecting to the trained forearm exhibited more profuse branching patterns than (contralateral) cells projecting to the untrained arm.

Rutledge et al. (96) have examined cortical dendritic morphology following behavioral conditioning using neocortical brain stimulation as a CS and foreleg shock as a UCS to condition leg flexion. Separate animals were given unpaired presentations of the CS and UCS to control for nonassociative effects of training. Neocortical pyramidal neurons from the stimulated region and from the contralateral hemisphere activated through collosal connections were examined. Results showed that material from conditioned animals exhibited an increased number of dendritic spines on vertical and oblique branches of apical dendrites when compared to material from control animals.

Vrensen and Cardozo (117) have trained adult rabbits on an operant, pattern discrimination task and subsequently examined several parameters of visual cortical synaptic morphology. Results showed that a) the frequency of complex synaptic grids increased substantially in trained animals; b) the surface area and number of dense projections per grid were smaller in trained animals, both for complex grids and for the total synaptic population; and c) there was a significant increase in the length of the postsynaptic terminals, number of vesicles, or other measures. Other observations by the authors led them to suggest that behavioral conditioning resulted in a "focalization" of synaptic grids that would act to increase synaptic efficacy. Thus, behavioral conditioning may be producing an enhanced activation of visual cortical neurons, possibly those neurons responsive to features of the visual patterns used as conditioning stimuli.

Spinelli and co-workers have recently reported experiments that extend earlier work on the effects of visual deprivation (104, 105). Although these studies were conducted in the developing animal, they are important in showing that, in addition to producing neuroanatomical changes, conditioning can result in functional alterations of the CNS that reflect specific aspects of the training conditions. Kittens raised under normal visual conditions were trained in an operant avoidance task requiring forearm flexion and were cued with either a horizontal or vertical visual pattern (104). The area of somatosensory cortex responsive to forearm stimulation was found to be greater in the hemisphere projecting to the trained forearm compared to the opposite hemisphere. The number of polymodal neurons responsive to visual stimuli in the forearm region had also increased significantly in the trained hemisphere and, more interesting, the majority of these cells exhibited an orientation preference for the stimulus that elicited the behavioral response. In a later study (105), the authors reported a substantial increase in dendritic branching of neurons within the cortical region demonstrating the physiological effects.

Finally, a recent study by Wenzel et al. (119) trained rats in a brightness discrimination task (using shock) and examined the number of synapses on apical dendrites of CA1 hippocampal pyramidal cells. Material from trained animals was compared with that of control animals that performed the identical response and received the same number of shocks, but shocks that were not correlated with brightness differences. Results showed that conditioning produced a 40% increase in the number of synapses in trained animals. While this difference decreased over days, a significant increase over control counts was still detectable fourteen days after training.

With respect to the four criteria used to evaluate environmental differences and visual deprivation, it is clear from the above studies that learning-induced neuroanatomical plasticity clearly occurs in the adult as well as in the developing animal. The studies reviewed here demonstrate the capability of the adult mammalian CNS to change structurally in response to behavioral conditioning.

Two issues arise with respect to functional significance. The first is whether the structural changes seen are related to behavioral learning or to other, nonassociative components of conditioning. As noted above, aspects of every conditioning procedure include elements unrelated to learning and mnemonic processes (such as changing levels of arousal or motor responding), and control procedures have been developed to differentiate between associative and nonassociative effects of behavioral training (e.g., (41)). Therefore, the degree to which neuroanatomical plasticity induced by conditioning procedures can be related specifically to associative aspects of training must be considered. The study by Rutledge et al. (96) included a control group given unpaired presentations of the CS and UCS, and results from those animals demonstrate the importance of control procedures in evaluating associative and nonassociative effects of conditioning. Specifically, both conditioned and control animals showed interhemispheric differences in the number of spines on terminal, apical dendrites. Only with respect to number of spines on vertical and oblique dendrites did the two groups differ. Thus, some morphological changes were specific to associative learning while others were not. The experiments of Vrensen and Cardozo (117), Larson and Greenough (67), and Wenzel et al. (119) all included separate groups of animals that controlled for such nonassociative effects of conditioning. All of the studies discussed above, then, are conclusive in showing that anatomical plasticity was related to associative learning and not to differences in the level of performance of stimulation of the experimental and control groups.

The second issue that arises with respect to functional considerations is whether the structural changes are actually the substrate for the changes in electrophysiological properties or behavior observed. For example, the results of Spinelli and Jensen (104) and Spinelli et al. (105) are encouraging in demonstrating that both learning-induced physiological and anatomical alterations occur within the same cortical region. Whether the two are actually related is still unknown. Even assuming that the increase in dendritic branching reported by Spinelli et al. (105) is indicative of an increased number of functional synaptic contacts, the available evidence does not allow the conclusion that the structural changes are the substrate for enlargement of forearm representation, or learned avoidance behavior. The same uncertainty exists with the results of other studies reviewed here. While their respective findings certainly suggest a relationship between learned behavior and the neuroanatomical plasticities they have documented, it is not certain that the structural changes are causally related to conditioned behavioral responses. Establishing such a relationship requires evidence on a number of different levels. For example, arguments for a causal relationship would be strengthened if results such as those reported by Vrensen and Cardozo (117) and Larson and Greenough (67) are restricted to appropriated neocortical regions, i.e., do not occur in nonvisual (117) or nonmotor, non-somatosensory (67) regions. Additional supportive evidence would come from examination of structural features of neurons strained intracellularly. With the latter technique, changes in electrophysiological properties could be related to changes in anatomical characteristics of the same neuron.

The permanence of learning-induced anatomical changes has yet to be addressed. The studies reviewed here are recent ones and are among the first to demonstrate that learning-related structural changes occur in the adult brain. The time course of their development, permanence, and possible reversal with extinction now become pertinent issues. Although in one sense, structural plasticity should be long-lasting to serve as a substrate for associative learning, any other constraints on time course would be premature. Temporal parameters of anatomical plasticity may vary depending on the learning process reflected by the anatomical change. Structural changes related to long-term storage may have a different time course than those underlying initial acquisition. Some structural changes may be associated with "learning to learn" (learning sets, see (51)) rather than learning a specific relationship between specific stimuli, and thus may be invariant in the course of learning and forgetting a number of problems within a learning set. Behavioral extinction may occur without total extinction at the neural level because relearning is almost invariably more rapid than initial learning (e.g., (97)), yet some neural extinction must reflect the behavioral extinction. These possibilities strongly suggest there may be various time courses of neuroanatomical change in relation to behavioral change. It is worth noting that a number of experiments have demonstrated that anatomical plasticity can occur with a rapid time course (24, 91, 114), so the capacity for rapid structural change exists.

With respect to the fourth criterion, the Vrensen and Cardozo experiment showed that only a subset of all structural dimensions of nerve cells are altered as a result of conditioning. This suggests either that only a subset of neuroanatomical features are capable of being modified by learning, or that the particular profile of features modified by conditioning is specific to the information learned. For example, training animals to respond differentially on the basis of other visual cues, such as color discrimination, may also result in neuroanatomical plasticity, but plasticity of structural features other than those of the synaptic grid. However, dendritic branching patterns and dendritic spine number may always be altered by conditioning, but the particular branching pattern and spatial location of new spines may vary with the particulars of each training condition, such as with the response system or stimuli used. From this perspective, it may be significant that maze training of rats alters dendritic branching patterns differently than exposure to enriched environmental conditions. For example, EC exposure is almost invariably associated with an increase in the number of higher-order branches on basal dendrites of neocortical neurons. In contrast, Greenough et al. (45) showed that maze training results in changes primarily on apical dendrites.

In summary, a considerable body of evidence from work on the effects of environmental differences, visual deprivation, and behavioral conditioning has demonstrated that the adult mammalian CNS changes structurally in response to changes in peripheral stimulation. In addition, the nature of these structural changes is compatible with the notion that neuroanatomical plasticity could serve as a mechanism for associative learning. In fact, recent studies analyzing neuroanatomical plasticity as a result of behavioral conditioning have been successful in documenting structural alterations that are specifically related to associative aspects of training.

Acknowledgements. In this brief review we have drawn heavily from our recent article in the Annual Review of Neuroscience (109). This work was supported in part by research grants from the National Science Foundation (NSF #BNS-81-17115) and the Office of Naval Research (ONR #N00014-83-K-0238) to R.F. Thompson, and from the National Science Foundation (NSF #BNS-80-21395) and the National Institutes of Mental Health (NIMH #MH-00343) to T.W. Berger.

REFERENCES

(1) Albus, J.S. 1971. A theory of cerebellar function. Math. Biosci. 10: 26-61.

(2) Berger, T.W. 1982. Hippocampal pyramidal cell activity during two-tone discrimination and reversal conditioning of the rabbit nictitating membrane response. Soc. Neurosci. Abstr. 8: 146.

(3) Berger, T.W., and Orr, W.B. 1983. Hippocampectomy selectively disrupts discrimination reversal conditioning of the rabbit nictitating membrane response. Behav. Brain Res. 8: 49-68.

(4) Berger, T.W.; Laham, R.I.; and Thompson, R.F. 1980. Hippocampal unit-behavior correlations during classical conditioning. Brain Res. 193: 229-248.

(5) Berger, T.W., and Thompson, R.F. 1978. Neuronal plasticity in the limbic system during classical conditioning of the rabbit nictitating membrane response. I. The hippocampus. Brain Res. 145(2): 323-346.

(6) Bernstein, L. 1973. A study of some enriching variables in a free-environment for rats. J. Psychosom. Res. 17: 85-88.

(7) Bingham, W.E., and Griffiths, W.J., Jr. 1952. The effect of different environments during infancy on adult behavior in the rat. J. Comp. Physiol. Psychol. 45: 307-312.

(8) Blakemore, C., and Cooper, G.F. 1970. Development of the brain depends on the visual environment. Nature 228: 477-478.

(9) Blakemore, C.; Garey, L.J.; Henderson, Z.B.; Swindale, N.V.; and Vital-Durand, F. 1980. Visual experience can promote rapid axonal reinnervation in monkey visual cortex. J. Physiol. 307: 25-26.

(10) Blakemore, C., and Van Sluyters, R.C. 1974. Reversal of the physiological effects of monocular deprivation in kittens: Further evidence for a sensitive period. J. Physiol. 237: 195-216.

(11) Blakemore, C., and Van Sluyters, R.C. 1975. Innate and environmental factors in the development of the kitten's visual cortex. J. Physiol. 248: 663-716.

(12) Borges, S., and Berry, M. 1976. Preferential orientation of stellate cell dendrites in the visual cortex of the dark-reared rat. Brain Res. 112: 141-147.

(13) Bromily, R.B. 1948. The development of conditioned responses in cats after unilateral decortication. J. Comp. Physiol. Psychol. 41: 155-164.

(14) Brons, J.F., and Woody, C.D. 1980. Long-term changes in excitability of cortical neurons after Pavlovian conditioning and extinction. J. Neurophysiol. 44: 605.

(15) Brooks, V.B. 1979. Control of the intended limb movements by the lateral and intermediate cerebellum. In Integration in the Nervous System, eds. H. Asanuma and V.J. Wilson, pp. 321-356. New York: Igaku-Shoin Press.

(16) Brooks, V.B.; Kozlovskaya, I.B.; Atkin, A.; Horvath, F.E.; and Uno, M. 1973. Effects of cooling dentate nucleus on tracking-task performance in monkeys. J. Neurophysiol. 36: 974-995.

(17) Brown, D.L., and Salinger, W.L. 1975. Loss of x-cells in lateral geniculate nucleus with monocular paralysis: Neural plasticity in the adult cat. Science 189: 1011-1012.

(18) Chang, F.F.; Wesa, J.M.; Greenough, W.T.; and West, R.W. 1981. Differential post-synaptic curvature in occipital cortex following differential rearing in rat. Neurosci. Abstr. 7: 772.

(19) Cohen, D.H. 1980. The functional neuroanatomy of a conditioned response. In Neural Mechanisms: Goal-directed Behavior and Learning, eds. R.F. Thompson, L.H. Hicks, and V.B. Shvyrkov. New York: Academic Press.

(20) Cohen, D.H. 1982. Central processing time for a conditioned response in a vertebrate model system. In Conditioning: Representation of Involved Neural Functions, ed. C.D. Woody. New York: Plenum Press.

(21) Coleman, P.D., and Riesen, A.H. 1968. Environmental effects on cortical dendritic fields. I. Rearing in the dark. J. Anat. 102: 363-374.

(22) Connor, J.R.; Melone, J.H.; Yuen, A.R.; and Diamond, M.C. 1981. Dendritic length in aged rats' occipital cortex: An environmentally induced response. Exp. Neurol. 73: 827-830.

(23) Cornwell, P., and Overman, W. 1981. Behavioral effects of early rearing conditions and neonatal lesions of the visual cortex in kittens. J. Comp. Physiol. Psychol. 95: 848-862.

(24) Cragg, B.G. 1967. Changes in visual cortex on first exposure of rats to light. Nature 215: 251-253.

(25) Creutzfeldt, O.D., and Heggelund, P. 1975. Neural plasticity in visual cortex of adult cats after exposure to visual patterns. Science 188: 1025-1027.

(26) Cummins, R.A.; Walsh, R.N.; Budtz-Olsen, O.E.; Konstantinos, T.; and Horsfall, C.R. 1973. Environmentally induced changes in the brains of elderly rats. Nature 243: 516-518.

(27) Deadwyler, S.A.; West, M.; and Lynch, G. 1979. Activity of dentate granule cells during learning: Differentiation of perforant path input. Brain Res. 169: 29-43.

(28) Diamond, M.C.; Johnson, R.E.; Ingham, C.; Rosenzweig, M.R.; and Bennett, E.L. 1975. Effects of differential experience on neuronal nuclear and perikarya dimensions in the rat cerebral cortex. Behav. Biol. 15: 107-111.

(29) Diamond, M.C.; Law, F.; Rhodes, H.; Lindner, B.; and Rosenzweig, M.R. 1966. Increases in cortical depth and glial numbers in rats subjected to enriched environment. J. Comp. Neurol. 128: 117-126.

(30) Disterhoft, J.F.; Kwan, H.H.; and Lo, W.D. 1977. Nictitating membrane conditioning to tone in the immobilized albino rabbit. Brain Res. 137: 127-144.

(31) Disterhoft, J.F., and Segal, M. 1978. Neuron activity in rat hippocampus and motor cortex during discrimination reversal. Brain Res. Bull. 137: 127-144.

(32) Donegan, N.H.; Lowry, R.W.; and Thompson, R.F. 1983. Effects of lesioning cerebellar nuclei on conditioned leg-flexion responses. Neurosci. Abstr. 9: 331.

(33) Dufosse, M.; Ito, M.; Jastrehoff, P.J.; and Miyashita, Y. 1978. Diminution and reversal of eye movements induced by local stimulation of rabbit cerebellar flocculus after partial destruction of the inferior olive. Exp. Brain Res. 33: 139-141.

(34) Eccles, J.C.; Ito, M.; and Sezentagothai, J. 1967. The Cerebellum As a Neuronal Machine. New York: Springer-Verlag.

(35) Fifkova, E. 1970. The effect of unilateral deprivation on visual center in rats. J. Comp. Neurol. 140: 431-438.

(36) Forgays, D.G., and Read, J.M. 1962. Crucial periods for free-environment experience in the rat. J. Comp. Physiol. Psychol. 55: 816-818.

(37) Garey, L.J., and Pettigrew, J.D. 1974. Ultrastructural changes in kitten visual cortex after environmental modification. Brain Res. 66: 165-172.

(38) Gilbert, P.F.C., and Thach, W.T. 1977. Purkinje cell activity during motor learning. Brain Res. 128: 309-328.

(39) Globus, A.; Rosenzweig, M.R.; Bennett, E.L.; and Diamond, M.C. 1973. Effects of differential experience on dendritic spine counts in rat cerebral cortex. J. Comp. Physiol. Psychol. 82: 175-181.

(40) Goddard, G. 1964. Functions of the amygdala. Psychol. Bull. 62: 89.

(41) Gormezano, I. 1972. Investigations of defense and reward conditioning in the rabbit. In Classical Conditioning. II. Current Research and Theory, eds. A.H. Black and W.F. Prokasy, pp. 151-181. New York: Appleton-Century-Crofts.

(42) Gormezano, I.; Schneiderman, N.; Deaux, E.; and Fentues, I. 1962. Nictitating membrane: Classical conditioning and extinction in the albino rabbit. Science 138: 33-34.

(43) Green, E.J.; Schlumpf, B.E.; and Greenough, W.T. 1981. The effects of complex or isolated environments on cortical dendrites of middle-aged rats. Neurosci. Abstr. 7: 65.

(44) Greenough, W.T.; Fulcher, J.K.; Yuwiler, A.; and Geller, E. 1970. Enriched rearing and chronic electroshock: Effects on brain and behavior in mice. Physiol. Behav. 5: 371-373.

(45) Greenough, W.T.; Juraska, J.M.; and Volkmar, F.R. 1979. Maze training effects on dendrite branching in occipital cortex of adult rats. Behav. Neural Biol. 26: 287-297.

(46) Greenough, W.T., and Volkmar, F.R. 1973. Pattern of dendritic branching in occipital cortex of rats reared in complex environments. Exp. Neurol. 40: 491-504.

(47) Greenough, W.T.; West, R.W.; and DeVoogd, T.J. 1978. Subsynaptic plate perforations: Changes with age and experience in the rat. Science 202: 1096-1098.

(48) Greenough, W.T.; Yuwiler, A.; and Dollinger, M. 1973. Effects of posttrial eserine administration on learning in "enriched" and "impoverished" reared rats. Behav. Biol. 8: 261-272.

(49) Gyllensten, L.; Malmfors, T.; and Norrlin, M.L. 1965. Effect of visual deprivation on the optic centers of growing and adult mice. J. Comp. Neurol. 124: 149-160.

(50) Haley, D.A.; Lavond, D.G.; and Thompson, R.F. 1983. Effects of contralateral red nuclear lesions on retention of the classically conditioned nictitating membrane/eyelid response. Neurosci. Abstr. 9: 643.

(51) Harlow, H.F., and Warren, J.M. 1952. Formation and transfer of discrimination learning sets. J. Comp. Physiol. Psychol. 45: 482-489.

(52) Hebb, D.O. 1949. The Organization of Behavior. New York: Wiley.

(53) Hilgard, E.R., and Marquis, D.G. 1940. Conditioning and Learning. New York: Appleton.

(54) Hirsch, H.V.B., and Spinelli, D.N. 1971. Modification of the distribution of receptive field orientation in cats by selective visual exposure during development. Exp. Brain Res. 13: 509-527.

(55) Hubel, D.H., and Wiesel, T.N. 1970. The period of susceptibility to the physiological effects of unilateral eye closure in kittens. J. Physiol. 206: 419-436.

(56) Hubel, D.H.; Wiesel, T.N.; and Le Vay, S. 1977. Plasticity of ocular dominance columns in monkey striate cortex. Phil. Trans. Roy. Soc. Lond. B 278: 377-409.

(57) Hymovitch, B. 1952. The effects of experimental variations on problem solving in the rat. J. Comp. Physiol. Psychol. 45: 313-321.

(58) Ito, M. 1970. Neurophysiological aspects of the cerebellar motor control system. Intl. J. Neurol. 7: 162-176.

(59) Ito, M. 1974. The control mechanisms of cerebellar motor system. In The Neurosciences, Third Study Program, eds. F.O. Schmitt and R.G. Worden. Boston: MIT Press.

(60) Ito, M. 1977. Neuronal events in the cerebellar flocculus associated with an adaptive modification of the vestibulo-ocular reflex of the rabbit. In Control of Gaze by Brain Stem Neurons, eds. R.G. Baker and A. Berthoz. Amsterdam: Elsevier Holland.

(61) Ito, M. 1982. Cerebellar control of vestibulo-ocular reflex: Around the flocculus hypothesis. Ann. Rev. Neurosci. 5: 275-296.

(62) Ito, M. 1982. Synaptic plasticity underlying the cerebellar motor learning investigated in rabbit's flocculus. In Conditioning: Representation of Involved Neural Functions, ed. C.D. Woody. New York: Plenum Press.

(63) Kaada, B.R. 1972. Stimulation and regional ablation of the amygdaloid complex with reference to functional representations. In The Neurobiology of the Amygdala, ed. B.E. Elftheriou, pp. 205-281. New York: Plenum Press.

(64) Kapp, B.S.; Gallagher, M.; Applegate, C.D.; and Frysinger, R.C. 1982. The amygdala central nucleus: Contributions to conditioned cardiovascular responding during aversive Pavlovian conditioning in the rabbit. In Conditioning: Representation of Involved Neural Functions, ed. C.D. Woody. New York: Plenum Press.

(65) Karamian, A.I.; Fanaralijian, V.V.; and Kosareva, A.A. 1969. The functional and morphological evolution of the cerebellum and its role in behavior. In Neurobiology of Cerebellar Evolution and Development: First International Symposium, ed. R. Ilinas. Chicago: American Medical Association.

(66) Kettner, R.E., and Thompson, R.F. 1982. Auditory signal detection and decision processes in the nervous system. J. Comp. Physiol. Psychol. 96: 328-331.

(67) Larson, J.R., and Greenough, W.T. 1981. Effects of handedness training on dendritic branching of neurons of forelimb area of rat motor cortex. Neurosci. Abstr. 7: 65.

(68) Lashley, K.S. 1929. Brain Mechanisms and Intelligence. Chicago: University of Chicago Press.

(69) Lavond, D.G.; McCormick, D.A.; Clark, G.A.; Holmes, D.T.; and Thompson, R.F. 1981. Effects of ipsilateral rostral pontine reticular lesions on retention of classically conditioned nictitating membrane and eyelid responses. Physiol. Psychol. 9(4): 335-339.

(70) Liang, K.C.; McGaugh, J.L.; Martinez, J.L., Jr.; Jensen, R.A.; Vasquez, B.J.; and Messing, R.B. 1982. Post training amygdaloid lesions impair retention of an inhibitory avoidance response. Behav. Brain Res. 4: 237-250.

(71) Lincoln, J.S.; McCormick, D.A.; and Thompson, R.F. 1982. Ipsilateral cerebellar lesions prevent learning of the classically conditioned nictitating membrane/eyelid response. Brain Res. 242: 190-193.

(72) Madden, J., IV; Haley, D.A.; Barchas, J.D.; and Thompson, R.F. 1983. Microinfusion of picrotoxin into the caudal red nucleus selectively abolishes the classically conditioned nictitating membrane/eyelid response in the rabbit. Neurosci. Abstr. 9: 830.

(73) Maffei, L., and Fiorentini, A. 1976. Asymmetry of motility of the eyes and change of binocular properties of cortical cells in adult cats. Brain Res. 105: 73-78.

(74) Marr, D. 1969. A theory of cerebellar cortex. J. Physiol. 202: 437-470.

(75) McCormick, D.A.; Clark, G.A.; Lavond, D.G.; and Thompson, R.F. 1982. Initial localization of the memory trace for a basic form of learning. Proc. Natl. Acad. Sci. 79(8): 2731-2742.

(76) McCormick, D.A.; Lavond, D.G.; Clark, G.A.; Kettner, R.E.; and Rising, C.E. 1981. The engram found? Role of the cerebellum in classical conditioning of nictitating membrane and eyelid responses. Bull. Psychonom. Soc. 18(3): 103-105.

(77) McCormick, D.A.; Lavond, D.G.; and Thompson, R.F. 1982. Concomitant classical conditioning of the rabbit nictitating membrane and eyelid responses: Correlations and implications. Physiol. Behav. 28: 769-775.

(78) Miles, F.A., and Lisberger, S.G. 1981. Plasticity in the vestibulo-ocular reflex: A new hypothesis. Ann. Rev. Neurosci. 4: 273-299.

(79) Mishkin, M. 1981. A memory system in the monkey. Phil. Trans. Roy. Soc. Lond. B 298: 85-95.

(80) Mishkin, M., and Oubre, J.L. 1976. Dissociation of deficits on visual memory tasks after inferior temporal and amygdala lesions in monkeys. Neurosci. Abstr. 2: 1127.

(81) Møllgaard, K.; Diamond, M.C.; Bennett, E.L.; Rosenzweig, M.R.; and Lindner, B. 1971. Quantitative synaptic changes with differential experience in rat brain. Intl. J. Neurosci. 2: 113-128.

(82) Movshon, J.A. 1976. Reversal of the physiological effects of monocular deprivation in the kitten's visual cortex. J. Physiol. 261: 125-174.

(83) Nachman, M., and Ashe, J.H. 1974. Effects of basolateral amygdala lesions on neophobia, learned taste aversions, and sodium appetite in rats. J. Comp. Physiol. Psychol. 87: 622-643.

(84) Norman, R.J.; Buchwald, J.S.; and Villablanca, J.R. 1977. Classical conditioning with auditory discrimination of the eyeblink in decerebrate cats. Science 196: 551-553.

(85) Oakley, D.A., and Russell, I.S. 1972. Neocortical lesions and classical conditioning. Physiol. Behav. 8: 915-926.

(86) O'Brien, J.H., and Quinn, J.K. 1982. Central mechanisms responsible for classically conditioned changes in neuronal activity. In Conditioning: Representation of Involved Neural Functions, ed. C.D. Woody. New York: Plenum Press.

(87) O'Brien, J.H.; Wilder, M.B.; and Stevens, C.D. 1977. Conditioning of cortical neurons in cats with antidromic activation as the unconditioned stimulus. J. Comp. Physiol. Psychol. 91: 918-929.

(88) Patterson, M.M. 1975. Interstimulus interval effects on a classical conditioning paradigm utilizing the hindlimb flexion reflex of acute spinal cat. Paper presented at the 5th Annual Meeting of the Society for Neuroscience, New York, November 1975.

(89) Patterson, M.M.; Cegavske, C.F.; and Thompson, R.F. 1973. Effects of a classical conditioning paradigm on hindlimb flexor nerve response in immobilized spinal cat. J. Comp. Physiol. Psychol. 84: 88-97.

(90) Pettigrew, J.D., and Freeman, R.D. 1973. Visual experience without lines: Effect on developing cortical neurons. Science 182: 599-601.

(91) Pettigrew, J.D., and Garey, L.J. 1974. Selective modification of single neuron properties in the visual cortex of kittens. Brain Res. 66: 160-164.

(92) Poltrew, S.S., and Zeliony, G.P. 1930. Grosshirnrinde und Assoziationsfunktion. Zoologie Biologie 90: 157-160.

(93) Pribram, K.H.; Reitz, S.; McNeil, M.; and Spevack, A.A. 1979. The effect of amygdalectomy on orienting and classical conditioning in monkeys. Pav. J. Biol. 14: 203-217.

(94) Rosenzweig, M.R., and Bennett, E.L. 1976. Enriched environments: Facts, factors, and fantasies. In Knowing, Thinking, and Believing, eds. J.L. McGaugh and L. Petrinovich, pp. 179-213. New York: Plenum Press.

(95) Ross, R.T.; Orr, W.B.; Holland, P.C.; and Berger, T.W. 1984. Hippocampectomy disrupts acquisition and retention of learned conditional responding. Behav. Neurosci. 98: 211-225.

(96) Rutledge, L.T.; Wright, C.; and Duncan, J. 1974. Morphological changes in pyramidal cells of mammalian neocortex associated with increased use. Exp. Neurol. 44: 209-228.

(97) Scavio, M.J., and Thompson, R.F. 1979. Extinction and reacquisition performance alternations of the conditioned nictitating membrane response. Bull. Psychonom. Soc. 13(2): 57-60.

(98) Schneiderman, N. 1972. Response system divergencies in aversive classical conditioning. In Classical Conditioning. II. Current Research and Theory, eds. A.H. Black and W.F. Prokasy. New York: Appleton-Century-Crofts.

(99) Shapiro, S., and Vukovich, K.R. 1970. Early experience effects upon cortical dendrites: A proposed model for development. Science 167: 292-294.

(100) Smith, A.M. 1970. The effects of rubral lesions and stimulation on conditioned forelimb flexion responses in the cat. Physiol. Behav. 5: 1121-1126.

(101) Smith, O.A.; Astley, C.A.; DeVit, J.L.; Stein, J.M.; and Walsh, K.E. 1980. Functional analysis of hypothalamic control of the cardiovascular responses accompanying emotional behavior. Fed. Proc. 39(8): 2487-2494.

(102) Spiegler, B.J., and Mishkin, M. 1981. Evidence for the sequential participation of inferior temporal cortex and amygdala in the acquisition of stimulus-reward associations. Behav. Brain Res. 3: 303-317.

(103) Spinelli, D.N.; Hirsch, H.V.B.; Phelps, R.W.; and Metzler, J. 1972. Visual experience as a determinant of the response characteristics of cortical receptive fields in cats. Exp. Brain Res. 15: 289-304.

(104) Spinelli, D.N., and Jensen, F.E. 1979. Plasticity: The mirror of experience. Science 203: 75-78.

(105) Spinelli, D.N.; Jensen, F.E.; and DiPrisco, G.V. 1980. Early experience effect on dendritic branching in normally reared kittens. Exp. Neurol. 68: 1-11.

(106) Squire, L. 1982. The neurophysiology of human memory. Ann. Rev. Neurosci. 5: 241-273.

(107) Thompson, R.F.; Berger, T.W.; Berry, S.D.; Clark, G.A.; Kettner, R.E.; Lavond, D.G.; Mauk, M.D.; McCormick, D.A.; Solomon, P.R.; and Weisz, D.J. 1982. Neuronal substrates of learning and memory: Hippocampus and other structures. In Conditioning: Representation of Involved Neural Functions, ed. C.D. Woody, pp. 115-130. New York: Plenum Press.

(108) Thompson, R.F.; Berger, T.W.; Cegavske, C.F.; Patterson, M.M.; Roemer, R.A.; Teyler, T.J.; and Young, R.A. 1976. The search for the engram. Am. Psychol. 31: 209-227.

(109) Thompson, R.F.; Berger, T.W.; and Madden, J., IV. 1983. Cellular processes of learning and memory in the mammalian CNS. Ann. Rev. Neurosci. 6: 447-491.

(110) Tolman, E.C., and Honzik, C.H. 1930. Introduction and removal of reward, and maze performance in rats. U. CA. Publ. Psychol. 4: 257-275.

(111) Tsukahara, N. 1981. Synaptic plasticity in the mammalian central nervous system. Ann. Rev. Neurosci. 4: 351-379.

(112) Tsukahara, N. 1982. Classical conditioning mediated by the red nucleus in the cat. In Conditioning: Representation of Involved Neural Functions, ed. C.D. Woody. New York: Plenum Press.

(113) Ulyings, H.B.M.; Kuypers, K.; Diamond, M.C.; and Veltman, W.A.M. 1978. Effects of differential environments on plasticity of dendrites of cortical pyramidal neurons in adult rats. Exp. Neurol. 62: 658-677.

(114) Valverde, F. 1971. Rate and extent of recovery from dark rearing in the visual cortex of the mouse. Brain Res. 33: 1-11.

(115) Voronin, L.L. 1971. Microelectrode study of cellular analogs of conditioning. Proc. Intl. Congr. Physiol. Sci. 9: 188-200.

(116) Voronin, L.L. 1980. Microelectrode analysis of the cellular mechanisms of conditioned reflexes in rabbits. Act. Neurobiol. Exp. 50: 335-370.

(117) Vrensen, G., and Cardozo, J.N. 1981. Changes in size and shape of synaptic connections after visual training: An ultrastructural approach of synaptic plasticity. Brain Res. 218: 79-97.

(118) Vrensen, G., and DeGroot, D. 1975. The effect of monocular deprivation on synaptic terminals in the visual cortex of rabbits: A quantitative electron microscopic study. Brain Res. 93: 15-24.

(119) Wenzel, S.; Kammerer, E.; Kirsche, W.; Matthies, H.; and Wenzel, M. 1980. Electron microscopic and morphometric studies on synaptic plasticity in the hippocampus of the rat following conditioning. J. Hirnforsch. 21: 647-654.

(120) Wiesel, T.N., and Hubel, D.H. 1963. Single-cell responses in striate cortex of kittens deprived of vision in one eye. J. Neurophysiol. 26: 1003-1017.

(121) Wiesel, T.N.; Hubel, D.H.; and Lam, D.M.K. 1974. Autoradiographic demonstration of ocular-dominance columns in the monkey striate cortex by means of transneuronal transport. Brain Res. 79: 273-279.

(122) Woody, C.D. 1970. Conditioned eyeblink: Gross potential activity at coronal-precruciate cortex of the cat. J. Neurophysiol. 33: 838-850.

(123) Woody, C.D. 1982. Neurophysiologic correlates of latent facilitation. In Conditioning: Representation of Involved Neural Functions, ed. C.D. Woody. New York: Plenum Press.

The Biology of Learning, eds. P. Marler and H.S. Terrace, pp. 509-531. Dahlem Konferenzen 1984. Berlin, Heidelberg, New York, Tokyo: Springer-Verlag.

Spatial Cognition and Memory in Captive Chimpanzees

E.W. Menzel, Jr.
Dept. of Psychology, State University of New York
Stony Brook, NY 11794, USA

Abstract. A thoroughgoing mechanistic approach to location and motion would, by definition, assume that the question of where a particular entity is located at any given instant and why it is there rather than elsewhere is, at least in principle, answerable purely in terms of immediate or "efficient" causation and the relevant universal laws of nature. Almost all current students of animal behavior would assume that, in addition to this, teleonomic principles (e.g., function, adaptation) that apply only to specialized classes of objects (i.e., living beings) are necessary in their discipline. And a very sizable and increasing number of these investigators would argue that, even beyond this, more specifically cognitive considerations are necessary. The analysis of how chimpanzees move about an outdoor field relative to one another and to hidden objects does not necessarily "force" one into the last position, but a cognitive vocabulary certainly seems to provide the most succinct, intelligible, and comprehensive account of the available data.

INTRODUCTION

Many, if not most, searches for unity among the sciences revolve in some fashion around problems of space, time, and motion. In keeping with this and with the goal of this workshop, the present paper shall focus on what is probably the most easily stated but generic such problem that might be of common concern to almost any learning theorist, naturalist, or philosopher of science: Given a particular animal at a particular point in the world out there, where will it go next, and why does it go there rather than elsewhere? As a subsidiary to this problem I shall also consider the question of what role teleological and mentalistic concepts "should" play in the formulation of one's experiments or observations and in the description and explanations of one's data. These are obviously not new problems. They date back to the days of Aristotle. But, on the other hand, research and debate upon them has seldom been any more active than it is today.

For chimpanzees (and, of course, for many other species) our naive spatial problem obviously gives rise almost immediately to questions regarding memory, for where chimpanzees go very often seems to depend upon where they have been before, if not upon the consequences of their past experiences, and it is difficult and in some cases seemingly impossible to reduce this description to one that relies solely upon an appeal to the immediate external causal factors that are affecting the animals at the moment. Analogously, the problem of cognition arises because in some cases the animals seem to be going beyond the information to which they have had direct access and, in effect, filling in some informational gaps themselves.

Should this sound too simplistic, one may, of course, go on to ask: What would we have to do and what factors or variables would we have to consider to answer our locational question to the satisfaction of most ethologists and psychologists? What do the data tell us about "the behavior of organisms" in general? What would it take to ascertain that some other animal (say, a rat vs. a chimpanzee) is doing "fundamentally the same thing"? How does our analysis differ from that which a student of physical mechanics would perform to describe the moment-to-moment locations of a planet in the sky or a rock that is rolling down a hill? Is it reducible to problems in physics as opposed to metaphysics - as Descartes, Jacques Loeb, and, in much more recent years, Gould and Gould (7) have argued? Among the many authors who have written on these or closely related questions, I recommend especially Lorenz (14, 15), Tolman (38, 39), Mayr ((17), Ch. 2), Lewontin (13), and Taylor (33).

REMEMBERING VS. REMEMBRANCES

Few hypotheses in the domain of animal learning have generated more research and counter-theorizing than Thorndike's: "The possibility is that animals have no images or memories at all, no ideas to associate. Perhaps the entire fact of association in animals is the presence of sense impressions with which are associated, by resultant pleasures, certain impulses, and that, therefore, and therefore only, a certain situation brings forth a certain act" ((35), pp. 108-109). By and large, animal researchers have gotten past this debate. Cognitively-inclined researchers are more apt simply to assume that animals have some form of "representational system" that allows their past experiences to affect their present behaviors, and then they go on to generate and test specific empirical predictions as to how the animals should behave if this working assumption were accurate (27).

The question of whether nonhumans have memories in some very strict sense, as opposed to "representations" in the very loose and generic fashion that Roitblat (27) uses this term is, however, by no means dead. It would not be difficult to find many current authors who take a very skeptical if not negative stance here (e.g., (1, 11, 16)), and by their definition and that of Thorndike's mentor, William James, they are surely right. According to James, "(human) memory proper" must be sharply distinguished from the mere recurrence of behaviors or even images, or their dependence upon past experiences. It entails "that 'warmth and intimacy' which were so often spoken of in the chapter on Self, as characterizing all experiences 'appropriated' by the thinker as his own ... a general feeling of the past direction in time ... a particular date conceived as lying along that direction, and defined by its name or phenomenal contents, an event imagined as located therein, and owned as part of my experience..." ((8), p. 288). This implies that the memory is conscious and that the subject has language and a formal concept of time and of Self.

I do not know of any evidence to date for memory in this sense for any nonhuman species, unless one takes seriously Patterson's (24) stories about conversing with gorillas in American Sign Language regarding what they did and felt on previous dates, and securing their apologies for their past misdeeds. Indeed, I wonder how much of our own ability to cope with present or future events on the basis of past experience would meet James' standards. This should not, however, be taken as an intended slur upon his distinction or upon the value of further animal research here.

LOGIC VS. "BIO-LOGIC"

As the foregoing sections suggest, animal memory has been approached in two rather different ways. On the one hand, many investigators have attempted to define beforehand in formal, logical terms what the presumed "essence" of this capacity (or some variety thereof) amounts to; to devise a concrete empirical test situation which could in principle be used with almost any species; to specify the exact performance criteria that any animal would have to meet before one would attribute it with this capacity; and finally, of course, to collect the necessary data and render a verdict.

On the other hand, other investigators might instead first of all watch what their animals already do in whatever situations one happens to find them (preferably, but not of necessity, under as naturalistic conditions as possible); seize on whatever apparent regularities in the animals' performances seem to suggest that their current behavior is inexplicable without taking into account what they did or experienced at some time in the past; and then (if other possible explanations cannot otherwise be ruled out), introduce the least amount of experimental change into this situation that will suffice to verify or falsify the investigators' hunches. This is not to say that formal concepts, definitions, and some degree of test standardization are irrelevant here, but only that they have been put in second rather than first place.

Over the past several decades, the presumed "upper limits" of memory and cognitive capacities in nonverbal animals have increased almost by leaps and bounds, and the notion that we are dealing here with unitary and species-general, as opposed to specialized and in some cases species-specific and situation-specific, capacities has for the most part crumbled. I would attribute a very large proportion of these changes to research that has followed the second approach outlined above.

Most of the work described below follows this approach. The parochial nature of the review stems largely from my decision to confine it to spatial problems that entail overt locomotion on the part of the animals. Actually, chimpanzee research that is devoted specifically to problems of memory outside of the confines of a Wisconsin-type apparatus (which involves manual reaching to one of a few objects that are within about .3 m of one another) is to date extremely meager - especially considering the amount of field research and the number of language-training projects that have been conducted in the past two decades, and the degree to which these settings would be ideal for the investigation of problems of memory. This is not, however, to say that students of memory who examine this literature will not find much useful information.

CHIMPANZEES

Simple observation and a sufficiently complete record of chimpanzees' travels and experiences in any given situation very often suffice to show that these animals must have very good object and spatial memory. Provided that the test situation is more than a small cage and is not totally familiar to them or empty of objects, whatever reinforcers are necessary to motivate their travels and activities are already "out there," thanks to natural selection, and whatever learning sets are necessary for one-trial learning and long-term retention have already been acquired by almost any juvenile chimpanzee that has been reared in a reasonably "normal" fashion (cf., (10, 18)).

Thus, for example, when a group of wild-born captive juveniles is released for a half-hour a day into an outdoor enclosure they might be observed (18, 19) to proceed to the identical clump of grass where one of them almost stepped on a snake on their last outing and to circle around it cautiously and with piloerection, possibly throwing a stick at it; to race for the spot in which they had discovered a blackberry bush or last left a favorite toy; or to spend most of the half-hour trying to pry the cage wire apart by hand or with a stick at the same point at which they had once escaped from the enclosure. If they have not as yet fully explored the entire enclosure, they usually commence each outing by returning to some of the places they have been before, and then gradually expanding into previously unentered places - becoming notably more cautious and apt to remain in close physical proximity to each other in these latter places. After they have explored the enclosure, almost any object that is moved from its previous location or orientation (e.g., a small tree that has fallen) and almost any new object anywhere within sighting distance will be investigated. It is as if the animals have learned, and remember from one day to the next, the visual appearances and relative positions of almost every reasonably-sized item in the enclosure, whether or not individual items have ever been associated with any obvious reinforcer such as food.

Tests of the above phenomena have also been conducted by experimentally introducing test objects and controls; for more detailed summaries, see (18, 19). Although humans were obviously involved in the tests, they may be thought of, at least in my own work, as ethological models or dummies of chimpanzees. Thus, for example, the social cues my co-workers and I gave to the chimps were for the most part modelled after those we had already seen the animals using with one another, and we carried the animals around the cage rather than letting them walk not only because this eliminated "response" cues (an important theoretical consideration in some noncognitive accounts of spatial learning and memory), but also because this simulated how a very young chimpanzee might in fact travel. Species in which infants are passively transported by their parents would be "natural" candidates for experiments that hope to demonstrate learning of spatial paths even when the animal is being passively transported.

Expectancies

The first convincing primate demonstration of behavioral "expectancies" was that of Tinklepaugh (36), who worked under the direction of Tolman and Köhler; logically speaking, my test procedures were variations on his and others' laboratory research (e.g., (3)). Thus in one test food, a snake, both of these, or nothing was hidden by an experimenter at any randomly-designated place in a 30 x 122 m field, in natural cover, and then a single member of a group of six chimpanzees was carried to this place, shown its contents, and returned to the group holding cage. On more than half of the trials the hidden objects were then removed. The experimenter left the enclosure, ascended an observation tower, and (within several minutes after the chimp had seen the object) opened the door of the holding cage by means of an attached cable.

Not only did the informed animal typically lead the group on a fairly direct course to the exact place where it had seen these things (other than the empty place), but also it reacted very differently on not finding them (e.g., searching the grass manually over a very limited region if it had seen food, but using a stick if it had seen a snake, and then walking all about the area and, in some cases, probing at old snake holes that were twenty meters or more away). Other members of the group seemed to have the same expectancies as to what was out there in the enclosure and (approximately) where it was located. Thus, for example, in the "food only" condition one or more of them ran ahead of the informed animal, sighted back periodically at its trajectory, and manually checked any possible hiding place toward which it looked and ran. But they often showed almost as much caution in the tests with snakes as their leader did, and if he or she made "cautious hoot" vocalizations or an "alarm call" while staring at the hiding place, even from a distance of up to ten meters, they frequently went over and threw something at that spot. It made no difference whether the snake was actually there or not.

Indirect Cues

In some tests the informed animal was not actually shown the contents of the pile, but instead the human informant made an imitation of a chimpanzee "food grunt" or a "cautious hoot" while staring and bodily pointing at a randomly selected spot, or otherwise stared at the spot and then shrugged and went on to another place (as if there were nothing there after all). The chimpanzees were, not surprisingly, not as excited when these indirect rather than direct, or social rather than nonsocial, cues were given; but the outcome usually matched the cues given, as described above (even when we had lied and given the "snake cue" where there was food, or vice versa). The only important exception occurred if we gave our cue while orienting to a place in which, obviously, nothing could easily be hidden - e.g., a bare patch of ground. For such a cue, as for the cue that we had intended to convey "nothing there," they typically did nothing.

Social cues did not have to be given right at a hiding place or when one was very close to the informed animal in order to get it to attend to, remember, and subsequently lead the group to that place. Thus, in the extreme case, the experimenter pointed from the observation tower to a distant spot and made either "food grunts" or "cautious hoots," and the animal that had been thus informed, while up to twenty meters away from the cue-giver and farther from the target spot, still responded with fair accuracy. Not surprisingly, its accuracy deteriorated as a function of the distance. If it did not know the exact location of hidden food, its groupmates as often as not found the food; and the averaged travel paths of the group-as-a-whole were at least as accurate as the path of the informed animal itself. Typically, as soon as the latter started to slow down and scan about rather than visually orient and run in a direct line (or otherwise act as if it did not know the exact spot), its followers fanned out more, as if they could judge its overall "angle of uncertainty," and sometimes climbed trees and scanned ahead from a height. An interesting variation of this test would be to give the test animal two different directional cues, from different angles, to determine how well it could triangulate from them to the exact target location. No formal experiments were conducted here, but my hunch is that chimpanzees, if not many other nonhuman species, already employ such strategies, especially, for example, when two or more of their groupmates simultaneously stare and possibly vocalize at approximately the same region in space from different positions and angles.

Tests with Two Informed Animals

Here, two chimpanzees were each shown a (different) hidden object. Sometimes these objects were identical and sometimes they differed, one being a more preferred object (e.g., food vs. an empty hiding place, fruit vs. vegetable, two vs. four pieces of fruit, fruit vs. toy). Each animal typically did whatever it took to recruit at least one follower (until, with age and experience, they grew bold enough to take off on their own in a different direction from everyone else) and shared the spoils. About 80 percent of the time the place that held the preferred goal was reached first and the informed animal going to it attracted the larger following, regardless of its status in the "leadership" hierarchy. Almost all so-called "errors" here involved following the more preferred of the two leaders (who also, not coincidentally, was more apt to share its food).

Is Travel to the Goal Object Necessarily Direct and Immediate?

If the Euclidean shortest-line path to a hidden object contained large pools of water or physical barriers, the chimpanzees typically anticipated this and took a roundabout path. (The classic experiments on chimpanzees' abilities in various classes of detour problems are, of course, those of Köhler (10).) Even when the informed animal had received only a distal social cue as to the approximate location of the hiding place, it could home in on the designated region from virtually any angle.

Other animals could also in some cases obviously constitute obstacles to the informed animal's getting the goal-object for itself, and in dealing with them it could show equal deviousness. Lawick-Goodall (12) has reported that in some cases wild chimpanzees that have seen a piece of food might withold responding to it for a half-hour or more, until animals that are clearly dominant leave the scene. Indeed, sometimes they actually get up, lead the other animals off somewhere, and then shake them off and return to the hidden food. I observed similar events on innumerable occasions, and under conditions where the locations of the food were varied at random and where (alternatively) very different behaviors could also be produced experimentally, at will. For example, if we had shown Gigi several piles of food and Shadow discovered one, he virtually never grabbed it up, but instead first looked around, and only if Gigi was more than about ten meters away and looking in some other direction would he reach for it, and then only slowly; and he would typically not eat it until having first carried it off some distance. If, however, Belle or Bandit were the informed animal, he simply grabbed the food and ate it on the spot (and they did not scream, chase, and beat up on him for taking "their" food, either). Or, again, if Belle were the informed animal, she herself came to sit down and groom herself ten meters or so away from the food, or to lead the group in a totally inaccurate direction if her nemesis, Rock, a much larger and relatively unfamiliar male who usually chased her from "her" food and took it all, was with the group; but, alternatively, she showed her typical "altruistic" and errorless performance if this male were locked in the holding cage for this trial. Not only did she know the nature and location of the hidden objects and perform behaviors that conveyed this information to her groupmates, but she could also, if necessary, withold much of this information.

Some of the findings in this and previous sections have been elaborated upon by other investigators (e.g., (26, 29, 31, 40)), in very different contexts and with emphasis upon the specifically communicative aspects of performance.

What Are the Invariants in the Above Behaviors?

Ordinarily, the informed animal's angle of departure from the release point was within ten degrees of the angle at which the hidden object was located and (over almost any block of ten or more trials) correlated .90 or higher with this datum. However, it would be fanciful to assume that a completely "mechanistic" account of the chimpanzees' locations and motions could be given, for this would be to imply that all causation was, in Aristotle's terminology, "efficient," i.e., that the data are in principle completely deducible from the antecedent and contemporaneous conditions, together with the relevant universal laws of nature.

What the relevant universal laws of nature are here is anyone's guess. It seems to me that relatively loose concepts and principles which apply at best to living beings, if not to a very limited class of living beings, are far more informative and pertinent; and it would be difficult to express them intelligibly or concisely without resorting to a teleonomic if not cognitivistic and purposivistic terminology. In other words, overall, the only invariant feature of the animals' travel paths and probably the most invariant feature of their other, more molecular behaviors was that these resulted in the animals' getting to whichever hidden objects they valued the most without at the same time having to get too far from their companions or risk losing all of the food to (or getting beat up by) their companions. Almost any behaviors of which juvenile chimpanzees are capable were used here more or less interchangeably. If one tried to state these facts from the outset in a purely "mechanistic" vocabulary most audiences, whether biologists or psychologists, would probably react by asking, "What are you getting at? What do the data mean?"

Length of delay was not varied systematically, but I would estimate that after direct sight of food, delays of up to 24 hrs would constitute no problem under these conditions (cf., (10, 41)), and that a social cue of direction could probably be retained for a half-hour or more. If the hidden objects affected the animals in a fashion that were analogous to the way in which a magnet might attract iron filings, one would obviously have to explain why only one of the iron filings in question (the so-called informed animal) was directly affected by the force field, and how the temporal as well as spatial "action at a distance" was possible. The old concept of "telotaxis" could of course be invoked, but it is an open question as to whether it predicts the behaviors in question rather than simply describing them post hoc.

The Travelling Salesman Problem

Perhaps the crucial issue so far as cognitive theorizing is concerned is how the animals are organizing the information that they pick up, and what level of integration we must invoke to explain their performances. From the standpoint of biological adaptation it is, of course, a matter of supreme indifference precisely how various animals solve a given sort of problem (e.g., by single, specialized receptor cells or by more complicated circuits; by a single sensory modality or by transfer of information between modalities; by nonsocial or social strategies); but cognition would ordinarily not be invoked unless the animal could in effect go beyond the information given and put together discrete bits of information, and from them come up with a solution that one could not readily predict from these "local facts" alone. The following test, which was based on those of Tinklepaugh (37) and Tolman (39), is one way of getting at this question.

A single member of a group of six chimpanzees was carried about the field and allowed to watch as an experimenter hid up to eighteen pieces of food in natural cover. On each trial a different set of hiding places and a different route was used. Following this procedure, which took about ten minutes, the experimenters turned the whole group loose, as usual. Not only did the informed animals find virtually all of the food (the uninformed group members scoring only by searching where the informed animal did, or begging from it directly), but they also took a route that bore no detectable relationship to that along which they had been carried, and which was not vastly less efficient in terms of its overall travel distance than that which they might have followed if all of the foods were visible at the time of response. Using computer simulation methods, the probability of selecting a route as short as the observed one (from among all possible itineraries between the several points) can be assessed separately for each trial. For no trial involving eighteen hiding places was this p value greater than one out of one thousand. It is most unlikely that the animals took into account either only a single hiding place at a time or all places simultaneously. Thus, for example, several hiding places that were clumped fairly close together were approached sooner than a single hiding place that lay in the opposite direction but was closer. And, in some cases, the animals actually stepped on one of the hiding places in the process of running to a different one, and then returned to it some time later.

If the informed animals found all of the food that we had shown them, they almost invariably quit forthwith and either laid down for a nap or commenced to play. And if they had themselves emptied any given place of its food, they very seldom returned to it. Places that they had searched without actually finding the food or that had already been emptied by a group-mate were, however, often re-searched. They very seldom searched any location that was more than a meter or two from an actual hiding place, whereas the uninformed controls did so on an uncountable number of occasions. Their most common "error" here was searching a hiding place that looked to us almost precisely like a "correct" one and was within a few meters of it. Obviously the chimpanzees did not know its exact location on the basis of a single cue system, but seemed to use a different set of cues for getting to its general locus in the field and for then homing in on it precisely. The concept of cognitive mapping is not the only possible explanation of the data, but I can think of none that seems any more succinct and able to encompass all of the details of performance. In other words, the animals seemed to know the nature and relative positions of the objects in question (including, from one moment to the next, their own).

Judging especially from our own difficulty in refinding the piles of food that we ourselves had hidden, even when we had a map that showed these locations to the nearest nine m^2, the animals' memory of exact locations seemed at least as good as our own. Tinklepaugh's (37) indoor laboratory study, which includes some comparative data, and Teleki's (34) attempts to match the skill of wild chimpanzees at locating termite mounds and the best places to "fish" on each, suggest the same thing. Possibly chimpanzees rely here on something like an eidetic image, as early investigators thought.

Recall

Perhaps the most impressive single performances I saw were the occasional instances when, after having eaten most of the food and lain down with their eyes closed for up to thirty minutes, an informed chimpanzee suddenly got up and, without any preliminary visual scanning about, walked straight to an unemptied hiding place that lay twenty meters or more away. If one may assume that it knew where the food was even while still flat on its back, with its eyes closed, where was the "immediate stimulus" for that? To be sure, I am talking here about a few isolated anecdotes; but if wild adult chimpanzees cannot be demonstrated to show such apparent recall in the morning in their sleeping nests, it is likely to be more a reflection of our own lack of investigative skills than in their lack of memory.

The ongoing research of S. Savage-Rumbaugh (personal communication) is highly pertinent here. She is testing an infant pygmy chimpanzee (Pan paniscus) in a heavily-wooded area of more than twenty-five hectares, and her animal, Kanzi, has learned many arbitrary symbols that refer to different classes of foods and other objects. All of Kanzi's rations are placed in the morning at various, widely distributed points in this area and are thus, of course, totally invisible from his sleeping quarters. Ordinarily, each type of food is in the same place from one day to the next. It is up to Kanzi to request a given type of food before a human companion takes him there. Not only can he do this, but even if his human companions do not themselves know where this place is located, he can direct them there. He rides on their shoulders, manually pointing down the appropriate trail at each of many and varied choice-points. From the standpoint of "memory proper" (see above), it would be most interesting to see if he can also indicate where he has been and what he has done after such an outing. I have never understood why a concept of "the present" should necessarily be thought of as any less sophisticated than concepts of the past or the future; indeed, the distinction between these terms rests on viewing "the present" as a single, infinitely divisible, static point rather than an interval - which is against the nature of all actual biological and psychological functioning (6, 8, 9) - but, regardless of how the performances in question might be interpreted, they will be of great interest.

Use of Televised Cues

On some occasions, the "uninformed" chimpanzees in my studies managed to poke a small peephole through the wall of their holding cage and get a view of us as we carried their groupmate over to a hiding place; this was instantly detectable by the fact that, instead of waiting to get their cues from others, they raced ahead in the appropriate direction on their own. What is the difference, in principle, between such a monocular and distal view of the scene and, say, the perception of a movie or photograph that might be shot from the same angle? And might it not be possible to use movies in the same fashion as students of vocal-auditory information processing conduct "playback" experiments, to analyze more precisely the nature of animals' visual systems and their representational capacities?

In the first such study (20), a wild-born infant chimpanzee (or, in an alternative condition, two of them) were held by an experimenter behind a small blind while a familiar caretaker held a pan of food in front of a window in the blind for a moment and then walked out into the enclosure and hid in the tall grass. Through the window the animals could, alternatively, either get a direct view of this procedure, a view of a small black-and-white television monitor which showed a "live" picture of the scene, or nothing. Even though the animals had had no known prior experience with television, their success in finding the hidden person on the basis of video information was (from the start) better than one would expect by chance, albeit much poorer than that in the "direct view" conditions. In subsequent food-finding (rather than person-finding) tests their performance initially deteriorated to chance, then eventually reached at least 85% "correct" for blocks of fifty or more trials. Premack and Premack ((25), p. 101) have more recently claimed that chimpanzees are incapable of solving precisely these same sorts of tests; but they fail to cite the above study or any other relevant data, and I therefore believe that they are mistaken.

That chimpanzees, or at least two very test-wise adult males, can utilize video information as a representation of a portion of their own body (cf. (4, 5)) as well as a portion of their living environment, has recently been demonstrated by Menzel, Savage-Rumbaugh, and Lawson (submitted for publication). The chimpanzees, Sherman and Austin, sat in one room and had visual access to a color monitor which showed them a picture of the backside of a door in an adjoining room. In this door there was a hole through which they could reach. A small spot of ink was placed at random anywhere within their arm's reach on the back of the door; they could see it only via the television. After they learned this problem (which was a very easy one for them), a variety of control and transfer tests were conducted. In one of them, the video picture was at random alternately reversed laterally, inverted 180 degrees, subjected to both of these distortions, or left "normal" - so that there was zero correlation between the perceived locus of the target and its actual locus, and the only way to get to it would be by tracking their hand on the video screen and moving it in whatever way it took to bring its image into contact with that of the target spot.

Both animals readily solved this problem. They also discriminated, from their first trial, between two simultaneously presented monitors, one of which showed a live image and the other of which showed a tape. If this tape was a commercial television program or their own last test session, they seldom even reached through the hole, but turned instead to the other monitor. Otherwise (e.g., if the tape showed the door alone), they stuck their hand through the hole for a second. If they had chanced to pick the monitor with the tape, they turned to the other screen, whereas if they had picked the live image, they proceeded to reach to the target spot. We tested ourselves on the same task and could not discriminate when we touched the spot except via visual information from the monitor.

Four rhesus monkeys whom we tried to "shape" to perform a much simpler version of this task, which used a high-quality mirror instead of video, readily recognized different targets (in their case, food or snake), but they never grasped where these objects were located in the real world, unless given a number of successive trials using only two locations - in which case, of course, they could learn the "spatial" aspects of the task by rote.

We interpret these data as further evidence (cf., (5)) for substantial differences between other primates vs humans and the great apes. We also concur with Köhler (10) and Lorenz (14) that the basis of these differences might perhaps best be disclosed by examining more precisely the organization of visual spatial perception and of the visual cortex. That differences as well as similarities between chimpanzees and humans are to be expected here seems obvious.

Can Chimpanzees Read "Real" Maps?

This question is, of course, a most anthropocentric one, but that does not make it any less interesting. One way to get at it would be to successively render a television picture into representations that are more and more abstract, reduced in size, and so forth. Woodruff and

Premack (as cited in (26)) have tried an even more straightforward technique: chimpanzees were confronted with two identical laboratory rooms, shown which of two containers in one of them contained food, and then permitted to enter the other room, which was baited in the same fashion. Then, once the animals were performing well here, a canvas was placed on the floor of the first room and treated as a "map" of the floor. It was gradually reduced in size until it was no larger than an ordinary map - the positions of the food containers also being moved in closer and reduced in their size.

The animals apparently learned to perform respectably here, but when the "map" was rotated and a transfer test involving a new map and a new room was given, their performance intitially dropped to chance. Premack and Premack (25) conclude that despite all of their training, the chimpanzees had really accomplished very little. It seems to me that this is judging them by rather anthropocentric standards.

At least in principle it should be quite possible for investigators to create for any visual animal situations in which a three-dimensional model or even a photograph is indiscriminable from the scene that is represented therein. While such perceptual objects are obviously not as such "representations" in a psychological sense, maybe they would be the best point from which to embark upon a study of how far a "shaping" study can go, and precisely where and for what reasons the performance of any given species eventually breaks down irreparably.

DISCUSSION

Are Cognitivistic Interpretations Necessary?

As Descartes said, "There are no Archimidean fixed points except insofar as we choose to treat them as fixed by our own thought" ((2), p. 266). Thus, where any given object or living being "really is," and why it is there rather than elsewhere, are indeterminate until one specifies the "fixed point" or (more generally) the point of view upon which one's descriptions or explanations are to be anchored. Which point of view one adopts depends in considerable measure upon what biological or psychological status one is willing to confer upon one's subject.

Since the time of Descartes, and even today (e.g., (7)), it has often been argued that animals should be accorded the status of "preprogrammed robots." No one to my knowledge has, however, been completely consistent here. No one today would, for example, explain a robot's actions in terms of its past evolutionary history or in terms of how its actions function to help it make its living in the world at large or to enhance the fitness of its genotype; explanations from such points of view are ordinarily reserved only for entities that we presume to be alive and to be products of Darwinian natural selection rather than overnight creation by an engineer. Nor would Darwin himself have been very likely to describe a robot's actions in mentalistic terms (or to call an animal's actions "robot-like") unless he was deliberately speaking in metaphors, for, in his own words, "Mind is a function of body - we must have some stable function to argue from." A Darwinian rationale for using mentalistic terms for the description of the actions of nonhuman animals rests, I presume, on three basic assumptions: a) Virtually all human observers believe that these terms are warranted in their own case. b) The apparent similarities between what they do and what other beings do are not merely hypothetical and metaphorical but stem in some measure from similarities in biological structures, functions, causal mechanisms, and developmental and (ultimately, and most importantly) evolutionary origins. c) Robot metaphors for the description of animal behavior may be justified insofar as they might generate novel predictions and empirical findings, which might not necessarily occur to one otherwise. The metaphors that cognitivists employ may, however, be justified on precisely the same grounds, and if anything, they probably come off better on this score. Indeed, from a biological point of view the various similarities that do exist between the actions of robots and the actions of living beings might be more tautological than profound, for the former were for the most part deliberately and consciously designed to simulate the latter, according to the designers' own (largely phenomenological) criteria of "similarity."

Speaking more generally, any animals that can make their living in the world at large in these days are almost by definition exceptional individuals, if not "geniuses" of some sort. The question so far as current ethology is concerned is, what sorts of problems do they face in the process and what strategies do they use for solving them? How cognitivistic an interpretation we place on this information is, in turn, largely a function of the perceived overall similarity between animals' actions and our own, and the relevant taxonomic criteria of overall similarity have yet to be fully biologized. There is no single point at which all investigators will feel "forced" to invoke cognitive concepts, and even individual thresholds in this regard are variable rather than constant. My own threshold in the case of chimpanzee "cognitive mapping" is a case in point, for it took any number of experiments to convince me, and it is probably impossible to rule out all other conceivable alternative interpretations.

What Is Special About Chimpanzees?

Students of nonhuman primates are notable for the frequency with which they infer from their own data that the differences between their subjects and humans are "merely matters of degree rather than of kind," and yet, at the same time, take umbrage at the suggestion of other researchers that the basic kinds of abilities to which they are referring are by no means unique to primates, but may also be seen in rats, pigeons, or bees. Even at the risk of fitting this stereotype, it is worth noting some of the similarities and differences between what chimpanzees do and what rats do on "radial maze" tasks (22, 23, 27, 32).

The radial maze task, like the chimpanzees' "travelling salesman problem," involves the placing of food in several different locations (in this case at the end of each of several - up to 17 - arms which radiate from a central, circular, choice-point like spokes from a wheel), and then examining how efficiently the animal in question organizes its travel path to these several locations and avoids going back to any given now-depleted food sources more than once without relying on any obvious intra-maze cue such as odor or any simple behavioral strategy (such as always responding to the various locations in a fixed sequence) that could not be said to entail "memory." Beyond this superficial characterization, there are innumerable differences in the chimpanzee task and Olton's task, which makes a detailed comparison of the data questionable. But there can be little doubt that the rats can be said to remember a large number of different locations and to organize their foraging route in a highly efficient and by no means stereotyped pattern. Anyone who has been convinced that chimpanzees' overt performances warrant our inferring cognitive mapping on their part would be less than pure empiricists if they hedged at making analogous inferences from the overt performances of Olton's rats. Thus, for example, in an eight-arm maze, rats typically make up to seven moves before going up any arm for the second time. Apparently, extra-maze cues are used as their landmarks here, for if, after visiting four arms, the rat is restrained on the center choice-point while the maze is rotated in space, it will revisit these arms, but only if they are in the same position (with respect to distal features of the room) as arms that have not yet been entered. No single such distal feature accounts for their orientation. Suzuki, Augerinos, and Black (32) demonstrated this quite clearly with an experiment in which a maze was surrounded by a dome on which various easily discriminable visual stimuli could be fixed to serve as potential extra-maze cues. In a series of carefully designed tests, this array was changed in various ways. Changes in any one stimulus element or in only a few of them produced little or no effect on the efficiency of the rats' performance, whereas large changes seemed to disorient them completely and cause them to act as if they were in a novel maze. It was as if the animals were using the whole array as a means of orienting their own representation of the test room.

This does not, of course, mean that rats and chimpanzees get about in the world in "fundamentally the same fashion." To say that both species "remember" or "have cognitive maps" could be saying little more than that our own verbal habits and behavioristic logic are such that we see no reason not to use the same vocabulary in each case, and what commonalities exist at the level of biological mechanisms remain for the most part to be determined. It is most unlikely, for example, that rats recognize their own images in reflections, read maps in a literal rather than metaphorical sense, or readily adapt to displaced visual images, as in the video experiment done with chimpanzees. It is an open question, too, how they would perform on a delayed response task that involved using a different set of locations on each and every trial, or even whether or not they would quit responding entirely to any arm of a radial maze once they had emptied all arms of food, rather than starting their whole routine all over again. For a substantial portion of the subtopics that I have discussed in this paper, I am not aware of any directly comparable studies on rats, let alone comparable performances. (The same thing could, of course, be said about the contrasts between chimpanzee and human studies of memory.)

This should not, however, be taken to imply that these species "should" be tested in the same fashion, or that the way in which chimpanzees solve their problems is somehow "better" than the ways that other species employ. Why is it ordinarily so much easier to get away with a cognitive interpretation of chimpanzees than of other species? In good part it is because of what we know or assume about how the animals in question solve their respective problems, how they came to be capable of their feats, and how well they could apply those same skills to a broad range of situations, including those for which they were not highly "prepared" (28, 30). However, no purely behavioral evidence is a sufficient index of cognition, independent of further information regarding the animals' natural history, biological makeup, and presumed phyletic status - otherwise, one might well have to attribute cognition to robots or even rocks. For most investigators, the question is not whether one should draw some a priori line between different classes of beings on the basis of nonbehavioral evidence, but principally where and how one should draw that line.

Are Cognitive Maps Necessary?

I prefer to use the term "cognitive mapping" rather than "cognitive maps" because it is not a noun and does not necessarily imply any internal object or entity that the animal in question must, figuratively speaking, scan in addition to looking at what it can perceive directly. As a shorthand way of stating that animals take into account the nature and relative positions of objects, including themselves, and can in effect deduce new S-S relationships and perform new S-R performances on which they were not specifically trained by the investigator, I do not see any better principle. It also generates many more specific questions, both at the behavioral and at the physiological level of analysis, that might not necessarily occur to one otherwise (see especially (21, 22)). At the same time, I would not be confident that any of the chimpanzee data would be inexplicable in terms of Gibson's (6) concept of the direct perception of structure. An obvious problem with map or mapping analogies is that once one indulges in them there is no reason in principle not to talk also about cognitive clocks, compasses, thermometers, lie-detectors, and all manner of other gadgets. Gibson was surely correct in pointing out that there must be a much simpler and more direct and monolithic way of getting around in the world than this. When asked to do so, many if not most human beings can, of course, sit down and introspectively conjure up a "map-like" image of where they are at present or where they were at some earlier date. It is, however, a completely open question as to what such an introspective ability has to do with the active process of locomoting about in the world, or vice versa (especially but not exclusively in the case of nonhuman species).

"Space" and "time" (and present, past, and future) are, by the same token, not conditions in which animals live but concepts that philosophers and scientists have invented to describe the conditions in which they and other animals live. Animals must obviously cope with these conditions if they are to survive and reproduce, but I would think that their capability to do so was a necessary precondition for the evolution of concepts of space and time, rather than vice versa. Unless one either believes in creationism (as Kant did) or attributes some sort of concepts of space and time to virtually every living being, the Kantian doctrine as to the a priori necessity of such concepts is thus debatable. So, too, is the Aristotelean and Jamesian doctrine that only those creatures that have a concept of time can possibly have "real" memory.

Is Memory Necessary?

Gibson argued in effect that memory is not necessary at all, and he and others of similar persuasion (9) have been highly astute in showing us how much information is in fact available "out there" in the present for animals to pick up by what he calls direct perception. (In this scheme of things "the present" is, of course, conceived of as an interval of time rather than a point, and it is an open question as to how long an interval one would have to have before leaving the domain of perception.) There is, however, a flaw or at least a gap in this argument. Insofar as Newton was right, one could argue on physical grounds alone that sufficient information for solving all of one's problems is available "out there," somewhere and somehow, at any given instant - for according to Newtonian theory no event fails to leave some trace; only the entire universe forms a closed system; and it runs according to deterministic, universal, and fundamentally simple laws. Gibson was quite correct, then, in saying that presently available information is sufficient in principle, and that memory is unnecessary if, in fact, we have the skill to pick up this available information. But to complete the account one would have to add that, by the same token, perception is unnecessary: any past moment in time also contains all the information that one could possibly ask for, and (if the universe is a totally deterministic system) there is, accordingly, no reason in principle why some creature could not make do purely on the basis of information that was coded into its genes by its ancestors. And, of course, we should not forget that one could make an analogous "in principle" argument as to the potential sufficiency of foresight and precognition, and the non-necessity of both perception and memory.

What is, in brief, necessary for all animals in general is not any particular mechanism of information processing, but some mechanism or set of mechanisms that suffices for the most important problems they face and is not overly costly, biologically speaking. Given what is known about "biological constraints" on various types of information processing mechanisms within particular species of animals, we may presume that in many cases mechanisms of direct perception are less costly than memory; but I know of no evidence that this would necessarily be true for all species - present, past, and future - in general. Whatever general rules apply here to DNA itself, in any and all of its manifestations, remain to be discovered.

REFERENCES

(1) Bennett, J. 1976. Linguistic Behaviour. Cambridge: Cambridge University Press.

(2) Descartes, R. 1955. The Philosophical Work of Descartes (translated by E.S. Haldane and G.R.T. Ross). New York: Dover.

(3) Fletcher, H.J. 1965. The delayed response problem. In Behavior of Nonhuman Primates, eds. A.M. Schrier, H.F. Harlow, and F. Stollnitz, vol. 1, pp. 129-165. New York: Academic Press.

(4) Gallup, G.G. 1970. Chimpanzees: self-recognition. Science 167: 86.

(5) Gallup, G.G. 1982. Self-awareness and the emergence of mind in primates. Amer. J. Primat. 2: 237-248.

(6) Gibson, J.J. 1979. The Ecological Approach to Perception. Boston: Houghton-Mifflin.

(7) Gould, J.L., and Gould, C.G. 1982. The insect mind: physics or metaphysics? In Animal Mind - Human Mind, ed. D.R. Griffin. Dahlem Konferenzen: Berlin, Heidelberg, New York: Springer-Verlag.

(8) James, W. 1890. Principles of Psychology. New York: Holt.

(9) Johansson, G.; Hofsten, G. von; and Jansson, G. 1980. Event perception. Ann. Rev. Psychol. 31: 27-63.

(10) Köhler, W. 1925. The Mentality of Apes. New York: Harcourt.

(11) Langer, S.K. 1972. Mind: An Essay on Human Feeling, vol. 2. Baltimore: Johns Hopkins University Press.

(12) Lawick-Goodall, J. van. 1971. In the Shadow of Man. Boston: Houghton-Mifflin.

(13) Lewontin, R.C. Darwin's real revolution. 1983. N.Y. Rev. Books 30(10): 21-27.

(14) Lorenz, K.Z. 1971. Studies in Animal and Human Behavior. Cambridge: Harvard University Press.

(15) Lorenz, K.Z. 1981. The Foundations of Ethology. New York: Springer Verlag.

(16) Marshall, J.C. 1982. A la representation du temps perdu. Behav. Brain Sci. 5: 382-383.

(17) Mayr, E. 1981. On the Growth of Biological Thought. Cambridge: Harvard University Press.

(18) Menzel, E.W. 1974. A group of young chimpanzees in a one-acre field. In Behavior of Nonhuman Primates, eds. A.M. Schrier and F. Stollnitz, vol. 5, pp. 83-153. New York: Academic Press.

(19) Menzel, E.W. 1978. Cognitive mapping in chimpanzees. In Cognitive Processes in Animal Behavior, eds. S.H. Hulse, H. Fowler, and W.K. Honig, pp. 375-422. Hillsdale, NJ: Erlbaum.

(20) Menzel, E.W.; Premack, D.; and Woodruff, G. 1978. Map reading by chimpanzees. Folia Primat. 29: 241-249.

(21) Nadel, L., and Willner, J. 1980. Context and conditioning: a place for space. Physiol. Psychol. 8: 218-228.

(22) O'Keefe, J., and Nadel, L. 1978. The Hippocampus As a Cognitive Map. Oxford: Clarendon Press.

(23) Olton, D.S., and Samuelson, R.S. 1976. Remembrance of places passed: spatial memory in rats. J. Exp. Psychol.: Anim. Behav. Proc. 2: 97-116.

(24) Patterson, F., and Linden, E. 1981. The Education of Koko. New York: Holt, Rinehart and Winston.

(25) Premack, D., and Premack, A.J. 1983. The Mind of an Ape. New York: Norton.

(26) Premack, D., and Woodruff, G. 1978. Does the chimpanzee have a theory of mind? Behav. Brain Sci. 4: 515-526.

(27) Roitblat, H.L. 1982. The meaning of representation in animal memory. Behav. Brain Sci. 5: 353-406.

(28) Rozin, P. 1976. The evolution of intelligence and access to the cognitive unconscious. In Progress in Psychobiology and Physiological Psychology, eds. J. Sprague and A. Epstein, pp. 245-280. New York: Academic Press.

(29) Savage-Rumbaugh, E.S.; Rumbaugh, D.M.; and Boysen, S. 1978. Linguistically-mediated tool exchange by chimpanzees (Pan troglodytes). Behav. Brain Sci. 4: 539-554.

(30) Seligman, M.E.P. 1970. On generality of the laws of learning. Psychol. Rev. 77: 406-418.

(31) Seyfarth, R.M.; Cheney, D.L.; and Marler, P. 1980. Monkey responses to three different alarm calls: Evidence for predator classification and semantic communication. Science 210: 801-803.

(32) Suzuki, S.; Augerinos, G.; and Black, A.H. 1980. Stimulus control of spatial behavior on the eight-arm maze by rats. Learn. Motiv. 6: 77-81.

(33) Taylor, C. 1964. The Explanation of Behaviour. London: Routledge and Kegan Paul.

(34) Teleki, G. 1974. Chimpanzee subsistence technology: materials and skills. J. Hum. Evol. 3: 575-594.

(35) Thorndike, E.L. 1898. Animal intelligence. An experimental study of the associative processes in animals. Psychol. Monogr. 2(8).

(36) Tinklepaugh, O.L. 1928. An experimental study of representative factors in monkeys. J. Comp. Psychol. 8: 197-236.

(37) Tinklepaugh, O.L. 1932. Multiple delayed reaction with chimpanzee and monkeys. J. Comp. Psychol. 13: 207-243.

(38) Tolman, E.C. 1938. Determiners of behavior at a choice point. Psychol. Rev. 45: 1-41.

(39) Tolman, E.C. 1948. Cognitive maps in rats and men. Psychol. Rev. 55: 189-208.

(40) Woodruff, G., and Premack, D. 1979. Intentional communication in the chimpanzee. Cognition 7: 333-362.

(41) Yerkes, R.M. 1943. Chimpanzees: A Laboratory Colony. New Haven: Yale University Press.

Standing, left to right:
Bob Bolles, Herb Terrace, Sam Revusky, Jean-Pierre Changeux, Kazimierz Zielinski, Emil Menzel, Tomasz Werka, Josef Rauschecker.

Seated, left to right:
John Gibbon, Wolf Singer, Peter Holland, Sara Shettleworth, Mort Mishkin.

The Biology of Learning, eds. P. Marler and H.S. Terrace, pp. 533-551. Dahlem Konferenzen 1984. Berlin, Heidelberg, New York, Tokyo: Springer-Verlag.

Biology of Learning in Nonhuman Mammals
Group Report

P.C. Holland, Rapporteur
R.C. Bolles
J.-P. Changeux
J. Gibbon
E.W. Menzel, Jr.
M. Mishkin
J.P. Rauschecker
S. Revusky
S.J. Shettleworth
W. Singer
H.S. Terrace
T.F. Werka
K. Zielinski

INTRODUCTION

Given the uncertainty about what was to be reconciled with what, our first order of business was to outline various notions of learning theory and of natural behavior, and then to consider questions of reconciliation or interaction. Subsequent discussions a) distinguished between learning as a general process and as a set of particular specialized adaptations, b) examined the study of cognitive processes and complex learning, c) considered the role of neurobiology in providing a common ground for learning theory and natural behavior approaches and in providing reasonable working mechanisms for learning and memory phenomena, and d) compared plasticity in development and learning.

WHAT IS LEARNING THEORY?

Not since the days of Hull (5) has there been a monolithic body of theory and data that could be simply identified as "learning theory." With some exceptions, modern learning theories tend to be constructed to account for relatively small sets of behavioral phenomena. Again with some exceptions, those theories make use of intervening variables couched in behavioral or cognitive rather than neurobiological terms.

Theories of learning have often been tied to the procedures of learning experiments. Thus, we have had theories of classical conditioning, theories of instrumental learning, theories of habituation, and so forth. Particular classifications of "kinds of learning" may encourage very different speculation about learning phenomena. In fact, many such catalogs gave fairly exclusive lists that tended to be as divisive as they were useful: some phenomena, such as behavioral sensitization and habituation, developmental changes, and imprinting were ignored earlier as not warranting the title of "learning," and by implication, not deserving serious analysis.

Before considering a few representative distinctions among theories of learning, we examined a number of classifications of learning procedures. The simplest one was based entirely on the operations of the experimenter. Three kinds of operations can be distinguished: the presentation of single events, the arrangement of relations among events, and the arrangement of relations between subjects' behavior and other events (10). Examples of traditional learning paradigms that fall in these broad categories are, respectively, habituation, in which a single stimulus is repeatedly presented by itself, classical or Pavlovian conditioning, in which one stimulus (say, a bell) is repeatedly paired with another stimulus, such as food, and instrumental or operant conditioning in which a stimulus such as food is presented only if the subject performs a particular behavior.

It is worth pointing out what is not specified by this classification. It does not specify how learning is to be assessed, it makes no distinctions about the nature of the events, relations, or behaviors involved, it does not identify outcome, and it does not classify process. Consequently, this classification covers a multitude of sins and rules out virtually nothing as an example of learning. Clearly, some sort of narrowing of such a classification scheme is necessary, but it is important to recognize that such narrowing may best come about after fairly extensive empirical and theoretical progress.

It is especially important to remember that procedure does not identify process. For instance, many theories of learning have distinguished between Pavlovian and operant conditioning at some level. But the arrangement of Pavlovian (stimulus-stimulus) relations does not guarantee that the subject's behavior will be governed by those relations, nor does the arrangement of instrumental (response-stimulus) relations insure that its behavior will be controlled by those relations (16). For example, restless activity of a dog during the conditioned stimulus in a Pavlovian conditioning experiment will be followed by the delivery of food. Thus, an unintended relation between the dog's behavior and the food has been arranged. Consequently, the dog may acquire such activity as part of its response to the conditioned stimulus because of an adventitious relation between that activity and food delivery. It is worth noting, however, that there is little evidence directly supporting such a view (12). Conversely, consider a rat that receives food at the end of an alley. Although the experimenter may assume he has arranged a response-stimulus (run-food) relation, the rat may be learning a Pavlovian stimulus-stimulus (end of alley-food) relation. If, as many data (Jenkins and Hearst, both this volume) suggest, rats approach stimuli paired with food as a Pavlovian-conditioned response, then our subject's "instrumental" behavior may well be Pavlovian in origin.

The point of these examples is that the question as to whether operant or Pavlovian conditioning has occurred cannot be addressed simply by pointing to the procedure used, but rather requires us to determine which relations govern the subject's performance. A variety of procedures for accomplishing this task have been used extensively in recent years. In many cases, the most accurate conclusion is that subject's behavior in conditioning experiments falls under the control of both sets of contingencies to varying degrees.

The importance of demonstrating which set of relations control a particular organism's behavior depends on the importance one assigns to the distinction between operant and classical conditioning. If they are viewed as two fundamentally different kinds of learning involving very different processes, then the question is essential. Increasingly, however, operant and Pavlovian conditioning are viewed as involving similar associative learning processes, but differing in the contents of that learning, that is, the elements that are associated (e.g., (1) and Jenkins, this volume). For instance, Pavlovian conditioning may involve S-S associations between internal representations of the CS and US, but operant conditioning may involve R-S associations between the response and the reinforcer and/or S-R associations between the discriminative stimulus and the response as well (1). Although considerable evidence suggests substantial commonality of behavioral mechanism between the associative processes involved, substantial disagreement still exists.

Most learning theories can be characterized as providing answers for three questions about learning (10). First, what are the conditions necessary for the establishment of learning? Second, what are the contents of learning, i.e., what is learned? And third, how is learning translated into the performance of observable behavior? Through the years learning theories have suggested a variety of answers to each of these questions. For instance, controversies raged over whether reinforcement was necessary for learning, whether organisms formed associations between stimuli and responses (S-R theories) or between stimuli and stimuli (S-S theories), and whether performance was generated by simple reflex action or with the intermediation of less easily specifiable but more flexible processes.

Rather than rehearse the history of these and other distinctions, it may be valuable to mention briefly some characteristics of the associative learning theories that are guiding much current behavioral research in learning. Recent theories seldom demand that particularly potent events normally termed reinforcers or unconditioned stimuli are necessary for learning to occur: there are many demonstrations of very substantial learning in the absence of such events. But most theories do demand that both elements to be associated (as, for example, a CS and US in classical conditioning) be well processed by the organism (1). This stricture is not trivial because many recent learning theories propose that experience-dependent changes in processing of conditioned and unconditioned stimuli are major determinants of variations in conditioning. More precisely, those theories assume that the extent to which an organism processes a CS or US event depends on the extent to which that event is predicted by other events, or on the extent to which that event predicts other events. Such an assumption has permitted the prediction of a wide variety of stimulus selection phenomena.

As discussed in more detail in the section on COGNITIVE PROCESSES AND COMPLEX LEARNING, a number of experimental operations distinguish between learned behavior that is mediated by internal representations of properties of events involved in conditioning procedures and that which is not. The question of S-S versus S-R is often recoded to mean, are properties of the reinforcer coded in learning to the extent that they mediate performance, or not ((1, 11) and Jenkins, this volume)? Consider a rat that receives pairing of a tone with food, and then later receives pairings of the food alone with a toxin, so that the rat will no longer consume the food. If the rat's performance of conditioned behavior to the tone is mediated by the rearousal of stored properties of the food, then the modification of those properties by the subsequent food-toxin pairings would produce a spontaneous reduction in conditioned behavior to the tone on the very first presentation of that stimulus after food-toxin trials. In a variety of such experiments, that outcome has been obtained (1, 11). Furthermore, some procedures generate learned performances that are immune to so-called post-conditioning reinforcer devaluations (1, 11). Thus, learned performance that is mediated by internal representations of detailed features of reinforcers and learning that is not so mediated can be distinguished. This distinction captures much of that embodied by the traditional S-S vs. S-R distinction without begging the question of the identification of stimuli and responses.

Two key concepts in modern learning theory are those of association and representation. It was clear that each of us had somewhat different uses of those terms. Concepts of association ranged from connections between neural units, to internal representations of relations that exist among environmental and organismic events, to an internal expression of the causal nature or texture of the environment, to a formal intervening variable which simply serves to integrate a variety of experimental data in conditioning. Some of us voiced the opinion that the de facto equation of learning with the formation of associations at any level of discourse limits us too much, especially by restraining our view to pairwise rather than multi-term relations among events. Nevertheless, the association remains the primary construct of most learning theories. The concept of representation is discussed in the section on COGNITIVE PROCESSES AND COMPLEX LEARNING (below).

A difference of opinion arose over whether lists of learning phenomena and theories of learning should be organized around procedures or around processes. One view was that behavioral constructs must be quite closely tied to procedures and outcomes, especially since they may have no intrinsic ties to neurobiological substrates. An opposing view was that we are, after all, trying to explain learning processes, and thus our theories should instead be classified by process or function. Several potential types of stimulus function were discussed, and the formulation of theories of elicitation, signaling, modulation, and other stimulus function described within procedure-based learning theories was urged. Finally, the view that neuroscience might provide a basis for "true" process theories of learning was expressed, but there was disagreement as to just what would be added by neurobiological analysis.

WHAT IS NATURAL BEHAVIOR?

It is important to recognize that we must reconcile like elements, i.e., learning theory and theory of natural behavior, and outcomes of laboratory studies of learning with behavior observed in the wild. Accordingly, we discussed both a number of natural behavioral phenomena that may be of interest to a wide spectrum of investigators and theoretical views of natural behavior.

Our theoretical discussion emphasized the framework provided by Tinbergen (15) within the ethological tradition. This framework identifies four broad classes of questions to be asked about natural behavior.

1 - What are the proximal (immediate) causes for behavior?
2 - What is the function of behavior for the animal in its world?
3 - How can evolution of a behavior be characterized?
4 - How does behavior develop?

Within this framework, the question of learning is subsumed by this last category.

We identified a number of ways in which a rapprochement between views of ethology and of learning theory is under way. First, ethologists have provided learning theorists with an impressive range of learning phenomena and processes for investigation (Gould and Marler; Shettleworth, both this volume). Deemed especially interesting by ethologist and learning theorist alike were behaviors such as song learning, imprinting, navigation, social behavior, communication, foraging, and behaviors that led to the notions of search images and spatial memory.

An important point was that at some point we must classify behavior not by what we do to the animal (that is, what events we administer to him) but by what he does. We considered some rough guides for the selection of natural behaviors for the study of learning. It would seem reasonable to study first those behaviors that seem most malleable by experience. Selection of those behaviors should be guided by a true comparative approach. Clearly, whether a particular behavior is influenced by learning contingencies will depend on the species studied, the place of a given behavior within a larger behavioral sequence, the ontogenetic period of the subjects studied, and so forth. That is, there may be no general rules such as "social behavior will always be less influenced by learning than feeding behavior." Nevertheless, guides for selection of behaviors for study may be readily obtained by careful comparative analyses of the naturally occurring behavior systems.

Second, the study of causation in ecologically relevant behavior systems may provide the learning theorist with more explicit performance rules. Psychologists have often been unconcerned with performance rules, equating learning and performance, as in some early S-R theories, or simply ignoring the question of performance, as in many recent cognitive theories of learning. However, a number of so-called behavior systems approaches have recently been proposed within traditional laboratory studies of learning (14). These approaches claim that the nature of learned behavior can best be understood by understanding the relatively preorganized molar behavior systems that are tapped in a conditioning experiment, and by seeing the relations between the artificial elements of the conditioning sequence and the sequence provided in an organism's natural use of these behavior systems. Similarly, many theories of behavior in operant conditioning experiments are theories of action, describing laboratory behavior in terms much like those used in more ethologically based descriptions.

In a related vein, it is clear that the most complete and detailed analysis of proximal cause has been carried out by learning theorists. Indeed, it is fair to say that such an analysis has been their major if not their only goal. It seems reasonable to expect that the application of the often quite sophisticated theories of cause in animal learning theory to more ecologically interesting behaviors may be fruitful. We should consider the role of processes assumed to mediate behavior, such as rehearsal, comparison, interference, attention, and so forth. Variables known to affect these processes may have effects on natural behaviors hitherto unanalyzed in these ways.

Third, performance rules may also be apprehended through a consideration of the functional aspects of the products of learning. At the micro level, Hollis (this volume) and others have suggested that the Pavlovian-conditioned responses that we observe in conditioning experiments are those that provide a useful function for the creature, i.e., increase fitness in a natural environment. At the more macro level, the generation of optimality functions within ethological traditions may provide rules for the apportionment of behavior in the learning laboratory. It is worth noting that both possibilities are past the point of speculation; several such research programs can be identified ((12) and Hollis, this volume).

A fourth possible interplay concerns the type of system investigated. Although learning theorists have often selected very limited response systems for investigation, the behaving organism is nevertheless confronted with a range of inputs and possesses a fairly wide range of adaptive capabilities, despite our choosing to ignore them. Selection of simpler preparations in which nature has effectively limited much of the input may provide us with simpler tasks, to which the strategies of learning theorists may be brought to bear. It was not clear to many of us, however, how such a strategy would be easily applied to the study of learning in mammals.

Fifth, the specification of feedback functions has been an important part of both learning theory and ethological approaches to the study of behavior (12). More frequent and detailed consideration of the nature of feedback functions in natural behavior may reveal new systematic relations in behavior.

Finally, we considered the possibility that the time-honored "building-block" approach could be taken more seriously. Can we in fact concatenate simple units to account for more complex, ecologically relevant behavior? Although psychologists generally reject such a notion today, we considered the possibility that neuroscience may provide the most appropriate units. Specific examples of unifying constructs within such a reductionistic approach are discussed in the section on NEUROBIOLOGICAL MECHANISMS OF LEARNING.

LEARNING AS A GENERAL PROCESS VERSUS LEARNING AS A SPECIALIZED ADAPTATION

There is very considerable evidence of common outcomes of a wide range of conditioning procedures over a wide range of species and response systems. For example, almost all systems show increasing responding with greater numbers of trials, decremental effects of prior exposure to either conditioned or unconditioned stimuli or both, a variety of interference effects, and many other phenomena (Revusky, this volume). There may also be general performance rules: approach to signals for appetitive events and/or withdrawal from signals for aversive events are found across a wide range of systems and species (Hearst, this volume), as is the classic Pavlovian principle that the CR often resembles the UR. The view was expressed that the emergence of a general learning process might in fact be expected, given that most creatures are exposed to similar causal sequences in their environments, and that many environments will provide pressures favoring learning with fairly constant properties.

On the other hand, individual animal species surely may have special needs. Similarly, a given animal may have very different needs in different behavior systems. So it is not unreasonable to suggest that specialized learning systems may have evolved. However, it is important to distinguish two usages of "specialization": first, special learning processes, and second, special purpose learning. Organisms may evolve learning processes which follow rules different from those of other learning systems. On the other hand, it is perfectly plausible that a special purpose learning system may follow general process laws. For example, flavor aversion learning has frequently been described as a special purpose learning system, but the rules of flavor aversion learning are qualitatively indistinguishable from those of other learning paradigms.

An issue that is frequently raised in discussions of general versus specific learning process is that of selective association, the finding that some pairs of events are more readily associated than others (Gould and Marler, Jenkins, and Revusky, all this volume). Frequently such findings are viewed as contradictory to the idea of a general learning process. But there is little evidence that the rules of association differ depending on the nature of the events associated. The failure of theories of association at the level of abstraction which characterizes learning theory today is that they do not give us rules for selectivity, i.e., they do not tell us a priori which events will be most readily associated.

But neither does learning theory tell us whether a particular visual stimulus will be more or less effective than a particular auditory event as a conditioned stimulus. Answers to these kinds of question may more likely come from other levels of analysis. For example, consideration of the adaptive significance of related natural behaviors may provide important guides (rats associate flavors more readily with illness than with pain because rats naturally use flavors in food selection but not in avoiding predators or other injury). Or, neurophysiological or anatomical evidence may provide a guide, e.g., the above relation occurs because neural representations of flavor and illness stimuli are more contiguous in the brain. Clearly, cooperation between the learning theorist, ethologist, and neurobiologist is necessary to specify such rules.

A final consideration was that specialized behavior functions may nevertheless reflect common, general learning processes. That is, evolution may act on performance rules rather than on associative processes directly. This notion is especially plausible in dealing with different response systems engaged in the same animal, perhaps at the same time (no conditioning situation yields only a single behavioral change). Thus, one response system may be able to afford to be conservative, showing sensitivity to contingency information while another system may not, without requiring a separate learning system for each behavioral consequence of learning.

COGNITIVE PROCESSES AND COMPLEX LEARNING

Many of the group felt that the primary contribution of learning theory to a science of behavior would involve the analysis of cognitive processes and complex learning phenomena (Lea, this volume). Rather than generate a catalog of cognitive processes, most of our discussion focused on the concept of representation. The notion has taken on diverse meanings, some highly specific and some more generic, and has been used at a number of levels, from the physiological to the behavioral to the conceptual. A relatively unspecific working definition of a behavioral representation as an organism-generated stimulus was considered. Most recent theories of conditioning assume that the organism stores some internal representation of the reinforcer that shares many of the properties of the reinforcer itself, and that much conditioned performance is mediated by that representation (1, 11). Consider a rat that experiences a pairing of a tone and a flavored food substance. According to these theories, one consequence of that pairing is that the tone acquires the ability to evoke a representation of the food, as in S-S learning theories. Given the appropriate motivational state, evocation of that representation by the tone produces some of the behaviors that presentation of the food itself produces. More importantly, the rat may operate on that representation as if its referent (the food) were physically present. Thus, subsequent pairing of that tone with an illness (but in absence of the food) results in the establishment of an aversion to the flavor of the particular food that had been paired with the tone (3). A number of analogous representation-mediated conditioning phenomena have recently been demonstrated (4).

Not all Pavlovian learning is mediated by such internal representations (Jenkins, this volume). Consider a simple second-order conditioning experiment in which rats first receive light-food pairings and then tone-light pairings. The tone acquires control over conditioned behavior. If performance of conditioned behavior to the tone were mediated by a representation of the light, then changing the value of the light, say, by pairing it with shock or extinguishing its relation to food, should eliminate the tone's tendency to evoke behavior. However, in a variety of experiments, second-order conditioned responses have been found to be unaffected by those manipulations (1, 11). One account is that those behaviors are not mediated by internal representations of stimuli, but instead are evoked more directly, as in S-R theories (1, 11). As noted in other contexts in this report, learned performance in conditioning is seldom purely mediated or unmediated by representations: depending on the selection of a number of specifiable features of the experimental situation, second-order conditioning that shows complete sensitivity, no sensitivity, or any intermediate sensitivity to such value changes can be obtained. Finally, it is worth noting that one invertebrate, Limax, shows a similar mediated or S-S type of second-order conditioning (Sahley, this volume). Thus, mediational functions assigned to such behavioral representations need not require an especially complex nervous system.

We next considered the question of animals' representation of serial order, a problem of interest to a number of investigators because of its relation to language acquisition and song learning. An experiment in which pigeons had to learn to respond to simultaneously presented stimuli in a particular serial order was considered (Terrace, this volume). After learning the correct four-term sequence, the subjects were tested on all possible two-term sequences. The quality of performance suggested that the subjects had learned more than simple pairwise associations of adjacent stimulus or stimulus-response units. Similarly, substantial savings in the pigeons' ability to discriminate the sequence which they had learned to produce from other sequences were observed. Together, the data suggested that the pigeons in some sense represented the serial order of the learned sequence. Analogous data from more natural behavior systems were noted: the song sparrow can develop its normal multi-phrase song structure, irrespective of whether or not multi-phrase models are present in its environment, as if some features of that song were already represented internally (Marler, this volume). The comparison of the representation of relatively fixed sequences such as that involved in the aforementioned case of song learning with that of more arbitrary sequences might provide a fruitful starting place in the examination of more complex representational processes.

Our discussion then turned to neurobiological conceptions of representation. We distinguished at least three senses of representation at this level. First, there is the neural representation of the receptor surface. It is essential to note that this level of representation is not isomorphic with the representation of a real stimulus. Second is the neural representation of a present stimulus, which forms the basis of perception and perhaps of some sorts of short-term memory. Third is a stored, permanent representation of a stimulus, which forms the basis of what could be called long-term memory.

In relation to these distinctions, we noted that the topographic organization of successive representations of receptors within the nervous system is progressively less like the organization of the receptors themselves, thus paving the way for more abstract representations at higher levels of processing. Also, we considered the notion that different kinds of behavioral representations might correspond to different levels within hierarchies of assemblies of cells, as discussed in the section on NEUROBIOLOGICAL MECHANISMS OF LEARNING. We also discussed Mishkin's (6) idea of neural event representations as convergences of neurons responsive to object features on individual higher-order neurons, such that those latter neurons can activate neural representations of the whole constellation of object properties, and the suggestion was made that spatial relations are represented neurally in quite a different manner and in quite a different place than are objects occurring in those relations with each other. Recent experimental evidence (7) which suggests the selective effects of various lesions on spatial abilities was mentioned, along with the thought that various behavioral concepts of representation might appear more credible to neurobiologists if more such selective effects of lesions on performances assumed to be mediated by those representations were discovered.

NEUROBIOLOGICAL MECHANISMS OF LEARNING

Three issues were discussed: the comparison of molluscan and mammalian conditioning, a neuroselection theory of higher nervous function, and wiring diagram approaches to a neuroscience of learning. The three issues were not incompatible. We were optimistic that a general process model of learning might be most attainable at the neurobiological level. An important feature of our discussion was the emphasis on the very different uses of similar terms by researchers from various traditions. The wisdom of the use of modifiers, as in <u>behavioral</u> inhibition or <u>neural</u> association, in interdisciplinary discussions became painfully obvious.

Molluscan Learning

There was considerable agreement that classical conditioning has been rigorously demonstrated in several molluscs (Quinn and Sahley, both this volume). At least one, Limax, has been pushed quite far behaviorally, that is, trained in fairly detailed behavioral conditioning paradigms (mentioned briefly in the section on COGNITIVE PROCESSES AND COMPLEX LEARNING). Most of our discussion, however, focused on Aplysia, in which extensive neurophysiological and neurochemical work has been carried out as well. Although interested by the data, we expressed some caution. Relatively little research on memory (i.e., storage of learned changes) has been conducted, and not all of the phenomena we discussed have been demonstrated in the behaving organism.

Considerable converging evidence suggests plausible mechanisms for learned changes in Aplysia (discussed by Menzel et al. and Quinn, both this volume). There is reason to believe that similar neurochemical and physiological mechanisms may be extended to several aspects of vertebrate learning as well. We noted that a single neurochemical mechanism can easily be given the power to mediate a variety of more complex learning processes by assuming a variety of wiring diagrams.

In this context, we spent considerable time discussing the nature of Hebb synapses, in which the gain of a synapse increases if activity on the presynaptic side is concomitant to activity on the postsynaptic side (2). On the whole, we stressed the importance of functional distinctions between alterations in receiving versus transmitting properties of a synapse rather than anatomical distinctions between pre- and postsynaptic changes. It seemed reasonable to consider a variety of possible changes dependent on the correlated activity of two neurons as mechanisms of neural plasticity both in molluscan and mammalian learning.

A Neuroselection Theory of Higher Brain Function

Many behavioral theories of instrumental learning attempt to provide rules for the generation of variation in behavior and the subsequent selection of behaviors that lead to reinforcement. Likewise, Changeux's neuroselection theory (this volume) attempts to provide mechanisms for the creation of diversity in neural units and the subsequent selection of those units that most closely match experience.

A key thought within neuroselection theory is that similar principles of plastic change may obtain at many levels of neuron organization. For instance, change occurs at the molecular level in the redistribution of receptor sites, at the cellular level in the activity-dependent loss of synaptic connections, and at the level of assemblies of cells in the analogous activity-dependent selection of assemblies. And, of course, these assemblies may themselves be stacked in hierarchical function and selected in comparable ways.

Neural representations of events are topologically organized and can be found in many places in the brain. Representations close to primary sensory areas may be characterized as more "concrete" or relating more to the receptor surface's organization, whereas those in more distant, association areas may be characterized as more "abstract" (see section on COGNITIVE PROCESSES AND COMPLEX LEARNING). These neural representations may be activated by corresponding events in the outside world, in which case we might call them percepts, or less directly (endogenously) activated, in which case we might call them memories, maps, or mental representations.

Another key point is that these neural representations have definite associative and combinatorial properties. Spatially and/or temporally contiguous neural representations can be associated, and such associated units can combine hierarchically. Thus, a mechanism to generate considerable diversity is assumed.

Learning, then, involves the comparisons of percepts with memories, i.e., exogenous and endogenous evocations of neural representations. Cells (or assemblies of cells) whose rhythmic activity match are said to resonate. It is this resonance which permits a higher-order grouping such as an assembly or assembly of assemblies to stabilize via neurochemical and biophysical mechanisms similar to those already studied at the cellular level. Thus, there will be a selection for those assemblies or higher-order units that match experience.

Of course, it must be kept in mind that development involves the growth of connections (or of their efficacy) prior to the selective loss of connections. And the environment does not play a uniform role at all levels and in all systems. Nevertheless, neuroselection theory provides organizing principles which may prove useful at several levels of discourse.

Wiring Diagram Approaches

We considered one particular wiring diagram approach to the underlying mechanisms of learning in some detail (6). Within this account neural representations of events are stored in the cortex, but only after work is performed in subcortical pathways, especially the limbic system. Most relevant from our perspective is the anatomical isolation of two separate learning systems. Although elements are thought to enter both systems simultaneously, the level of coding and mode of operation of the two systems are quite different. One system is involved in more cognitive tasks and stores representations of stimuli, responses, and reinforcers, whereas the other system does not. That is, the second system apparently runs off behavior without spontaneous rearousal of neural representations of those events. It is worth noting that this distinction corresponds quite closely to the historically prominent split between S-S or "cognitive" and S-R "behavioral" theories of learning described earlier (see also section on COGNITIVE PROCESSES AND COMPLEX LEARNING). Furthermore, just as modern learning theory notes that S-S and S-R learning can exist side by side, this theory also claims that a full account of behavior must recognize that learned behavior must be considered as engaging an amalgam of both kinds of processing. It is interesting to note that this theory also runs afoul of the same problems affecting learning theories in the account for performance. Although the simpler, less richly represented system apparently generates behavior through simple reflex action, as did behavioral S-R theories, the more cognitive system has no immediately obvious route through which to generate behavior. However, rather than leaving the organism buried in thought, a neurobiological theory simply leaves the representation lost in the brain. Perhaps consideration of elicitation mechanisms for various behavioral units (perhaps those identified by ethologists) may provide the clues to the wedding of activity in the cognitive system to performance at the behavioral level. Indeed, substantial progress in tracing these more complex wiring diagrams has been made.

RELATIONS BETWEEN DEVELOPMENT AND LEARNING

Most of our discussion concerned the parallels at a descriptive level between filial imprinting in chicks and the development of the visual system, as reflected in studies of the degree of binocularity and selectivity for orientation of cortical cells in cats as a function of early visual experience (described in (8) and by Singer, this volume). Those parallels include time courses, relative irreversibility, relation of amount of experience to the amount of retention, the self-terminating nature of the changes (the shortening of sensitive periods once an adequate stimulus has occurred, or the extension of such periods if adequate stimuli are not presented), and the selective aspects of the changes (the process can be engaged only by stimuli that activate assemblies with appropriate resonance characteristics or prestabilized combinatorial mechanisms, that is, preselected stimuli). Furthermore, we saw a continuum between experience-independent change in embryology, through experience-dependent change in ontogeny, to the adaptive changes we call learning, and expressed optimism that similar cellular mechanisms of plasticity might be found. It is fair to say that in some cases the similarities broke down on more detailed analysis. Nevertheless, on the whole the analogies were compelling.

We first considered plasticity in the developing cat's visual system. Once again, the concept of Hebb synapse played an important role. Variations in the activity of afferents from the eyes are induced in these experiments by the occlusion of one eye. The efficacy of excitatory transmission for afferent pathways from the eye increases if they are active in temporal contiguity with the postsynaptic target. Conversely, the gain decreases when the postsynaptic target is active while the presynaptic terminal is silent. Finally, no changes in gain occur when the postsynaptic path is inactive, regardless of the status of presynaptic terminals.

Within this system, it also seemed quite clear that the adaptive processes are gated: a number of experimental manipulations including the depletion of norepinephrine and a variety of other operations that can be described as preventing the cats from attending to visual input (e.g., muscle paralysis and the disruption of visuomotor coordination by reorienting the eye) prevented plastic changes. Consistent with existing evidence is the hypothesis that the gating mechanism might involve the entry of Ca^{2+} into the dendrites of cortical neurons, a mechanism except for locus identical to that proposed by a number of investigators for a wide range of plastic changes of neuronal interaction in a number of systems.

The end product of the whole process is the selective removal of fibers ascending to particular areas of cortex, affecting the organization and, presumably, the function of those areas. In fact, behavioral data show the kittens to be behaviorally impaired in ways to be expected from the structural abnormalities. The mechanisms identified so far are especially appealing since they may be viewed as providing suitable mechanisms for the formation of assemblies within Changeux's neural selection theory. The rules that Changeux (this volume) proposes to govern cellular changes resemble those governing the action of groups of cells in this situation.

Learning theorists among us were struck by the resemblance of the procedures and outcomes of many experiments in this preparation with those in learning experiments. For example, the task of the organism may be viewed as an associative one in which the receptive fields of two eyes must be coordinated, or subunits must be connected together to form an oriented receptive field. More compelling were the results of "competition" experiments in which the prior acquisition of dominance in one eye for a particular feature prevented the subsequent acquisition of binocular sensitivity to that feature when both eyes were opened, but not to other features on which no dominance was preestablished (9). That outcome as well as more detailed aspects of the data correspond formally to the variety of blocking phenomena found in the learning laboratory: pretraining a CS A with one US prevents the acquisition of responding to a CS B when A and B are jointly paired with the same US, but not if they are paired with a different US (see Jenkins, this volume).

Our discussion of filial imprinting in chicks followed similar lines. The behavioral properties of this system were more the province of another group at this workshop and the reader is referred to that report (Bateson, Immelmann, and Kroodsma et al., all this volume). The most intriguing analogy beyond the behavioral similarities is the existence of a potentially similar neural mechanism. Imprinting to a visual cue will occur only when a particular brain structure, the intermediate region of the medial hyperstriatum ventrale, is intact. Further, exogenous stimulation of that region at a particular pulse rate generates preferences for light flashes at the same rate at a later time. A working hypothesis within the imprinting literature is that stimulation of that neuronal region acts as an enabling condition so that neurons in visual pathways can be modified (Bateson, this volume). A mechanism analogous to that proposed in the visual development case is plausible, although little supporting data are as yet available.

RECONCILIATION OF LEARNING THEORY AND NATURAL BEHAVIOR?

Throughout this report, we have suggested potentially valuable realms for the interaction of learning theorists, neuroscientists, and ethologists. Both the learning theorist and the neuroscientist have been turning toward the study of more ecologically relevant activities, and ethologists are using sophisticated tools of the neurosciences. Neurobiological theories seem to be increasingly useful in revealing possible process similarity in a variety of cases of behavioral plasticity. Although to many, learning theories have become increasingly esoteric, it must be said that they are able to describe increasingly wide-ranged and detailed phenomena, and they have suggested a number of heuristically useful psychological processes.

Collaboration among investigators from a variety of traditions is already under way and is likely to expand substantially in the next few years, in part as a result of gatherings like this one. Clearly, attitudes have changed in the past few years. Conspicuously absent from our discussions were several topics which formerly evoked considerable disagreement and hostility among researchers from the traditions represented here. There were no champions of extreme positions in classic nature-nurture controversies, in the relative importance of field and laboratory studies, in the use and selection of "representative species" in the investigation of learning processes, or in other issues. Researchers generally showed considerably more understanding of the methods, goals, and attitudes of other approaches than was the case a decade ago.

Our "wish list" comprises two kinds of wishes, one for particular sorts of data or theory, and another set of wishes that the other investigator would do his research differently. Wishes of the first kind included the discovery of anatomical correlates of behavioral processes inferred in recent learning theories, especially those of various sorts of representation; the development of systematic theories of natural behavior; the discovery of general process markers in natural behavior; and the discovery in traditional laboratory experiments of not-specially-adapted natural behavior markers which indicate common functions of natural and laboratory behaviors. Wishes of the second kind included hopes that ethologists would try out more sophisticated behavioral processes in their analysis of causation, hopes that learning theorists would be more sensitive to the function and evolution of behavior, and hopes that all would tie their theoretical speculation to concepts of neuroscience.

REFERENCES

(1) Dickinson, A. 1980. Contemporary Animal Learning Theory. Cambridge: Cambridge University Press.

(2) Hebb, D.O. 1949. The Organization of Behavior. New York: Wiley.

(3) Holland, P.C. 1981. Acquisition of representation-mediated conditioned food aversion. Learn. Motiv. 12: 1-18.

(4) Holland, P.C. 1983. Representation-mediated overshadowing and potentiation of conditioned aversions. J. Exp. Psychol.: Anim. Behav. Proc. 9: 1-13.

(5) Hull, C.L. 1943. Principles of Behavior. New York: Appleton-Century.

(6) Mishkin, M. 1982. A memory system in the monkey. Phil. Trans. R. Soc. Lond. B 298: 85-95.

(7) O'Keefe, J., and Nadel, L. 1978. The Hippocampus As a Cognitive Map. Oxford: Clarendon Press.

(8) Rauschecker, J.P., and Singer, W. 1981. The effects of early visual experience on the cat's visual cortex and their possible explanation by Hebb synapses. J. Physiol. 310: 215-239.

(9) Rauschecker, J.P., and Singer, W. 1983. Competition and orientation-dependent recovery from monocular deprivation in the kitten's striate cortex. Dev. Brain Res. 10: 305-308.

(10) Rescorla, R.A., and Holland, P.C. 1976. Some behavioral approaches to the study of learning. In Neural Mechanisms of Learning and Memory, eds. E. Bennett and M.R. Rosenzweig. Cambridge, MA: MIT Press.

(11) Rescorla, R.A., and Holland, P.C. 1982. Behavioral studies of associative learning in animals. Ann. Rev. Psychol. 33: 265-308.

(12) Staddon, J.E.R., ed. 1980. Limits to Action: The Allocation of Individual Behavior. New York: Academic Press.

(13) Staddon, J.E.R., and Simmelhag, V. 1971. The "superstition" experiment: A reexamination of its implications for the principles of adaptive behavior. Psychol. Rev. 78: 3-43.

(14) Timberlake, W. 1983. The functional organization of appetitive behavior: Behavior systems and learning. In Advances in the Analysis of Behavior: Biological Factors in Learning, eds. M.D. Zeiler and P. Harzem, vol. 3. Chichester: Wiley.

(15) Tinbergen, N. 1951. The Study of Instinct. Oxford: Clarendon Press.

(16) Zielinski, K. 1979. Extinction, inhibition, and differentiation learning. In Mechanisms of Learning and Motivation, eds. A. Dickinson and R.A. Boakes. Hillsdale, NJ: Erlbaum.

The Biology of Learning, eds. P. Marler and H.S. Terrace, pp. 553-584. Dahlem Konferenzen 1984. Berlin, Heidelberg, New York, Tokyo: Springer-Verlag.

Biological Predispositions to Learn Language

L.R. Gleitman
Dept. of Psychology, University of Pennsylvania
Philadelphia, PA 19104, USA

INTRODUCTION

Language learning clearly is an outcome of specific exposure conditions, but just as clearly requires specific biological adaptations. There is no controversy about this claim as stated, for it is obvious to the point of banality. To believe that special biological adaptations are a requirement, it is enough to notice that all the children but none of the dogs and cats in the house acquire language. To believe that language is nevertheless learned, it is sufficient to note the massive correlation between living in France and learning French, and living in Germany and learning German. Controversy does arise, however, on the issue of whether language knowledge is based on a specific and segregated mental faculty or, instead, utilizes the same machinery in the head that is implicated in the acquisition of all complex cognitive functions. Many linguistic theories postulate not only a distinct mental representation or faculty of language (a "language organ," in Chomsky's wording, functioning as autonomously as, say, the liver), but a highly modularized system internal to language itself (13). Proponents of such positions expect that language learning will be largely maturationally determined, that the maturation functions may be quite separate from those in other cognitive domains, and that different modules within the language system may mature quasi-independently. In clear contrast, most developmental psycholinguists hold that language acquisition is best described by a global learning procedure that is responsible for the acquisition of, e.g., knitting, arithmetic, and ancient history as well as, say, English (e.g., (3, 50)).

They expect that the unfolding of language will be jointly dependent on specific opportunities to receive relevant data and on cognitive development.

Rudimentary current evidence about learning is insufficient to adjudicate these distinct claims about how biology and experience interact to produce language knowledge. Exacerbating these difficulties, there are at present many contending descriptions of grammar - of what is finally learned. Given disagreements on what is learned, it is hard to devise an adequate description of how it is learned (for example, see (45, 55) for quite disparate formal models of a learning procedure based on correspondingly disparate conjectures about the grammar that is attained). But in my view, the presently available evidence tips the balance of plausibility toward a biologically preprogrammed learning procedure specific to language.

In the present discussion, I first present a schematic description of the language learning task, followed by a sketch of the kinds of argument often put forward in favor of significant and relatively autonomous biological preprogramming supporting that learning. Thereafter, I summarize the kinds of investigation that are being carried out by our own group of investigators.

ANALYSIS OF THE LANGUAGE LEARNING PROBLEM

Chomsky (11) and other investigators (44, 55) have provided a schematic analysis of language learning; roughly, it looks like this:

1) The learner receives some sample utterances from the language.

2) He simultaneously observes situations: objects, scenes, and events in the world.

3) These utterance/situation pairs constitute the input to language learning.

4) The learner's job is to project from these sample utterance/situation pairs a system, or grammar, that encompasses all sentence/meaning pairs in the language. This job includes, of course, learning the meanings and forms of words, for these affect the meanings of sentences. But it also includes learning the syntactic structures, for these affect the meaning of sentences even when the component words are held constant (i.e., Caesar killed Brutus and Brutus killed Caesar mean different things).

5) For learning to proceed, the learner must have some means for representing the utterances and the situations to himself in a linguistically relevant way. For example, though the learner receives utterances in the form of continuously varying sound waves, he must be disposed to represent these as sequences of discrete formatives such as phone, syllable, word, phrase, and sentence. And though he receives impressions of some single object, he must be disposed to represent it in various ways, such as "Fido," "dog," "mammal," "physical object," and not in other ways, such as "undetached dog-parts."

6) The learner must also have some strategies for manipulating these data in the interest of extracting the regularities that bind them.

7) He must have some perspective on the kinds of descriptive devices, or rule systems, that he is willing to countenance as statements of the regularities. To put this another way, different language learning "machines" (particular representational systems, with particular computational procedures) will construct different grammars, based on the same data.

As presented, this analysis is neutral about many issues. For example, it is possible that tutors are required for some of these steps to be taken successfully. Perhaps the learner requires the caretakers to say only very simple sentences, or sentences of restricted kinds, early in the learning period, and perhaps the learner requires reinforcement for correct performances and correction of errors he makes along the way. Depending on whether these additional conditions are met, quite different kinds of machine will be able to learn a human language. Therefore, it is of some interest to examine the real input circumstances of children and the early generalizations they draw from these inputs. This may help to disentangle the kinds of internal and external resources learners recruit to crack the language code.

A SIGNIFICANT INNATE BASIS FOR LANGUAGE LEARNING

Three main kinds of argument favor the supposition that language and its learning are biologically preprogrammed. The first two derive from empirical study of learning: a) language learning proceeds in uniform ways within and across linguistic communities, despite extensive variability of the input provided to individuals, and b) the character of what is learned is not simply related to the input sample. A third argument is logical: c) the child acquires many linguistic generalizations which experience could not have made available.

Uniform Learning

Inquiry into language learning is constrained by one main principle. The right theory has to cope with the fact that everybody does it, by specifying a learning device guaranteed to converge on the grammar of any language to which it is exposed, in finite time, and in fair indifference to particulars of the sample data received (54). This principle is based on the real world facts of the matter. Under widely varying environmental circumstances, learning different languages under different conditions of culture and child rearing, and with different motivations and talents, all nonpathological children acquire their native tongue at a high level of proficiency within a narrow developmental time frame. (This does not mean there are no differences in final attainment, but these differences pale into insignificance when compared with the samenesses.)

Moreover, there are very interesting similarities in the course this learning takes. Isolated words appear at about age one, followed by two-word utterances at about age two. Thereafter, sometime during the third year of life, there is a sudden spurt of vocabulary growth accompanied, coincidentally or not, by elaboration of the sentence structures. By about four years of age, the speaker sounds essentially adult, though his sentences tend to be quite short because the use of embeddings is limited, and though some item-specific information continues to come in through age eight or nine. (By item-specific information, I have reference particularly to some features of derivational morphology and significant growth of vocabulary.)

Summarizing, similarities in the pattern of learning are observed across individuals and across linguistic communities. Lenneberg (36) was perhaps the first to argue that these uniformities in course of learning, despite differences in experience, are a beginning indicant that language learning has a significant biological basis. He provided some normative evidence that the achievement of basic milestones in language learning are predictable from the child's age and seem to be intercalated quite closely to developments that are known on other grounds to be maturationally dependent (e.g., the appearance of sitting, standing, walking, and jumping). Other findings were used by Lenneberg to argue that there is a sensitive (or critical) period for language learning. For example, "foreign accents" are typical in adult learners but not child learners of second languages even when time of exposure and use are equated. Downs Syndrome individuals that he studied seemed to acquire language in the same way as normals, only slower, and their learning seemed to stop at about puberty, whatever their current competence with the language. Finally, Lenneberg maintained that recovery from brain injury that implicates language is likely in children, but rare in adults. On such evidence, he conjectured that the capacity for language learning tends to whither away as the brain matures.

Not all of Lenneberg's findings have withstood subsequent review too well, however. For example, in later discussion I will present a rather different picture of learning in Downs Syndrome individuals. Moreover, the results about the course of learning (from Lenneberg and others) are so fragmentary as to be consistent with quite distinct conjectures about the processes that underlie this learning: though there certainly are gross similarities among children whose age is the same, there may be detailed distinctions among them consequent on distinctions in their exposure conditions. Certainly that is true for specific vocabulary learning, and who is to say, and on what grounds, that this detail is "unimportant"? Symmetrically, the differences in environment may have been overestimated. Indeed, no two mothers will say exactly the same sentences to their offspring. But still the set of input utterances may be constrained in various ways (they may be especially "simple" sentences), and they may be presented under special conditions (e.g., accompanied by corrections or relevant didactic comments, said only in the presence of "easily interpretable events," etc.). Under these improved conditions, perhaps any open-minded all purpose inductive device would generate the same uniformities in the course of learning as is observed for young children acquiring their native tongue. However, as I will now try to show, closer inspection of patterns of development gives more weight to the view that special dispositions in the learner are guiding language acquisition.

Character of Learning: Disparities Between Input and Output

The character of language knowledge at various developmental moments is hard to reconcile with the superficial properties of the input data. To be sure, the child learns from what he hears. However, he does not directly copy these heard sentences, but makes systematic "errors." These errors can be understood, but only by claiming the learner filters the input data through an emerging system of rules of grammar, rules to which he is never directly exposed. (No one explains the language rules to the children. One reason is that the mothers do not explicitly know the rules. The other reason is that the children would not understand the explanations.)

Noncanonical sentences in; canonical sentences out. Some convincing examples were developed by Bellugi (2). One of her cases concerned interrogative structures. In simple sentences of English, auxiliary verbs appear after the subject noun-phrase but before the verb, e.g., I can eat pizza. But in yes/no questions, the auxiliary precedes the subject, e.g., Can I eat pizza?. And in so-called wh-questions, this subject/auxiliary inversion appears again, e.g., What can I eat?, this time without the object noun-phrase (pizza) and with an initial "wh-word" (e.g., what or when or who) instead. The learner of English is exposed by his caretakers to many such wh-questions (about 10% of all the utterances he hears on one estimate (41)). Nevertheless, young learners generally do not reproduce these forms that they hear so often. Instead they produce a form that is virtually never spoken by adults, namely, What I can eat? or sometimes even What can I can eat?. A related finding, also from Bellugi (2) is that, while over 90% of maternal auxiliaries in declaratives are contracted (e.g., We'll go out now), the child uses only the uncontracted forms early in the learning period (We will go out now). In light of such findings, the sense in which the child is learning the language from the presented environment of utterances is evidently quite abstract.

There is a generalization that predicts these errors, as Bellugi pointed out: The child is biased toward "canonical" surface structure formats for his utterances (see also (27, 52)). In the canonical declarative sentence of English, the subject does precede the whole verb-phrase, including its auxiliary. An abstract "movement transformation" reorders these elements in questions.[1] If it is supposed that the child acquires formation rules that underly the declarative first and countenances movement rules only later, this particular error is predictable. Similarly, it is possible to suppose that only the full canonical forms (will) of words like 'll are entered into the learner's mental lexicon. Contraction is achieved by a rule that operates on these abstract lexical representations under restricted circumstances. In sum, it is certainly possible to explain why the young child does not behave exactly like his models, who contract and invert. To explain this, one can invoke a bias toward canonical forms in the language being learned. But this, in turn, implies that the young learner has an ability to reconstruct the canonical forms for questions and for words like 'll. These canonical forms are related to the utterances he actually hears, but only by covert rules. Considering the tender age of the language learning humans who apparently can perform the complex data manipulations required to recover these rules - quite effortlessly and unconsciously - it is likely that significant biological dispositions are guiding their analyses.

Open class/closed class. An even more general disparity between input and output, as Brown (8), Gleitman and Wanner (27), and many other have discussed, has to do with the differential pattern of acquisition of the so-called open class and closed class stock of morphological items (usually called content and functor words by psychologists). The distinction between these two classes is not easy to state formally (and, in fact, is partly controversial, see (32) for the clearest explication). But technicalities aside, it is fairly clear. It has been known for some time that there are lexical categories that admit new members freely, i.e., new verbs, nouns, and adjectives are being created by language users every day (hence "open class"). Other lexical categories change their membership only very slowly over historical time (hence "closed class"). The closed class includes the "little" words and affixes, the conjunctions, prepositions, inflectional and derivational suffixes, relativizers, and verbal auxiliaries. These examples given, it is obvious that the closed class and open class

---

[1]Certain linguistic characterizations (see, e.g., (11) for an early description) assume that interrogative sentences are at some level represented in a format whose phrase organization is just like that of declarative sentences, with the subject noun-phrase preceding the auxiliary and the object noun-phrase following the verb. A rule obligatorily applies to such structures, inverting the order of the subject and the auxiliary. But for those interrogatives beginning with a so-called "wh-word," there is an additional complication: The object noun-phrase is a question morpheme (wh-) joined with a pronoun (e.g., -at or -en, as in what or when), i.e., I can eat what or, following the inversion, Can I eat what. A further obligatory rule applies to such structures. The wh-word is moved to the left, yielding What can I eat? One virtue of such an analysis is that it conforms transparently to the semantics of questions: What can I eat seems to be the way of querying ("wh?") that unknown thing (-at), labelled by a noun-phrase, that can be eaten. Another virtue is that it materially simplifies the description of sentence form. For example, a reasonable generalization about many verbs, e.g., rely on is that they must be followed by a noun-phrase: John relied on sounds strange because is does not have such a following noun-phrase. But this generalization seems to be defeated in wh-questions where, indeed, rely on does occur without a following noun-phrase: What did John rely on? The solution is that, at some stage of derivation, this sentence did have the noun-phrase, namely wh-thing or what, but that that noun-phrase was subsequently "moved." As I am now arguing, the facts about the speech of two- and three-year-olds lend independent plausibility to such conjectures about the mental representation of sentences. Thus, I believe, they bear on the psychological relevance of grammatical descriptions that do or do not countenance movement rules (compare, e.g., (45, 55)).

morphemes are different in many ways. The closed class items are restricted in semantic content (e.g., nobody's name is a preposition) and in syntactic function (as follows from the fact, stated above, that they belong to only certain lexical classes). They differ phonologically as well, i.e., closed class morphemes are in most usages unstressed, some are subsyllabic, and many more become subsyllabic by contraction (e.g., will contracts to 'll).

Perhaps the most striking fact about early speech and the context in which it is learned is that open class and closed class materials are made available to the learner simultaneously, but these subcomponents of the morphological stock are incorporated into the child's speech at different developmental moments. The mother's speech consists of simple sentences like "The book is on the table," including the closed class items the, is, and on. But there is a well-known stage of language learning, the so-called two-word or telegraphic stage, at which the output is "Book table," with all the closed class items omitted. Thus, the most primitive learners seem to have certain devices that allow them to filter out the closed class.

It is not obvious how to explain these developing speech patterns without begging the questions of language learning. For example, even if it were possible to say that the omitted items were the meaningless ones, which it emphatically is not, one would not want to claim that the child examines the semantics of closed class words and, on the basis of this, decides to omit them all. This would beg the question at issue: how the semantics of such items are arrived at in the first place.

Ditto for a claim based on the syntactic functioning of closed class items. Gleitman and Wanner (27, 28) conjectured that it is the special phonology of the closed class items that renders them opaque to youngest learners - they are subsyllabic items or unstressed syllables. Differential attention to the unit stressed syllable in the incoming speech wave represents, we believe, one of the significant biases of young learners.[2] Because

[2]The specific acoustic correlates of primary stress in English include longer duration, higher fundamental frequency, and intensity. For the arguments that follow to hold, it would be necessary to show that related acoustic properites are available and exploited to mark phrase boundaries in the nonstress-accent languages, and that these are the properties to which infants are sensitive and which they reproduce in their first utterances (see (17) for a review of supportive evidence).

of the child's selective attention to stressed syllables, the pattern of acquisition does not mirror the environment directly, but mirrors the environment only as it is mediated by these preexisting biases as to how to represent the sound wave. (See Fernald (17) for a full discussion of acoustic-perceptual dispositions in infants and the position that the filtering effect of these predispositions plays a role in language learning.)

It is very interesting to notice that the distinction between the open class and closed class morphological components is an organizing factor in language use even after learning is complete. For example, in adults, speech errors differ for the two classes and almost never involve an exchange between an open class and a closed class item (21); the patterns of long-term language forgetting differ for the two classes (14); within-sentence code switching is constrained differently for the two classes (31); judgmental performance differs for the two classes (23, 24) as do properties of reading acquisition (26, 48); finally, the very definition of the distinction between Broca's and Wernicke's aphasia involves a differential dissolution of these two subcomponents of speech performance ((7, 38), but see also (37) for evidence that the language faculty itself may be intact in Broca's aphasics).

Lexical selections and argument roles. So far I have noted learning biases of the child in the syntactic and morphophonological domains that act to predict a difference between what goes into the child's ear and what first comes out of his mouth. The case for lexical category acquisition is just as clear. Evidently, a learner hears verbs and adjectives as well as nouns early in life. But the child's earliest words are overwhelmingly nouns that encode simple concrete objects; verbs that encode activities tend to appear later, and verbs that encode mental states later still; finally, adjectives, that encode properties of things, are later than all these others (e.g., (15, 22, 40)). Similarly, certain syntactically - rather than lexically - encoded semantic-relational categories (e.g., "agent of the action," for example, John in John eats peas) uniformly appear early and others (e.g., "instrument of the action," for example, knife in John eats peas with a knife) are uniformly found later, again though instances of all of them appear in the child's data base - his mother's speech - from the beginning (6).

In sum, the child is clearly learning from what he hears (English from English, French from French), but the detailed properties of the development are hard to describe as arising very simply or directly from the environment of heard utterances. Likely, global aspects of cognitive development will account for many of these learning patterns, e.g., attention to concrete objects and physical activities before mental states and properties is unlikely to have specifically linguistic sources. But the morphophonemic and syntactic choices of novice language learners are less likely to be explained as deriving from general properties of cognitive development.

Language Knowledge That Experience Could Not Provide

Another kind of argument for innate language learning capacities comes from logical analyses by Chomsky (12). This has to do with the learning of certain language properties that experience could hardly have made available. Taking one of his examples, a distinction between higher and lower clause in a phrase structure configuration determines certain properties of movement transformations and of the reference of pronouns. (The characterization of the linguistic facts are perforce rough in this discussion, but will have to do for a sketch.) Specifically, consider again the movement of certain material in English yes/no questions. In the sentence Is the man a fool?, the is has moved to the left from its canonical position in declarative sentences, The man is a fool. But can any is in a declarative sentence be moved to form a yes/no question? It is impossible to judge from one-clause sentences alone. The issue is resolved by looking at more complex sentences, that contain more than one clause: It is the is in the higher clause, never the is in the lower clause, that moves to form yes/no questions from structures underlying, e.g., The man who is a fool is amusing and The man is a fool who is amusing. This generalization explains the acceptability of yes/no questions such as Is the man who is a fool amusing? and Is the man a fool who is amusing? but the absence of *Is the man who a fool is amusing? and *Is the man is a fool who amusing? Whether is moves depends on its structural position in the sentence.

Notice that an alternative analysis, namely, serial position of the two is's in the string of words, as opposed to structural position of the two is's in a clause hierarchy, could not explain the facts about English structure: In the examples just given, the moving is was once the first is in the sentence, but once the second. Learners apparently know that movement rules are structure-dependent, not simply serial-order dependent. To my knowledge, no child ever makes the mistake of saying, "Is the man is a fool who amusing?" The important point here is that it is hard to conceive how the environment literally gives the required information to the learner. Surely only the correct sentences, not the incorrect ones, appear in the input data. But the generalization required for producing new correct sentences is not directly presented, for no "hierarchy of clauses" appears in real utterances - only a string of words is directly observable to the listener. And certainly there is no instruction about clauses. Even if mothers knew something explicit about these matters, which they do not, it would not do much good for them to tell the aspiring learners that: "It's the is in the higher clause that moves."

Many generalizations about sentence form turn on the same or related configural distinctions (though again the characterization being used here is rough and underestimates the complexity of the descriptive facts). To give one more example, co-reference is possible when the pronoun precedes its antecedent, but only if the pronoun is in a lower clause than the antecedent. For example, the man mentioned in the second clause could be the one who arrived in the sentence When he arrived, the man danced, but not in the sentence He arrived when the man danced. Even in these hard cases where the pronoun precedes its potential antecedent, learners come to interpret co-reference correctly, without special tutoring, and without an opportunity directly to experience the hierarchical structures.

As a final example, to explain what is finally learned, it has been necessary for Chomsky (and indeed, all syntacticians, to my knowledge) to postulate certain ghostly "null elements" in the representation of sentences. For example, a sentence I took up earlier was What can I eat? Many linguists will transcribe this sentence as What can I eat ϕ, as though some "trace" of its canonical phrase structure, with object noun-phrase following the verb were somehow "still there" in the question form. Many descriptive facts about English speech can be explained only by postulating such soundless, tasteless, odorless entities in the mental representation of sentences. As one example, note that the item is, which can contract in many positions in English sentences, e.g., What's your name?, in other positions cannot contract: One can say I wonder who he is? but not I wonder who he's. Suppose that a "trace" of the moved noun-phrase constituent appears in the underlying structure of this sentence, e.g., I wonder who he is ϕ. It can now be postulated that contraction is prohibited preceding the deletion site. This claim is useful if and only if a large number of superficially distinct restrictions on contraction can be subsumed under the same generalization, and indeed this turns out to be the case.

To give one more example, the same principle explains why one can use the wanna contraction of want to in This is the rabbit who I wanna banish, but not in *This is the rabbit who I wanna vanish. Under an analysis similar to that given in the preceding section, there is a missing noun-phrase in each of these sentences: in each case a rabbit has disappeared. But that noun-phrase is missing from different places in the underlying representations of these sentences, a consequence of the fact that vanish is an intransitive verb (whose subject is missing) while banish is a transitive verb (whose object is missing). That is, the mental structure of these sentences is plausibly transcribed as This is the rabbit who I want to banish ϕ and This is the rabbit who I want ϕ to vanish. In each case, the trace, ϕ, appears where once there was a rabbit. But only in the second case does the trace intervene between want and to. If it is contiguous want and to that contract to wanna, the real facts about restrictions on contraction are explained. As a further demonstration of the power of the analysis, notice that This is the rabbit who I want to visit is ambiguous (between I want to visit the rabbit and I want the rabbit to visit) while This is the rabbit who I wanna visit is not: Only the transitive reading of the verb visit remains under the contraction, for the same reasons as stated above. Young children honor these very abstractly described restrictions on contraction and have never been observed to err during the learning period. This implies that mentally they are in possession of a descriptive device that involves something like these null elements though they never "heard" them.

Summarizing, it is hard to imagine how the environment instructs the learner that not serial order simply, but the hierarchical arrangement of clauses, determines properties of sentences and their interpretations, or that spirit-like null elements determine whether contraction is possible and, given the contraction, what the meaning must be of potentially ambiguous sentences. Yet errors on these properties are so rare that no one, to my knowledge, has so far succeeded in observing one made by a child. In short, errorless learning of structural properties of the language not transparently offered to experience - the poverty of the stimulus information - forms still another sort of argument that learning biases, this time biases in the formation and interpretation of configural structures underlying sentences, are required as explanations of how the language organization is achieved by the child.

THE DEPRIVATION PARADIGM

During the last several years, my associates and I have looked at some natural cases where some of the components of the language learning situation are varied. We have operated by taking to heart the view that utterance/situation pairs are required input to language learning, as described in the schema for learning with which I began. Therefore we have looked at situations where this information is changed. Certain populations of learners allow us to see what happens if there is less or different information about the utterances; other populations allow us to see what happens if there is less or different information about the world. Symmetrically, certain populations are exposed to normal input data but differ from the normal in their current or final mental state. In these latter cases, we are asking what happens when different learning devices are exposed to the same data.

Varying the Language Samples

There seems to be quite general agreement that a learner exposed to random samples of the sentences of a language would be unable to converge on the grammar. The main difficulty is the richness of the incoming data, which would seem to support so bewildering a variety of generalizations, including wrong and irrelevant ones, that we would expect learners to vary extremely in the time at which they hit on the grammar of their language. It is usually assumed, following findings from Brown and Hanlon (9), that negative feedback, or correction, from the environment cannot be relied on to solve this problem. This is not so much that feedback for ungrammaticalness is never given, though Brown and Hanlon have shown that it is rare. The main problem is that for a variety of structures and contents - such as those described in the preceding section - the child never errs in the first place so no opportunity to correct him, or to describe his learning as a consequence of such correction, arises. Another problem, as Brown and Hanlon showed, is that correction is given very often for matters other than grammaticalness of the child's utterances, for example, their truth or moral propriety. Should the child construe all corrections, then, as grammatical corrections, he might falsely conclude that ink-on-the-wall is an outlawed phrase, rather than an outlawed act. In light of these difficulties, most investigators assume that language learning is on positive examples only. Formally speaking, it is a lot easier to develop a mechanical procedure for learning from presented sample utterances if it assumed that that procedure receives negative feedback when it makes a false generalization, i.e., if it is told that some new sentence it tries out is not a correct sentence of the language being learned (for that demonstration, see (29)). Hence, some proposals have considered whether restrictions or simplifications of sentences presented to novices can substitute for overt corrections.

Effects of maternal simplification: The Motherese Hypothesis. The paradox so far is that language learning is hard to describe given positive data only (a sample of the correct sentences), and yet all the real learners seem to do very well even though they receive only, or almost only, such positive data. A very popular response to this problem among developmental psycholinguists has been to suppose that the environment provides detailed support for learning by ordering the input utterances (see the collection of articles edited by Snow and Ferguson (53), for many papers adopting this position). We have called this "the Motherese Hypothesis." It holds that caretakers present linguistic information in a set sequence, essentially smallest sentences to littlest ears. And in fact there is no doubt that adults speak quite differently to children and to adults, so the utterances heard by the children are not random selections from the adult language. The utterances to youngest learners are very short, slow in rate, and the like. Some investigators propose that this natural simplification from caretakers (whatever its source or motivation) plays a causal role in learning. Evidence favoring this hypothesis comes from Fernald (16, 17) who has shown that infants prefer to listen to the prosodically exaggerated forms of Motherese, which is used in every known culture: Apparently, as Fernald states, Motherese is "sweet music to the species." Gleitman and Wanner (27, 28) have proposed that these exaggerated prosodic cues can help an appropriately preprogrammed learner to reconstruct a global parse of the input sentences, a reconstruction that materially simplifies subsequent steps in the language learning task.

But evidently there are limits on the work that Motherese can do for the learner. Particularly it is hard to maintain the view that the preselection of syntactic types by caretakers can bear materially on the acquisition of grammar. Though restricting the sentence types may exclude certain hypotheses from being considered early, they may also make available hypotheses that would be insupportable given the full range of the language structures. As I described in the preceding section, the child would be unable to distinguish the "string movement" hypothesis from the "structural movement" hypothesis if all the input sentences were uniclausal (for discussion, see (12, 55)); if more complex sentences are offered, the string movement hypothesis fails on data. However, there is at least some surface plausibility to the idea that the mother first teaches the child some easy structures. After he learns these, she moves on to the next lesson. To help this idea go through, we would have to grant the caretaker some implicit metric of syntactic/semantic complexity so that in principle she could choose judiciously the sentences that might be good to say to learners. Here too there are some initial supportive findings: Caretakers' speech changes to some degree, in correspondence with the learner's age (41).

In our studies of the Motherese hypothesis (24, 42), we first collected extensive samples of maternal speech to young learners (age range 15 to 27 months). Rather to our surprise, the properties of this speech did not seem promising as aids to learning syntactic forms. The mothers' speech forms were only rarely (less than 10% of the time) canonical sentences, and they were neither uniform syntactically, nor more explicit in how they mapped onto the meanings, than the sentences used among adults. They <u>were</u> short and clear, but this hardly suggests anything very specific about how they reveal the syntax of the language. But it is still possible that some less obvious properties of maternal speech are especially useful to learners. To study this, we revisited the original mother/child pairs six months after the first measurement (42). Analyzing the child's speech at these two times, we were in a position to compute growth scores for each child on many linguistic dimensions. The question was which properties of the mother's speech, at time one, had predicted the child's rate of growth on each measure, explaining his status at time two. (The correlational analysis used a partialling procedure that removed baseline differences among the learners.)

One interesting outcome of these studies was that a number of dimensions of learning rate were utterly indifferent to large differences in the speech pattern of the mothers. For example, the child's increasing tendency to express predicates (as verbs) and their obligatory arguments (as nouns) was not predictable from the particular speech forms presented. On the contrary, in the age range studied, the child's progress with the closed class morphology was a rather strict function of maternal speech style. For example, almost all the variance in rate of learning the English auxiliary verbs is predicted by the preponderance of yes/no questions in maternal speech (for a replication of this finding, see (20)). The effect of these is to place the closed class items in first serial position, with stress, and without contraction (e.g., <u>Will you pass the salt?</u> rather than <u>You will pass the salt</u>, <u>You'll pass the salt</u>, or <u>'Ll you pass the salt?</u>). Thus either, or both, positional biases or biases toward stressed syllables (as conjectured by Gleitman and Wanner (27, 28)) can be postulated for the child, and mothers whose usage gets through these child filters have children who learn the closed class materials the faster.

Recall from the earlier discussion, however, that these environmental factors do not say much about what is learned. For one thing, as just stated, no known special properties of Motherese explain the learning patterns for open class materials or their organization in the child's sentences. Furthermore, the evidence is clear that children first learn to say declaratives with auxiliaries in medial unstressed position even though the environment favoring learning how to do so is hearing those auxiliaries in initial stressed position.

In sum, certain universal properties of natural languages (expressing the predicates and arguments of propositions, for example) seem to emerge in the child at maturationally fixed moments and are insensitive to the naturally occurring variation among mothers. But elements and functions of the closed class, for children in this age range, seem to be closely affected by specifiable facts about the input. Even here, however, the environment exerts its influence only as the information it provides is filtered through the child's learning biases. For example, the serial position, but not the frequency, of maternal auxiliary use affected the learning rate (for a general discussion of the Mothereses Hypothesis and its limitations, see (25)).

Unfortunately, while these studies preclude certain strong forms of the Motherese hypothesis, they leave almost everything unresolved. First, the limited effects of environment on language learning that we found may be attributable to threshhold effects of various sorts, to the attenuated sample, or to the measures or analyses used. These complaints are fair even though they lose some force given the positive findings for the closed class component. Nonetheless, there was clear impetus for looking at cases in which the child's environment was more radically altered.

The creation of language: isolated deaf children. We therefore next studied a population grossly deprived of formal linguistic stimulation (15). These were six deaf children of hearing parents who had decided to educate their children "orally," by having them taught to vocalize and lip read. Accordingly, in advance of the planned training period, the parents made no attempt to teach a manual language. More important, these parents did not know a manual language, so they were not in a position to present the easiest sentences first, the harder ones later. It has been observed that children in these circumstances develop an informal system of communicative gestures, called "home sign." It was the genesis of this system that we wished to study. Though many questions arise about how precisely we could analyze this exotic communication system, it is fair to say that the interpretive puzzles we faced are not materially different from those confounding the study of two- and three-year-old English speakers, by adult English-speaking psycholinguists. In each case, one has to try to interpret the child's messages relying heavily on their real-world context of use (cf. (5)). In doing so, one encounters the same perils and pitfalls as the language learner himself. We settled for using the methods traditionally emplyed in studying normal language learning, and we achieved about the same results, for early stages.

These linguistic isolates began to make single gestures (invented by themselves) at the same developmental moment that hearing learners of English speak one word at a time. Two and three sign sequences, encoding the same semantic/relational roles, appeared at the same age as hearing learners speak in two and three word sentences. To the (rough) extent that the words in these primitive sentences are serially ordered by young hearing learners according to these semantic roles, similar serial ordering of the same categories described the self-generated gesture system. It seems then that even if the environment provides no sample sentences, the child has the internal wherewithal to invent forms himself, to render the same meanings.

These results become more interesting when compared to the findings mentioned earlier, concerning the hearing learners. To the degree that the propositional forms and meanings appeared in indifference to variations in maternal input, these same properties appeared at the same time in the deaf learners, exposed to no formal language input at all. The closed class subcomponent, responsive at this stage to variations in maternal input, did not appear at all in the signing of the isolated youngsters. The first suggestion here is that the closed class is laid down later, and in a different developmental pattern than other properties of the language system, an argument I made earlier based on quite different observational evidence (see section on A SIGNIFICANT INNATE BASIS FOR LANGUAGE LEARNING). The second suggestion is that the one subcomponent of the language system is more environmentally dependent than the other and may not appear at all in some exposure conditions.

The creation of language: creoles from pidgins. A fascinating line of research (4, 49) concerns the process of language formation among linguistically heterogeneous populations: pidgins and their creolization. This work shows that there are very interesting overlaps between the rudimentary first attempts of young children learning an elaborated natural language and the devices that appear early in the history of a new language. For example, at the first stages of both, the sentences are uniclausal, have rigid canonical phrase order, etc. (for an admirable discussion that makes these connections to language acquisition research, see (51)).

Most interesting of all in the present context, this work suggests that the final (phonological) steps in creating a closed class morphology may be carried out by five to eight year old youngsters exposed to a pidgin as their first language. The pidgin itself characteristically contains only impoverished closed class resources, again a property shared with the speech forms of all very young learners. The learners who hear a pidgin refine, grammaticize, and expand upon the open class resources of the pidgin in late stages of their learning. In a final step, these new resources are phonologically reduced by the learners, reproducing the destressed (and often contracted) closed class morphology that is characteristic of fully elaborated languages (see (56) for discussion of closed class items and their distribution in languages of the world).

A very similar development has been observed by Newport and Supalla ((43) and in preparation). They study adults who learned formal sign language (ASL) in early childhood from their deaf parents and have shown that this learning is virtually identical to learning of spoken languages. But they also study mature deaf individuals who were isolated from ASL (and spoken language) during early childhood, the normal language learning period, either because of the oralist beliefs of their caretakers or because they acquired deafness later in life. That is, these investigators studied subjects like those of Feldman et al. (15) when they grew up. I have already noted that these isolated individuals develop a pidginized form of language, one that lacks complex embedding devices, closed class items, and the like. Newport and Supalla have shown that, when these individuals are finally exposed to formal ASL, if at ages later than six or seven years, they again learn a form of that language that is highly deficient in the ASL equivalents of closed class morphology. They drop their own home sign pidgin, but they create a pidgin from the elaborated language to which they are now exposed. This is common in adult learners of any language.

Now the most fascinating result from these investigators is for deaf children of such first-generation deaf individuals. Keep in mind that these deaf signers who acquired the language relatively late are the ones who use a rather pidginized form of ASL. Their sentences, of course, form the basis for the second-generation deaf child's induction of ASL. Now (unlike the younger deaf isolates) these learners at approximately four and five years of age refine, expand, and grammaticize certain open class resources and create a closed class mophology in the course of their learning. In a nutshell, for both the spoken pidgin of Sankoff and LaBerge and the gestural pidgin of Newport and Supalla, the first language-learning situation, carried out at the correct maturational moment, creates new resources out of the air, resources that are abstract and are the very hallmarks of fully elaborated natural languages.

Summary comments. The evidence reviewed suggests that certain syntactic properties, though not all (those which involve the closed class being the exception), appear in the learner in the same way even though the utterance samples vary. The studies of the deaf isolates suggest that there is no requirement for an experienced tutor who presents the easy sentences first to secure the first principles of natural language syntax. On the contrary, should the environment for first-langugage learning be deficient in the sample utterances, the learner will improve the language in the course of learning it.

Varing the Interpretive Information

There is more to the child's input than a sample of utterances - presumably it would be impossible to learn language just from listening to the radio. Specifically, in the logic sketched earlier, it was asserted - in agreement with most investigators, whatever their theoretical persuasions - that the child requires a real world context that accompanies the speech events: some situation that he can interpret. In fact, many investigators assert that there is little mystery left in the language-learning feat once it has been acknowledged that the child can interpret the extralinguistic world meaningfully (e.g., (1)). However, it is not so easy to state just how "relying on meaning" succeeds in helping a child learn language.

One difficulty is that every object in the world can be described by many different kinds of words, a fact I alluded to earlier: The same object out there can be called Felix, a cat, a mammal, etc. On seeing the object then, the child still has a problem in determining the intended meaning of a word used to refer to it (the analysis of "basic level categories," e.g., (47) is sure to be of use in approaching such problems). Similarly, any given scene or event in the world can be described by many different sentences. For instance, scenes suitable to The cat is on the mat are just as suitable to The mat is under the cat and Get that damn cat off the new mat. Thus there is a considerable distance between meaningfully interpreting a scene and catching just how a heard sentence relates to it. To maintain the position that the scene helps the child learn, it will be necessary to provide the natural (perceptual and/or cognitive) analysis of the world that biases the learner to see cats-on-mats, not mats-under-cats, plus a conspiratorial agreement between mother and child such that the mother refer to scenes only in ways that match these biases, whatever they will turn out to be (for very useful discussions of the relations between language and perception, see (30, 39)). In addition, there will have to be further conspiracy for marking specially any other intents, for, after all, the grammar that is ultimately learned has to allow for the saying and comprehension of mat-under-cat.

A related problem has to do with how the utterance is to be analyzed, even assuming that the coconspirators have figured out how to be united on the interpretation. For example, suppose the mother says "Rabbit jumps" when the learner can see a rabbit jumping. And suppose, with Pinker (45), that the child believes things are to be the nouns, actions are to be the verbs. Even in these very favorable circumstances there seem to be at least two choices the learner can make. He can suppose that English is a noun or subject-first language, in which case rabbit is the required noun, or he can suppose English is a verb or predicate-first language, in which case rabbit is the required verb[3]. Given all this, it is hard to know how the child gets his act off the ground.

[3]A comeback to this supposed difficulty, of course, is that, over many utterances that dissociate rabbit and jumping, a distributional analyzer can make the choice. But the semantic bootstrap notion has been put forward as a crucial step that precedes and renders possible subsequent distributional analysis; moreover, its very purpose is to relieve the learner of the burdensome tasks of storing and manipulating large quantities of data so as to dissociate, over the corpus as a whole, rabbits from jumping situations and jumping activities from rabbit objects.

Summarizing once again, all parties agree that language is learned in partial dependence on the real scenes that are there to be interpreted in the world. However, to my knowledge, nobody has succeeded in providing the required cognitive-perceptual analysis of how scenes are to be interpreted against heard utterances.

Our approach (34, 35) to these problems has been to look once more at differing environments in which language is learned. In the studies I have mentioned, the learner was in some ways deprived of information about language forms. What happens if the child is deprived of some opportunities to interpret heard utterances against the world of real objects and events? Surely, a blind learner suffers some such deprivations. Though he can hear and touch objects, so can a sighted learner. A claim in the literature (10) is that a child learns which words refer to what - and hence their meanings - because, as he listens, he follows his mother's gaze and pointing gestures. Even supposing (falsely) that the mother of a blind child names objects only when the child is holding them, in what sense could this be equivalent to gazing and pointing, in directing reference making?

In the light of these limitations on the blind learners' opportunities to discern the referents of many heard words and sentences, we have been surprised to discover that blindness hardly delays language onset; moreover, after the first few words are said, the pattern of linguistic development is virtually identical for sighted children and for neurologically intact blind children. This includes both the development of a lexicon, used appropriately to map onto the world, and the development of syntactic structure and the semantic/relational categories this describes. Apparently, receiving different, and less, interpretive information has no dramatic effect on overall acquisition rate or the character of that learning.

Some details of the blind child's learning are quite interesting. We expected to find the largest differences between blind and sighted learners in acquiring the visual vocabulary, words like look and see, for here the information base is maximally different from the normal. However, our blind subject used these words as early as do sighted children. In the education literature, such uses are called verbalism, often said to be detrimental to the child, who should be discouraged from use of the sighted vocabulary lest he or she fall victim to "loose thinking." But, on the contrary, the meanings the blind child came up with seem quite appropriate, though, of course, they map onto a different sensory world. A sighted child told to "Look up" will tilt his head and orient his eyes upward - even if he is blindfolded during the testing. But our blind subject raises her hands, keeping the head immobile. It is not that the blind child simply conflates *look* and *touch*. For one thing, she responds to "Touch the doll, but don't look at it" by a tap or scratch on the doll, and then to "Now you can look at it" by exploring it manually. And in response to "Touch behind you" she touches her back, but in response to "Look behind you" she searches the space behind her.

On this and much related evidence, we think the blind English-speaking child has developed a distinction as made in French, between *toucher* and *tater*, between manual contact and apprehension by manual exploration. The question is whether the maternal contexts of use of the sighted terms is special, providing a basis on which the child could develop her special construals. Our finding is that no very superficial description of the contexts explain the learning. For one thing, the caretakers use *look* and *see* to their blind offspring in a surprising way: just as they do to their sighted offspring, i.e., to mean "perceive" on some occasions (e.g., "Look at this boot"), to mean "consider an event or state of affairs" on other occasions (e.g., "Let's see if granny's home," said while dialing the telephone), and to mean "resemble" on others (e.g., "Oh, you look like a kangaroo in those overalls"). As these examples begin to show, it is not even possible to say that the mother of a blind child reserves the use of *look* and *see* to occasions when the listener has a relevant object in hand or close to hand - a generalization that potentially could explain how she settled on the interpretation "explore or apprehend manually."

To be sure, there is a correlational effect here: Usually the mother speaks of *looking* when her blind child is near some target object. But the trouble is that this situational factor is not very informative. This is because the mother says a goodly variety of simple verbs (e.g., *have*, *play*, *give*, *put*, *hold*, *say*, *be*) under the same circumstances: Since the child is blind, the mother most often talks of things nearby. But this means that many verbs cannot be discriminated from each other in terms of the nearbyness of things talked about. Rather, our conclusion in the investigations of the blind is that these children recruit several sources of information that jointly can be informative about which verb has which meaning. A contribution is made by the situational factors (e.g., *look* but not *get* or *come* is usually used when a target object is nearby). But a separate contribution is made by examining the constraints on the syntactic forms in which the different verbs participate. Space forbids reproducing a full description here. But as an example, notice that give is a verb that takes three noun-phrases, the first of which expresses the agent of the action, e.g., John gives Mary the ball. But look, an inalienable perceptual activity, can express no agent, i.e., it is semantically incoherent and syntactically anomalous to say *John looks Mary the ball. We take the position that a child disposed by nature to analyze these syntactic formats can extract and differentiate the verb meanings, while a learner dependent on observation of linguistic circumstances alone has an insufficient basis for making these inductions. Considering the intricacy of the syntactic analyses required even of blind toddlers, we take their success as another argument for significant biological support of language learning.

Varying the Endowment of the Learner

The literature just sketched is consistent with a maturationally-driven acquisition process, heavily dependent on specific linguistic and perceptual representations, with progress relatively independent of exposure time or type. If this position is correct, then organisms differently endowed should not be able to learn under anything like the same exposure conditions. For cats and dogs, however, there are many arguments much weaker than their lack of a "language faculty" that will serve to explain why they do not learn English. The case has been made more interestingly for primates by Premack and Premack (46), for they have shown that chimpanzees have certain general conceptual wherewithal in common with humans: But that still does not allow them to function with syntactic categories like those of a human language. My colleagues and I have begun to look at special human populations to pursue this kind of issue.

Language learning in Downs Syndrome (DS) retardates. Fowler (18) first examined the linguistic functioning of a small group of Down's Syndrome adolescents who for three years had shown no further linguistic development, i.e., had arrived at some steady final state. They were selected for homogeneity on several measures of cognitive function (e.g., MA about six years) and an anchor measure of language function (mean length of utterance, MLU, 3.0 - 3.5). This is the level usually achieved by normals between ages 2 and 3. It is important to note that these individuals differ extensively from one another in other aspects of cognitive functioning, e.g., some of them were vastly better than others at primitive arithmetic, but this did not predict differences among them in language skill. Not only their gross language level as assessed by MLU, but also the internal properties of their language knowledge, as assessed by a variety of standard instruments used by developmental psycholinguists, were found to be the same as that of 2 1/2 year old normal controls. This similarity in the course and character of early language learning between normal and DS children has also been found by many others (e.g., (33, 36)). Hence, it looks as if DS individuals may be a diminished case of the normal endowment. (In contrast, nonDS retardates we have studied differ both from the DS individuals and from each other in linguistic developmental patterning, making them a less likely group to study for the present questions).

The important issue to Fowler, Gelman, and Gleitman (in preparation) had to do with the course of language development in the DS population, and so we instituted a longitudinal study of individuals whose IQ was about the same as the original adolescent group. This study is still underway, and the number of subjects is very small, but a fewer generalizations are already apparent. These individuals began to speak very late, at about 5 years of age. But once language was manifest at all, their rate of growth was normal for some succeeding time. Correcting for the onset-time difference, they traversed Brown's (8) first four "stages" of language learning in the same absolute period of time required by normals. The internal structure of the knowledge at each interim measurement was virtually identical to that of much younger normals traversing the same stages. (All these data pertain to spontaneous speech; measures of comprehension, including knowledge of word meaning, during this period give the retardates the advantage.) However, at this point (MLU about 3.5), the learning of the retardates in the IQ range studied came to a halt. Perhaps the halt is permanent. It certainly extends for over a year in the few individuals we have so far observed, but we have to wait to see if they may start to learn once more. The adolescent population mentioned above suggests, however, that at least some DS individuals at this IQ level reach just a ceiling of attainment, equivalent to that of two year old normals, and then learn no more.

Summarizing, the interest of the longitudinal findings have to do with two main points: a) learning is not slow, but at normal pace, until some ceiling is achieved, often at a point very early in life, and b) the character of knowledge is the same as for normals at the same stage of language development. The progress of the DS individuals, constrained as it is, suggests to us that a very low-level, automatic process is at work to determine the rockbottom aspects of linguistic function.

Other populations. So far the findings I have discussed suggest that language learning survives intact despite many differences in the exposure conditions, e.g., exposure to a pidgin or an elaborated language, to speech which varies among individual mothers, to speech with diminished opportunity to observe the world (the blind case), or to no speech at all (the isolated deaf children). In contrast, a change in endowment has dramatic consequences for language learning. For the retardates, there is a very low ceiling on accomplishment. This motivates a search for yet other populations that would allow us to disentangle effects of biological status and effects of exposure.

One such population we are studying is children who differ in gestational age at birth (Landau and Gleitman, in preparation). This work is still underway, and it has many technical problems of which the worst is that prematurity is often accompanied by neurological defects that may be relevant to language attainment. We have attempted to control such variables, e.g., by choosing only individuals whose birthweight was normal for their gestational age and who had no observable neurological abnormalities, but one should still be wary about the generalizations that can be drawn. Acknowledging this, our current findings do suggest that there is a stable effect of prematurity on language onset, i.e., onset time is better predicted by time since conception (neurological status) than time since birth (exposure time).

Another potential source of evidence for the contribution of biology to language learning is the character of the learning process in those who are exposed late to a new language, for example, child second-language learners. If biological status bears a significant burden of explanation for the character of language knowledge, independent of exposure, we might expect both the rate and the patterns of learning to differ for second-language learners. This would help explain a very striking phenomenon. A four year old foreign child, transported to America, requires only one year's exposure to speak English like a native five-year old, who has had five years of exposure and practice. Of course, another interpretation has to do with the fact that the emigre has priorly learned some other language. But a related, and again very large and striking phenomenon that escapes this defect has to do with native bilingualism. Many children are brought up in homes where two or three languages are spoken. Anecdotal evidence - but rather voluminous anecdotal evidence - suggests that the two or three languages are learned as fast as one; namely, at the level of peers learning a single one of the two or three languages. This is a pretty queer kind of learning, it seems.

It can handle twice as much data without apparent strain - and handle the additional problem of disentangling two data bases which, if confused, would yield an incoherent system. A theory in which induction from information provided in the environment is not the limiting factor on rate of acquisition could handle these facts (if they are real facts) rather easily: It would be the present expressive power of the learning machinery that is the limiting factor in language growth.

Experimental evidence on this topic is thin. Though there is an enormous literature on child bilingualism and second-language learning, generally it has not focussed on the kinds of issue I have considered in this paper. There is only one study I know of that seems to attack them directly in the way that is required - that is Newport and Supalla's ongoing study, mentioned earlier, of deaf individuals learning ASL at different ages. Because of "oralist" teaching methods and because deafness is often acquired late, there are cases of deaf individuals learning a first language at ages ranging from infancy to the late forties! They are left without a formal language until put in a situation where ASL is used. Newport and Supalla's preliminary findings are that the character of final knowledge of the manual language is predictable from the age of the learner at first exposure, independent of the number of years the individual subsequently used it, e.g., as stated in an earlier section of this paper, late learners fail to acquire the closed class ASL morphology even after decades of exposure and everyday use. Such findings strengthen the case that the neurological status of the learner is dramatically implicated in what he can learn. In this case, evidence was provided supporting the lay impression that young children are better language learners than adults. Whether that evidence is strong enough to support Lenneberg's claim for a "critical period" roughly equivalent to that involved in duck imprinting or birdsong learning remains for further investigation to determine.

CONCLUSIONS

I have tried to describe some of the complex facts about language and its learning that I suppose are at least within calling distance of an explanation just in case there are task-specific, biologically given predispositions in humans to support them. I know of no extant learning theory specific enough in its claims about the human representational system and learning strategies to explain these same facts and to subsume as well learning and knowledge of other human cognitive systems. Possibly, such a global learning theory can be developed and can be successful.

Until or unless such a theory is developed, such language-specific learning devices as proposed, e.g., by Wexler and Culicover (55), are the closest thing we have to an account of how language is acquired. I myself therefore do not take Chomsky's postulation of an autonomous "language faculty" with principles of learning all its own as an approach that makes extravagant claims on fragmentary evidence. On the contrary, I take that claim as representing appropriate scientific modesty given the state of the art in describing either language or its learning. At their best, schematic models offered by linguists and learnability theorists go some small way toward describing the awesomely complex facts about the learning of human language. They go no distance whatsoever in describing "all learning" or "all human knowledge." Therefore, at present, they are best interpreted as interim descriptions of language and nothing else. It will be a victory for psychology, but one I scarcely anticipate in the near future (say, within a millenium), if that task-specific view turns out to be too modest after all.

Acknowledgements. This paper summarizes work I have done over a period of years with many collaborators who are cited in the text. Many - indeed, most - of the theoretical and experimental ideas described are attributable to these collaborators. Particularly, the paper is organized around a position that postulates a bottom-up learning process mediated by the child's distinctive encoding of open and closed class members of the morphological stock; these ideas were developed in their present form in collaboration with Wanner (28). Another presentation of the ideas expressed herein, in a closely related format, appeared in (19). The paper in its present form also appeared in Language Learnability and Concept Acquisition, eds. A. Marras and W. Demopoulos, and is reprinted here with their permission and that of the Ablex Publishing Corporation, Norwood, NJ. I thank H. Gleitman for significant help in reading earlier drafts of this paper and for offering many detailed suggestions for improvement. I also thank the National Foundation for the March of Dimes, which supported much of the research described as well as the writing of this paper.

REFERENCES

(1) Bates, E., and MacWhinney, B. 1982. Functionalist approaches to grammar. In Language Acquisition: State of the Art, eds. E. Wanner and L.R. Gleitman. Cambridge, MA: Cambridge University Press.

(2) Bellugi, U. 1967. The acquisition of negation. Unpublished doctoral dissertation. Harvard University.

(3) Bever, T.G. 1982. Some implications of the nonspecific bases of language. In Language Acquisition: State of the Art, eds. E. Wanner and L.R. Gleitman. Cambridge, MA: Cambridge University Press.

(4) Bickerton, D. 1975. Dynamics of a Creole System. New York: Cambridge University Press.

(5) Bloom, L. 1970. Language Development: Form and Function in Emerging Grammars. Cambridge, MA: MIT Press.

(6) Bloom, L.; Lightbrown, P.; and Hood, L. 1975. Structure and variation in child language. Monog. Soc. Res. Child. Dev. 40 (Serial No. 160).

(7) Bradley, D.C.; Garrett, M.F.; and Zurif, E.G. 1979. Syntactic deficits in Broca's aphasia. In Biological Studies of Mental Processes, ed. D. Caplan. Cambridge, MA: MIT Press.

(8) Brown, R. 1973. A First Language: The Early Stages. Cambridge, MA: Harvard University Press.

(9) Brown, R., and Hanlon, C. 1970. Derivational complexity and order of acquisition in child speech. In Cognition and the Development of Language, ed. J. Hayes. New York: Wiley.

(10) Bruner, J.S. 1974/75. From communication to language: A psychological perspective. Cognition 3: 255-287.

(11) Chomsky, N. 1965. Aspects of the Theory of Syntax. Cambridge, MA: MIT Press.

(12) Chomsky, N. 1975. Reflections on Language. New York: Random House.

(13) Chomsky, N. 1981. Lectures on Government and Binding. Dordrecht: Foris Publications.

(14) Dorian, N. 1978. The fate of morphological complexity in language death. Language 54(3): 590-609.

(15) Feldman, H.; Goldin-Meadow, S.; and Gleitman, L. 1978. Beyond Herodotus: The creation of language by linguistically deprived deaf children. In Action, Symbol, and Gesture: The Emergence of Language, ed. A. Lock. New York: Academic Press.

(16) Fernald, A. 1982. Acoustic determinants of infant preference for "motherese." Unpublished Ph. D. dissertation, University of Oregon.

(17) Fernald, A. 1983. The perceptual and affective salience of mothers' speech to infants. In The Origins and Growth of Communication, eds. C. Feagans, C. Garvey, and R. Golinkoff. New Brunswick, NJ: Ablex Publishing Corp.

(18) Fowler, A. 1981. Language learning in Downs Syndrome children. Manuscript, University of Pennsylvania.

(19) Friedman, S., and Klivington, K., eds. 1984. The Brain, Cognition and Education. New York: Academic Press, in press.

(20) Furrow, D.; Nelson, K.; and Benedict, H. 1979. Mothers' speech to children and syntactic development: Some simple relationships, J. Child. Lang. 6: 423-442.

(21) Garrett, M.F. 1975. The analysis of sentence production. In The Psychology of Learning and Motivation, ed. G.H. Bower, vol. 9. New York: Academic Press.

(22) Gentner, D. 1982. Why nouns are learned before verbs: Linguistic relativity vs. natural partitioning. In Language Development: Language, Culture, and Cognition, ed. S. Kuczaj. Hillsdale, NJ: Erlbaum.

(23) Gleitman, H., and Gleitman, L.R. 1979. Language use and language judgment. In Individual Differences in Language Ability and Language Behavior, eds. C.J. Fillmore, D. Kempler, and W.S-Y. Wang. New York: Academic Press.

(24) Gleitman, L.R., and Gleitman, H. 1970. Phrase and Paraphrase. New York: Norton.

(25) Gleitman, L.R.; Newport, E.L.; and Gleitman, H. 1984. The current status of the Motherese hypothesis. J. Child. Lang. 11(1): 43-80.

(26) Gleitman, L.R., and Rozin, P. 1977. The structure and acquisition of reading I: Relations between orthographies and the structure of language. In Toward a Psychology of Reading, eds. A. Reber and D. Scarborough. Hillsdale, NJ: Erlbaum.

(27) Gleitman, L.R., and Wanner, E. 1982. Language acqusition: The state of the state of the art. In Language Acquisition: The State of the Art, eds. E. Wanner and L.R. Gleitman. New York: Cambridge University Press.

(28) Gleitman, L.R., and Wanner, E. 1984. Current issues in language learning. In Developmental Psychology, ed. M. Bornstein. Hillsdale, NJ: Erlbaum, in press.

(29) Gold, E.M. 1967. Language identification in the limit. Inform. Control 10: 447-474.

(30) Jackendoff, R. 1983. Semantics and Cognition. Cambridge, MA: MIT Press.

(31) Joshi, A. 1983. On code switching. Information and Control, in press.

(32) Kean, M.L. 1979. Agrammatism: A phonological deficit? Cognition 7(1): 69-84.

(33) Lackner, J.R. 1976. A developmental study of language behavior in retarded children. In Normal and Deficient Child Language, eds. D.M. Morehead and A.E. Morehead. Baltimore: University Park Press.

(34) Landau, B. 1982. Language learning in blind children. Unpublished Ph.D. dissertation, University of Pennsylvania.

(35) Landau, B., and Gleitman, L.R. The Language of Perception in Blind Children. Harvard University Press, in press.

(36) Lenneberg, E. 1967. Biological Foundations of Language. New York: Wiley.

(37) Linebarger, M.C.; Schwartz, M.F.; and Saffran, E.M. 1983. Sensitivity to grammatical structure in so-called agrammatic aphasics. Cognition 13(3): 361-392.

(38) Marin, O.; Saffran, E.; and Schwartz, M. 1976. Dissociations of language in aphasia: Implications for normal function. Ann. N.Y. Acad. Sci. 280: 868-884.

(39) Miller, G.A., and Johnson-Laird, P.N. 1976. Language and Perception. Cambridge, MA: Harvard University Press.

(40) Nelson, K. 1979. Structure and strategy in learning to talk. Monog. Soc. Res. Child Dev. 38 (Serial No. 149): 1-2.

(41) Newport, E.L. 1977. Motherese: The speech of mothers to young children. In Cognitive Theory, eds. N.J. Castellan, D.B. Pisoni, and G. Potts, vol. 2. Hillsdale, NJ: Erlbaum.

(42) Newport, E.L.; Gleitman, H.; and Gleitman, L.R. 1977. Mother, I'd rather do it myself: Some effects and noneffects of maternal speech style. In Talking to Children: Language Input and Acquisition, eds. C.E. Snow and C.A. Ferguson. Cambridge, MA: Cambridge University Press.

(43) Newport, E.L., and Supalla, T. 1980. The structuring of language: Clues from the acquisition of signed and spoken language. In Signed and Spoken Language: Biological Constraints on Linguistic Form, eds. U. Bellugi and M. Studdert-Kennedy. Dahlem Konferenzen. Weinheim/Deerfield Beach, Fl./Basil: Verlag Chemie.

(44) Pinker, S. 1979. Formal models of language learning. Cognition 7: 217-283.

(45) Pinker, S. 1982. A theory of the acquisition of lexical interpretive grammars. In The Mental Representation of Grammatical Relations, ed. J. Bresnan. Cambridge, MA: MIT Press.

(46) Premack, D., and Premack, A.J. 1983. The Mind of an Ape. New York: W.W. Norton.

(47) Rosch, E.; Mervis, C.B.; Gray, W.D.; Johnson, D.M.; and Boyes-Braem, P. 1976. Basic objects in natural categories. Cog. Psychol. 8: 382-439.

(48) Rozin, P., and Gleitman, L.R. 1977. The acquisition and structure of reading II: The reading process and the acquisition of the alphabetic principle. In Toward a Psychology of Reading, eds. A. Reber and D. Scarborough. Hillsdale, NJ: Erlbaum.

(49) Sankoff, G., and Laberge, S. 1973. On the acquisition of native speakers by a language. Kivung 6: 32-47.

(50) Slobin, D.I. 1973. Cognitive prerequisites for the development of grammar. In Studies of Child Language Development, eds. C.A. Ferguson and D.I. Slobin. New York: Holt, Rinehart and Winston.

(51) Slobin, D.I. 1977. Language change in childhood and in history. In Language Learning and Thought, ed. J. Macnamara. New York: Academic Press.

(52) Slobin, D.I., and Bever, T.G. 1982. Children use canonical sentence schemas: A crosslinguistic study of word order and inflections. Cognition 12(3): 229-266.

(53) Snow, C.E., and Ferguson, C.A., eds. 1977. Talking to Children: Language Input and Acquisition. New York: Cambridge University Press.

(54) Wexler, K. 1982. A principle theory for language acquisition. In Language Acquisition: State of the Art, eds. E. Wanner and L.R. Gleitman. Cambridge, MA: Cambridge University Press.

(55) Wexler, K. and Culicover, P. 1980. Formal Principles of Language Acquisition. Cambridge: MIT Press.

(56) Zwicky, A.M. 1976. On clitics. Paper read at the Third International Phonologie-Tagung at the University of Vienna, Sept. 2, 1976.

The Biology of Learning, eds. P. Marler and H.S. Terrace, pp. 585-616. Dahlem Konferenzen 1984. Berlin, Heidelberg, New York, Tokyo: Springer-Verlag.

On Perceptual Predispositions for Human Speech and Monkey Vocalizations

M.R. Petersen* and P.W. Jusczyk**
*Dept. of Psychology, Indiana University
Bloomington, IN 47405
**Dept. of Psychology, University of Oregon
Eugene, OR 97403, USA

Abstract. Early models of infant speech perception posited the existence at birth of special speech-processing mechanisms. More recent research with both adults and infants has questioned the fundamental claim that speech sounds require special sensory coding strategies. Instead, if a speech-specific perceptual mode exists at all, it probably takes the form of certain attentional and interpretive processes enlisted when treating speech signals as linguistic messages. Interestingly, this view is also consistent with some recent research on vocal perception mechanisms used by monkeys. A shift in emphasis from speech-specific sensory codes to interpretive specializations leads to a new model of the development of speech perception capacities.

INTRODUCTION

Although learning can be (and has been) defined in a variety of ways, a common goal of all theories of learning is to account for how associations are formed between stimuli and other events. Some theorists might place heavy emphasis on the role of stimulus-stimulus associations while others highlight the importance of stimulus-response associations, but all show some measure of concern for how an organism learns to connect specific stimuli with other aspects of its environment and behavior.

Until quite recently most workers in the field of animal learning tacitly assumed that, aside from obvious hedonic and intensity differences, all stimuli were approximately equivalent to one another with respect to ease of associability. Seligman and Hager (66) term this the equipotentiality perspective and point out that it held sway over nearly all the research conducted on basic learning processes in animals until the late 1960s and early 1970s. Beginning about that time, a number of studies reported that the nature of the stimulus was an extremely important variable inasmuch as it could either impede or facilitate performances in simple learning tasks. For example, in a classic study Garcia and Koelling (20) found that when food flavor and an auditory-visual cue were paired with radiation-induced illness, only the food flavor became aversive. But when the same stimuli were paired with electric shock to the feet, only the auditory-visual signal became aversive. A wealth of reports of this general type have flooded the literature since the publication of Garcia and Koelling's seminal finding and all point in the same direction: stimuli for which organisms are biologically adapted to attend or associate produce different patterns of behavior than stimuli which are, in an evolutionary sense, arbitrary or irrelevant. One of the more hotly contested issues to arise out of this new research is the question as to whether the character of the stimulus fundamentally alters the learning process itself or whether general learning processes are simply responding to the high salience of biologically relevant cues (see Hollis and Revusky, both this volume). But one could just as well ask whether the stimuli themselves undergo specialized perceptual processing or whether they receive the same, general treatment accorded any stimulus. On this view, the outcome of either special or general perceptual processes could conceivably serve as input to either a special or general set of learning processes. In any event, the central point is that the stimulus itself might be accorded special consideration, irrespective of the nature of the learning process.

Although most traditionalists in the animal learning field have only recently come to appreciate this point, their counterparts in the areas of human learning and cognition have been wrestling with distinctions between special and general processes for nearly three decades. This is especially evident in the research and theories generated about human language processes, most particularly with regard to the perception of spoken language. On the basis of a vast array of studies comparing speech perception with ordinary auditory perception, it has been concluded by many that speech and non-speech sounds enlist distinct perceptual processes. For instance, early evidence showed that speech is perceived categorically while non-speech is processed in a rather continuous fashion.

Furthermore, it was shown that important speech contrasts are processed primarily by the left cerebral hemisphere, whereas non-speech signals evoke either right-hemisphere or bilateral processes. Thus, it was suggested, whereas a very general set of auditory mechanisms are adequate for most acoustic signals, speech requires a mode of processing specially adapted for decoding various peculiarities of the highly complex speech signal (41, 42). This concept was then extended to the problem of early language learning to argue that if this special processor is part of the infant's innate endowment, then perhaps it makes a crucial contribution to language acquisition by treating speech differently from other sounds even before the child is able to comprehend the meaning of the utterances (13).

So, we see that the concerns of animal and human learning researchers have begun, in a sense, to converge on a common issue: Do organisms make use of special perceptual processes when attending to biologically significant stimuli? This is, of course, an important question that must be addressed eventually if we ever hope to understand the nature of the perceptual contribution to learning processes.

As hinted above, the area of human speech perception - in both adults and infants - has long been preoccupied with the specialization question. Consequently, there is a substantial history of research and theory in the field which can be used to illustrate the sorts of issues that emerge in the search for special versus general perceptual processes. Inasmuch as the charge of this paper is to summarize the evidence for perceptual predispositions in human infants, one objective will be to examine the ongoing controversy over whether the human infant possesses language-specific perceptual capacities. A set of questions parallel to those raised about human speech perception have also been addressed in recent studies of how animals perceive their vocal communication signals, so some of that work shall be discussed in the spirit of providing a comparative perspective. Finally, a model will be outlined which attempts to relate the infant's initial speech perception capacities and proclivities to the early phases of language acquisition wherein the child's perception of speech begins to approximate that of the adult.

THE CASE FOR SPEECH-SPECIFIC PERCEPTUAL PROCESSES IN INFANTS

The search for speech-specific perceptual capacities in infants has, from the start, been closely tied to research and theory about adult speech perception. So, it is important to appreciate why some have seen a need to posit the existence of special processes in adults. Early attempts to study adult speech perception confronted a thorny problem which to this day remains one of the major, unresolved issues in the field. The essence of the problem is the apparent lack of direct correspondence between the acoustic realization of a speech sound and the percept it elicits in a listener (42). For example, although most English-speaking adults readily percieve the [d] in /di/, /da/, and /du/, extensive spectrographic analysis - the principal tool for acoustic analysis of speech - has failed to uncover an invariant acoustic cue for [d] across the different vowel contexts. Faced with this general type of problem, researchers at Haskins Laboratories suggested that the invariance might instead lie in the rules used by the speaker to produce the sound (42). If it is assumed that the listener has knowledge of these rules, then perhaps he perceives the speech sound by calculating how he would have produced it. This view was termed the motor theory of speech perception, and it represented an explicit suggestion that there is something unique about speech perception: other acoustic stimuli could be processed by a general set of auditory system processes, but certain speech sounds involved special pathways that depend critically on language-specific knowledge. This bold claim instigated a whole new line of research which sought to pinpoint the ways in which the perception of speech and non-speech could be distinguished. One of the most important findings to emerge out of this was evidence that speech appeared to be perceived categorically, whereas non-speech was perceived continuously.

Data on the categorical perception of speech comes from studies in which listeners are asked to perform categorization and discrimination tasks with speech signals from a computer-synthesized continuum (12, 43). Two or more stimuli from different phonemic categories (e.g., [b], [d], and [g]) are selected to represent endpoints and a continuum of stimuli that spans the different categories is created by varying a single acoustic feature. Listeners typically a) categorize the continuously varying speech signals into discrete phonemic classes and b) show extremely poor discrimination of different stimuli from the same class, but excellent discrimination of stimuli belonging to different classes. Thus when discrimination performance is plotted for successive pairs of stimuli along a speech continuum, one observes marked discontinuities that correspond to the boundaries between the different phonemic categories measured in the categorization task. This pattern of results was distinct from that obtained with non-speech acoustic continua (e.g., pure tones varying in frequency) (61, 62): a) discrimination functions are typically smooth and monotonic, and b) although listeners might sort acoustically different signals into the same category, they are still able to discriminate between different exemplars within a category.

The apparent differences in the perception of speech and non-speech signals led to proposals for two distinct types of processing: a categorical mode specialized for perceiving certain types of speech sounds (especially consonants) and a continuous mode designed for general-purpose auditory perception. Subsequent findings that vowel continua (e.g., [i], [I], [E]) (19) are perceived continuously, together with the results on categorical perception of consonants, were interpreted as support for the motor theory of speech perception. Since stop consonants are produced in a discontinuous fashion by constricting discrete portions of the vocal tract whereas vowels are produced by continuous changes in the position of articulators, the motor theory expects categorical perception of the first and continuous perception of the latter. Thus, although the categorical mode was reserved for speech sounds, the continuous mode played a role in both speech and non-speech perception.

Findings of this sort set the initial backdrop for studies of infant speech perception: Do linguistically inexperienced infants come to the world already endowed with the specialized categorical mode of processing? The pioneering study of this question was conducted by Eimas and his colleagues (16) who examined 1- and 4-month-olds' abilities to discriminate synthetic /ba/ and /pa/ syllables. The acoustic distinction between these sounds resides in a feature known as voice onset time (VOT), which is a measure of the point in an utterance when voicing (laryngeal pulsing) begins. For English /ba/ voicing is present throughout the syllable (VOT = Omsec), but it is delayed by about 25msec in /pa/ (VOT = 25msec). Extensive research with adults has shown that when this feature is systematically varied to produce a VOT continuum, the continuum is categorically perceived with a boundary in the vicinity of +30msec. The infants were tested in a habituation paradigm with several different pairs of stimuli from a VOT continuum and showed discrimination results quite comparable to those obtained with adults. Pairs of stimuli drawn from the same phonetic category were indiscriminable from one another, but pairs selected to straddle the category boundary were easily discriminated. Thus, the results indicated the existence of a categorical mode of processing in infants who had not received extensive linguistic experience. This finding led many researchers to conclude that categorical perception was an innate trait of the human infant and, since the categorical mode was thought to reflect a speech-specific process, that infants were biologically predisposed to perceive speech sounds (13). In the years since Eimas et al.'s study, infants have been tested on numerous other phonetic contrasts and, by and large, the studies indicate that infants are capable of distinguishing virtually all of them (see Aslin, Pisoni, and Jusczyk (3) for a thorough review).

Even though the infants in the Eimas et al. study were too young to use language, they doubtless received some exposure to the English /ba/-/pa/ contrast in their homes before they were laboratory-tested. So, it might be argued that this moderate, passive exposure could have influenced the results. However, this possibility was discounted by subsequent reports that infants from non-English-speaking environments, where the VOT boundaries actually differ substantially from those for English, showed sensitivities in the same VOT regions as Eimas et al.'s subjects (40, 72). In addition, others have shown that these cross-language parallels extend to contrasts other than VOT, so that the finding seems to represent a general characteristic of language perception (73). Interestingly though, Werker and Tees (76) have recently shown that although 6- to 8-month-olds from English-speaking homes are capable of discriminating contrasts from foreign languages, by 8-10 months of age, discrimination performance begins to deteriorate, and by 10-12 months it is even more impaired. Finally, when tested at 4 yrs of age, children are unable to discriminate the same foreign-language contrasts they were capable of distinguishing at 6-8 months. Therefore, although specific language experience appears not to affect performance at the very early ages that Eimas et al. dealt with, it may well play a role in helping to maintain discriminative capacities in children as they grow older.

The evidence reviewed to this point has all been consistent with the view that speech engages a unique perceptual process - the categorical mode - and that infants appear to be innately endowed with this same capacity. On this basis one might be tempted to conclude that the categorical mode represents an important means by which the apparently overwhelming task of learning a first language is reduced to manageable levels by providing the infant with an automatic processor specifically attuned to complex speech signals. However, the notion that categorical perception represents a processing mode unique to speech, and any claims associated with it, has come under heavy criticism from a number of directions over the last few years.

The most serious challenges to the "speech is special" perspective have come from: a) the demonstration that certain classes of complex, non-speech sounds are categorically perceived and b) the finding that animals show evidence of categorically discriminating human speech sounds. For instance, Pisoni (60) and Miller et al. (51) constructed continua of non-speech sounds that contained the important acoustic cues for VOT, but in a nonphonetic context (i.e., tones and/or noise). When faced with these sorts of signals, adult human listeners categorized and discriminated them in a speech-like, categorical manner. Moreover, Jusczyk et al. (28) subsequently showed that the non-speech signals used by Pisoni (60) were also categorically discriminated by infants. Thus, instead of being specific to speech, it would seem that the categorical mode employed by both infants and adults could be accounted for in terms of general auditory processing capabilities. Further support for such a conclusion comes from studies that tested animals with the very speech continua that elicit categorical processing in humans. For example, Kuhl and her colleagues have shown in a series of experiments that animals (chinchillas and monkeys) both categorize and discriminate speech sounds in a way that is virtually indistinguishable from humans (34-37). As a consequence of findings of this sort, most workers have abandoned the idea that categorical perception reflects a speech-specific mode. But, at the same time, a new wave of studies have proffered a different set of phenomena as indicative of the unique character of speech perception (41, 64).

One of these phenomena has been studied in both adults and infants and was interpreted as indicating specialized speech processing by both. Eimas and Miller (15) found that the infant's discrimination of formant transition duration differences used to signal a contrast between [ba] and [wa] depended upon contextual information in the form of syllable duration, even though the information for syllable duration came well after the transition information. The argument that the infant's behavior in this setting is indicative of specialized processing mechanisms rests on certain assumptions drawn from a study with adults by Miller and Liberman (50). In their study, Miller and Liberman observed a similar effect of syllable duration on the locus of the perceptual boundary between [ba] and [wa]. Whereas the [ba]-[wa] formant transition duration boundary for syllables with overall durations of 80 msec was 32 msec, it was 47 msec for syllables with durations of 296 msec. In other words, the adult listener appears to compensate for the overall duration of the utterance in the course of deciding what phonetic segment is spoken. However, the compensation is not based solely on duration; rather, the specific nature of the information contained in the longer duration utterance plays a critical role. Miller and Liberman observed that increasing the duration of the steady-state portion of a vowel, as might happen when speaking rate slows down, had the effect of shifting the perceptual boundary towards longer transition duration values. On the other hand, an equivalent increase in syllable duration produced by adding a final stop consonant to the vowel (which would tend not to slow down speaking rate) actually produced shifts in the transition duration boundary towards shorter values. Thus, it was not just the syllable duration but the nature of the syllable structure that determined the location of the perceptual boundary. For this reason, Miller and Liberman argued that the phonetic boundary location was dependent on estimates of speaking rate. Eimas and Miller (15) employed similar logic in arguing for phonetic processing effects in their study. However, unlike Miller and Liberman (50), they were not able to assess the consequences of substituting an additional consonant in place of an increased vowel duration for their long duration syllables. Hence, there is no way of knowing whether, for infants as well as adults, the effect was dependent on the nature of the syllable structure rather than overall duration.

However, the most telling argument against the view that any specialized speech processing capacity is responsible for the effects observed in these studies comes from recent research with non-speech contrasts. Carrell, Pisoni, and Gans (10) have demonstrated that all the effects observed by Miller and Liberman with adults can be obtained with non-speech stimuli. Furthermore, Jusczyk et al. (27) have found that 2- to 3-month-olds show shifts in the discrimination of frequency transitions with changes in overall stimulus duration, paralleling the results of Eimas and Miller (15). Therefore, it appears that the type of compensation observed by Miller and Liberman is a general feature of human auditory processing rather than a specific response to a change of speaking rate.

Equally important as the ability to discriminate phonetic contrasts is the ability to recognize the same phonetic segment when spoken by different speakers or with a different inflection. The acoustic characteristics of speech sounds vary greatly from speaker to speaker, yet the adult listener is able to ignore such differences in recognizing the identity of a given word. In effect, perception of the phonemic segments is invariant across these differences. Kuhl and her co-workers have looked at the infant's capacity to ignore irrelevant differences in speaker's voice and intonation patterns in making phonetic discriminations. They first trained infants to discriminate between single tokens of two different syllables spoken by the same speaker. Then, in successive phases of the experiment, they introduced new tokens of the syllables spoken by different speakers and with varying intonation contours. The infant was deemed to have achieved some degree of perceptual constancy for the phonetic segments being tested if he or she could successfully maintain the discrimination between the two types of segments in the face of the irrelevant changes introduced by adding new tokens varying the intonation patterns and speaker's voice. Kuhl found evidence that six-month-old infants are able to ignore changes in intonation patterns and speakers' voices for both vowel (32, 33) and fricative (22) segments.

One might be tempted to suggest that the finding of perceptual constancy across differences in speakers' voices represents evidence of a speech-specific process. In particular, it might be argued that the only commonality that exists in tokens of the same syllable uttered by different speakers is phonetic rather than acoustic. While it is true that studies such as that of Peterson and Barney (59) indicate that there is a great deal of acoustic variation in tokens produced by different speakers, it is also the case that nonhuman mammalian species such as the dog (5) and the chinchilla (9) are apparently capable of adjusting to variations in speaker's voice and intonation contour. Hence, the mechanisms that extract constancies of this sort appear to be generally available in the mammalian auditory system, suggesting a basis in some measure of overall acoustic similarity rather than an analysis into speech-related component dimensions.

In summary, an examination of the possible grounds for attributing the infant with specialized speech processing mechanisms reveals very little support for the notion. Instead, it appears that the existing body of data from studies of infant speech perception can be explained in terms of general processes and mechanisms of the human auditory system.

IS THERE A SPECIAL SPEECH MODE IN ADULTS?

Having dismissed the claims for the existence of specialized speech processing mechanisms in infants, it might seem as though a wholesale rejection of the claim for a specialized speech mode is in order. However, there currently exists a substantial body of evidence from research with adults that is very difficult to account for without assuming that speech sounds undergo some form of specialized processing. In particular, there are a number of studies that demonstrate that the same sounds can be processed in quite different ways depending upon the listener's attentional set to hear them as speech or non-speech signals.

One experimental paradigm which has been employed in studies of speech and non-speech processing involves the dichotic presentation of different portions of a speech syllable (44). For example, the third formant transition may be played to one ear while the remaining portion of the syllable is played simultaneously to the other ear. Under such testing conditions subjects report hearing the acoustic signal as both speech and non-speech simultaneously. The so-called "duplex perception" is one of hearing both a speech syllable and a chirp. Thus, the information in the third formant transition contributes to the perception of the whole syllable as well as its standing alone as a chirp. Liberman et al. (44) showed that listeners were able to make independent judgements about the speech and non-speech qualities of the stimuli. For example, varying the intensity of the third formant transition affected only judgements about the perceived loudness of the chirp, and not the overall syllable. The implication is that the third formant transition undergoes two modes of processing simultaneously, and that one of these modes is used in the perception of speech.

Further evidence for the view that speech sounds undergo special processing comes from studies that have employed ambiguous stimuli (4, 8). Bailey et al. (4) created a set of non-speech stimuli by replacing the formant structure of synthetic speech syllables with frequency- and amplitude-modulated sinewaves. Of most interest was their finding that perceptual boundary shifts occurred when subjects were instructed to hear the stimuli as speech rather than non-speech sounds. A more recent investigation by Best et al. (8) upholds Bailey et al.'s original finding of differences under speech and non-speech expectations. More specifically, Best et al. observed a trading relation between two cues - the onset value of the first formant and the duration of a silent closure interval following an initial fricative sound - only when subjects perceived the sinewaves as speech. The trading relation with the sinewaves mirrors the one which occurs with speech sounds, where less silence is needed to change "say" to "stay" when the first formant has a low onset than when it has a high onset. No subjects who interpreted the sinewaves as non-speech stimuli gave any evidence of employing a trading relation between the first formant onset value and silent closure interval. Thus, the trading relation emerged only under conditions in which subjects analyzed the sinewaves as speech. This finding suggests that subjects are employing different criteria in evaluating speech and non-speech signals.

To the extent that listeners do weight the information available in the acoustic signal differently when they are set to interpret it as speech, it becomes sensible to refer to a special mode of perception for speech. The studies involving duplex perception and ambiguous sinewave stimuli are suggestive of such specialized processing. There are at least two possible ways by which specialized speech processing might come about: a) the speech signal might be processed by some specialized perceptual mechanisms, or b) special interpretive strategies may be used when treating the acoustic signal as a linguistic message (phonological categorization). As noted earlier, the available data from human infants and adults and from animals would seem to favor the second sort of explanation. By this latter view, the speech mode of perception develops as a consequence of trying to attach linguistic meaning to speech. The speech mode is not the result of a set of specialized innate perceptual mechanisms; rather, it is an interpretive schema for weighting the acoustic information gained via general auditory processing mechanisms. In other words, the difference between speech and non-speech modes of perception lies solely in the weightings assigned to various aspects of the acoustic signal. A similar view of the differences between speech and non-speech processing has been expressed by Oden and Massaro (54).

COMMUNICATIVE VALENCE AS A DETERMINANT OF SPECIAL PROCESSING IN MONKEYS

A series of studies conducted by Petersen and his co-workers on vocal perception in monkeys has converged on an interpretation of specialized processing which is remarkably consistent with that offered for human speech perception (6, 56, 57, 59, 79). We present them in the present context in the hope of broadening the perspective from which researchers and theorists ordinarily view the perception of human speech processes. Work with nonhuman species is beginning to alert us to the fact that some of the characteristics of the speech perception process are not all that unique to human language per se but may represent general solutions to the problems presented by any complex acoustic communication system. In this sense, the consideration of monkey vocal perception mechanisms helps put the study of speech-related processes into a biological context.

Primate vocalizations were long viewed as nothing more than non-semantic markers of emotional and motivational states, but recent research (11, 67, 68) suggests that they can, in fact, serve rudimentary semantic functions. For instance, Seyfarth, Cheney, and Marler (67, 68) have shown that the vervet monkey uses acoustically distinct calls to warn other group members of the approach of each of the major classes of animals that prey upon them. Even more striking, though, is the fact that without seeing the predator itself, vervets will respond to each of the calls with a unique response. For example, upon hearing a snake call a vervet will stand up and look down at the ground around it, whereas upon hearing a leopard call it will take to the trees. Follow-up work has extended these observations to other vocalizations (11) and other species (70). Thus, a view currently held by many (see (71) for several recent reviews) is that monkey vocalizations are like human speech sounds in the sense that any single utterance carries many pieces of information. Some features are of the "paralinguistic" sort, allowing determination of individual identity, sex, age, emotional state, and so on; but other features are likely to carry semantic-like information about the environment, an individual's relative status, future intentions, and the like. A major objective of much research in this area is to dissect monkey calls and identify the correspondences between functions and acoustic features. Of course, this entails conducting perceptual studies with the monkeys to determine how they process these cues. It was in the course of doing just that that Petersen and his collaborators uncovered perceptual specializations that are strikingly similar to some observed in human speech.

The study species was the Japanese macaque. Green (21) examined the vocal repertoire of this animal and developed a rather exhaustive taxonomy of its lexicon. The Japanese macaque produces upwards of 80-90 different vocalizations, of which two were selected for intensive study in various perceptual tests. The calls chosen were the smooth early high (SE) and smooth late high (SL) coos which, according to Green, occur in distinct behavioral contexts. Both calls are highly tonal, frequency modulated (FM) signals that can be differentiated on the basis of the location, within the call, of the FM segment: in SE's it appears during the first two-thirds of the call, in SL's it occurs in the last third. Interestingly, though, there is considerable variability in the precise location of the FM segment both across and within individual monkeys. Additionally, the calls vary in many other respects including duration, fundamental frequency, presence of harmonics, extent of modulation, and so on. Thus, if the Japanese macaques recognize these variants as belonging to two call classes, they presumably normalize this variability to extract a more or less constant percept. The first test, then, was to determine whether they exhibited perceptual constancy of the sort observed for human speech. Using Green's acoustic criteria, several exemplars of the two call types were selected from a pool of vocalizations recorded in the field. Several Japanese macaques were trained, using operant techniques, to respond differentially to a single token of each class, and then their ability to generalize to novel instances of the calls was measured by progressively enlarging the number of SE's and SL's they were required to discriminate. To control for the possibility that the animals were simply learning a discrimination imposed upon them by the structure of the experiment, rather than exhibiting a natural tendency to differentiate among the two call types, several non-Japanese comparison monkeys were also tested on the identical task. The Japanese monkeys responded correctly virtually immediately to the novel exemplars, but the comparison monkeys required extensive training before they finally mastered the task. This hinted that the Japanese monkeys were prepared to attend selectively to the FM segment of the calls while the comparison monkeys were not. This was verified in a subsequent experiment which required all the monkeys to differentiate the same calls on the basis of fundamental frequency (high or low) instead of the temporal location of the FM segment. In this case, the comparison monkeys quickly learned the discrimination but the Japanese monkeys had great difficulty. This suggested that the Japanese monkeys were especially attentive to the temporal location of the FM segment, a strategy which facilitated their performance on the first task but interfered with performance of the second. The comparison monkeys showed the opposite tendency and the reverse performance pattern.

Interestingly, the pattern shown by the Japanese monkeys - rapid learning of a discrimination which required attention to a communicatively important dimension and slow learning when forced to ignore the same cue - is precisely the result obtained by Kuhl and her colleagues for the discrimination of phonemes by infants (32, 33). To be sure, infants are able to readily ignore irrelevant variations in intonation contour and speaker's voice in solving phonetic discriminations for both vowel and fricative segments. But when the tables are turned and they must ignore the phonetic feature and attend to the other cues, the phonetic cue interferes, resulting in poorer performance.

It is especially noteworthy that although the Japanese and comparison monkeys learned the FM segment task at different rates, the latter did eventually master the task. This suggested that although both groups of monkeys had the necessary sensory capacities for performing the discrimination, since the FM feature was of no communicative value to the comparison animals it was not preferentially attended to. In fact, once the animals had reached asymptotic performance on the FM segment task they were tested for generalization to 26 novel calls and no species-differences were observed. Thus, the contention that the comparison and Japanese monkeys both had the requisite sensory processes was supported. The main difference among them was that the FM feature was communicatively significant to the Japanese macaques and not to the others, so the Japanese macaques learned the initial discrimination more readily.

A more recent study in this series suggests that, although the Japanese and comparison animals are both attending to the same FM feature once they master the task, they appear to be using different neural processing strategies (58). This evidence comes from a study which permitted measurement of lateralization processes while the animals discriminated among the SE and SL calls. Stimuli were presented alternately, on a random basis, to one ear at a time. By comparing the discrimination performances of the individual ears a measure of neural lateralization was obtained. Of course, the typical finding with adult humans is that speech sounds produce a right ear performance advantage (REA), indicative of left hemispheric dominance. The Japanese and comparison monkeys were tested for several weeks on a discrimination involving 15 calls and then, to ascertain whether they were all attending to the same acoustic feature (the FM segment), they were tested for generalization to 26 novel calls. The generalization test confirmed that both groups of monkeys were listening to the same feature, but the laterality data showed that only the Japanese macaques had an REA; the comparison animals exhibited no significant ear performance advantage. Hence, it appears that, at some level, the neural processes employed by the monkeys for whom the sounds have some meaning are different than those used by animals for whom the calls are communicatively inconsequential.

Now, it is an open question as to whether this difference among monkey species is largely a consequence of genetic predispositions as opposed to the social communication environments in which they are reared. But it is interesting to note that a parallel of sorts has been noted for humans with different native languages. In Thai, for example, changes in fundamental frequency serve a critical linguistic function by underlying certain semantic distinctions (1, 2). In English, though, fundamental frequency changes serve only a paralinguistic function by providing information about a speaker's emotional state, sex, or age. This difference in feature usage has a correlate in lateralization; namely, Thai speakers show an REA for fundamental frequency whereas English speakers do not (39). It seems highly unlikely that genetic differences among the respective human populations underlie this neural processing difference - especially in light of the evidence presented above on the human infant's remarkable ability to discriminate the contrasts of every foreign language with which they have been tested. Rather, it seems probable that one's language experience results in a weighting of those features distinctive to the language - a process which might be reflected in the lateralization of language processes. The obvious question, then, is whether the acquisition of communicative competence by Japanese macaques might follow a similar path. Only ontogenetic studies can tell for sure. But, for the moment at least, it seems worth giving serious thought to the possibility that the ontogenetic processes in humans and monkeys might be highly similar.

THE RELATIONSHIP BETWEEN INFANT SPEECH PERCEPTION CAPACITIES AND LANGUAGE ACQUISITION

The research that we have reviewed to this point suggests that when it approaches the task of acquiring a language the human infant is already in possession of a number of important speech perception capacities. For example, categorical discrimination and perceptual constancy provide the infant with a means of coping with the variability that exists between speech tokens of the same type when produced by different speakers or by the same speaker on different occasions and at different speaking rates. The fact that such categories exist for infants reduces what otherwise would be an infinite variety of different types of speech sounds down to a manageable few. Were it the case that languages differ only in which subset of categories they include in their sound structure, and <u>not</u> in where the boundaries for the same categories are drawn, then the task of acquiring the sound structure of the language might be reduced to determining the correct subset of the categories and the appropriate rules for combining them. However, the available data suggest that the situation is more complicated than this. Although research on this problem has been extremely limited, there is some evidence that differences do exist in the location of the perceptual boundary for certain phonemic contrasts. For example, Williams (77) has shown that the perceptual boundary for VOT differences occurs at about +4 msec for

Spanish speakers, but at about +25 msec for English speakers. These findings with adult listeners stand in marked contrast to those obtained for infants. The latter indicate a relatively uniform perceptual boundary for discriminating VOT differences in infants of different language backgrounds (e.g., English (16), Spanish (40), and Kikuyu (72)). Together the findings for infants and adults suggest that infants must learn not only which categories their language includes, but also how the category boundaries are drawn in the language. Thus, any attempt to relate early speech perception abilities to the process of language acquisition must take into account the fact that these capacities are themselves shaped by the sound structure of the language being learned.

Putting aside for the moment the fact that the sound properties of the language being learned will ultimately affect the nature of the perceptual categories, what can we say about the nature of the infant's earliest representations of the sound properties of language? At first glance, it might seem reasonable to assume that the earliest representations of the sounds of language are directly equivalent to categories based on the finest discriminations that infants are capable of making. Although this possibility cannot be rejected out of hand, there are a number of reasons to be skeptical of a claim that the underlying representations of speech sounds can be equated with the limits of perceptual capacities determined in some test setting. The data from infant speech perception studies are collected under conditions which, given present knowledge and available methodological tools, present the infant with the best opportunity to display the full extent of his or her perceptual capacities. Usually, only a single contrast is presented at a time, and the infant is exposed to multiple presentations of the sounds to be discriminated. Great efforts are taken to minimize any distractions which might compete for the infant's attention. Such test conditions are rarely, if ever, duplicated in the everyday world of the infant. Instead, there are competing sources of information available, and seldom any systematic presentation of information regarding contrasting speech sounds. Hence, a representation equivalent to the infant's perceptual limits is an ideal unlikely to be realized in the normal experience of the infant.

Even if one were able to counter the suggestion that the test conditions in infant speech perception experiments are seldom achieved in the natural setting of the infant, it would not necessarily follow that the infant's representation of speech sounds would be equivalent to units corresponding to the minimal contrast discriminable between any two utterances. Thus, the finding that infants are able to discriminate between two utterances that differ by only a single phonetic feature does not necessarily imply that they represent the utterances as collections of features, or even as a series of phonetic segments. A much more global representation of the utterances is also a possibility, provided that it serves to differentiate them. In fact, discrimination data alone will not permit us to determine what units the infant uses to represent speech. For, although these data might inform us of the fact that the infant is able to discriminate the occurrence of a given phonetic segment, say, [b], in all possible syllable contexts from every other type of phonetic segment, it does not necessarily follow that the infant perceives the [b]'s in [ba], [bo], [bi], [bu], etc., to be the same element. The only way to determine the latter is from studies that explicitly test categorization. Unfortunately, attempts to do so directly have been unsuccessful up to now, and what indirect evidence there is, is at best equivocal (e.g., (29)).

There seems to be some agreement at present in the field of infant speech perception that the stretch of information with which the infant deals is apt to be something on the order of a syllable (3, 7, 14, 26). The chief grounds for arguing in favor of the syllable comes from two sources. First, there is the failure to produce clear-cut evidence that infants recognize the identity of phonetic segments in different syllable context - e.g., the [b] in [bi], [ba], [bu], [bɛ] (Jusczyk and Derrah, in preparation; see also (29)). Second, Bertoncini and Mehler (7) found that infants were better able to discriminate speech patterns that conformed to lawful syllable structures than those which violated such structures (e.g., when the stimuli consisted only of clusters of consonants).

Given our starting assumption that the pre-linguistic infant's representation of speech takes the form of syllables, what can we say about the role of speech perception capacities in language acquisition? As an initial step in this direction, it would be helpful to consider the nature of the task that confronts the infant. One fundamental prerequisite for acquiring a spoken language is to be able to identify the different words of the language and to be able to discriminate them from one another. In other words, the infant must begin by acquiring some sort of vocabulary in the language. To be sure, the infant will also have to learn the rules for lawfully relating the words of the language to one another in complex utterances, but the first step has to involve learning to recognize the vocabulary items that will be related to one another. Moreover, another consideration is the speed with which the words can be recognized. The skilled listener is normally able to understand speech spoken at rates of 3 or 4 words per second. Thus, a constraint on the language learner is that he or she arrive at some sort of organization that makes it possible to understand speech produced at such rates. Clearly, a process for recognizing speech that involved exhaustively comparing the incoming signal serially to all elements in some mental lexicon would not be a plausible procedure given the time constraints. Hence, one expects that there is pressure on the infant to organize the items-to-be-learned in such a way as to facilitate the rapid recognition of words. Undoubtedly, the nature of the basic speech perception capacities is a critical factor in determining how the infant organizes the items in his recognition vocabulary. Moreover, it is likely that the sound properties of the language being learned also constrain the form of the organization that the listener employs. Indeed, it may be that the attempt to find the optimal organization for processing words in a particular language is ultimately responsible for the cross-language differences in perceptual boundaries that have been observed for adult listeners. Finally, an additional point to consider is that within a given language, words often take on multiple forms, and that the relationships between these forms are generally lawful (e.g., the relationship between singular and plural forms of the same word). For the average listener such variations apparently pose no great difficulty, even on first hearing a new variant of some familiar lexical item (provided that the context is appropriate). Thus, the underlying organization for the word recognition process should be structured so as to permit these generalizations naturally.

With these conclusions in mind, let us speculate as to the role of infant speech perception capacities in the acquisition of a particular language. From the point of view of the perception of language, when the infant learns a particular word, then he or she must store some form of representation of that word that can be used to check against information available from the auditory analysis of incoming utterances. When there is a satisfactory match between the output of the analysis and the stored representation for a given word, then that particular word will be recognized.

There are two important points to be made here. First, it should be obvious that the auditory analysis of utterances goes on even before the infant has "learned" a word. This type of analysis is available for all the utterances that the infant hears. In fact, it is this analysis that is the source of the infant's discrimination of speech sound contrasts in the typical speech perception experiments. Second, the information that is stored as the representation of a word is only an approximation of that available from the auditory analysis. Thus, the infant might begin by storing only a few highly salient properties of the word such as its prosodic structure and overall acoustic shape. Later the representation would be elaborated by the addition of other information that would help to distinguish it from other words with similar sound properties. In the end, the information contained in this representation would constitute a prototype of the sound properties of the word.

At this point it is useful to consider the nature of the information available in the analysis of the incoming acoustic signal. Processing at this first or pre-phonetic level is apt to be continuous as opposed to being segmented into processing units. The speech signal itself is transformed in various ways by the peripheral auditory system. The transformations involved are a consequence of such things as the temporal resolution of the auditory system, the sensitivity of the ear to different frequencies, auditory masking, and the like. Analytic processing routines applied to this transformed signal provide a description of the important acoustic characteristics present in each syllable. In essence, these analytic processing routines comprise the dimensions on which any acoustic stimuli - speech or non-speech - might be compared. Categorical decisions are based on the outputs of each of these analytic processing routines. The routines are part of the innate endowment of the infant; a fact which explains the similarities observed in the early processing of speech sounds by infants from different language backgrounds.

Thus, the analytic processing routines yield a description of the incoming acoustic signal which forms the basis for any representation of a word that the infant might develop and store. At first, new words might be stored as separate entities with no particular organization relating them, other than that they are language items. Consequently, whenever the infant attempts to identify a particular word, it would be necessary to search exhaustively through the set of stored items until a match was obtained. Whereas such a procedure might be workable as long as the number of items to be searched remained small (say, less than 50), it would soon become unwieldy as the number of items greatly increased. At this point, some sort of systematic organization of the representations of the vocabulary items would be required. It is reasonable to suppose that this organization is based on characteristics of the information available in the representations. One suggestion (cf. (26, 31)) is that information about the spectral characteristics of the onsets of words could provide the basis for the organization. According to this view, words having similar spectral onset characteristics would occupy positions relatively close to one another. The obvious reason for choosing an organization on the basis of onset characteristics is that this matches the temporal sequence of the input string and would allow the perceiver to begin processing the incoming utterance before it has been completed. An organization of this sort, whereby processing can begin immediately, is important for the listener to be able to cope with the speed with which successive words are presented under normal speaking conditions. By organizing things along these lines, the perceiver essentially narrows the size of the set of the items to be searched during word recognition to a smaller set having similar onset characteristics. Given our original assumption of a perceptual unit on the order of a syllable, it would follow that syllables having similar onset characteristics would be grouped together, and especially that multisyllable words having the same initial syllable would occupy adjacent locations. For multisyllabic words, the onset characteristics of non-initial syllables would further constrain the location of these words in the growing recognition network. Although we have posited an organization based on the spectral onset characteristics of syllables, it is certainly plausible that there is some other basis that is derived from the information available in the analytic processing routines. In any event, it is likely that the information available through such routines does serve to further constrain the organization of the network of stored word representations.

At this point it is worth considering in a bit more detail the nature of the representations employed in the network. As noted earlier, the initial representations are likely to be quite global, incorporating a limited number of salient features of the acoustic form of the word. In this respect, our view is similar to the position taken by a number of investigators studying the acquisition of phonology. These researchers have suggested that the initial phonological distinctions occur between global features of whole words, with a gradual progression towards more fine-grained units such as phonemes (e.g., (17, 49, 53)). By our view, as the size of the vocabulary grows, there will be more pressure to make the form of the representation more specific to differentiate it more readily from other similar words that the infant has learned. It is at this point that the specific nature of the sound structure of the language that is being learned begins to have an impact on the relationship between the acoustic characteristics of a word and its representation by the infant. Specifically, the task of the language-learner is to devise a representation for a particular word so that it has the greatest number of properties in common with the various tokens of that word, yet the fewest number in common with utterances of different words. In other words, the language-learner's representation will take the form of a prototype (23, 63, 65). As Osherson and Smith (55) have noted, representations in the form of prototypes are particularly well suited for identification procedures that require rapid decisions about category membership.

The nature of the prototype for a given word will depend on what other words it must be distinguished from - i.e., the characteristics of the sound structure of possible words in the language. Hence, from the information available about a word produced by the analytic processing routines, the infant would be inclined to select that most germane to distinguishing between the possible words in that language. This is apt to be a gradual process dependent on a certain amount of experience with the language. Hence, the prototype associated with a particular word would undergo continual refinement as the infant mastered the phonological structure of the language.

What might be involved in refining a prototype in this way? One possibility is that the infant learns to weight the information available through the analytic processing routines in certain ways. That is, certain types of information will be deemed as more important indicants of distinctions between words in the language than others. In effect, the perceiver prioritizes which sources of information are to be checked in word recognition and the relative importance of each. In some cases, it is possible that the analytic processing routines themselves undergo some modification (e.g., the criterion value used to determine which of the possible output messages that the routines will return would be reset from some initial default value to another one given the language context). Thus, in order to refine the prototype in an appropriate manner, the infant develops a scheme or prescription for weighting the information to be used in recognizing the words in the language. This scheme, which is specific to the language being learned, will be directly reflected in constraints that guide the formation of the prototypes to be used in recognition.

Let us consider a possible example of the way in which the phonological structure of a given language may influence the weighting scheme that the infant may develop. Voicing differences are used to signal contrasts between words in many languages. However, as Lisker (46) has noted, there are many different acoustic correlates of voicing information. It would not be unreasonable to assume that languages might differ in the exact manner in which such correlates are used to signal voicing changes. In other words, different languages might select for different aspects of the acoustic signal. A case in point concerns voicing changes in English and Spanish. In Spanish, the determination of voicing contrasts between words is largely dependent on whether voicing is present at the onset of a segment, whereas in English, many so-called voicing contrasts are cued by differences in the amount of aspiration (voiced sounds being unaspirated and voiceless ones generally aspirated as Ladefoged (38) has noted). Furthermore, Macken (47) has observed that Spanish speakers substitute a stop-spirant distinction for voicing contrasts in utterance initial positions 30-40% of the time, and that Spanish-learning children initially employ the stop-spirant distinction for this purpose in their early productions as well. Therefore, given the present view, one would anticipate that English and Spanish speakers have developed different means of weighting information relating to voicing contrasts. The available perceptual evidence is certainly consistent with this view, since it is known that the category boundary for voiced-voiceless contrasts differs for the two languages (77, 78). Thus, it follows from the present view that the cross-language differences found in studies of speech perception with adults are a direct consequence of the development of the weighting schemes incorporated into the prototypes. A further implication is that bilingual speakers would develop different weighting schemes for each language. Some support for this notion is present in the results of a study by Williams (77) who found that fluent bilingual speakers show two sets of category boundaries along the voicing continuum, depending on the language context that they are prepared to receive.

Not only does the present view account for cross-language differences in speech perception, but it also helps to account for similarities and differences that occur in speech and non-speech processing. For example, since categorical perception is attributable, at least in part, to the analytic processing routines, it is not surprising that it will be present for certain non-speech contrasts as well as for speech contrasts. Moreover, since the information available through such routines will be weighted specially in the prototypes that occur in speech contexts, some differences might be expected in the way in which the same acoustic information is perceived in speech and non-speech contexts. Thus, results such as those of Best,

Morrongiello, and Robson (8) or Bailey, Summerfield, and Dorman (4) which demonstrate that the same acoustic information is categorized differently, depending on one's expectations to hear it as either speech or non-speech, are explicable as a consequence of employing different weightings in the two settings. In effect, the specialized phonetic processing of speech that has been observed in various experimental settings (e.g., (41, 64)) is the natural outcome of developing a weighting scheme for recognizing words in a particular language.

Having considered some aspects of the way in which the structure of the representations of the individual words develop, we turn to a further examination of the word recognition network and its relation to phonological development. One important issue concerns the way in which the ability to analyze words into component phonemes arises. Recall that the perceptual units that we postulated are of syllable length. Again the organization of the word recognition network helps to explain how the ability for phonemic segmentation might arise. Words that share common initial segments would tend to be located in close proximity to each other owing to similarities in their acoustic characteristics at onset. The pressure to implement some further segmentation of the perceptual units would conceivably lead in the direction of looking for commonalities between items in the same general vicinity in the network. Common characteristics shared by a large number of items in the same general vicinity and not shared by more distant items might serve as an initial basis for developing representations corresponding to phonemic units. Once again, this would amount to developing a prototype, only in this instance for a phonemic segment. The phonemic prototype in this instance would be used either to match the output from the analytic processing routines in the case of new words to be analyzed or to match the information encoded in the prototype for the whole word when decisions regarding previously stored items are required. One important caveat is necessary here. Although phonemic segmentation might arise as a result of operations performed on the word recognition network, it would not necessarily follow that the network would be reorganized so as to provide a segment-by-segment analysis of words during the course of on-line processing. Instead, a segment-by-segment analysis might constitute an entirely independent process occurring only in special circumstances (e.g., when encountering a non-word or an unfamiliar word).

It is difficult to say exactly when the child begins to engage in phonemic segmentation. It is improbable that this type of process is one that infants would be likely to use. There is some suggestion that the ability to perform phonemic segmentation is one that arises in conjunction with learning how to read (e.g., (45, 74)). In fact, Morais et al. (52) found that illiterate adults were unable to learn a task that involved segmentation into phonemes, even though they were able to master the same task when it involved manipulating syllables.

Although we have not presented an explicit time-frame for the development of the word recognition network that we have discussed, it is likely to be a rather lengthy process, perhaps beginning with children as young as 9 months of age (76), lasting several years, and covering much, if not all, of the period studied in investigations of the acquisition of phonology. In this regard, it is worth noting that the present proposal is compatible in a number of respects with much recent theorizing about phonological development. Although earlier theories in this area tended to treat the first stages of phonological acquisition in terms of developing oppositions between phonemic segments (e.g., (25)), the more recent trend has been to view initial stages of this process as involving contrasts between larger word and syllable units with only a gradual progression to smaller cluster and phoneme units (e.g., (17, 49, 53, 75)). Thus, Ferguson (17) notes, "... opposites in the child's system are at first in terms of words, and only gradually do the partial similarities and differences between words and within words come to be in terms of segmental identities and oppositions." (p. 287). The support for his claim comes from observations of the child's early productions of words. In particular, he believes that this contention is supported by the fact that words vary greatly in the stability with which they are pronounced, suggesting that a detailed representation in terms of phonemic segments is not available to the child at this stage. Additional evidence comes from Ferguson and Farwell's (18) observations of prosodic interchanges within words (whereby the child perceives some phonetic characteristic but locates it in the wrong place or spreads it throughout the entire word in production). Of course, this view that the child starts with a global representation of words and only later derives a segmental one is consistent with our position on the development of the word recognition system.

Another shift in focus in studies of phonological development concerns the role assigned to rules in description of the child's behavior. Initially, there was a tendency to provide a description of the child's behavior in terms of the rules that the child was employing and to discuss these rules in relation to ones in descriptions of adult-based phonologies. More recently, there has been a recognition of the need to separate rules that relate to output constraints on what the child is able to articulate from those that relate to generalizations about the sound structure of language (e.g., (24, 30, 48)). Much of the early behavior of the child is best viewed as the acquisition of rules for pronunciation, rather than as acquiring rules regarding generalizations about structural properties of the language. In this respect, it is interesting to note that changes in pronunciation tend to occur first for individual words, and only later does there appear to be an attempt to generalize the potential rule to other words in the child's vocabulary (18). The latter tendency is apparently responsible for the instances of overgeneralization and regression in pronunciation that have often been observed (e.g., (30, 69)). Again the notion is that there is a tendency for words to undergo an increasing sort of phonological organization during the course of development.

It is quite likely that developments in pronunciation rules will have consequences with respect to the word recognition network that is proposed here. For example, the effort to generalize rules for pronunciation may be a factor that leads to the search for commonalities between different words, resulting in the development of representations for individual phonemic segments. More importantly, knowledge gained about phonological regularities in the language might be directly incorporated into the network in ways that might help to facilitate word recognition (e.g., by providing shortcuts or alternative routes as Klatt (31) has suggested). One consequence of reorganizing the network in this fashion would be the fact that during on-line processing, it would not be necessary to postulate the explicit application of phonological rules. Instead, the application of such rules would be a by-product of the way in which the recognition system is structured - i.e., the rules would be "pre-compiled" into the network. The effort involved in organizing the network in this way might be considerable, yet the payoff would come in terms of the increase in speed that it would permit in on-line speech processing.

Finally, we suspect that the type of representation of words that we have described in conjunction with the acquisition of vocabulary items, and subsequently in word recognition, is the representation that underlies the child's production of words. The difference between this underlying representation and the actual word that the child produces is a consequence of the articulatory routines that the child has at his or her disposal at the time.

In conclusion, the model that we have presented here is highly speculative and preliminary. There are a number of issues that we did not address, and for the ones that we did deal with, the data that exist are by no means conclusive. Nevertheless, we believe that the present account is a reasonable way to view the transition from pre-linguistic to post-linguistic speech perception capacities.

REFERENCES

(1) Abramson, A.S. 1975. The tones of central Thai: Some perceptual experiments. In Studies in Thai Linguistics in Honor of William J. Gedney, eds. J.G. Harris and J.R. Chamberlain, pp. 1-16. Bangkok.

(2) Abramson, A.S. 1978. Static and dynamic acoustic cues in distinctive tones. Lang. Speech 21: 319-325.

(3) Aslin, R.N.; Pisoni, D.B.; and Jusczyk, P.W. 1983. Auditory development and speech perception in infancy. In Handbook of Developmental Psychology, ed. H. Mussen. New York: Academic Press, in press.

(4) Bailey, P.J.; Summerfield, A.; and Dorman, M. 1978. On the identification of sinewave analogues of certain speech sounds. In Status Report on Speech Research. SR-51/52. New Haven, CT: Haskins Laboratories.

(5) Baru, A.V. 1975. Discrimination of synthesized vowels [a] and [i] with varying parameters in dog. In Auditory Analysis and the Perception of Speech, eds. G. Fant and M.A.A. Tatham. London: Academic Press.

(6) Beecher, M.; Petersen, M.; Zoloth, S.; Moody, D.; and Stebbins, W. 1979. Perception of conspecific vocalizations by Japanese monkeys (Macaca fuscata). Brain Behav. 16: 443-460.

(7) Bertoncini, J., and Mehler, J. 1981. Syllables are units in infant speech perception. Infant Behav. Devel. 4: 247-260.

(8) Best, C.; Morrongiello, B.; and Robson, R. 1981. The perceptual equivalence of two acoustic cues for a speech contrast is specific to phonetic perception. Perc. Psych. 29: 191-211.

(9) Burdick, C.K., and Miller, J.D. 1975. Speech perception by the chinchilla: discrimination sustained /a/ and /i/. J. Acoust. So. 58: 415-427.

(10) Carrell, T.; Pisoni, D.; and Gans, S. 1980. Perception of the Duration of Rapid Spectrum Changes: Evidence for Context Effects with Speech and Nonspeech Signals. Paper presented at the meeting of the Acoustical Society of America, Los Angeles.

(11) Cheney, D., and Seyfarth, R. 1982. How vervet monkeys perceive their grunts: Field playback experiments. Anim. Behav. 30: 739-751.

(12) Cooper, F.S.; Delattre, P.; Liberman, A.; Borst, J.; and Gerstman, L. 1952. Some experiments on the perception of synthetic speech sounds. J. Acoust. So. 24: 597-606.

(13) Eimas, P. 1975. Speech perception in early infancy. In Infant Perception, eds. L. Cohen and P. Salapatek, pp. 193-231. New York: Academic Press.

(14) Eimas, P.D. 1983. On infant speech perception and the acquisition of language. Paper presented at Symposium on Invariance and Variability of Speech Processes, M.I.T.

(15) Eimas, P.D., and Miller, J. 1980. Contextual effects in infants' speech perception. Science 209: 1140-1142.

(16) Eimas, P.; Siqueland, E.; Jusczyk, P.; and Vigorito, J. 1971. Speech perception in infants. Science 171: 303-306.

(17) Ferguson, C.A. 1978. Learning to pronounce: The earliest stages of phonological development in the child. In Communicative and Cognitive Abilities: Early Behavioral Assessment, eds. F.D. Minifie and L.L. Lloyd. Baltimore, MD: University Park Press.

(18) Ferguson, C.A:, and Farwell, C.B. 1975. Words and sounds in early language acquisition: English initial consonants in the first fifty words. Language 51: 419-439.

(19) Fry, D.; Abramson, A.; Eimas, P.; and Liberman, A. 1962. The identification and discrimination of synthetic vowels. Lang. Speech 5: 171-189.

(20) Garcia, J., and Koelling, R. 1966. Relation of cue to consequence in avoidance learning. Psychonomic Science 4: 123-124.

(21) Green, S. 1975. The variation of vocal pattern with social situation in the Japanese monkey (Macaca fuscata): A field study. In Primate Behavior, ed. L. Rosenblum, vol. 4. New York: Academic Press.

(22) Holmberg, T.L.; Morgan, K.A.; and Kuhl, P.K. 1977. Speech Perception in Early Infancy: Discrimination of Fricative Consonants. Paper presented at the 94th Meeting of the Acoustical Society of America, Miami.

(23) Hyman, R., and Frost, N. 1975. Gradients and schema in pattern recognition. In Attention and Performance, eds. P.M.A. Rabbitt and S. Dornic. New York: Academic Press.

(24) Ingram, D. 1974. Phonological rules in young children. J. Child. Lan. 1: 46-64.

(25) Jakobson, R. 1968. Child Language Aphasia and Phonological Universals. The Hague: Mouton.

(26) Jusczyk, P.W. 1983. On characterizing the development of speech perception. In Neonate Cognition: Beyond the Blooming, Buzzing Confusion, eds. J. Mehler and R. Fox. Hillsdale, NJ: Erlbaum Associates, in press.

(27) Jusczyk, P.W.; Pisoni, D.B.; Reed, M.A.; Fernald, A.; and Myers, M. 1983. Infants' discrimination of the duration of a rapid spectrum change in nonspeech signals. Science 222: 175-177.

(28) Jusczyk, P.W.; Pisoni, D.B.; Walley, A.; and Murphy, J. 1980. Discrimination of relative onset time of two-component tones by infants. J. Acoust. So. 67: 262-270.

(29) Katz, J., and Jusczyk, P.W. 1980. Do Six-month-olds Have Perceptual Constancy for Phonetic Segments? Paper presented at the International Conference on Infant Studies, New Haven, Connecticut.

(30) Kiparsky, P., and Menn, L. 1977. On acquisition of phonology. In Language, Learning and Thought, ed. J. McNamara. New York: Academic Press.

(31) Klatt, D.H. 1979. Speech perception: A model of acoustic-phonetic analysis and lexical access. J. Phonetics 7: 279-312.

(32) Kuhl, P.K. 1979. Speech perception in early infancy: Perceptual constancy for spectrally dissimilar vowel categories. J. Acoust. So. 66: 1668-1679.

(33) Kuhl, P.K. 1983. Perception of auditory equivalence classes for speech in early infancy. Infant Beh., in press.

(34) Kuhl, P., and Miller, J. 1982. Discrimination of auditory target dimensions in the presence or absence of variation in a second dimension by infants. Perc. Psych. 31: 279-292.

(35) Kuhl, P.K., and Miller, J.D. 1975. Speech perception by the chinchilla: Voiced-voiceless distinction in alveolar plosive consonants. Science 190: 69-72.

(36) Kuhl, P.K., and Padden, D.M. 1982. Enhanced discriminability at the phonetic boundaries for the voicing feature in macaques. Perc. Psych. 32: 542-550.

(37) Kuhl, P.K., and Padden, D.M. 1983. Enhanced discriminability at the phonetic boundaries for place of articulation in macaques. J. Acoust. So. 73: 1003-1010.

(38) Ladefoged, P. 1971. Preliminaries to Linguistic Phonetics. Chicago: University of Chicago Press.

(39) Lancker, D. van, and Fromkin, V. 1973. Hemispheric specialization for pitch and "tone": Evidence from Thai. J. Phonetics 1: 101-109.

(40) Lasky, R.; Syrdal-Lasky, A.; and Klein, R. 1975. VOT discrimination by four to six and a half month old infants from Spanish environments. J. Exp. C. Psy. 20: 215-225.

(41) Liberman, A. 1982. On finding that speech is special. Am. Psychol. 37: 148-167.

(42) Liberman, A.; Cooper, F.; Shankweiler, D.; and Studdert-Kennedy, M. 1967. Perception of the speech code. Psychol. Rev. 74: 431-461.

(43) Liberman, A.; Harris, K.; Hoffman, H.; and Griffith, B. 1957. The discrimination of speech sounds within and across phoneme boundaries. J. Exp. Psychol. 54: 358-368.

(44) Liberman, A.M.; Isenberg, D.; and Rakerd, B. 1983. Duplex perception of cues for stop consonants: Evidence for a phonetic mode. Perc. Psych., in press.

(45) Liberman, I.Y.; Shankweiler, D.; Fisher, F.W:; and Carter, B. 1974. Explicit syllable and phoneme segmentation in the young child. J. Exp. C. Psy. 18: 201-212.

(46) Lisker, L. 1978. In qualified defense of VOT. Lang. Speech. 21: 375-383.

(47) Macken, M.A. 1980. Aspects of the acquisition of stop systems: A cross-linguistic perspective. In Child Phonology, eds. G.A. Yeni-Komshian, J.F. Kavanagh, and C.A. Ferguson, vol I: Production. New York: Academic Press.

(48) Menn, L. 1980. Phonological theory and child phonology. In Child Phonology, eds. G.A. Yeni-Komshian, J.F. Kavanagh, and C.A. Ferguson, vol. I: Production. New York: Academic Press.

(49) Menyuk, P., and Menn, L. 1979. Early strategies for the perception and production of words. In Studies in Language Acquisition, eds. P. Fletcher and M. Garman. Cambridge, MA: Cambridge University Press.

(50) Miller, J., and Liberman, A. 1979. Some effects of later-occurring information on the perception of stop consonant and semi vowel. Perc. Psy. 25: 457-465.

(51) Miller, J.D.; Wier, C.C.; Pastore, R.; Kelly, W.J.; and Dooling, R.J. 1976. Discrimination and labeling of noise-buzz sequences with varying noise-lead times: An example of categorical perception. J. Acoust. So. 60: 410-417.

(52) Morais, J.; Cary, L.; Alegria, J.; and Bertelson, P. 1979. Does awareness of speech as a sequence of phones arise spontaneously? Cognition 7: 323-331.

(53) Moskowitz, A.I. 1973. Acquisition of phonology and syntax: A preliminary study. In Approaches to Natural Language, eds. G. Hintikka, J. Moravesik, and P. Suppes. Dordrecht: Reidel Publishing Co.

(54) Oden, G.C., and Massaro, D.W. 1978. Integration of featural information in speech perception. Psychol. Rev. 85: 172-191.

(55) Osherson, D.N., and Smith, E.E. 1981. On the adequacy of prototype theory as a theory of concepts. Cognition 9: 35-58.

(56) Petersen, M.; Beecher, M.; Zoloth, S.; Moody, D.; and Stebbins, W. 1978. Neural lateralization of species-specific vocalizations by Japanese macaques (Macaca fuscata). Science 202: 324-327.

(57) Petersen, M.R. 1982. The perception of species-specific vocalizations by primates: A conceptual framework. In Primate Communication, eds. C. Snowdon, C. Brown, and M. Petersen, pp. 171-211. New York: Cambridge University Press.

(58) Petersen, M.R.; Beecher, M.D.; Zoloth, S.R.; Green, S.; Marler, P.R.; Moody, D.B.; and Stebbins, W.C. 1983. Neural lateralization of vocalizations by Japanese macaques: Communicative significance is more important than acoustic structure. Behav. Neurosci., in press.

(59) Peterson, G.E., and Barney, H.L. 1952. Control methods used in a study of vowels. J. Acoust. So. 24: 175-184.

(60) Pisoni, D. 1977. Identification and discrimination of the relative onset time of two-component tones: Implications for voicing perception in stops. J. Acoust. So. 61: 1352-1361.

(61) Pollack, I. 1952. The information in elementary auditory displays. J. Acoust. So. 24: 745-749.

(62) Pollack, I. 1953. The information in elementary auditory displays II. J. Acoust. So. 25: 765-769.

(63) Posner, M.I., and Keele, S.W. 1968. On the genesis of abstract ideas. J. Exp. Psy. 77: 353-363.

(64) Repp, B. 1982. Phonetic trading relations and context effects: New experimental evidence for a speech mode of perception. Psychol. B. 92: 81-110.

(65) Rosch, E. 1978. Principles of categorization. In Cognition and Categorization, eds. E. Rosch and B.B. Lloyd. Hillsdale, NJ: Erlbaum Associates.

(66) Seligman, M., and Hager, S. 1972. Biological Boundaries of Learning. New York: Appleton-Century-Crofts.

(67) Seyfarth, R.; Cheney, D.; and Marler, P. 1980. Monkey responses to three different alarm calls: Evidence of predator classification and semantic communication. Science 210: 801-803.

(68) Seyfarth, R.; Cheney, D.; and Marler, P. 1980. Vervet monkey alarm calls: Semantic communication in a free-ranging primate. Anim. Behav. 28: 1070-1094.

(69) Smith, N.V. 1973. The Acquisition of Phonology. Cambridge: Cambridge University Press.

(70) Snowdon, C.T. 1982. Linguistic and psycholinguistic approaches to primate communication. In Primate Communication, eds. C. Snowdon, C. Brown, and M. Petersen. New York: Cambridge University Press.

(71) Snowdon, C.T.; Brown, C.H.; and Petersen, M.R., eds. 1982. Primate Communication. New York: Cambridge University Press.

(72) Streeter, L. 1976. Language perception of 2-month-old infants shows effects of both innate mechanisms and experience. Nature 259: 39-41.

(73) Trehub, S.E. 1976. The discrimination of foreign speech contrasts by infants and adults. Child Dev. 47: 466-472.

(74) Walley, A.C.; Smith, L.B.; and Jusczyk, P.W. 1980. Classification of CV syllables by readers and pre-readers. In Research on Speech Perception, Progress Report No. 6, Indiana University.

(75) Waterson, N. 1971. Child phonology: A prosodic view. J. Linguist. 7: 179-211.

(76) Werker, J.F., and Tees, R.C. 1983. Cross-language speech perception: Evidence for perceptual reorganization during the first year of life. Infant Beh., in press.

(77) Williams, L. 1977. The perception of stop consonants by Spanish-English bilinguals. Perc. Psy. 21: 289-297.

(78) Williams, L. 1977. The voicing contrast in Spanish. J. Phonetics 5: 169-184.

(79) Zoloth, S.; Petersen, M.; Beecher, M.; Green, S.; Marler, P.; Moody, D.; and Stebbins, W. 1979. Species-specific perceptual processing of vocal sounds by Old World monkeys. Science 204: 870-873.

The Biology of Learning, eds. P. Marler and H.S. Terrace, pp. 617-628. Dahlem Konferenzen 1984. Berlin, Heidelberg, New York, Tokyo: Springer-Verlag.

Human Learning and Memory

W.K. Estes
Dept. of Psychology and Social Relations, Harvard University
Cambridge, MA 02138, USA

Abstract. Salient trends in the transition from animal to human learning are: a) a progressive constriction of the scope of elementary conditioning and associative learning processes and subordination of these to higher, mainly verbal, processes; and b) a greatly extended time scale for the relation between acquisition and use of information. The current practice in cognitive psychology is to study human learning wholly within an information processing framework. This strategy has yielded useful results, yet questions can be raised concerning the possible advantages of also looking at biological aspects of human information processing, especially with regard to the development of learning ability, readinesses and sensitive periods for particular forms of learning, and biological constraints on learning.

THE ROUTE TO HUMAN LEARNING

Preparatory to discussing the relevance of biological ideas to the interpretation of human learning, it will be useful to characterize in a general way how human learning resembles and how it differs from forms of animal learning that have been much more intensively studied in relation to their biological settings and to underlying neurophysiological processes. I shall start with a quick scan of lower and higher animal forms, pointing up a few salient aspects of learning at different levels.

Learning in Lower Animals

Simple habituation and conditioning (in the sense of sensitization) are demonstrable even in lower invertebrates, for example, Aplysia. However, fully controlled demonstrations of associative learning become available only as research progresses to somewhat higher animal forms. The prototype is classical conditioning, in which a response originally evoked only by some unconditioned stimulus comes to be evoked by a new, conditioned stimulus that has been paired with it. The usual arrangement is for the new stimulus to precede the unconditioned one on training trials and the result is that the animal comes, in effect, to anticipate the unconditioned stimulus (usually a biologically significant event) when the conditioned stimulus occurs.

At the time of Pavlov and Watson, conditioning was typically viewed as an extremely elementary and even mechanical form of learning, but continuing research has revealed many complexities (3). For example, it has become clear that, in the higher animals, simple contiguity of a conditioned and an unconditioned stimulus is not always sufficient for learning to occur. Rather, it may be necessary that the conditioned stimulus convey some information relevant to predicting the unconditioned stimulus above that previously available to the animal in the situation. Again, it was believed earlier that conditioning can produce associations between arbitrary pairs of stimuli so that which event is chosen to become a signal for an unconditioned stimulus is essentially immaterial. Here, continuing research has shown that conditioning in animals is not completely flexible, and that, rather, the learning builds on species-specific behavioral and physiological systems in such a way that different stimuli belonging to the same system may readily become associated through conditioning, whereas stimuli belonging to different systems may be very difficult to associate. (Thus, a solution with a particular taste may readily become a conditioned signal for a gastric disturbance originally evoked by a poisonous substance, whereas a visual or auditory stimulus is refractory to such conditioning.)

Although classical conditioning appears phenotypically similar in basic properties from lower invertebrates to higher subhuman vertebrates, there is a distinct shift in its role in the animal's economy. In the lower forms, habituation and simple classical conditioning generally appear to constitute the whole of the animal's learning repertory, whereas in higher animals, these forms of learning are confined mainly to visceral processes and reflexes. The overt behavior of higher animals appears to be guided primarily by its consequences, in accordance with principles of instrumental or operant conditioning. But even here there is a distinct phylogenetic trend. In lower animals the effect of a reward or other reinforcing event appears to be a direct strengthening of the particular rewarded action; in contrast, for animals closer to man, the process seems better described as one in which animals learn what rewarding or punishing consequences follow various combinations of situations and actions and then are guided in their performance by feedback from signals indicating whether they are moving toward or away from sources of reward.

What are taken to be higher forms of learning, having to do with learning common responses to categories of stimuli, also appear to depend strongly on species-specific characteristics. Thus, prolonged and arduous training may be necessary to teach animals at the level of rats or pigeons to respond differentially to, say, circles versus triangles, but with little identifiable practice, pigeons come to respond to categories of stimuli that have some natural biological significance for them, for example, bodies of water or types of vegetation (9).

Learning in Man

Habituation and classical conditioning can be demonstrated in man, but they are narrowly confined to vegetative processes and simple reflexes and even here are often subordinated to higher, mainly verbal processes. Anticipating significant events continues to be an important aspect of learning, but it occurs largely at the level of verbal associations. Learning to profit from the consequences of actions has some of the properties of operant conditioning in animals, but the coupling of response and reinforcement is looser. In fact, most normal human learning cannot be described in terms of increasing or decreasing tendencies to repeat rewarded or punished actions; it is better characterized as a process of acquiring information about the probabilities of various event sequences and their utilities and then using this information in decision making.

In general, human learning rarely appears to be limited to the formation of associations between particular stimuli and responses. Categorization is the rule. Category, or concept, learning in human beings is little constrained by species-specific factors. Rather, it seems possible for the human learner to form arbitrary categorizations of events or objects that have only common functional significance (for example, furniture, tools) as readily as those that might seem to have more biological significance (for example, vegetables, weeds).

HUMAN LEARNING AS INFORMATION PROCESSING

Although, as seen in the preceding sketch, continuities can be discerned between some aspects of human learning and learning in lower forms, the differences, taken together, suggest such a gap that the currently prevailing view is to describe human learning in a quite different conceptual framework from animal learning (6). Perhaps the most striking distinction between animal and human learning has to do with the time scale of the acquisition and use of information. In animals, learning mainly has to do with information relevant to their current problems, though there are some exceptions (for example, acquiring information about broad features of an environmental locale or territory). In human learning, it is the rule rather than the exception for information acquisition to have little to do with current problems or tasks. Sometimes a long-term purpose for information being acquired is discernable, but sometimes information appears to be gathered and stored in memory for its own sake - as seems to be true of much of the acquisition of knowledge in schools or from general reading. The major questions about these characteristic forms of learning from the viewpoint of most psychologists have to do not with what instigates or maintains the learning, but rather how it proceeds. The prevailing descriptive framework views man as a special kind of information processing system, rather than as a simple extrapolation of the sequence of pictures of animal learners at successive phylogenetic levels.

Organization of Memory

Though cognitive theorists differ on some details, a modal picture of the time course of human information processing of a stimulus input is as follows (12). The sensory input is first subjected to perceptual processing, the result being to determine what objects are present in the input (whether these are material objects, like people or trees, or linguistic objects, like printed letters or spoken syllables). A representation of the constellation of objects and events perceived is then maintained for a short time in what is termed primary memory. Roughly speaking, the individual is aware of the contents of primary memory but has not yet determined their significance.

During the few seconds when a representation of the input is available in primary memory, attentional processes select some elements for passage into short-term working memory, a subsystem of memory in which items are typically encoded in terms of visual or auditory features and can be kept in an active state by rehearsal processes. It is assumed that this stage is the point in processing at which motivation makes itself felt, by determining the criteria for passage of items into working memory.

While items are in working memory they are subject to cognitive operations, such as comparison, that may lead to the detection of categorical relationships and thus to reorganization of the information in accord with task demands before it is stored in long-term memory. Long-term storage takes two forms. A representation of the events perceived at a particular time together with the context in which they occurred is termed an episodic memory. The episodic system provides a temporally organized record of an individual's experience. If a newly formed memory representation incorporates, instead, relations among items or events that are independent of the original context (for example, a person's name, a chemical formula), it is said to belong to semantic, or categorical, memory. This system provides the factual knowledge base for processes of reasoning and problem solving.

The information processing that leads to learned categorizations or concepts may proceed at different levels. At what might be termed the most primitive level, individual instances or exemplars of a to-be-learned category (for example, particular animals if one is learning to distinguish dogs from cats) are stored in memory together with their category labels. Another, and more distinctively cognitive, route to categorization is for the learner to notice which particular features or attributes of a class of exemplars have higher probabilities for one category or another and to begin to classify new instances in accord with the feature or attribute probabilities.

Finally, a major component of human information processing that apparently is not shared with lower animals is the acquisition of information from linguistic inputs. Here, learning is not a matter of remembering the input itself, but rather of remembering what is represented by the symbols detected in the input. Perhaps the currently most influential view is that when a human being hears or reads a sentence, the detailed form of the sentence is quickly lost from memory and what is retained is the information conveyed by the sentence, transformed into a standard propositional format. Thus the result of such learning is not a collection of memories of particular words or sentences, but rather a network of associations among remembered propositions.

In contemporary cognitive psychology, the investigation of human information processing and the acquisition of propositional knowledge is characteristically carried on wholly without reference to properties of the human learner as a biological organism. Clearly, research carried on in this framework can make progress toward understanding complex learning. But the question remains whether this orientation is optimal. Are there any reasons to believe that deeper understanding of human learning processes would result if one took into account not only analysis in information processing terms but also consideration of their biological basis?

A Model for the Learning of Fuzzy Categories

For the biologically oriented investigator, it may be difficult to imagine the form of a learning theory formulated strictly at a psychological level. As a preliminary to discussing the degree to which such a theory could be self-sufficient, or even a useful component of a larger theory, it may be helpful to summarize a particular example. A particularly suitable model for this purpose is the model for category learning put forward by Medin and Schaffer (13), which was formulated explicitly with a view to maintaining continuity with related models of animal discrimination learning.

The task of this model is to provide a quantitative account of the way people learn to categorize collections of objects or events when the categories are not necessarily defined by any simple rules that could be discovered by a learner, as in the everyday examples of games and ethnic groups. It is assumed that one learns such a classification on the basis of experience with a series of exemplars by forming memorial representations of the individual exemplars and grouping these in memory in accord with their category assignments. In an experiment to which the model was applied successfully, the stimuli to be categorized were photographs of faces of women clipped from college yearbooks, the individual photos varying in hair color, shirt color, length of hair, and similar attributes (14). It is assumed that when a learner perceives an individual exemplar, in this case a photograph, the first step in information processing is to encode the stimulus in memory in terms of its values on relevant attributes. In the simplest case, only two values on each attribute are recognized and can be denoted by 0's or 1's, as in Table 1. The table could be taken to represent an experiment involving three exemplars of each of two categories, with each exemplar being encoded by the learner on four attributes. Thus, exemplar A_1 of category A has the value 1 on the first and second attributes and 0 on the third and fourth, and so on. After these dimensional representations of experienced exemplars have been stored in memory, the learner categorizes new exemplars by computing their similarities to the representations already

TABLE 1 - Representation of a simple experiment on categorization.

	Exemplar					
	Category A			Category B		
Attribute	A_1	A_2	A_3	B_1	B_2	B_3
1	1	0	1	0	0	1
2	1	0	1	0	1	0
3	0	1	1	0	0	1
4	0	1	1	0	1	0

in memory. The computed similarity is directly related to the number of attributes on which the new and the remembered exemplar have the same value, and the probability that a test exemplar will be assigned by the learner to a given category is equal to the sum of its similarities to the remembered exemplars of that category divided by the sum of its similarities to that and any alternative categories (for a simplified presentation of the quantitative details, see (7)).

The principal successes of the model have been in providing accounts of the relative difficulty of different classification problems, and of the learning of different individual exemplars within a particular problem. According to the model, these relative difficulties depend simply on the degrees of similarity of exemplars within and between categories and on the saliences of the attributes on which exemplars are encoded in the learner's memory system.

From a broader standpoint than that of assessing the quantitative adequacy of the model to predict properties of the data, given information about similarities and saliences, it is a severe limitation that the model does not speak to questions having to do with the relative difficulty of learning classifications of objects or events in a natural environment. It is possible that these could always be accounted for simply on the basis of similarity and stimulus salience, but there is room for doubt. Another possibility is that some classifications are easier for human learners than others owing to differences in biological relevance of the categorizations. A conspicuous feature of present-day research on human learning in the information processing framework is that questions of the latter sort are rarely raised.

POINTS OF CONTACT BETWEEN HUMAN LEARNING AND BIOLOGY

Motivation

In the current cognitively oriented approach to human learning, considerations of the relation between motivation and learning are almost a no-man's-land. The focus of research is on the properties of learning, and especially its products, with essentially no attention being given to questions of why the learning occurs or what maintains it. In the rare instances when the term motivation occurs at all in treatments of information processing, motives are presumed to take the form of verbal plans, or the equivalent, that relate actions to goals in the learner's memory system (2). Presumably some events or states of affairs serve as goals for any given instance of learning, but questions are seldom raised concerning their biological basis.

In the more classical learning theories that build closely on research on animal learning, it is assumed that learning occurs because the necessary activities on the part of the organism are maintained and directed by the reinforcing consequences of actions (3). Some events are assumed to be innately reinforcing for an organism, and other stimuli that occur in temporal contiguity with these acquire the same reinforcing potency by a conditioning process. Whether any such process is responsible for the potency of the events that instigate and direct most ordinary human learning is entirely unknown. Considering some of the conspicuous general features of human learning, the possibility seems unlikely. Since much human learning is a matter of acquiring information that may have only long-term, if any, implications for the learner's behavior, it seems most implausible that the step-by-step cognitive operations responsible for the information acquisition could depend on identifiable reinforcing consequences. It is an interesting possibility that the accrual of information might be inherently reinforcing for human beings, but there is little in the way of relevant evidence.

Developmental Aspects of Learning

The elaborate information processing capabilities of the adult do not come into play full blown, but rather result from a lengthy developmental process. The nature of this process is little understood. One view that has been extremely influential during the formative period of cognitive psychology is that the various component information processes that constitute adult learning and thinking appear in the child in a fixed developmental sequence. In the extensive theorizing of Piaget, which has been largely responsible for this deterministic conception, there has been little effort to account for the developmental sequence in terms of biological factors. There has been speculation that differential rates of myelinization of nerve fibers in different parts of the brain might be implicated, but sources of direct evidence on this and related hypotheses have not yet come to hand (11).

A conspicuous overall trend in the development of human learning was pointed up by Hebb (8) in his influential treatment of the neurophysiological substrate of behavior. Hebb emphasized, in particular, the progressive shift from passive and slow to actively directed and fast learning, both in phylogeny from lower to higher organisms, and in the ontogeny of individual species, including man. For animals lower on the phylogenetic scale, and even for human infants, learning is mainly a matter of forming associations among relatively simple stimuli and responses. This form of learning seems always to be slow and incremental, even when it occurs in the human adult. Learning in the adult, especially the adult human being, however, is characteristically a matter of associating complex patterns of stimulus information often of a symbolic character, and complexly organized behavioral routines.

A start on an interpretation of the shift from slow to fast learning in human development might take the following form. The essence of fast learning in the adult is the capability of mentally manipulating symbolic representations of relatively complex patterns of stimulus or response information, selecting those to enter active working memory by attentional processes, and then, by means of rehearsal, generating multiple opportunities for new associations relative to a current task to be formed. The formation of the complex stimulus and response units that are critical to adult information processing is slow for several reasons. Perhaps most importantly, the learning must almost necessarily be an automatic process carried on incidentally to ordinary day-to-day activities, for a very large number of these units must be formed, and there is no way that the child could foresee the long-term relevance of such units and thus select those to be learned by active attentional processes. The lengthy childhood during which the developing human being is largely protected from the need to cope with environmental exigencies may be critically important in making possible the vast amount of slow learning leading to the formation of the cognitive units that serve adult learning and problem solving. Whether unitization is truly automatic given a normal biological state of the child or whether it requires triggering by some as yet unidentified kinds of experiences is largely unknown.

A related question is whether sensitive periods or readinesses for particular kinds of learning (for example, reading readiness) depend simply on achieving a certain amount of progress in forming the relevant kinds of cognitive units or whether there is some independent biological basis.

Biological Constraints on Human Learning

A major theme in the study of animal learning during the past dozen years has been the emerging recognition of a major role of species-specific biological constraints on learning processes (1, 10). Thus rats regularly learn to run from one end of a shuttle box to another to avoid shock, but only with great difficulty learn to turn a wheel with the same reinforcement; however, these animals readily learn to turn the wheel to obtain food. Pigeons readily learn to peck a key that operates a food mechanism, but only with great difficulty, if at all, can they be trained to peck a key to turn off a shock. The accumulation of large numbers of similar examples for animals at various phylogenetic levels has made it abundantly clear that to understand animal learning of any particular animal species, one must study not only the learning itself but also its relation to its biological and ecological setting.

The prevailing view of human learning is that it is almost wholly general-purpose in character and can be understood without reference to biological or ecological considerations. Whether there will at some point be a shift in the Zeitgeist comparable to that which occurred earlier for animal learning is a matter for speculation. One view, and perhaps the dominant one in current cognitive psychology, is that adult learning is able to cope with almost any form of information that may prove relevant to adapting to a problem situation. Investigators who have looked at cross-cultural comparisons might add the qualification that adult learning capabilites are not wholly general, but rather tend to match the domain of problems for which the individual is prepared by early experiences in a particular environment and culture (5).

The general-purpose conception is sometimes more directly challenged in particular settings. A notable case is the acquisition of language, where the very influential theorizing of Chomsky (4) presumes an innate correspondence between learning capacities of the child and the properties of the language the child is prepared to acquire. Unfortunately, experiments comparable to those that have elucidated the roles of experiential and biological factors in the acquisition of bird songs are not feasible with human children, and it is not yet apparent what alternative sources of evidence might be brought to bear on this issue in the human case. Looking in another direction, if one grants some credibility to the idea of innate preparedness for particular forms of learning in the case of language, then the question arises whether directing as close attention to properties of what is learned in other domains as Chomsky did for grammar might uncover other constraints on human learning that have not been apparent to casual observation.

Acknowledgement. Supported in part by NSF Grant BNS 80-26656.

REFERENCES

(1) Bolles, R.C. 1975. Learning, motivation, and cognition. In Handbook of Learning and Cognitive Processes: Introduction to Concepts and Issues, ed. W.K. Estes, vol. 1, pp. 249-280. Hillsdale, NJ: Lawrence Erlbaum Assoc.

(2) Bower, G.H. 1975. Cognitive psychology: An introduction. In Handbook of Learning and Cognitive Processes: Introduction to Concepts and Issues, ed. W.K. Estes, vol. 1, pp. 25-80. Hillsdale, NJ: Lawrence Erlbaum Assoc.

(3) Bower, G.H., and Hilgard, E.R. 1981. Theories of Learning, 5th ed. Englewood Cliffs: Prentice-Hall.

(4) Chomsky, N. 1965. Aspects of the Theory of Syntax. Cambridge: M.I.T. Press.

(5) Cole, M., and Scribner, S. 1974. Culture and Thought: A Psychological Introduction. New York: John Wiley.

(6) Estes, W.K. 1978. On the organization and core concepts of learning theory and cognitive psychology. In Handbook of Learning and Cognitive Processes: Linguistic Functions in Cognitive Theory, ed. W.K. Estes, vol. 6, pp. 235-292. Hillsdale, NJ: Lawrence Erlbaum Assoc.

(7) Estes, W.K. 1983. Categorization, perception, and learning. In Perception, Cognition, and Development: Interactional Analyses, eds. T.J. Tighe and B.E. Shepp, pp. 323-351. Hillsdale, NJ: Lawrence Erlbaum Assoc.

(8) Hebb, D.O. 1949. The Organization of Behavior: A Neurophysiological Theory. New York: John Wiley.

(9) Herrnstein, R.J. 1979. Acquisition, generalization, and discrimination reversal of a natural concept. J. Exp. Psychol.: Anim. Behav. Proc. 5: 116-129.

(10) Hinde, R.A. 1973. Constraints on learning: An introduction to the problems. In Constraints on Learning, eds. R.A. Hinde and J. Hinde, pp. 1-19. New York: Academic Press.

(11) Kendler, T.S. 1983. Labeling, overtraining, and levels of function. In Perception, Cognition, and Development: Interactional Analyses, eds. T.J. Tighe and B.E. Shepp, pp. 129-162. Hillsdale, NJ: Lawrence Erlbaum Assoc.

(12) Lachman, R.; Lachman, J.L.; and Butterfield, E.C. 1979. Cognitive Psychology and Information Processing: An Introduction. Hillsdale, NJ: Lawrence Erlbaum Assoc.

(13) Medin, D.L., and Schaffer, M.M. 1978. Context theory of classification learning. Psychol. Rev. 85: 207-238.

(14) Medin, D.L., and Schwanenflugel, P.J. 1981. Linear separability in classification learning. J. Exp. Psychol. Hum. Learn. Mem. 7: 355-368.

The Biology of Learning, eds. P. Marler and H.S. Terrace, pp. 629-642. Dahlem Konferenzen 1984. Berlin, Heidelberg, New York, Tokyo: Springer-Verlag.

Models of Language Acquisition

D.N. Osherson* and S. Weinstein**
*Center for Cognitive Science, M.I.T.
Cambridge, MA 02139
**Dept. of Philosophy, University of Pennsylvania
Philadelphia, PA 19104, USA

How do we come to believe what we do? This question has for millenia engaged the interest of philosophers inclined to reflect on human nature and has in this century been the primary impetus for much research in psychology. More recently, the attention of philosophers and psychologists alike has focussed not only on belief formation and fixation per se, but also on the acquisition of competences which have been argued to be founded on systems of beliefs apparently inaccessible to conscious survey and control. Formal learning theory attempts to provide a framework within which various precise models of belief formation and fixation can be elaborated. The investigation and comparison of such models can provide insight into some of the questions which philosophers and psychologists have raised about the fixation of belief.

The type of situation to which the framework of formal learning theory is applicable may be roughly described as follows. A subject is placed in an environment about which she comes to have certain beliefs or within which she develops certain competences. The environment is viewed as a potentially inexhaustible fund of data upon which the formation of the subject's beliefs or the development of her competences might depend. At any given time, the environment makes available to the subject only a finite piece of the data which is potentially available. As ever more data becomes available to the subject over time, her beliefs or competences may become fixed. This fixation consists in the subject's attaining a stable state of belief or competence as the environment further unfolds. In situations where learning is in question, this stable state of belief or competence must in some way be a correct representation of the environment.

Various phenomena instantiate the type of situation just described. In the first part of this paper we will focus our attention on language acquisition and will make explicit the details of the framework which formal learning theory provides for constructing models of this process. In the second part of the paper we will indicate how this framework may be used to clarify several issues which have been of interest to philosophers, psychologists, and linguists concerned with language acquisition. The final section of the paper will consider the extent to which formal learning theory can shed light on questions about rationality.

We will begin by describing in detail one of the models of language acquisition which has been elaborated within the framework of formal learning theory. This model was formulated by Gold (2) and is called language identification in the limit from arbitrary text. In this model the environments for learning mentioned above are construed as texts for the language. A text for a language is an arbitrary enumeration of the sentences of that language. The learner, when placed in such an environment, has available to her ever longer initial segments of the text. The learner is viewed as conjecturing for each such initial segment of the text a grammar which is construed as her guess about the character of her linguistic environment, or better, as that which underlies her current linguistic competence. The learner's competence stabilizes in a given environment just in case there comes a time after which she always conjectures the same grammar. She is said to have learned in the given environment in case the grammar to which she converges generates the language for which the environment is a text. A learner is credited with the capacity to learn a given language just in case she learns it in any environment for that language.

In the description of Gold's model of language acquisition just given, the learner is construed as a function from finite pieces of possible linguistic environments to grammars which represent the learner's linguistic competence at each stage of development within an environment. The model defines for each such function the collection of languages which that function can acquire. In Gold's model attention is restricted to recursively enumerable languages and computable learning functions. We will identify languages with recursively enumerable sets of natural numbers and grammars for a language with the indices of Turing machines that accept that language. Texts are then infinite sequences of natural numbers and learning functions are computable mappings from finite sequences of natural numbers to natural numbers.

Before proceeding on to discuss how Gold's model of language identification can be generalized to other models of language acquisition in the context of formal learning theory, it might help to pause and consider some examples of collections of languages which can and others which cannot be identified with respect to the Gold model. First, it is useful to observe that any single language is identifiable with respect to the Gold model - this because each constant function identifies the language for which its uniform guess is a grammar. This example shows that the learnability of single languages is not an issue of interest from the point of view of the models we will be considering. What is of interest are the variety of collections which can be learned. A second example of a collection which can be identified is the set of all finite languages. A learning function which identifies this collection may be described roughly as follows - it conjectures a grammar which generates all and only the data it has seen so far. Given a text for a finite language there comes a point where all the data the learning function has seen are all the data it will ever see, and at this point the learning function described stabilizes to a correct conjecture about its environment.

A host of examples of collections of languages which are not identifiable is provided by the following result of Gold - no collection of languages which properly contains the collection of finite languages is identifiable. It should be remarked that this last example of collections of languages which cannot be identified played some role in shaping the character of further research on formal models of language acquisition. Among the collections of languages which are demonstrably not identifiable on the basis of the above result are the regular languages, the context free languages, the context sensitive languages, and, of course, the collection of all recursively enumerable languages - that is, all collections in the Chomsky hierarchy. On this basis it was argued by some that Gold's model of language acquisition is inadequate. For, the argument went, the natural languages are certainly located somewhere in the Chomsky hierarchy and they can all be acquired. Therefore, Gold's model of acquisition is inadequate and some change, for example, enrichment of the linguistic environment to include data not only about sentences of the language but also about the circumstances in which those sentences might be uttered, is required. We do not intend to take up the various proposals that have been made along this line but would only like to observe at the moment that the argument adumbrated is a non sequitur. That the natural languages fall into the Chomsky hierarchy may well be true in the sense that every natural language is, let us say, context-sensitive. But this is certainly not to say that every context-sensitive language is a (possible) natural language and hence that the entire collection of them can be acquired. In particular, it is usually argued that no finite language is natural, and hence the failure of a model of language acquisition to provide a learning function which identifies all the finite languages and then some cannot be regarded as a defect of that model vis-a-vis its usefulness in application to the study of the acquisition of the natural languages.

We will turn now to the question of how the model of identification in the limit from arbitrary text may be generalized. The exposition of Gold's model suggests that there are three dimensions along which we may vary the model to obtain other models of language acquisition. These dimensions are the environment for learning, the criterion for successful learning, this combining the two components of a criterion for stabilization and a criterion of correctness, and the learning strategy. We will explain how variation can be achieved along each of these dimensions in turn and describe some of the consequences of such variation in terms of the collections of languages that can be acquired with respect to the resulting models.

In the Gold model described above, the linguistic environments for a given language, L, are all the texts for L, that is, all the sequences which enumerate L. A learner must succeed in identifying L in each such environment if she is to be credited with the capacity to learn L. Alternative construals of linguistic environment amount to alternative definitions of the relation, t is an environment for the language L. Two types of alternative to the construal of environment for L as arbitrary text for L may be distinguished. The first type, which we will call a text theory of linguistic environments, leaves the general character of the total evidential states unchanged. They are still characterized as texts, that is, infinite sequences of sentences. But, the particular class of texts which are counted as being environments for L may no longer be all and only the enumerations of L. Two kinds of variation here are of some interest from the point of view of modelling the process of language acquisition. First, the actual linguistic environments in which language is acquired are not obviously free from the intrusion of strings which are not sentences of the language to be learned. Nor is it clear that such environments should be construed, even ideally, as including all sentences of the language to be learned. This suggests considering the performance of learning functions on texts which contain sentences not in the target language and texts which omit sentences from that target language. If we consider texts with only finite omissions or intrusions, we arrive at the notions of incomplete or noisy environments. Since each of these environments for a language strictly includes the arbitrary texts for that language, it is reasonable to conjecture that these environments render learning more difficult in the sense that not every collection identifiable with respect to arbitrary text is identifiable with respect to noisy or incomplete text. Indeed, as is obvious, the class of finite languages is not identifiable with respect to either of these environments. Somewhat more interesting is the fact that there are collections of languages each of which is infinite and disjoint from every other member of the collection which are identifiable by computable learning function with respect to arbitrary text but not with respect to noisy text and other such collections not identifiable with respect to incomplete text. These results give an initial indication of the extent to which enlarging the class of environments in which success must be achieved restricts the collections of languages which can be identified.

A second way of altering the environments for learning yet remaining within a text theory is to consider restrictions on the order in which data is made available to the learner. The set of arbitrary texts for a language includes many bizarre orderings of the data which are unlikely to take part in the course of child language acquisition. It might be thought that a learner should be credited with the capacity to identify a language even if she cannot succeed in such unnatural environments. Indeed, in the study of child language acquisition the hypothesis has been advanced that the linguistic environment of children is specially controlled by their caretakers so as to make the task of language acquisition easier. The idea behind this so called motherese hypothesis is that by ordering the data for language acquisition in a particular way the caretaker can suggest easy generalizations about the syntax of the language to the child - generalizations that would otherwise be missed in an arbitrary ordering of the data. Work of Newport, Gleitman, and Gleitman (3) suggests that the motherese hypothesis as usually stated is dubious, that is, the sort of regularities hypothesized in caretaker's speech to children are generally absent. Nonetheless, the situation here is less than settled, and it would be useful to contruct environments which embody the kinds of restrictions on text orderings envisioned by the motherese hypothesis and to consider their effect on the classes of languages which can be identified. The results available from the side of formal learning theory do not as yet bear directly on this issue. Gold showed that if we restrict environments to primitive recursive texts for a language, then the collection of all the recursively enumerable languages is identifiable. On the other hand, it is a corollary of work reported in Blum and Blum (1) that if a collection of languages is identifiable by a computable learning function operating on recursive texts, then it is identifiable by a computable learning function operating on arbitrary texts.

An important feature of text theories is that they construe the linguistic environment as providing only positive information about the language to be learned. This restriction on available information does restrict the collections of languages which can be identified. There is, however, considerable evidence that in the course of language acquisition children receive little or no data about the ungrammaticality of strings which are not sentences of the language which they are engaged in learning. Thus, text theories are generally regarded as the proper way of modelling the process of child language acquisition.

We will now proceed to consider alternatives to the criterion of successful learning posited by Gold's model. According to the Gold model, children confronted with a natural language, L, eventually conjecture some grammar, g, for L and then never abandon g thereafter. The grammar, g, is entirely adequate for L, generating exactly its sentences, and stabilization to g is eventually perfect encompassing no further deviations, even temporary. Gold's version of identification has seemed too restrictive to some. Actual language learners, it is thought, achieve only an approximation to the input language, and stability in Gold's stringent sense is not attained. Within formal learning theory several alternatives to Gold's criteria of correctness and stabilization have been investigated.

A criterion of correctness determines the degree of accuracy with which a learner's conjectures must represent the target language if she is to be credited with the capacity to identify that language in given environments. The Gold criterion requires that the learner's conjectures eventually generate exactly the target language. There are many weaker criteria which could be argued to do better justice to the facts of child language acquisition. One criterion which has been studied within the framework of formal learning theory is the finite difference criterion. With respect to this criterion, a grammar, g, is regarded as a correct representation of a language, L, if the symmetric difference between L and the language generated by g is finite.

A criterion of stabilization specifies the degree of freedom the learner has in changing her conjectures in the limit while still being credited with having an abiding linguistic competence. An obvious alternative to Gold's criterion of stabilization, i.e., that eventually the learner must vary not at all in the grammars she conjectures, is to require only that eventually all the learner's conjectures are conjectures of grammars for the same language. The motivation for such a weakening of the criterion for stabilization is as follows. In acquiring a language a child eventually reaches a state of linguistic competence. If we identify that competence with the child's possession of a grammar for the language she is learning, then eventually the child must have available at each moment such a grammar. But there is nothing in this description to require that the child eventually has the same grammar available from moment to moment. For example, the learner might discover how to lower the processing time of a subset of sentences already accepted, albeit inefficiently, by her latest conjecture; the new processing strategy might require modification of her grammar without changing the language accepted. Alternatively, the learner might come upon an unfamiliar construction from a grammatically reliable source. Rather than verify that her current conjecture accepts this construction, she might modify her grammar to ensure its acceptance. If her old conjecture already accepted this construction, then her new grammar will be equivalent to it. This situation could arise indefinitely often for a sufficiently cautious learner. Osherson and Weinstein (4) established relations among the language classes which are identifiable with respect to combinations of various criteria of correctness and stabilization.

The third parameter on which the class of identifiable collections of languages depends is the learning strategy - that is, the set of learning functions which are regarded as possible models of learners. In the Gold model, the learners are only constrained to be computable functions. This choice reflects the currently popular computational theory of mental processes but is certainly not, in itself, a very strong constraint on models of human learners. More restrictive learning strategies represent hypotheses about the character of the human learning function. Formal learning theory provides a framework for investigating in detail the effect of such restrictions on the identifiable collections of languages. In so far as such restrictions have empiricial support as hypotheses about human language learners, the information provided by formal learning theory could then lend support to conjectures about features of the natural languages.

We will now give examples of some of the learning strategies that have already been investigated and some of the results that have been discovered in the course of this research. For present purposes we will assume that the environments are arbitrary texts and that the learning criterion is the one Gold proposed, i.e., intensional identification in the limit.

As remarked before, it is generally believed that each natural language is an infinite collection of sentences. This appears to follow from the consideration that no natural language contains a longest sentence. If this apparently universal feature of natural languages corresponds to an innate constraint on children's linguistic hypotheses, then they would appear to be barred from conjecturing a grammar for a finite language. This constraint on the space of grammatical conjectures may be formulated as a learning strategy. Call a learning function nontrivial if all its conjectures are grammars which generate infinite languages. It is clear that this strategy is restrictive in terms of the collections of languages it renders identifiable since no such collection will contain any finite language. What is more surprising is that nontriviality even restricts the collections of infinite languages that are identifiable. That is, there is a collection of languages, L, such that L is identifiable by a computable learning function and every member of L is inifinite, but no nontrivial computable learning function identifies L. Thus, even so weak a hypothesis about the learner, that she conjectures only grammars for infinite languages, renders unlearnable collections of infinite languages which would otherwise be identifiable.

If a learning function identifies a language, then it must converge on each text to some grammar for that language. But it is not required that on different texts for the same language it converge to the same grammar. It might be thought that this degree of freedom enjoyed by arbitrary computable learning functions would substantially increase the collections of languages they could identify. Different orders of presentation of the data might well lead to different sequences of generalizations about the data and hence to differing grammars in the limit. A theory of grammar which provided just one grammar for any language for which it provided a grammar at all might be thought severely to constrain the collections of identifiable languages. We will call a learner who is constant over differing orders of presentation an order independent learner. A learning function is order independent just in case for each language it identifies there is a single grammar to which it converges on all texts for that language. It is a corollary of a result of Blum and Blum (1) that order independence is not in fact a restrictive strategy. That is, if a collection of languages, L, is identifiable by a computable learning function, then L is identifiable by an order independent computable learning function.

One model of the place of grammatical theory within the study of language acquisition is the following. The grammatical theory imposes restrictions on grammars in such a way that the class of grammars admissible by the theory includes grammars of all and only natural languages. The natural languages are identified with the languages that can be acquired by normal human infants under casual conditions of access to linguistic data.

This suggests another kind of learning strategy which we will call prudence. A learning function is prudent just in case all its conjectures are grammars of languages which it in fact can identify. The above envisioned connection between grammatical theory and language acquisition suggests the following question. Is prudence a nonrestrictive or a restrictive strategy? That is, can every identifiable collection of languages be identified by a learner who stores only grammatical hypotheses for languages which she can in fact learn, or do some identifiable collections require a learner to conjecture hypotheses for languages she cannot identify? This is an open problem in formal learning theory.

Among the issues which have attracted the attention of philosophers, linguists, and psychologists concerned with language acquisition is the debate between rationalist and empiricist theories of learning. With the help of the framework developed above, we would like to distinguish two different issues which seem to get run together in some of the literature addressed to this controversy. The first issue, what we will call the rationalist-empiricist debate proper, is, in effect, a question about learning strategies. Suppose we have fixed on empirical grounds the nature of the environments in which natural languages are acquired and the criterion of success to which children conform in the natural case. The rationalist's gambit is then the following. If we employ a learning strategy whose space of conjectures is severely limited by purported principles of universal grammar, we can account for the known facts about language acquisition. For example, we can explain how each of the 5000 or so natural languages known to us can be acquired, and we can account for the speed at which they are acquired and for intermediate stages in the process of their acquisition. The empiricist rejoinder is that the assumption that the human learning function is of the type described by the rationalist is unnecessary to account for such facts, for learning functions which implement what might be called broadly empiricist learning strategies can model the actual process of language acquisition just as well. Sometimes such strategies have been termed all-purpose general learning strategies. By this is presumably not meant that the strategies are general in the sense that they can learn anything. That would make the empiricist too easy a prey, for as we have seen above, there is no learning function computable or not which can identify all languages. Presumably, some specific proposal about the character of the learning strategy is intended such as its lack of domain specificity or its being based on associationist principles. Some such learning strategies have been investigated which embody generate-and-test models of induction. These prove to be rather weak in terms of the collections of languages they can identify. This is not by any means to say that all the evidence is in on the side of rationalism. Rather, the power of various specific proposals which embody empiricist learning strategies needs to be investigated. Whether or not it is fruitful to pose the debate about language acquisition as one between rationalism and empiricism is an open question, given the wide variety of learning strategies one can formulate and the limited extent to which any of them have been studied.

A different, but related, issue is what one might call innatist vs. environmentalist theories of language acquisition. The innatist maintains that the course and outcome of the acquisition process is largely independent of perturbations in the linguistic environment whereas the environmentalist claims that the process of acquisition is largely driven by the environment. If we fix the notion of successful learning and fix a given learning function, we may view the collection of languages acquired as a function of environments. We may then view the innatist-environmentalist debate as an issue about the stability of the collections of languages acquired in varying linguistic environments by the human learning function. For example, the innatist might suppose that the human learning function would acquire the same collection of languages up to the natural criterion of success in each of the environments arbitrary text, noisy text, incomplete text, and imperfect text (that is, texts with both omissions and intrusions), whereas the environmentalist might expect the collections acquired in these environments to differ widely. The framework of formal learning theory makes it possible, then, to study the connection between innatism and rationalism on the one hand and empiricism and environmentalism on the other. Without further study, there is little reason to believe that a learning function is rationalist if and only if innatist and empiricist if and only if environmentalist.

Another aspect of formal learning theory is the alternative approach it gives to addressing issues about general features of the class of natural languages. This class, as noted above, is sometimes defined as the collection of languages which can be acquired by normal human children under casual conditions of access to linguistic data. In so far as one can settle the character of linguistic environments, criterion of successful learning, and human learning strategy on empirical grounds, formal learning theory provides a basis for studying the collection of natural languages which may supplement the linguist's methods of projecting universals from the extant natural languages. An interesting question which has recently been investigated from the point of view of formal learning theory regards the number of natural languages. Theories of universal grammar elaborated by Chomsky and Bresnan in the past few years seem to imply that there are only finitely many natural languages. Recently, various assumptions formulated within formal learning theory have been shown to have this same consequence. We will discuss these assumptions briefly and informally in order to illustrate the way in which this theory may provide insight into issues of linguistic interest.

The argument that there are finitely many natural languages is based on the following six assumptions. The first four are hypotheses about strategies which the human learner is supposed to implement. The first is that the human learner is subject to some form of memory limitation, that is, there is a fixed upper bound on the amount of data the learner can store beyond what she encodes in her conjecture. The second is that the human learner is prudent as defined above, that is, she conjectures only grammars for languages she can learn. Third, it is supposed that if the learner conjectures a grammar that accounts for the data, then she does not abandon that conjecture for a grammar which is of arbitrarily greater complexity. Fourth, it is supposed that if the learner can change from conjecturing a grammar, g, for a language, L, to a grammar, g', for L', then she may change to a grammar, g' for L' which is of complexity comparable to the complexity of g. The final two assumptions concern the performance of the learner in various environments. The first of these is that the learner identifies target languages in finite intrusion environments, that is, texts which contain only finitely many occurrences of non-sentences of the target language. The second is that the learner converges on, though perhaps incorrectly, noisy texts for the languages she acquires. Any learning function which satisfies these conditions identifies at most finitely many languages. Moreover, for each n there is a learning function satisfying these six conditions which identifies exactly n languages.

In conclusion, we will consider briefly two ways in which formal learning theory may be used to study questions about rationality. The first of these deals with what might be called canonical forms for learning while the second deals with the existence of ideal learning strategies.

It might be thought that certain learning strategies represent canons of rationality - any learning function which falls under such a strategy is to that extent a rational strategy, any which falls outside it irrational. Two such strategies are conservation and consistency. A learning function is conservative just in case it abandons a conjecture only when that conjecture fails to account for the data. A learning function is consistent just in case all its conjectures account for the data. In so far as a strategy is deemed rational, it is interesting to ask whether such a strategy provides a canonical form for learning, that is, whether every collection which is identifiable by any learning function whatsoever is identifiable by a learning function which falls under that strategy. Neither conservatism nor consistency provides canonical forms for learning in this sense. That is, there are collections of languages, L and L', each identifiable by a computable learning function, such that L is not identifiable by any conservative computable learning function and L' is not identifiable by any consistent computable learning function.

Were there rational learning functions, one argument for adhering to such might be that they were in some sense ideal. As has already been noted, no single learning function identifies all the effectively enumerable languages, so that is too stringent a sense of ideality perhaps. Another common reading of ideality is maximality, that is, an ideal learning function is one than which no other does strictly better. Unfortunately, there are very few ideal learning functions in this sense either, for it can be shown that if a learning function, F, identifies an infinite language, then there is a learning function, F', such that the collection of languages identified by F is strictly included in the collection of languages identified by F'. Thus, the only ideal learning functions in this sense are those which identify exactly the collection of finite languages.

Other notions of ideal learning function connected with the time and space required for computations on the data and with the amount of data required before convergence are currently under study. These notions of ideality seem, however, to have little connection with notions of rationality.

One promising avenue for further research in formal learning theory is its application to problems of concept acquisition in both human and infrahuman species. Jenkins (this volume) notes that one of the prominent features of contemporary learning theory is its recognition that there are biological constraints on the range of concepts which an animal can acquire, and that these constraints vary from one species to another. The framework of formal learning theory provides the tools to study the limits on concept acquisition exhibited across species and to clarify and assess explanations for these limits. In order for formal learning theory to contribute in this way, however, considerable progress needs to be made on another of the major problems discussed at this workshop, namely, the nature of the representations of the concepts which are acquired (cf. Terrace, this volume). What is required is a theory of the representation of concepts analogous to the theories of syntactic representation developed by Chomsky and others over the past two decades. Only in the context of such a rich representation of concepts, of "what is learned," could the framework of formal learning theory be fruitfully applied to questions of scientific interest.

REFERENCES

(1) Blum, L., and Blum, M. 1975. Toward a mathematical theory of inductive inference. Inform. Contr. 28: 125-155.

(2) Gold, E.M. 1967. Language identification in the limit. Inform. Contr. 10: 447-474.

(3) Newport, E.; Gleitman, H.; and Gleitman, L.R. 1977. Mother, I'd rather do it myself: some effects and non-effects of maternal speech style. In Talking to Children, eds. C.E. Snow and C.A. Ferguson. Cambridge, England: Cambridge University Press.

(4) Osherson, D.N, and Weinstein, S. 1982. Criteria of language learning. Inform. Contr. 52: 123-138.

The Biology of Learning, eds. P. Marler and H.S. Terrace, pp. 643-665. Dahlem Konferenzen 1984. Berlin, Heidelberg, New York, Tokyo: Springer-Verlag.

Brain Mechanisms of Language

J.W. Brown
Dept. of Neurology, New York University Medical Center
New York, NY 10016, USA

Abstract. Our knowledge of the brain mechanisms of language depends heavily on evidence derived from aphasia study, supplemented by electrocortical and metabolic investigations. Clinical observation reveals definite patterns of language change with focal lesions. These patterns are also apparent with stimulation of regions in the left frontal and temporo-parietal areas. The symptoms of aphasia reflect processing stages in normal language, and areas identified with these symptoms correspond to stages in forebrain evolution. A consideration of this material leads to a microgenetic or unfolding model of language representation in the brain.

INTRODUCTION

Aphasia is a natural interface between studies in psychology and neuroscience, but psychologists and neuroscientists have not always looked to the aphasia material for data relevant to issues in their respective fields. This is because of some misconceptions about the meaning of clinical pathology. One misconception is that since aphasia depends on a pathological state of the brain, it is irrelevant to studies of normal function. Aphasia is noise in a damaged machine. Yet, there are links between aphasic language and normal speech. For example, one can see almost the full range of aphasic errors in normal sleeptalking (1). A close study of the different types of errors that occur with brain damage in aphasia or other neuropsychological disorders shows that errors are not random disruptions but patterned and predictable decompositions which, it can be argued, cleave psychological processes apart along their natural lines.

Another misconception is that anatomical regions isolated by symptom correlations are insufficiently fine-grained for physiological analysis. While this is true in a general sense, the importance of the aphasia material is that it suggests that the relevant units or "chunks" of brain for psychological processing are not the command cells and columns of behavioral neurophysiology but fields or cell populations. Aphasia study is consistent with the idea that the "function" of an area arises out of a network dynamic rather than from a mosaic of component structures.

In my view, psychology will not be wedded to neuroscience unless it first passes through neuropsychology. This does not mean that one should borrow paradigms from normal psychology and try them out on aphasic populations. Neuropsychology is not psychology plus neuroscience. The purely psychological or neurological approach misses the point of neuropsychological study. The methods and concepts of neuropsychological research should be generated directly from the pathological material, and the best place to begin is with the nature of the symptom and the process of symptom formation.

WHAT IS A SYMPTOM?

One cannot go from the loss of a performance to an understanding of the process that is impaired. The loss is only an outcome, and the same outcome can be achieved in different ways. Teuber (65) concluded that "positive symptoms may be more revealing than the absence of symptoms." One could say that positive symptoms have a greater specificity than negative ones, while the absence of symptoms gives little information concerning function in an area. A positive symptom is an event, it points to a change in processing. It is not like a deficit where something is missing in a performance. It is an active phenomenon that can be approached directly.

Consider a paraphasic error. When a patient with brain damage or stimulation in language cortex names a chair a "table," what is the meaning of this symptom? One can say that the paraphasia does not simply reflect the disarray of a lexical-semantic device, since the error is linked to the target item. The device, the lexical store or the semantic representation, is not destroyed since recovery is always a possibility and normal performances alternate with defective ones. Luria's idea (42) of a weakened associative strength does not account for the variety of error types nor the systematic relation to lesion site. Is a paraphasia the response of an aphasic right hemisphere? While the effects on language of a second mirror lesion in the right hemisphere are imprecisely known, the attribution of a function to right hemisphere mechanisms does not gain us a better understanding but only shifts the problem to another hemisphere.

Another possibility is that the symptom reflects a moment in a processing sequence. For example, in the situation where the patient says "table" for chair, the error would point to a stage in lexical retrieval where these items are relatively covalent with respect to the target. On this view the paraphasia results when the process leading to the lexical representation is attenuated prior to full differentiation. In this approach the symptom takes on a new meaning. It is a slice of cognition actualizing in a moment of psychological time. Similarly, the effect of a brain lesion is not to remove an operation from the repertoire of behavior but to display a processing stage normally concealed beneath a surface performance. If this interpretation is correct, the symptoms of pathological behavior assume a greater significance for psychological theory. A study of symptom change in relation to brain damage provides the basis for a model of normal processing in relation to neural structure.

In the following section, the different forms of aphasia and the evidence for their anatomical localization are considered from the standpoint of such an approach. According to this approach, the anterior and posterior language zones consist of distributed systems laid down in the evolution of the forebrain. Stages in normal language production and representation, as inferred from aphasic errors, map onto structural levels in these systems (11, 12).

ANTERIOR APHASIAS

Various disorders of speech and purposeful action follow damage to parts of mesial frontal cortex and areas on the convexity. The regions of most importance are anterior cingulate gyrus, supplementary motor area, and the inferior premotor cortex. This section describes impairments in speech which result from damage to these regions.

Anterior Cingulate Gyrus

A bilateral lesion of the anterior cingulate gyrus leads to akinetic mutism. In this disorder, the patient sits or lies motionless as if in a catatonic state. There may be no definite changes in motor tone or reflexes. Nielsen and Jacobs (50) described the first case of this type, but other reports have followed (e.g., (18, 35)). The clinical picture differs from akinetic mutism with periacqueductal (midbrain) lesions in that the akinesia is punctuated with bouts of excitement, like the excited phases in the course of a catatonic stupor. Mutism with excitation after bicingulate lesions has also been observed in monkey (MacLean, personal communication). The patient with anterior cingulate lesions is not only speechless, but aphonic with no attempt at vocalization, nor is there evidence for preserved language comprehension. With recovery, speech returns through whispering and hoarseness rather than dysarthria or aphasia. Mutism has been reported following callosal section in man (7), where damage to cingulate gyrus or cingulum is likely. These patients also recover through a phase of hoarseness. Small bilateral stereotactic lesions of anterior cingulate gyrus (cingulotomy) give placidity with little in the way of language, motor, or cognitive deficit (23). This effect may be a partial expression of the more pronounced akinetic disorder with extensive destructions.

In monkey, stimulation studies indicate that anterior cingulate gyrus is part of a limbic and brain stem system mediating vocalization (34). In man, stimulation of anterior cingulate gyrus produces integrated motor behavior and affective change (Fig. 1). Bancaud et al. (2) concluded that the anterior cingulate region mediates a primitive or archaic level in behavior.

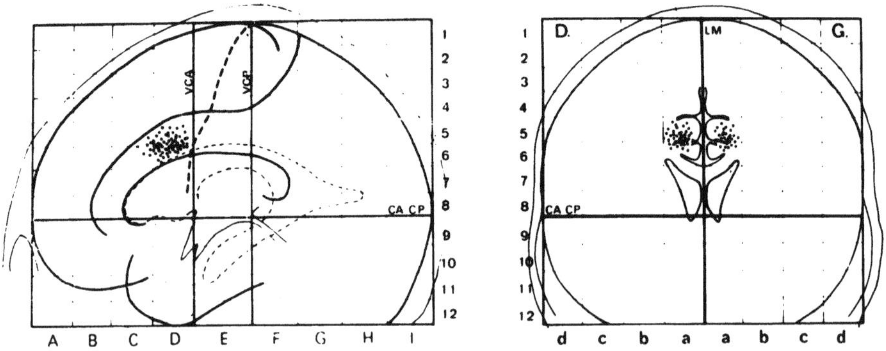

FIG. 1 - Results of 116 stimulations in 80 subjects (2). Responses include alerting without object orientation, simple integrated movements of the hands and legs, synergic oral movements, and affective responses.

Supplementary Motor Area (SMA)

Interest in SMA has been sparked by recent findings of a bilateral increase in cerebral blood flow during language tasks (40), though it has been known for some time (53) that the area is involved in speech production. Stimulation gives (motor) speech arrest or vocalization, hesitation or distortion of speech, and anomia. Palilalic iteration, cries, and vocalizations have been reported with irritative lesions of either side, though clinical deficits seem to occur only with left hemisphere lesions (33).

The usual picture is mutism at onset resolving into transcortical motor aphasia (11, 60). The patient evolves from mutism to a state of little or no spontaneous speech with some repetition and naming possible. Repetition may be agrammatic, or if normal, agrammatism may appear in spontaneous speech and/or writing as these performances return. Repetition is usually deliberate, not echolalic. Speech may have few aphasic errors, with difficulty largely for initiation, such that repetition and naming survive as simpler performances. The patient with a left SMA lesion has a type of partial mutism, i.e., impaired initiation of spontaneous speech, but also a partial akinesia affecting primarily the upper limbs. Ambulation is usually good, though inertia or slowness resembling Parkinsonism may occur.

There is often marked difficulty initiating upper limb action, especially for the contralateral arm. When asked to lift or lower the arm, the patient may require initiation of the movement by the examiner. A touch of the patient's finger in the direction of the movement often suffices for it to be completed. There may be an impression that the patient is uncooperative, for strength and tone are normal, and he may stare at the examiner without attempting to carry out the command. Of interest is that stimulation in SMA gives contralateral or bilateral movements involving the proximal (e.g., shoulder) musculature (68), while hypertonia at the shoulders is reported as a consequence of unilateral ablations in monkey (67).

The effects of bilateral SMA lesions are unknown. Presumably, a persistent deficit occurs since unilateral cases tend to show recovery. In monkey, unilateral lesions do not produce paresis, and bilateral lesions lead to changes in posture and tonus (67). The SMA has been surgically extirpated in epileptic cases (39, 54) with few residual motor or speech impairments.

Left Premotor Area
The limits of Broca's area are imprecisely known, since the area is defined by its association with Broca's aphasia, and there is little agreement on what constitutes this disorder. Two major components of Broca's aphasia are a phonetic-articulatory defect - with sequencing errors, "verbal apraxia," perseverations, and "cortical anarthria" - and agrammatism. These components may refer to different anatomical substrates: for example, agrammatism to the premotor cortex around posterior inferior F3 and misarticulation to focal lesions of posterior inferior F3 (Broca area proper).

Agrammatism. Agrammatism (telegrammatism) describes utterances in nonfluent aphasics consisting of strings of content words with a relative lack of small grammatical or function words. Articulation may be good without dysarthria or phonemic paraphasia. There is difficulty with inflections and auxilliaries and a dropping out of unstressed syllables - especially initial unstressed ones - and simplification of syntax. The normal intonation contour is lost. There is a monotony of speech and a reliance on lexical stress in production. Some studies of syntactic deficits in anterior aphasics are reviewed in Caramazza and Zurif (21).

There are few studies of the anatomical pathology of agrammatism, but the impression from a few case studies is that the responsible lesion is usually in the premotor area anterior or superior to pars opercularis of F3 (29, 42). A case has recently been described with a restricted inferior premotor lesion (48). It is of interest that agrammatism occurs with lesions of the right hemisphere, mainly frontal, in crossed aphasic dextrals with a frequency surprising in view of its rarity in left frontal cases. Agrammatism is the major form of aphasia in young children with a left hemisphere lesion involving frontal or temporal regions (9).

Clinically, there are links between agrammatism and transcortical motor aphasia. Agrammatics show better (less agrammatic) repetition, an observation responsible for its earlier interpretation as a type of speech economy. There is the appearance of agrammatism in children and crossed dextrals out of an initial stage of mutism. The difficulty initiating actions with the upper limbs typical of transcortical motor aphasia with SMA lesions now appears as a limb apraxia, where the movement is initiated properly but is derailed as it unfolds toward the target. Similarly, instead of inability to initiate vocalization, there is now oral apraxia with substitutions, omissions, and poorly executed orofacial movements.

Broca's area and Broca's aphasia. Lesions restricted to the foot of F3 (area 44) give phonetic-articulatory defects in utterances which may be syntactically normal. The misarticulation does not show a clear word class effect, and the intonation pattern is closer to normal. A phonetic disturbance in motor aphasics has been documented in studies by Blumstein (6). Luria (42) described impairments in initiation and sequencing of speech sounds in motor aphasia. Speech production errors in patients with Broca area lesions have been best described in surgical and traumatic cases (31, 42). In these populations, a phonetic-articulatory defect has been demonstrated with various subclassifications (kinetic, kinaesthetic aphasia, verbal apraxia, phonetic disintegration, etc.), depending on the dominant error type. Though vascular cases have been described with involvement of Broca's area, selective defects are usually not observed. There are cases with inferior pre-central and/or premotor lesions where misarticulation has been the major residual (31, 44, 49, 66).

Broca aphasics also have oral and limb apraxia (26, 58), though perhaps not the same type as in patients with agrammatism. In misarticulation cases, one has the impression of a dyspraxic impairment, with clumsiness in the distribution of the distal (oral and limb) innervation rather than para-praxic substitutions, though there are no studies which look at these differences within motor aphasics.

Interpretation of Frontal Aphasias

Impairments of speech and action with frontal lesions have traditionally been viewed as disruptions of, or disconnections between, components in a complex functional system. On this view, each deficit reflects a normal mechanism situated in the damaged area, or an interruption of information flow between these areas. More recently, the different clinical syndromes have been proposed to reflect processing stages in the unfolding of the utterance or the action (11).

According to this view, a base level (motor envelope) elaborates an archaic level in speech and motility, combining the incipient vocal and somatic movement in a space centered on the body axis. The action is organized about the axial and proximal musculature, linked to respiratory and other rhythmic automatisms, close to motivational and drive-like states. As it develops, the action undergoes a specification of its motor components, with an isolation of limb, body, and vocal motility. This phase in the action is mediated by frontal paralimbic formations, anterior cingulate gyrus, and SMA. Damage to these structures gives akinetic mutism and the SMA syndrome. This deep rhythmic structure is derived through left generalized (premotor) cortex to an oscillator elaborating the speech melody, disruption of which leads to a prominence of stress-bearing content words (agrammatism). Subsequently, there is a transformation through left inferior premotor cortex to the fine temporal program of sound sequences. Disruption exposes a stage of phonological (phonetic) encoding (Broca's aphasia) and its articulatory realization through inferior precentral cortex (phonetic disintegration) (Fig. 2).

As the utterance undergoes progressive specification, limb action develops from a global, axial movement to independent upper limb motility (impaired initiation), through a stage of the sequential laying down of the movement (apraxia) to the fine distal innervation (dyspraxia).

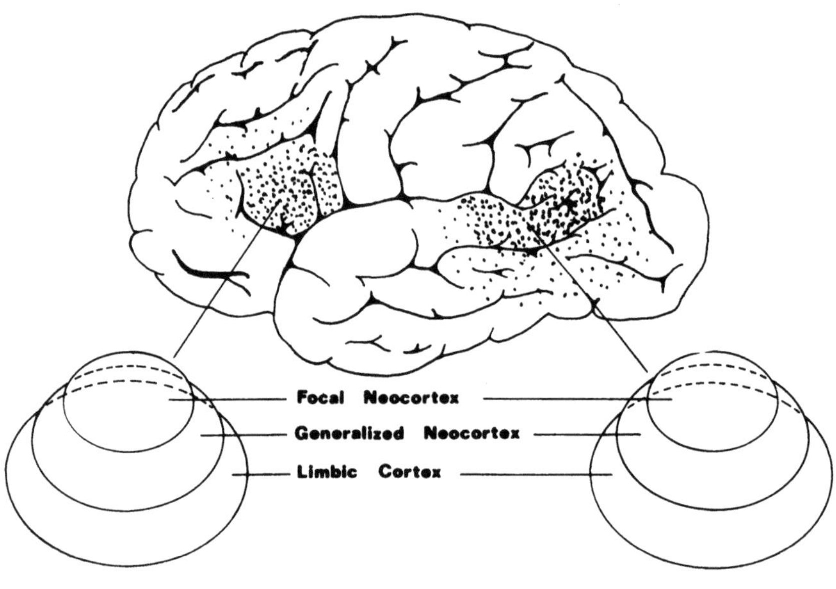

FIG. 2 - The Broca and Wernicke zones develop through a process of regional or core specification within the background "integration" neocortex. The result is a system of evolutionary layers, or growth planes, in the anterior and posterior language areas, through which the motor and perceptual components of the utterance unfold.

POSTERIOR APHASIAS

Damage to posterior temporal neocortex and contiguous structures give rise to various types of fluent or Wernicke's aphasia. Although the fluent aphasias are linked to comprehension deficits (receptive, sensory aphasia), the different subtypes are defined largely by the relative occurrence of two major errors in production: lexical-semantic and phonological (phonemic) errors. There is some evidence that lexical errors occur primarily with lesions of inferolateral temporal (T2, T3) and parietal "integration" cortices, and that lexical processing is more bilaterally organized, while phonological errors occur with lesions of the left Wernicke area proper (posterior T1 and supramarginal gyrus).

The clinical data is consistent with studies of cortical stimulation (52), indicating that phonological processing is linked to posterior T-1 and supramarginal gyrus, and lexical processing is related to more lateral regions. This distinction is supported by CT studies of posterior aphasics (Fig. 3) (20). There is also evidence from regional blood flow studies (38, 45) for left posterior temporal increases on phonological but not lexical tasks. Studies in our laboratory are examining these regional effects more directly with positron emission tomography (PET) using a phoneme-monitoring paradigm. Preliminary results indicate activations of temporal cortex bilaterally and implicate Insula in language function, with cortical asymmetries less prominent than anticipated. The Insula may figure importantly in aphasic disorders (12, 29), though precise clinical data are lacking.

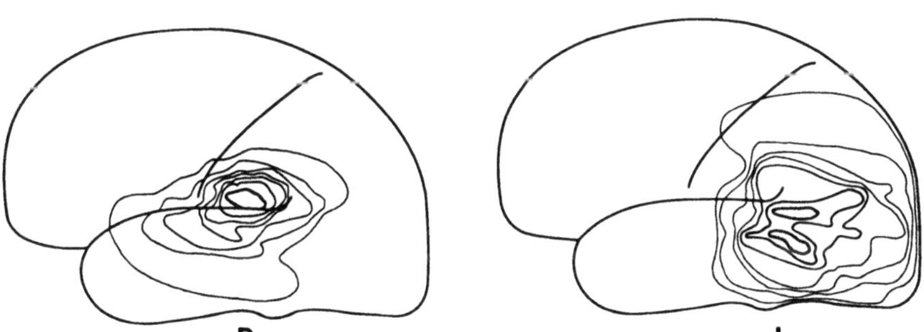

FIG. 3 - Outline of CT lesions in cases of posterior aphasia with predominantly phonemic (P) and lexical (L) errors. From Cappas et al. (20).

Limbic Structures and Inferotemporal Cortex

Bilateral lesions of the limbic system including hippocampus, perhaps dorsomedial nucleus, septum and mammillary bodies (69), or inferolateral temporal cortex (41), give rise to an amnestic (Korsakoff) syndrome with memory deficits and confabulation. Confabulation is defined as a fabrication or filling in of memory gaps, but there are important links with posterior aphasia where a difficulty in the comprehension of word meaning is the central impairment.

In confabulation, lexical substitution (i.e., semantic paraphasia) is unusual but may occur. The problem is at an inter-sentential level with conceptual derailments which are determined by symbolic and affective relationships (3, 63). For example, a story of a rape is recalled as a woman burned in a fire. Florid confabulation may approach aphasia, as in the following example of a patient recalling the story of the "Hen and the Golden Eggs":

> "It was about a horse that bred one year or something like that, but then didn't breed anything because there was no contact."

Bilateral lesions of inferolateral temporal cortex (T2, T3) may also give rise to semantic jargon. In this disorder the semantic deviance penetrates intra-sententially to affect naming and repetition, as well as spontaneous speech. Naming errors may show thematic links, as in:

pipe → "smoking mob"
glove → "you can't wear bread,"

or may be distant in meaning from presumed targets:

thumb → "envelope"
key → "featherhair."

An example of the spontaneous speech of a patient with mild semantic jargon is the following:

> "perhaps there was a possibility that I reach, reach a photograph for someone who teaches a photograph of some sort. I may have asked for it and I would have assumed my own vantage of yes and no,"

and a more severe case:

> "Face of everything. This guy has got to this thing, this thing made out in order to slash immediately to all of the windpails."

Patients with semantic jargon have a memory disorder embedded in the aphasia and on recovery are amnestic for the jargon episode. As with

Korsakoff cases, there is confusion or euphoria during the confabulatory or jargon period. Conversely, Korsakoff patients are often aphasic acutely, with lexical errors predominating (Victor, personal communication), and show semantic encoding problems (22).

The pathological anatomy of semantic jargon is not well established, though there are sufficient case reports to suggest the importance, especially in young and middle-aged subjects, of bilateral, probably subtemporal lesions. A unilateral left temporal lesion can produce semantic deviance, though usually not the anomaly of the bilateral case.

Temporo-parietal "Integration" Cortex

This refers to posterior T2 and inferior parietal lobule, especially angular gyrus. A left-sided lesion of this region gives lexical-semantic errors and/or word finding difficulty in speech, naming, and repetition. Speech tends to be empty with circumlocution, and semantic paraphasia is prominent on naming tests. Errors tend to be in the same semantic category (e.g., chair → "table"), though occasionally errors more distant in meaning may occur. Word frequency determines the likelihood of success, but not the nature of the naming error (e.g., red → "fuchsia").

Right-sided lesion of posterior "integration" cortex gives an impairment in the comprehension of word meaning in perhaps 20-40% of patients. This was first demonstrated by Eisenson (27) and Critchley ((24); see also (55)). Deficits in phonology and/or syntax are rare, if they occur at all, in right-handers. This finding, together with evidence of good lexical-semantic capacity in the "isolated" right hemisphere (70), the lack of rCBF and evoked potential asymmetry with lexical activation procedures, and the need for bilateral lesions to produce semantic jargon are consistent with the idea that lexical-semantic processing develops out of a system that is bilaterally represented, to one that is relatively left lateralized ("integration" cortex).

Lexical substitutions may coexist with anomia, in which omission errors predominate. Anomia does not correlate well with focal left hemisphere lesions but, when severe, does suggest a left parietal or temporal convexity lesion. It also occurs as an initial stage in dementia and other diffuse conditions. In this respect, anomia is intermediate between confabulation and semantic jargon, with bilateral pathology, and phonological disorders, with discrete focal left-sided lesions (see below). The anomic error is interpreted as a failure to fully realize a reasonably well specified lexical item. Anomic patients behave as if in a "tip-of-the-tongue" state, though there is evidence for more diffuse semantic categories, and anomics do poorly on TOT judgments.

Summary. Bilateral inferotemporal lesions give confabulation or marked semantic deviance, whereas lesions of left posterior "integration" cortex give lexical errors which tend to be in the same abstract class as the target item, or omission errors in situations (TOT state) where the patient appears to have adequate lexical representations. These relationships suggest that the pathology disrupts successive processing stages in the course of lexical retrieval. This process consists in a progressive specification of the lexical representation over strata in the evolutionary growth of the forebrain.

Wernicke's Area (Posterior T1 and Supramarginal Gyrus)
Lesions restricted to posterior T1 and its "parietal" continuation as supramarginal gyrus give rise to phonological errors in otherwise well specified lexical targets (Figs. 2 and 3). Various types of errors can occur, including substitution, deletion, and transposition (metathesis):

kite	→	"dite"
green	→	"reen"
president	→	"predisent."

The disturbance is typically present in conversational speech, as well as naming and repetition. When mild it is referred to as conduction aphasia. All cases of conduction aphasia have involvement of posterior T1 and/or supramarginal gyrus (10, 25, 30).

The phonological error affects content words preferentially, and lexical targets are usually recognizable. Blumstein's (5) finding that phonemic errors respect distinctive feature distance may not be the rule for posteriors if one includes cases with neologism. There are instances where the phonemic error seems preferentially close to, and others where it seems distant from, targets. One has the impression of a "zeroing-in" on the target phoneme comparable to the specification of lexical representations in the lexical-semantic phase of retrieval.

When the phonological disturbance extends into function words, the disorder becomes so profound that the listener can no longer segment the stream of speech sounds into word-like units. An example of this "phonemic jargon" is the following (from Perecman and Brown (56)):

(oo et voony a piuvwivwenhy espida etsabaforfor orlilik boughts).

The pathological localization is uncertain. This case had bilateral Wernicke area lesions, but a patient has been reported with a large Sylvian lesion (57).

Lesions of the left Wernicke area may extend to neighboring "integration" cortex. In such cases, the phonological disturbance is combined with a lexical-semantic disorder resulting in neologism or neologistic jargon:

glasses → (karera)
chair → (patchit)

The view of neology as a two-level (phonemic + semantic) deficit is consistent with its pathological localization (36) and the pattern of clinical recovery (8). However, there are cases of neologism due to a severe phonological disorder, as perhaps in the Perecman-Brown case above, and some authors have postulated "device-generated" neologisms to fill empty anomic segments ((19); see also Buckingham (16)).

Summary. Focal left-sided lesions of the Wernicke area give phonological errors in recognizable lexical targets. Bilateral Wernicke area lesions give phonemic jargon. Extension of a left Wernicke area lesion to surrounding "integration" cortex gives a combined phonemic-semantic deficit, neologistic jargon. These observations are consistent with the idea that lexical realization is completed through a stage of phonemic representation mediated by the Wernicke zone.

Interpretation of the Posterior Aphasias

The process of lexical retrieval begins with a stage of conceptual knowledge built up on perceptual, experiential, and affective relationships. The configuration that develops out of this stage is transformed through levels in a semantic representation to an abstract category of the lexical target. The specification of the target corresponds with an evocation of its phonological shape.

Brain damage displays successive moments or loci in this process. Bilateral lesions of evolutionarily more archaic systems display early processing, whereas left focal lesions of more recent evolutionary structures display end point or surface processing.

Thus, in parallel with the anterior system, the posterior sequence leads from a bilateral stage of limbic and inferotemporal cortex where the conceptual content is aroused in memory (amnestic syndrome) leading to a semantic operation. Disruption at this point gives retrieval deficits in which symbolic and affective links to targets are prominent (confabulation). Subsequently, the lexical representation is specified through a series of progressively narrowing fields by way of dominant posterior "integration" cortex. Disruption gives semantic paraphasias, leading from wide or distant, to narrow or categorical, relationships, finally to the correct lexical representation and inability to evoke its phonological form (anomia). Phonological encoding of the lexical representation through dominant focal (Wernicke) neocortex leads to a stage of phonemic representation, disruption of which gives conduction aphasia.

BRAIN MECHANISMS OF LANGUAGE

Components and Continua

The idea of a discrete representation of language subfunctions, as implied in older models as well as current componential or modular theories, finds only limited support in the aphasia material. While there is evidence for an anterior phonetic and posterior phonemic component, and a posterior lexical-semantic component, focal lesions do not dissect these components with the selectivity predicted by some psychological models. There are patterns of destructuring within components but not a loss of specific elements.

For example, within the lexical component, the evidence for category-specific stores is extremely weak. The evidence for a loss of specific semantic categories (e.g., foods or colors) or a selective preservation of some categories (e.g., cities, body parts) with a loss of others, should be taken with caution. One has to consider the object space, the relation to modality, the degree of impairment in a perceptual modality, the skill or familiarity with, or the frequency of, a class of lexical items. Individual lexical items or concepts are not abolished. Word frequency is important in establishing the level of difficulty, but not in predicting the nature of response errors. In contrast, the problematic notion of "semantic distance" appears to have value in the analysis of lexical errors.

Another difficulty is that lexical errors tend to be accompanied by disorders of attention, mood, and awareness which differ from those which are present with other types of aphasia, suggesting that a component may not have a representation distinct from other cognitive or paralinguistic features. Still, the idea of a separate lexical-semantic and phonological component is central to the interpretation of the aphasias, and the data strongly support the inference that these components are represented through different brain regions. The evidence for a separate syntactic component is less compelling.

Connectivity Between the Language Areas

The concept of discrete centers or modules implies an interaction between them through conduction pathways. In language study, a prime target for such interaction is the arcuate fasciculus, a fiber bundle projecting between the Wernicke and Broca region. Lesion of this pathway is said to give rise to conduction aphasia, an interruption in flow between the two centers. However, there are no documented cases of aphasia with lesions restricted to this pathway, there are cases of conduction aphasia with lesions sparing the pathway, and there are instances where the pathway is involved without aphasia. Puusepp (59) reported cases of section of the arcuate fasciculus without aphasia, and Penfield is reputed to have sectioned it without consequence. Stimulation in Wernicke's area in a single patient did not produce electrocortical effects in Broca's area (Ojemann, personal communication). Moreover, Ojemann (52) reports simultaneity in electrocortical events across the anterior and posterior sectors. This recalls findings in owl monkey of simultaneous activation of multiple cortical visual areas (47) and argues against posterior to anterior conduction.

If language cortex is neither generalized and equipotential, nor arranged into centers which interact through association bundles, like cities and thoroughfares on a geographic map, how are the brain mechanisms of language to be conceived? One possibility is that specialized regions (Broca, Wernicke areas) emerge over time within more generalized fields, such that there is a (qualitative) gradient from the periphery of a zone to its core. Inconsistencies in clinical localization might be explained through the degree of encroachment on the core, or the degree to which the field is specialized. According to this view, the degree of regional specification of the Broca and Wernicke zones would vary from one individual to another.

Regional Specification

The concept of a progressive differentiation of the Wernicke and Broca zones has developed out of evidence for a relationship between aphasia type and age. For example, a lesion of Wernicke's area produces a different aphasia, actually four or five different types of aphasia, depending on the age of the individual. This observation suggests that changes in functional representation reflect the continuing growth of the language areas over the life span (14, 15, 51). This interpretation is consistent with findings of dendritic growth in the aged brain (17) and greater dendritic arborization (growth) in left than in right Broca area (62).

This concept is related to the idea of diffuse and focal representation and evidence that the right hemisphere is more "diffusely" organized than the left hemisphere (32, 64). PET studies provide support for this hypothesis in that right hemisphere activation with nonverbal stimuli tends to be diffuse while stimulation with language material gives focal increases in the left hemisphere ((46), Mazziota et al., personal observations). These results could reflect either material-specific or hemispheric differences. Given a diffuse:focal/right:left hemisphere analogy, one might predict that the degree of lateral asymmetry for a performance corresponds with the degree of regional specification for the processing underlying that performance. In other words, regional specification is the means through which lateralization and the transition from a diffuse to focal organization occurs.

The concept of a progression from a diffuse to a focal organization and the regional specification of the language area, though developed independently in neuropsychology, has appeared in different forms in other areas of brain research. Core differentiation has been proposed as a fundamental aspect of forebrain evolution (61). In neural development, there is a change from generalized to specific innervation in the maturation of sensory and other systems (28). Electrophysiological studies in animals demonstrate a restriction of the cortical locus for sensory evoked potentials with age (4, 43). Studies of purposeful movement show that diffuse cortical activation (Bereitschaftspotential) precedes focal activation in the motor region (37). Even the effects of sulcus principalis lesions in monkey have been interpreted in terms of a focus-field concept. Thus, there is good evidence that a diffuse to focal shift occurs in development, and some evidence that this may continue into late life. What are the implications of this data for theories of brain and language?

The Broca and Wernicke zones, which presumably undergo regional specification, mediate phonological processing, as inferred from aphasia study, metabolic, and electrocortical investigations. Phonological processing is assumed to occur as an end stage in the processing sequence that lays down an utterance. Phonological processing is more strongly asymmetric (left hemisphere) than lexical-semantic processing. These observations indicate an association between a late-developing brain area, lateral asymmetry, and a late processing stage. Specifically, there seems to be an inner relationship between the direction of cerebral growth and the direction of language processing.

Structure and Process

These findings and speculations have been expanded to a general theory of brain-behavior relations (11, 13), but whatever one's theory, they do indicate a need for a more dynamic view of brain and localization. Structure is not fixed and immutable but undergoes change even apart from the obvious change that occurs in early development, pathology, and decay. Structure is more like a process in a continual state of growth. Language maps onto structure viewed in this way. The various components of language, or for that matter any neuropsychological performance, are not absolutely wired into specific brain areas but, as Monakow suggested long ago, undergo "chronogenic localization" in relation to age and experience.

The view of language as a microtemporal unfolding over evolutionary growth planes, with regional specification of language areas reflecting a sustained evolutionary growth trend, captures this dynamic quality and helps to explain many long-standing problems not resolved in traditional accounts. The centers of classical aphasiology are not sites for mechanisms but phyletic systems through which a configuration is transformed one stage further in a processing continuum. A center is a process at a certain moment in the realization of an utterance.

Acknowledgement. Portions of this paper have been published in: Brown, J.W., Frontal Lobe Syndromes. In Handbook of Clinical Neurology, 2nd ed. Clinical Neuropsychology, ed. J. Frederiks. Amsterdam: North Holland, in press.

REFERENCES

(1) Arkin, A.M., and Brown, J.W. 1972. NREM sleep speech, drowsy speech, aphasic and schizophrenic speech. Psychophysiol. 9(2): 210.

(2) Bancaud, J.; Talairach, J.; Geier, S.; Bonis, A.; Trottier, S.; and Manrique, M. 1976. Manifestations comportementales induites

par la stimulation électrique du gyrus cingulaire chez l'homme. Rev. Neurol. 132: 705-724.

(3) Betlheim, S., and Hartmann, H. 1951. On parapraxes in the Korsakoff psychosis. In Organization and Pathology of Thought, ed. D. Rappaport. New York: Columbia University Press.

(4) Bignall, K.E.; Imbert, M.; and Buser, P. 1966. Optic projections to nonvisual cortex of the cat. J. Neurophysiol. 29: 396-409.

(5) Blumstein, S. 1973. A Phonological Investigation of Aphasic Speech. The Hague: Mouton.

(6) Blumstein, S. 1981. Phonological aspects of aphasia. In Acquired Aphasia, ed. M. Sarno. New York: Academic Press.

(7) Bogen, J. 1976. Linguistic performance in the short-term following cerebral commissurotomy. In Studies in Neurolinguistics, ed. H. Whitakers, vol. 2. New York: Academic Press.

(8) Brown, J., ed. 1981. Jargonaphasia. New York: Academic Press.

(9) Brown, J., and Hecaen, H. 1976. Lateralization and language representation. Neurology 26: 183-189.

(10) Brown, J.W. 1972. Aphasia, Apraxia and Agnosia. Springfield, IL: Charles C. Thomas.

(11) Brown, J.W. 1977. Mind, Brain and Consciousness. New York: Academic Press.

(12) Brown, J.W. 1979. Language representation in the brain. In Neurobiology of Social Communication in Primates, eds. H. Steklis and M. Raleigh. New York: Academic Press.

(13) Brown, J.W. 1983. The microstructure of perception: physiology and patterns of breakdown. Cog. Brain Theory 6(2): 145-184.

(14) Brown, J.W., and Grober, E. 1983. Age, sex and aphasia type. J. Nerv. Ment. Dis. 171: 431-434.

(15) Brown, J.W., and Jaffe, J. 1975. Hypothesis on cerebral dominance. Neuropsychologia 13: 107-110.

(16) Buckingham, H. 1981. Where do neologisms come from? In Jargonaphasia, ed. J.W. Brown. New York: Academic Press.

(17) Buell, S., and Coleman, P. 1979. Dendritic growth in the aged brain and failure of growth in senile dementia. Science 206: 854-856.

(18) Buge, A.; Escourolle, R.; Rancurel, G.; and Poisson, M. 1975. Mutisme akinetique et ramollissement bi-cingulaire. Rev. Neurol. 131: 121-137.

(19) Butterworth, B. 1979. Hesitation and the production of verbal paraphasias and neologisms in jargon aphasia. Brain Lang. 8: 133-161.

(20) Cappas, S.; Cavallotti, G.; and Vignolo, L. 1981. Phonemic and lexical errors in fluent aphasia: correlation with lesion site. Neuropsychologia 19: 171-177.

(21) Caramazza, A., and Zurif, E., eds. 1978. Language Acquisition and Language Breakdown. Baltimore: Johns Hopkins University Press.

(22) Cermak, L., and Butters, N. 1973. Information-processing deficits of alcoholic Korsakoff patients. Q. J. Stud. Alchol. 34: 1110-1132.

(23) Corkin, S.; Twitchell, T.; and Sullivan, E. 1979. Safety and efficacy of cingulotomy for pain and psychiatric disorder. In Modern Concepts in Psychiatric Surgery, eds. E. Hitchcock et al. Amsterdam: Elsevier.

(24) Critchley, M. 1962. Speech and speech-loss in relation to the duality of the brain. In Interhemispheric Relations and Cerebral Dominance, ed. V. Mountcastle. Baltimore: Johns Hopkins Press.

(25) Damasio, H., and Damasio, A. 1980. The anatomical basis of conduction aphasia. Brain 103: 337-350.

(26) De Renzi, E.; Pieczuro, A.; and Vignolo, L. 1966. Oral apraxia and aphasia. Cortex 2: 50-73.

(27) Eisenson, J. 1962. Language and intellectual modifications associated with right cerebral damage. Lang. Speech 5: 49-53.

(28) Goldman, P. 1976. Advances in the Study of Behavior, vol. 7, pp. 1-90. New York: Academic Press.

(29) Goldstein, K. 1948. Language and Language Disturbances. New York: Grune and Stratton.

(30) Green, E., and Howes, D. 1977. The nature of conduction aphasia. In Studies in Neurolinguistics, ed. H. Whitakers, vol. 3. New York: Academic Press.

(31) Hecaen, H., and Consoli, S. 1973. Analyse des troubles du langage au cours des lesions de l'aire de Broca. Neuropsychologia 11: 377-388.

(32) Hecaen, H., and Sauguet, J. 1971. Cerebral dominance in left-handed subjects. Cortex 7: 19-48.

(33) Jonas, S. 1981. The supplementary motor region and speech emission. J. Comm. Disord. 14: 349-373.

(34) Jurgens, U., and Pratt, R. 1979. The cingular vocalization pathway in the squirrel monkey. Exp. Brain Res. 34: 499-510.

(35) Jurgens, U., and von Cramon, D. 1982. On the role of the anterior cingulate cortex in phonation: a case report. Brain Lang. 15: 234-248.

(36) Kertesz, A. 1981. The anatomy of jargon. In Jargonaphasia, ed. J.W. Brown. New York: Academic Press.

(37) Kornhuber, H. 1974. Cerebral cortex, cerebellum and basal ganglia. In The Neurosciences: Third Study Program, eds. F.O. Schmitt and F.G. Worden. Cambridge, MA: MIT Press.

(38) Knopman, D.; Rubens, A.; Klassen, A.; and Meyer, M. 1982. Regional cerebral blood flow correlates of auditory processing. Arch. Neurol. 39: 487-493.

(39) Laplane, D.; Talairach, J.; Meininger, V.; Bancaud, J.; and Bouchareine, A. 1977. Motor consequences of motor area ablations in man. J. Neurol. Sci. 31: 29-49.

(40) Larsen, B.; Skinhoj, E.; and Lassen, N. 1978. Variations in regional cortical blood flow in the right and left hemispheres during automatic speech. Brain 101: 193-209.

(41) Lhermitte, F., and Signoret, J. 1972. Analyse neuropsychologique et différentiation des syndromes amnésiques. Rev. Neurol. 126: 161-178.

(42) Luria, A. 1970. Traumatic Aphasia. The Hague: Mouton.

(43) Marty, R. 1962. Développement post-natal des responses sensorielles du cortex cerebral chez le chat et le lapin - aspects psychologiques et histologiques. Arch. Anat. Micr. 51: 129-264.

(44) Masdeu, J., and O'Hara, R. 1983. Motor aphasia unaccompanied by faciobrachial weakness. Neurology 33: 519-521.

(45) Maximilian, V.A. 1980. Cortical blood flow asymmetries during monoaural verbal stimulation. In Functional Changes in the Cortex during Mental Activation, ed. V. Maximilian, pp. 103-121. Malmo: University of Lund.

(46) Mazziotta, J.; Phelps, M.; Carson, R.; and Kuhl, D. 1982. Tomographic mapping of human cerebral metabolism: auditory stimulation. Neurology 32: 921-937.

(47) Merzenich, M., and Kass, J. 1980. Principles of organization of sensory perceptual systems in mammals. Progr. Psychobiol. Physiol. Psychol. 9: 1-42.

(48) Miceli, G.; Mazzucchi, A.; Menn, L.; and Goodglass, H. 1983. Contrasting cases of Italian agrammatic aphasia without comprehension disorder. Brain Lang. 19: 65-97.

(49) Mohr, J.; Pessin, M.; and Finkelstein, S.; et al. 1978. Broca's aphasia: pathological and clinical. Neurology 28: 311-324.

(50) Nielsen, J., and Jacobs, L. 1951. Bilateral lesions of the anterior cingulate gyri. Bull. LA Neurol. Soc. 16: 231-234.

(51) Obler, L.; Albert, M.; Goodglass, H.; and Benson, D. 1978. Aphasia type and aging. Brain Lang. 6: 318-322.

(52) Ojemann, G. 1983. Brain organization for language from the perspective of electrical stimulation mapping. Behav. Brain Sci. 6(2): 189-230.

(53) Penfield, W., and Roberts, L. 1959. Speech and Brain-Mechanisms. Princeton, NJ: Princeton University Press.

(54) Penfield, W., and Welch, K. 1949. The supplementary motor area of the cerebral cortex of man. Trans. Am. Neurol. Assoc. 74: 179-184.

(55) Perecman, E. 1983. Cognitive Processing in the Right Hemisphere. New York: Academic Press.

(56) Perecman, E., and Brown, J.W. 1981. Phonemic jargon: a case report. In Jargonaphasia, ed. J.W. Brown. New York: Academic Press.

(57) Peuser, G., and Temp, K. 1981. The evolution of jargonaphasia. In Jargonaphasia, ed. J. Brown. New York: Academic Press.

(58) Poeck, K., and Kerschensteiner, M. 1975. Analysis of sequential motor events in oral apraxia. In Cerebral Localization, eds. K. Zulch et al. Heidelberg: Springer Verlag.

(59) Puusepp, L. 1937. Alcune considerazioni sugli interventi chirurgici nelle malattie mentali. Gior. Accad. med. Torino 100: 3-16.

(60) Rubens, A. 1975. Aphasia with infarction in the territory of the anterior cerebral artery. Cortex 11: 239-250.

(61) Sanides, F. 1975. Comparative neurology of the temporal lobe in primates including man with reference to speech. Brain Lang. 2: 396-419.

(62) Scheibel, A.; Fried, I.; Paul, L.; et al. 1982. A histological substrate for speech-related cortical asymmetry. Ann. Neurol. 12: 76.

(63) Schilder, P. 1951. Studies concerning the psychology and symptomatology of general paresis. In Organization and Pathology of Thought, ed. D. Rappaport. New York: Columbia University Press.

(64) Semmes, J. 1968. Hemispheric specialization: a possible clue to mechanism. Neuropsychologia 6: 11-26.

(65) Teuber, H.-L. 1964. The riddle of frontal lobe function in man. In The Frontal Granular Cortex and Behavior, ed. J. Warren and K. Akert. New York: McGraw-Hill.

(66) Tonkonogy, J., and Goodglass, H. 1981. Language function, foot of the third frontal gyrus and Rolandic operculum. Arch. Neurol. 38: 486-490.

(67) Travis, A. 1955. Neurological deficiencies following supplementary motor area lesions in Mucaca mulatta. Brain 78: 174-198.

(68) Van Buren, J., and Fedio, P. 1976. Functional representation on the medial aspect of the frontal lobes in man. J. Neurosurg. 44: 275-289.

(69) Victor, M.; Adams, R.; and Collins, G. 1971. The Wernicke-Korsakoff Syndrome. Philadelphia: Davis.

(70) Zaidel, E. 1977. Lexical organization in the right hemisphere. In Cerebral Correlates of Conscious Experience, eds. P. Buser and A. Rougeul-Buser. New York: Elsevier.

The Biology of Learning, eds. P. Marler and H.S. Terrace, pp. 667-685. Dahlem Konferenzen 1984. Berlin, Heidelberg, New York, Tokyo: Springer-Verlag.

The Neuropsychology of Memory

L.R. Squire
Veterans Administration Medical Center
San Diego, CA 92161 and
University of California, San Diego
La Jolla, CA 92093, USA

Abstract. Neuropsychology aims to describe how the brain accomplishes learning and memory, in a way that speaks both to cognitive psychology and neuroscience. This paper presents a summary of presently available information about the neuropsychology of human memory, emphasizing three ideas: a) The neural substrate of memory continues to change for a long time after initial learning. This change (memory consolidation) is distinct from the changes underlying forgetting and involves the medial temporal region of the brain. b) The nervous system honors the distinction between two kinds of learning and memory (procedural vs. declarative). The former is spared in amnesia and does not depend on the integrity of the particular brain regions that when damaged cause amnesia. c) Animal models of human amnesia in the monkey are now available. These models should permit those brain regions damaged in amnesia to be identified and should lead to more detailed neurobiological study of these regions.

INTRODUCTION

The neurobiological study of memory implies first of all an effort to understand how neurons function and how neuronal function can change. This analysis has properly proceeded, in ever increasing detail, at the level of cells and synapses with biochemical, physiological, and anatomical techniques. To understand how the brain accomplishes learning and memory, one must also obtain information at the brain systems or neuropsychological level. For example, it is important to know whether there is one kind of memory or many, how memory changes over time, what brain regions or brain systems are involved in memory functions, and how they are involved in memory.

Neuropsychological studies potentially connect cognitive psychology and philosophical inquiries about cognition to fundamental neuroscience. Fodor wrote "it was widely held that philosophers ought to provide a survey of the conceptually coherent options, and that there are, in fact, fewer of these than might be supposed" (12). Neuropsychology aims to discover the particular way in which the brain accomplishes memory storage. This endeavor is empirical, inferential, and aims at a level of abstraction that is biologically meaningful. Sometimes, hypotheses developed in cognitive psychology can be tested in neuropsychological experiments, thereby assessing the biological reality of psychological theory. Conversely, by linking behavior to brain function, neuropsychology can sometimes provide functional significance for neurobiological observations.

One strategy for addressing neuropsychological questions about memory is to study human amnesia. It has turned out that the human amnesic syndrome can sometimes occur as a relatively circumscribed entity in the absence of other cognitive deficits, and such cases can be instructive about how memory functions are organized in the brain (for reviews, see (3, 6-8, 41, 42, 50)). To the extent that the neuropathology of these cases is known, and especially to the extent that human amnesia can be mimicked in experimental animals like the monkey, it then also becomes possible to learn something about the identity of the brain regions that are involved in memory. In this paper, I draw on work from our laboratory to consider three aspects of amnesia that bear on questions about memory and the brain: retrograde amnesia and memory consolidation, the phenomenon of spared learning in amnesia, and attempts to achieve an animal model of human global amnesia in the nonhuman primate.

Before considering these three topics, it will be useful to identify the brain regions where damage is known to cause amnesia in man (Fig. 1).

The medial temporal region became associated with memory functions primarily because of the noted patient H.M. (39). In 1953 in an effort to relieve severe epilepsy, this individual sustained bilateral removal of hippocampal gyrus, amygdala, and the anterior two thirds of the hippocampus. Although it has traditionally been held that the hippocampal

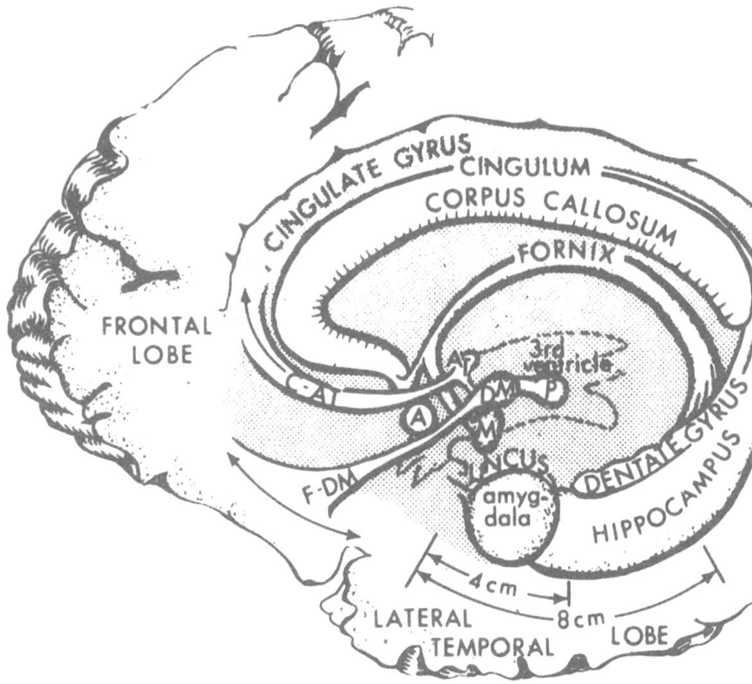

FIG. 1 - Schematic drawing of the medial surface of the human brain showing structures implicated in diencephalic and medial temporal amnesia.

removal was the critical aspect of the neurosurgery that caused amnesia, recent work with monkeys has raised the possibility that the amygdala may play an important role as well (29). However, many questions remain (23, 45), and further work with animal models of human amnesia will be needed to settle this question.

The diencephalic region became linked to amnesia through detailed neuropathological study of the Korsakoff's syndrome (for reviews, see (24, 47)). It is known that damage to the mammillary bodies and the dorsomedial thalamic nucleus correlates highly with the amnesic deficit in these patients, but it is not yet clear which specific structure or structures must be damaged to produce the syndrome. Again, animal models of human amnesia will be needed to settle this issue. Considerable success has been achieved in recent years in establishing animal models (see (30, 45)), and this work will be summarized in the final section of this paper. Bilateral damage to either diencephalic midline or the medial temporal region is sufficient to cause pronounced amnesia. Although the impairment in the two circumstances is similar in many ways, there are differences between the two kinds of impairment. The critical structures may be part of a common circuit (30), though diencephalic and medial temporal structures contribute differently to memory functions (40).

MEMORY CONSOLIDATION

Here I consider the concept of memory consolidation (26) and the role of the medial temporal region in consolidation. Memory consolidation refers to the idea that memory changes with the passage of time after learning (aside from the changes related to normal forgetting). Consolidation has been much discussed in the disciplines of experimental psychology, physiological psychology, and neuropsychology, though there has seldom been good agreement on what exactly it is or how long it lasts. My colleagues and I have proposed that memory consolidation best refers to a long-lived, dynamic process (43). The basic idea that memory changes for a long time after learning was stated clearly many years ago:

> "In normal memory a process of organization is continually going on, - a physical process of organization and a psychological process of repetition and association. In order that ideas may become a part of permanent memory, time must elapse for these processes of organization to be completed" (5).

Our reasons for adopting this view of memory consolidation are based on three facts of retrograde amnesia. First, electroconvulsive therapy (ECT) typically causes a temporally-limited retrograde amnesia, covering the period a few years prior to treatment without affecting the period prior to that time. This has been shown most clearly using tests based on remote memory for one-season television programs. Thus the susceptibility of memory to disruption can decrease for as long as a few years after learning.

Second, a recent study of experimental mice receiving electroconvulsive shock (ECS) has also demonstrated that retrograde amnesia can be temporally graded over a long time period (44a). Though a large literature involving primarily rats and mice has shown that retrograde amnesia is usually brief (seconds to minutes) following a single ECS (27), in this study the cumulative effects of four ECS treatments were sufficient to cause retrograde amnesia when the treatments were given one day to three weeks after training. Four ECS treatments were given at hourly intervals on one of several different days (1 to 70) after training in the one-trial step-through passive avoidance task, a standard laboratory test much used in experimental studies of memory in rodents. Testing for retention always occurred two weeks after ECS. In untreated mice, memory persisted for more than twelve weeks. The susceptibility of memory to disruption by ECS gradually diminished during the weeks after training. In humans, memory for one-season television programs persisted for more than sixteen years, and the susceptibility of memory to disruption by ECT diminished during the first few years after the programs had been broadcast. Thus, for both humans and for mice, retrograde amnesia can be temporally graded and can cover a significant portion of the lifetime of a memory (Fig. 2).

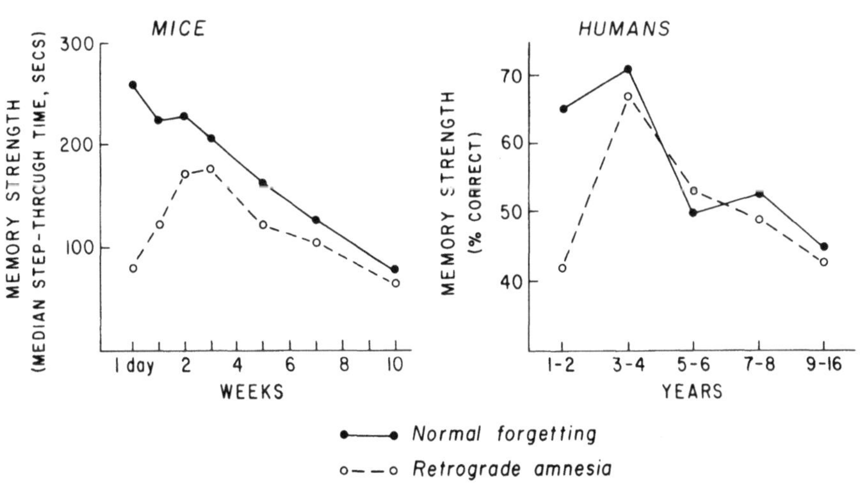

FIG. 2 - Long gradient of retrograde amnesia in mice and humans.

Third, case H.M., an example of medial temporal amnesia, has consistently been described as having a retrograde amnesia of a few years and good memory for earlier time periods. The facts of case H.M. connect ideas about memory consolidation to the medial temporal region. The medial temporal region cannot be a site of storage for all memories, since H.M. recalls memories that were acquired long before the onset of his amnesia. It appears, instead, that the medial temporal region has a necessary role in the formation and development of memory. We have proposed that at the time of learning the medial temporal region establishes a relationship with memory storage sites elsewhere in the brain, primarily in the neocortex (43). This interaction is required for as long as a few years after learning in order for memory to develop and be maintained in a normal way. There need be nothing special or fixed about this length of time, since many or most memories do not endure as long as a few years. Accordingly, the role of the medial temporal region in the consolidation of memory is presumed to continue so long as information is being forgotten, during the reorganization and stabilization of what remains. This gradual process strengthens and makes more coherent those elements of memory that are not lost through forgetting, thereby enabling stable memory representations to be maintained across time.

We have supposed that the length of the slow process in which the medial temporal region participates, i.e., memory consolidation, is not predetermined at the outset nor is this process relentlessly gradual or automatic. Rather, it is understood as dynamic, affected by rehearsal and association and by subsequent memory storage episodes. These events would influence the fate of recent, and unconsolidated, memories by remodeling the neural circuitry underlying the original representation. During consolidation, some parts of the original representation would be lost through forgetting, while those that remain would become more stable and more coherent. The representation of an event thereby changes over time and eventually is able to support memory storage and retrieval in the absence of the medial temporal region.

PRESERVED LEARNING

Knowing what amnesic patients can succeed at is as important to questions about memory and the brain as knowing what they cannot learn. This kind of information is needed to define the scope and limits of amnesia, and it helps to characterize the function of the affected brain regions. It is now known that amnesia is a selective deficit. Accordingly, the role of the medial temporal region and of the diencephalic midline in memory functions is narrower than once believed.

The traditional view of amnesic patients had been that they can acquire motor skills or perceptual-motor skills (e.g., specific feats of eye-hand coordination) but cannot learn much else. During the last ten years, however, investigators have compiled a considerable list of tasks that are less clearly perceptuo-motor, yet can be accomplished by amnesic patients, sometimes in a normal fashion (9, 49, 53). A particularly good example is mirror-reading (i.e., reading words reversed in a mirror) which has been studied extensively in normal subjects by Kolers (20, 21). Working with three kinds of amnesic patients, Cohen and I showed that the capacity to learn the mirror-reading skill developed at an entirely normal rate over a period of three days and then was retained at normal levels for three months (11). Yet, many of the patients did not remember having worked at the task before, and all were amnesic for the particular words that they read.

Recently, we have shown that this same mirror-reading skill can be taught prior to ECT and that it then survives the course of treatment (44). In this case, two weeks after ECT, the patients who received ECT showed as much savings of the mirror-reading skill as control subjects; yet unlike control subjects, the patients who received ECT could not recall their previous testing experience nor recognize the words that they had read. Thus the mirror-reading skill survives both anterograde and retrograde amnesia.

These results have suggested a distinction between information based on skills or procedures and information based on specific facts or episodes. Both diencephalic and medial temporal patients are capable of acquiring the skill of mirror-reading, but they cannot acquire facts about the world that would ordinarily be acquired by using these skills, i.e., the fact that they had been tested or the ability to recognize as familiar the words that they had read. The view that amnesia selectively spares skills has also been proposed recently by Moscovitch (34). This distinction drawn here is similar to others that have been proposed between kinds of knowledge (e.g., knowing-how and knowing-that (38); memory without record and memory with record, (4); procedural and declarative knowledge (52)).

These ideas have considerable relevance to questions about how memory is organized in the brain (42). The distinction between skills or procedures and facts or episodes appears to be honored by the nervous system, in that the brain accomplishes memory storage so as to distinguish prominently between two kinds of knowledge. Procedural knowledge can be acquired independently of the brain regions affected in amnesia. The medial temporal region and the diencephalic structures affected in amnesia are not needed for its acquisition and maintenance. Considering what is known about the cellular or synaptic basis of information storage in simpler nervous systems (18), it is worth raising the possibility that procedural learning is phylogenetically primitive, and that it occurs as "on-line" changes or tuning in already existing neural circuitry. Procedural knowledge is implicit and accessible only by engaging in or "running through" the skills in which the knowledge is embedded. By this view, procedural knowledge may be stored in, and expressable only through, activation of brain systems engaged in the learning tasks. As such, it may be relatively specific and inaccessible to other systems. It is not yet known whether phylogenetically primitive forms of learning can be acquired normally in amnesia, though studies of classical conditioning in amnesic patients make this an interesting possibility (51).

By contrast, declarative knowledge requires the participation of the diencephalic midline and the medial temporal region and is subject to gradual consolidation. The sites of information storage appear to maintain a necessary functional link to these critical structures during the period of consolidation. It is possible that this link is what permits declarative, but not procedural, memory to be accessible and not bound to the system in which learning occurred. Declarative knowledge includes the facts and data of conventional memory experiments. It affords recognition that an event happened previously. It includes both specific information about time and place (episodic knowledge) as well as general information (semantic knowledge), which though not specific to time and place is nevertheless obtained in the course of specific experiences. Indeed, one of the consequences of consolidation might be to extract and integrate the semantic knowledge acquired through different time and place experiences.

The idea that procedural knowledge might be so specific as to include certain kinds of classical conditioning (and the attendant CS-US contingencies) implies that the procedures that can be acquired in amnesia can be quite specific. The process of activation (25) or perceptual fluency (16) is relevant to this issue. Activation, as studied in priming and certain other tasks, can be distinguished from recognition of a word as one previously presented. Rozin (37) suggested that activation, or what he termed a "hot-tubes effect," might occur normally in amnesia, and Warrington and Weiskrantz (48) have shown that completion of word stems by amnesic patients can be normal.

It turns out that the results depend critically on the instructions given to subjects (14). Amnesic patients and control subjects saw 3-letter word stems that could be completed to form any of ten common words. When the instruction was to complete each word stem with the word from the recently studied list, amnesic patients performed more poorly than control subjects. They also performed poorly on other conventional memory tests, i.e., free recall and recognition. However, different results were obtained when the subjects studied a list, but then at retest were not explicitly told that they were in a memory experiment. Instead, they were asked to complete the word stems with the first simple word that came to mind. In this condition, amnesic and control subjects performed identically, completing about 50% of the word stems to match target words. The probability of completing word stems by chance to form target words was 9.5. percent (Fig. 3). This tendency or bias to complete the word stems in a particular way declined at the same rate in amnesic and control subjects, reaching chance after two hours. In other studies, we have found that in both amnesic patients and control subjects this bias is reduced when testing occurs in a modality other than the one in which the information was originally presented.

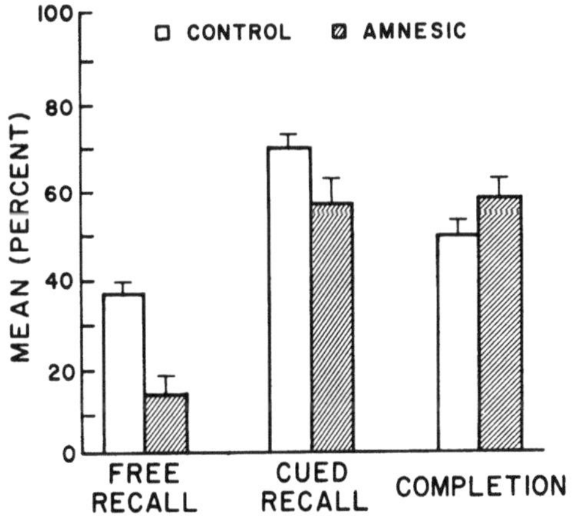

FIG. 3 - Impaired free recall and impaired cue recall of word stems by amnesic patients, but normal completion of word stems.

In all cases, despite their deficit on conventional memory tests, amnesic patients exhibited a normal influence of the effects of previously presented words on word completion. This specific effect of prior experiences on behavior occurred without any awareness by the patients about what they had learned. Similar findings for amnesic patients have been reported by others (17, 48). Thus both skill acquisition and priming effects are spared in amnesia. In these cases, it is supposed that they occur by virtue of plasticity inherent in particular processing structures or procedures, without establishing the kind of record in memory that provides access to the outcomes of having engaged these processing structures or procedures.

Perhaps the most dramatic example of the learning of a specific skill by amnesic patients comes from recent work by Cohen. Inspired by the work of Simon with the Tower of Hanoi problem (2), Cohen noted that the solutions acquired by many normal subjects seemed implicit, in the sense that they had difficulty telling what they had learned. This raised the possibility that the knowledge acquired might have some of the characteristics of a skill or procedure, like mirror-reading. In this problem subjects find the optimal solution to a puzzle that requires discovery of iterative procedures for transferring a stack of five blocks from one peg to another. The puzzle can eventually be solved in thirty-one moves. Case N.A. and case H.M. were able to learn this puzzle over daily sessions to an optimal solution, despite claiming little or no knowledge at the beginning of each session as to what was required of them (9, 10).

The question inevitably arises as to whether the problem of amnesia is a problem of storage or a problem of access. My colleagues and I have favored the view that the amnesic deficit is a failure to develop and maintain some part of the synaptic connectivity that ordinarily constitutes memory representations (41, 43). In this sense we view amnesia as a storage deficit, and not just a problem of access. The deficit cannot be an access problem only, since in medial temporal amnesia patients have access to premorbid memory acquired long before the onset of amnesia. This issue has been discussed in more detail elsewhere (42).

The storage/retrieval question is larger and harder to settle than it might first appear to be. Indeed, the storage/retrieval dichotomy has its roots in a long-standing debate about whether normal forgetting itself is a matter of changes in storage, i.e., actual reversal of some of the synaptic events that originally subserved memory, or changes in retrieval or accessibility. It seems reasonable that the synaptic events underlying both forgetting in normal subjects and rapid forgetting in amnesia may be similar. I suggest that both normal forgetting and abnormally rapid forgetting depend at least in part on an actual loss or reversal of synaptic change. In Aplysia, forgetting of habituation is reflected in a return to baseline of the synaptic changes originally responsible for habituation (18). Neurobiological evidence of this kind was needed to show conclusively that for Aplysia loss of habituation is a matter of a change in storage. In other animals the storage/retrieval issue will also require neurobiological evidence to settle entirely.

ANIMAL MODELS OF HUMAN GLOBAL AMNESIA

Work with monkeys can contribute a good deal when it is done against a background of the neuropsychological facts obtained from human amnesia. These two domains of investigation are mutually facilitatory. Human amnesic cases can provide extensive information about the nature of memory disorders. Monkeys can be prepared with precise lesions and can be studied with tests that are rigorously analogous to the human tests. In this way one can hope to bridge neurology and neuroanatomy and to relate function to brain structure with a precision that would be very difficult to obtain from human cases àlone.

Until recently, the effects of medial temporal and diencephalic lesions in experimental animals like the monkey seemed difficult to relate to work with human patients. Uncertainty about which tasks to use to assess memory in the monkey almost certainly contributed to the difficulty. Because the amnesic syndrome in man is now known to apply to a narrower domain of learning and memory than previously supposed, one must select only particular tasks to study amnesia in monkeys. Several tasks are available that appear to test memory just as human memory is tested in studies of amnesia. The best studied of these tasks is trial-unique, delayed matching and nonmatching-to-sample (13, 31). Here the monkey is given sets of paired trials, always with unique stimuli. First, the monkey sees a novel object which he displaces to obtain food. Then, seconds, minutes, or even hours later, the monkey sees two objects side by side - the original one and a new one. To obtain food reward, the monkey must displace the familiar (match-to-sample) or the novel (nonmatch-to-sample) of the two objects. Normal monkeys learn to perform this task and can then demonstrate differential behavior toward old and new objects (i.e., recognition memory) across long intervals. This task requires delayed recognition of a familiar object, and human amnesic patients cannot master it (Squire and Zolar-Morgan, unpublished observations).

With tasks like these, considerable headway has been made in recent years towards identifying the brain regions that are damaged in amnesia. In humans, the neuropathology is seldom known in the same patients who are extensively studied, and the lesions that the patients have are often the result of disease (e.g., Korsakoff syndrome) or brain injury (e.g., case N.A.), where undoubtedly some damage occurs to brain regions in addition to those responsible for the memory impairment.

In the case of medial temporal amnesia, Mishkin (29) suggested that conjoint hippocampus-amygdala damage is required to produce the syndrome that has been studied so extensively in case H.M. and other patients. Only monkeys with the combined lesions exhibited a severe impairment. In addition, an impairment following combined lesions was found in the tactile modality (36). If correct, the view that combined damage is required would force a revision of the traditional and widespread belief that damage to hippocampus alone can cause amnesia. So far, this idea is based on work with monkeys and the delayed nonmatching-to-sample task, and more studies using other tasks are needed. Other investigations have found that hippocampal lesions alone can produce a decided impairment in memory as measured by concurrent learning (35), retention of simple object discriminations (22), and even the delayed nonmatching-to-sample task (23). The critical questions are whether amygdala damage always add substantially to the deficit caused by hippocampal lesions alone, whether amygdala damage alone causes as much impairment as hippocampal lesions, and whether the impairment reported for either amygdala or hippocampal lesions alone is clinically meaningful, i.e., severe enough to be called amnesia if the same deficit occurred in humans. Further studies of this important issue are needed across several tasks and with histological verification of lesions.

Horel (15) suggested that neither the hippocampus nor amygdala is involved in amnesia. He reviewed the existing data, arguing that the traditional view is not so well supported as might be thought. Instead he suggested that the critical area is the temporal stem, a band of white matter lying adjacent to hippocampus. This view can now be discounted, however, because monkeys with temporal stem lesions are not amnesic (56).

This same investigative approach can be used to address the question of diencephalic amnesia. It has recently been shown that monkeys with bilateral lesions involving dorsomedial nuclei and anterior nuclei of thalamus exhibit a pronounced amnesic deficit (1). Moreover, monkeys with dorsomedial thalamic lesions, who exhibited amnesia, showed normal rates of forgetting (54), like human patients with diencephalic amnesia (41). It seems reasonable to expect that the present uncertainty about amygdala and hippocampus, about dorsomedial nucleus and mammillary bodies, and about the possible role in memory of other thalamic nuclei should be resolved within the next few years.

Work with monkeys in the last few years has come very close to achieving an animal model of human amnesia. Although the neural structures which when damaged produce amnesia have not been identified unambiguously, a group of anatomically related candidate structures has been found, and the behavioral deficit associated with damage to these structures has been found to match closely what is observed in human amnesia (33, 45). Monkeys with conjoint hippocampus-amygdala lesions perform progressively more poorly on the delayed nonmatching-to-sample task as the delay between sample trial and matching trial is increased from five seconds to ten minutes (29, 56). They also fail on three other tasks: delayed response (but again, only on long delays), concurrent discrimination (eight pairs given during forty trials per day), and acquisition and retention of simple object discriminations (55). These tasks are known to be sensitive to human amnesia. Indeed, they are failed by amnesic patients when administered to them in the same way as they are given to monkeys.

At the same time, the pattern of behavioral sparing after these lesions is similar to what would be required by the classification scheme for two memory systems outlined in the previous section. The monkeys are entirely normal at learning and retaining perceptual-motor skills (55). They also can succeed at visual pattern discrimination tasks (45, 55, 56). This task has been analyzed previously, with the suggestion that it consists primarily of skill learning (45). The small impairment observed during acquisition of this task may be attributable to a small part of the necessary information being acquired as a fact or proposition. Some evidence supporting this suggestion has also been obtained (55). Further discussion of the notion of two memory systems in monkeys can be found elsewhere (32, 45).

The findings with humans and monkeys, taken together, have led us to a tentative formulation concerning the existence of two memory systems and their organization. A late-evolving memory system dependent on limbic structures may have appeared in mammals, and possibly in all vertebrates. This system is fast, adapted to one-trial learning, and stores representations declaratively, such that they are accessible to information-processing systems other than ones initially engaged during learning. By this account, the phylogenetically more primitive system is slow, adapted to incremental learning, and stores information as changes in the processing structures initially engaged in learning. The many feats of learning that can be exhibited by nonmammalian vertebrates and by invertebrates (see the many relevant papers in this volume, especially those by Gould, Quinn, Sahley, Lea, and Revusky) may provide opportunities to test more precisely these ideas about two memory systems, which can be stated only generally at the present time. Might it be possible in these animals to dissociate some examples of learning from others by interventions? Should we expect that most or all examples of learning and memory demonstrated by invertebrates can also be accomplished by amnesic patients? Might those examples of learning by invertebrates that amnesic patients could not accomplish (e.g., rapid acquisition of spatial information about food sources as demonstrated by honey bees) be different in some principled way from how mammals would approach the same task? For example, in the honey bee the learning may reflect a specialized adaptation for learning, in that what is learned is minimally accessible to other response systems.

Much remains to be understood about the nature of the two memory systems proposed here. Nevertheless, we are encouraged that we may have uncovered two categories of learning and memory that are biologically meaningful and which may prove useful to those working on species other than man and monkey.

SUMMARY

Based on these and other neuropsychological studies, considerable progress is being made toward identifying the brain structures important in memory and toward describing the roles of these structures in accomplishing memory storage. A great deal is also being learned about the cellular changes that subserve memory, about the anatomy and physiology of several of the candidate brain structures, and about how their neural circuitry can exhibit various forms of plasticity (for reviews, see (19, 28, 46)). Neuropsychological studies can illuminate these neurobiological studies, which are conducted at more elemental levels of analysis, by establishing a clear connection between such studies and functional questions about the organization of memory in the brain. At the same time, neuropsychological investigations can identify for the neurobiologist particular brain structures involved in memory functions that can then be studied in a more detailed way; moreover, if things go well, the findings from neuropsychological work should be able to offer to the neurobiologist some testable hypotheses about the function of the identified brain structures.

Acknowledgement. Supported by the Medical Research Service of the Veterans Administration and by NIMH Grant, MH24600. S. Zola-Morgan, N. Cohen, and P. Graf collaborated on one or more of the studies and contributed to the ideas discussed here.

REFERENCES

(1) Aggleton, J., and Mishkin, M. 1983. Visual recognition impairment following medial thalamic lesions in monkeys. Neuropsychol. 21: 189-197.

(2) Anzai, Y., and Simon, H.A. 1979. The theory of learning by doing. Psychol. Rev. 86: 124-140.

(3) Baddeley, A. 1982. Implications of neuropsychological evidence for theories of normal memory. In Philosophical Transactions of the Royal Society of London, eds. D.E. Broadbent and L. Weiskrantz, pp. 59-72. London: The Royal Society.

(4) Bruner, J.S. 1969. Modalities of memory. In The Pathology of Memory, eds. G.A. Talland and N.C. Waugh, pp. 253-259. New York: Academic Press.

(5) Burnham, W.H. 1903. Retroactive amnesia: illustrative cases and a tentative explanation. Am. J. Psychol. 14: 382-396.

(6) Butters, N., and Cermak, L.S. 1980. Alcoholic Korsakoff's Syndrome: An Information Processing Approach to Amnesia. New York: Academic Press.

(7) Butters, N., and Squire, L.R., eds. 1984. The Neuropsychology of Memory. New York: Guilford Press.

(8) Cermak, L.S. ed. 1982. Human Memory and Amnesia. Hillsdale, N.J.: Lawrence Erlbaum Associates.

(9) Cohen, N.J. 1981. Neuropsychological evidence for a distinction between procedural and declarative knowledge in human memory and amnesia. Unpublished Dissertation, University of California, San Diego.

(10) Cohen, N.J., and Corkin, S. 1981. The amnesic patient H.M.: Learning and retention of a cognitive skill. Soc. Neurosci. Abstr. 7: 235.

(11) Cohen, N.J., and Squire, L.R. 1980. Preserved learning and retention of pattern analyzing skill in amnesia: dissociation of knowing how and knowing that. Science 210: 207-209.

(12) Fodor, J. 1981. Representations. Cambridge, MA: MIT Press.

(13) Gaffan, D. 1974. Recognition impaired and association intact in the memory of monkeys after transection of the fornix. J. Comp. Physiol. Psychol. 86: 1100-1109.

(14) Graf, P.; Squire, L.R.; and Mandler, G. 1984. The information that amnesic patients do not forget. J. Exper. Psychol.: Learn. Mem. Cog. 10: 164-178.

(15) Horel, J.A. 1978. The neuroanatomy of amnesia: a critique of the hippocampal memory hypothesis. Brain 101: 403-445.

(16) Jacoby, L.L. 1982. Knowing and remembering: some parallels in the behavior of Korsakoff patients and normals. In Human Memory and Amnesia, ed. L.S. Cermak, pp. 97-122. Hillsdale, NJ: Lawrence Erlbaum Associates.

(17) Jacoby, L.L., and Witherspoon, D. 1982. Remembering without awareness. Can. J. Psychol. 32: 300-324.

(18) Kandel, E. 1976. Cellular Basis of Behavior. New York: Freeman.

(19) Kandel, E. 1977. Neuronal plasticity and the modification of behavior. In Handbook of Physiology, eds. J.M. Brookhart, V.B. Mountcastle, E.R. Kandel, and S.R. Geiger, vol. 1, pp. 1137-1182. Bethesda, MD: American Physiological Society.

(20) Kolers, P.A. 1976. Pattern-analyzing memory. Science 191: 1280-1281.

(21) Kolers, P.A. 1979. A pattern-analyzing basis of recognition. In Levels of Processing in Human Memory, eds. L. Cermak and F.I.M. Craik, pp. 363-384. Hillsdale, NJ: Lawrence Erlbaum Associates.

(22) Mahut, H.; Moss, M.; and Zola-Morgan, S. 1981. Retention deficits after combined amygdalo-hippocampal and selective hippocampal resections in the monkey. Neuropsychologia 19: 201-225.

(23) Mahut, H.; Zola-Morgan, S.; and Moss, M. 1982. Hippocampal resections impair associative learning and recognition memory in the monkey. J. Neurosci. 2: 1214-1229.

(24) Mair, W.G.P.; Warrington, E.K.; and Weiskrantz, L. 1979. Memory disorder in Korsakoff's psychosis: a neuropathological and neuropsychological investigation of two cases. Brain 1023: 719-783.

(25) Mandler, G. 1980. Recognizing: The judgment of previous occurrence. Psychol. Rev. 87: 252-271.

(26) McGaugh, J., and Gold, P. 1976. Modulation of memory by electrical stimulation of the brain. In Neural Mechanisms of Learning and Memory, eds. M.R. Rosenzweig and E.L. Bennett, pp. 549-560. Cambridge, MA: MIT Press.

(27) McGaugh, J., and Herz, M.M. 1972. Memory Consolidation. San Francisco: Albion.

(28) McGaugh, J.; Lynch, G.; and Weinberger, N., eds. 1984. Conference on the Neurobiology of Learning and Memory. New York: Guilford Press.

(29) Mishkin, M. 1978. Memory in monkeys severely impaired by combined but not by separate removal of amygdala and hippocampus. Nature 273: 297-298.

(30) Mishkin, M. 1982. A memory system in the monkey. In Philosophical Transactions of the Royal Society of London, eds. D.E. Broadbent and L. Weiskrantz, pp. 85-95. London: The Royal Society.

(31) Mishkin, M., and Delacour, J. 1975. An analysis of short-term visual memory in the monkey. J. Exp. Psychol. 1: 326-334.

(32) Mishkin, M.; Malamut, B.L; and Bachevalier, J. 1984. Memories and habits: Two neural systems. In Conference on the Neurobiology of Learning and Memory, eds. J.L. McGaugh, G. Lynch, and N.M. Weinberger. New York: Guilford Press.

(33) Mishkin, M.; Spiegler, B.J.; Saunders, R.C.; and Malamut, B.L. 1982. An animal model of global amnesia. In Alzheimer's Disease: Report of Progress, eds. S. Corkin, K.L. Davis, J.H. Growdon, E. Usdin, and R.J. Wurtman, pp. 235-247. New York: Raven Press.

(34) Moscovitch, M. 1982. Multiple dissociations of function in amnesia. In Human Memory and Amnesia, ed. L. Cermak, pp. 337-370. Hillsdale, NJ: Lawrence Erlbaum Associates.

(35) Moss, M.; Mahut, H.; and Zola-Morgan, S. 1981. Concurrent discrimination learning of monkeys after hippocampal, entorhinal, or fornix lesions. J. Neurosci. 1: 227-240.

(36) Murray, E.A., and Mishkin, M. 1981. Role of the amygdala and hippocampus in tactual memory. Soc. Neurosci. Abstr. 7: 237.

(37) Rozin, P. 1976. The psychobiological approach to human memory. In Neural Mechanisms of Learning and Memory, eds. M.R. Rosenzweig and E.L. Bennett, pp. 3-46. Cambridge, MA: MIT Press.

(38) Ryle, G. 1949. The Concept of Mind. London: Hutchinson.

(39) Scoville, W.B., and Milner, B. 1957. Loss of recent memory after bilateral hippocampal lesions. J. Neurol. Neurosurg. Psychiat. 80: 11-21.

(40) Squire, L.R. 1981. Two forms of human amnesia: an analysis of forgetting. J. Neurosci. 1: 635-640.

(41) Squire, L.R. 1982. The neuropsychology of human memory. Ann. Rev. Neurosci. 5: 241-273.

(42) Squire, L.R., and Cohen, N.J. 1984. Human memory and amnesia. In Conference on the Neurobiology of Learning and Memory, eds. J.L. McGaugh, G. Lynch, and N. Weinberger. New York: Guilford Press.

(43) Squire, L.R.; Cohen, N.J.; and Nadel, L. 1984. The medial temporal region and memory consolidation: a new hypothesis. In Memory Consolidation, eds. H. Weingartner and E. Parker. Hillsdale, NJ: Lawrence Erlbaum Associates.

(44) Squire, L.R.; Cohen, N.J.; and Zouzounis, J. 1983. Preserved memory in retrograde amnesia: sparing of a recently acquired skill. Neuropsychologia 22: 145-152.

(44a) Squire, L.R., and Spanis, C.W. 1984. Long gradient of retrograde amnesia in mice: continuity with the findings in humans. Behav. Neurosci. 98: 345-348.

(45) Squire, L.R., and Zola-Morgan, S. 1983. The neurology of memory: the case for correspondence between the findings for man and non-human primate. In The Physiological Basis of Memory, 2nd ed., ed. J.A. Deutsch, pp. 200-268. New York: Academic Press.

(46) Thompson, R.; Berger, T.; and Madden, J. 1983. Cellular processes of learning and memory in the mammalian CNS. Ann. Rev. Neurosci. 6: 447-492.

(47) Victor, M.; Adams, R.D.; and Collins, G.H. 1971. In The Wernicke-Korsakoff Syndrome, eds. F. Plum and F.H. McDowell. Philadelphia, PA: Davis.

(48) Warrington, E.K., and Weiskrantz, L. 1978. Further analysis of the prior learning effect in amnesic patients. Neuropsychologia 16: 169-177.

(49) Weiskrantz, L. 1978. A comparison of hippocampal pathology in man and other animals. In Functions of the Septo-Hippocampal System, CIBA Foundation Symposium, No. 58. Oxford: Elsevier.

(50) Weiskrantz, L. 1982. Comparative aspects of studies of amnesia. In Philosophical Transactions of the Royal Society of London, eds. D.E. Broadbent and L. Weiskrantz, vol. 298, pp. 97-109. London: The Royal Society.

(51) Weiskrantz, L., and Warrington, E.K. 1979. Conditioning in amnesic patients. Neuropsychologia 17: 187-194.

(52) Winograd, R. 1975. Frame representations and the declarative-procedural controversy. In Representation and Understanding, eds. D. Bobrow and A. Collins. New York: Academic Press.

(53) Wood, F.; Ebert, V.; and Kinsbourne, M. 1982. The episodic-semantic memory distinction in memory and amnesia: clinical and experimental observations. In Human Memory and Amnesia, ed. L. Cermak, pp. 167-193. Hillsdale, NJ: Lawrence Erlbaum Associates.

(54) Zola-Morgan, S., and Squire, L.R. 1982. Two forms of amnesia in monkeys: rapid forgetting after medial temporal lesions but not diencephalic lesions. Soc. Neurosci. Abstr. 8: 24.

(55) Zola-Morgan, S., and Squire, L.R. 1984. Preserved learning in monkeys with medial temporal lesions: sparing of motor and cognitive skills. J. Neurosci. 4: 1072-1085.

(56) Zola-Morgan, S.; Squire, L.R., and Mishkin, M. 1982. The neuroanatomy of amnesia: amygdala-hippocampus vs. temporal stem. Science 218: 1337-1339.

Standing, left to right:
Scott Weinstein, John Morton, Jason Brown, Mike Petersen.

Seated, left to right:
Larry Squire, Bill Estes, John Marshall, Walter Huber.

(Not shown): Tom Bever, Klaus Grossmann.

The Biology of Learning, eds. P. Marler and H.S. Terrace, pp. 687-705. Dahlem Konferenzen 1984. Berlin, Heidelberg, New York, Tokyo: Springer-Verlag.

Biology of Learning in Humans
Group Report

J.C. Marshall and J. Morton, Rapporteurs
T.G. Bever
J.W. Brown
W.K. Estes
K.E. Grossmann
W.G.H. Huber
M.R. Petersen
L.R. Squire
S. Weinstein

If there were a (reasonably) well worked out biology of learning in humans, what might it look like? To what overall structure would such a theory conform? The account should presumably specify: a) what is learned by humans, that is, the nature of the representational domains that can be acquired qua knowledge or skill structures; b) the nature of the processing machinery whereby these domains are learned; c) what "innate" structure must be presupposed if domains_{1-n} are to be learned in accordance with given learning procedures.

Given a concern with the biology of learning, a functionally specified learning theory should be linked to an account of the particular neural instantiations of the postulated mechanisms, stated in neuroanatomical, neurophysiological, and neurochemical language.

To what extent does classical learning theory (derived in large part from the study of animals other than ourselves) meet the adequacy criteria we have proposed? In particular, is it plausible to suppose that the principles of classical and instrumental conditioning, defined over a simple vocabulary of (quasi-) observable stimuli and responses, can mediate the acquisition of human knowledge and skill? Although some basic habits are no doubt formed in such a fashion in infants and young children, the (traditional) vocabulary of learning theory is too impoverished to describe the acquisition of knowledge and the cognitive activities involved in problem solving. It is true that the classical concept of "association" is ubiquitous in human cognitive psychology (everything is represented in "associative networks"), but the explanatory power of the notion is debatable. Some members of the group argued as follows: A folk term has been elevated to theoretical status, yet all it denotes is any form of linkage (however indirect) between any pair of elements in any representational system. The original "laws" of association (derived from Aristotle), whereby any elements can be linked by spatial or temporal contiguity (as in classical conditioning), suggests the width of the category (never mind the quality); association by "similarity" begs the question of what counts as similar for a particular organism in a particular domain; and the claim (with respect to language acquisition) that grammatical subjects must be "associated" with predicates fails to indicate the structural nature of the association. Nonetheless, current association theory is more sophisticated than the original Aristotelian version; "associations" are treated as only one component of cognitive models and the concept of association itself has been enriched and extended from simple inter-item connections to associative networks that may have quite complex, hierarchical structures. Much research is addressed to testing hypotheses about the details of these structures.

"Association" thus provides an element of unity in otherwise heterogenous segments of theory in that associative relations with similar properties are assumed to hold among units at different levels of complexity, abstractness, and representational type (stimulus/response, "concepts," and "images"). In some influential theories (e.g., (9)), associations of different qualitative types ("labelled connections") appear.

Given such an enrichment of associationism and the development of new methodologies, some of the circularity of old-style theory disappears. Thus "similarity" can be assessed and quantified with procedures distinct from those used to test for associative memory; association has ceased to be (quasi-) synonymous with learning and memory with the demonstration that some forms of short-term visual and auditory memory are nonassociative. Nonetheless, it must be noted that even elaborated notions of association involve but one methodological option in theory construction. Associationist theories still place the explanatory burden on structural relationships between elements in mental representations. There remains a strong possibility that such accounts encourage a belief in spurious similarities across domains of knowledge that would look very different under theoretical accounts which place more stress on process.

For those committed to operant conditioning, reinforcement is a tautological notion (if it is learned, it must have been reinforced); for the cognitively inclined, "knowledge of results" (or "information feedback") looks like a more useful concept. That is, a basic principle of human learning may be the TOTE unit (Test-Operate-Test-Exit) of Miller, Galanter, and Pribram (17), although this formulation also finesses the question as to the representational format in which hypotheses are couched and evaluated. Nonetheless, the TOTE is an example of a concept that may help supply the property of goal-directedness to theories of human learning that have been provided for animal learning theories by the notions of drive and reinforcement.

Returning to our first question (what is learned?), how should we cut the cake of "natural behaviors" for humankind? A first slicing might distinguish between sensory and/or motor skills and the representation of "knowledge." This cut seems a priori reasonable in that visual object recognition, e.g., is by definition specific to the visual (input) system, at least up to the 2 1/2-D (and perhaps the 3-D) representation (16); drawing is by definition specific to some muscular output system (once the internal code reaches a level of the neural system where it becomes "motoric"). But our "knowledge" that Paris is the capital of France is presumably supra-modal (or multi-modal?) and can be expressed by a variety of output modalities, despite each individual having acquired this information via a particular input modality. What further modularization could be justified at input or output (beyond the eyes and ears or hands and feet being different organs)?

At input, object recognition and the identification of individual conspecifics (face recognition) are often regarded as distinct functional components. How would such a modularization be justified, over and above the fact that human faces are typically distinct in form and color from trees? In this case, the (mere) fact that the structural representations assigned to faces and trees are different does not in itself mandate the postulation of distinct analyzing devices qua discrete mechanisms. (That is, the structural descriptions assigned to "stimuli" must be controlled, in part at least, by the shape of the stimuli presented.) Presumably, however, one would become (a little) more confident of the modular story if the parameters of learning to distinguish within and between trees and faces were distinct - that is, grossly distinct. Motivational considerations could modulate different rates of learning, but very early smiling responses to faces (and schematic dinner plates) presumably implicate some "innate template" that catches onto a biologically important feature of the environment. One notes that it is the specificity of the configuration that suggests modularity (see (10)); even tree recognition may presuppose "innate" figure/ground discrimination, edge detection, color discrimination (fine-tuned by experience). Similar arguments apply at output. Walking and skating are not distinct skills because they involve distinct patterns of motor coordination, but because of their ontogenetic history.

Within the neuropsychological tradition, double dissociation of functions (24) has also been taken as prima facie evidence of modular components (20). Teuber's insight was that, given two tasks and two patients with focal brain damage, strong claims for the distinctness of the function(s) that mediate(s) the tasks follow from a pattern of performance in which on task A, patient A is more impaired than patient B, but on task B, patient B is more impaired than patient A. The situation is formally analogous to crossover interactions in an analysis of variance and hence avoids the interpretive problems associated with "intrinsic" task difficulty. When double dissociation is found, it does indeed follow that each task has (at least) one functional component which is not held in common with the other task ("weak modularity"); it does not follow that the two tasks are functionally (or anatomically) distinct throughout their entire sequence of respective processing stages ("strong modularity").

Since the number of plausible ways of categorizing cognitive functions is clearly large, a sensible heuristic would be to value most highly those solutions suggested by a number of different sources of evidence. If (approximately) the same domain is isolated by, say, representational format, distinct learning history, and discrete neuroanatomic instantiation, one might feel justified confidence that the fractionation is biologically meaningful. One "natural behavior" of humans for which fair convergence exists is language. Language capacity is clearly species-specific. Animals (even smart ones) do not talk (or learn American sign language, especially its syntactic properties); people (even unintelligent ones) do. How should the capacity to acquire (and use) a language be characterized? Two general approaches are often distinguished: the Organ Model and the Cognitive Model. We will illustrate the distinction from the study of language, which has proved a valuable "test bed" for studies of learning in a biologically prepared mode. The nature of "what is learned" is reasonably well worked out, thereby enabling one to raise issues concerned with the respective roles of endogenously specified constraints, environmental triggering, and communicative functions in the acquisition of conceptual and linguistic form, as well as the pragmatic uses to which the system can be put in human interaction.

The concept of a "language organ" (6), analogous (qua organ) to the visual system or the liver, stresses the centrality of the computational principles unique to linguistic communication. That is, the formal nature of grammars (rules of or constraints upon phonology, syntax, and logical form) is regarded as the critical, distinctive attribute of linguistic communication. In current theory (7) the number of core grammars is fixed and finite. In the organ approach (6), the principles that govern core grammar are attributed to the human genotype; the "growth" of a particular grammar for a specific language is then triggered by environmental evidence that sets the parameters left open in core grammar; exposure to linguistic input has the function of fixing whether the language of the child's environment is, e.g., configurational or non-configurational, pro-drop or not, and the base order of constituents (subject-verb-object or subject-object-verb, etc.).

By contrast, the "cognitive model" characterizes language as a complex adaptation to a set of problems posed by social life (how to communicate efficiently with a symbol system) and by individual problem solving (how to represent tasks in a notation that facilitates their accomplishment). On this account, language acquisition and use recruits other (more basic) faculties of mind to its own ends; in this assembly of cognitive capacities, some of which may be unique to humans, the distinctiveness of language lies in the mix, not the ingredients.

On either of these models, the language system seems to emerge as a functional module. The evidence comes from examining the course of language acquisition, grammatical structure itself (i.e., the end product of the learning), and anatomic specialization. Briefly, we review these sources of evidence below.

Language growth takes place in a fixed (maturational?) sequence, perhaps bounded by a sensitive period for acquisition. Gleitman (this volume) presents evidence for a biological clock that sets the acquisition process into motion. On the other hand, changes and in some cases massive deprivation of linguistic input (e.g., isolated deaf youngsters) and opportunities to interpret certain of these inputs against real-world events (e.g., blind children learning vision-related terms such as look or green) have only highly restricted effects on the rate or character of learning. On the other hand, differences in mental status have predictable effects on the timing of language growth (e.g., in children born prematurely) and in the character of what is learned (e.g., in Downs Syndrome retardates). Thus, facts about mental status more than facts about specific exposure conditions seem to predict the course and internal structure of learning.

Moreover, Gleitman argues that the rapid pace of learning, as well as similarity of learning despite differences in the talent and motivation of learners, and differences in their cultural circumstances as well as linguistic input, further support a maturational interpretation of the course of learning. However, it is fair to ask, "Rapid with respect to what?" and to remark, "So what is not learned from a degenerate environment?" In answer, rapidity is, in comparison to learning of the same linguistic materials by adults, presumed to be beyond some sensitive period for language learning (e.g., young children learn about seven new vocabulary items a day, every day, every week, every month, during the period from about 18 months to 6 years old; all of them, even the dullest, do so; the evidence that adults cannot perform like feats is quite good).

But much more interesting is the sense in which the environment for language learning seems to be impoverished and degenerate, presenting something of a conundrum for learning-theoretic approaches to answering the big question: How does the child extract the essence of his native tongue all the same? Particularly, the strongest evidence for significant biological supports for language learning come from the logic of what is learned under the real circumstances that obtain: certain linguistic properties, e.g., null elements and the rules for anaphora (cf. (5, 6), and Gleitman, this volume, for a discussion) are never modelled directly in the input to the child but are acquired errorlessly. Moreover, as Gleitman also notes, the course of acquisition does not seem to be generally understandable as a direct modelling of the input. For example, though open and closed class vocabulary is presented to the learner from the beginning in the speech of the mother, the open class is acquired early, the closed class appearing only a year or more later; furthermore, canonical declarative sentences in the data base are rare, but they are the first forms acquired by the learner.

Despite such evidence from the course of learning and reasoning from the product of learning, it might seem possible to show that the child's decisive moves toward his native tongue are much less the outcome of some innate specification for language and much more the outcome of an inductive machinery (albeit no doubt a smart one) than Gleitman, for one, has argued. Speaking particularly, it could be that the child's environment for learning is not so restrictive and degenerate as Gleitman claims. For example, Gleitman, in common with such theorists as Wexler and Culicover (26), conjectures that learning must be largely from positive evidence. If that is so, the case for an internally, biologically driven acquisition process becomes stronger. This is because, as Gold (12) and Wexler and Hamburger (27) have shown formally, a transformational grammar cannot be acquired by an open-minded inductive device, based on positive instances (grammatical sentences) alone, though it could be acquired given both positive instances and negative instances (so marked).

The position taken by Gleitman and others, that negative evidence is simply not available to the learner to any material extent, is based in part on certain findings in the literature of developmental psycholinguistics: work by Brown and Hanlon (3) showing that mothers rarely correct their children's grammar. If that is so, then the child <u>does</u> receive positive instances (i.e., the grammatical sentences spoken by the mother) but does <u>not</u> receive negative instances (i.e., corrections of the false attempts by the learner). And if <u>that</u> is so - given Gold's theorem - then the learner cannot be proceeding by open-minded induction but must be contributing to his own learning by bringing to bear certain preexisting (biologically given) biases about the principles of this system to which he is being exposed.

However, an apparently reasonable counterattack is that there are other, more subtle, kinds of correction provided to the learning child. For example, a mother* may scowl at non-sentences or fail to understand them, she might correct her own mistakes, or even explicitly correct a child older than those studied by Brown and Hanlon; such responses may be good clues for the learner that something is amiss in the way he talks. One objection to this rejoinder is that numerous investigators have sought such properties in the interaction between learning child and teaching mother but have failed to find them. If the child learner <u>does</u> find them, we have to conclude that he is a more exquisite interpreter of the ongoing scene than the average child language investigator; but perhaps that is likely. A more important objection to the view that the child learns from correction is that he or she, to profit from such subtle corrections, would have to disentangle them from the huge number of corrections that are made by mothers, all too unsubtly, and are obvious from inspection of the mother/child dyad; namely, corrections on grounds of propriety and truth (see (3)). But most important is that the position of "subtle correction" misses a logical point. For the vast number of linguistic generalizations, the child never errs at all and so the mother never gets a chance to correct him either overtly or subtly. To take a trivial example, the child has to err and say, "House the is red," if the mother is to get her chance to scowl and thus teach that determiners in English precede nouns; as a less trivial example, the child has to say, "Is the man who a fool is interesting?" before the mother can teach that only is in the main clause can be fronted in yes/no questions (or at least, the mother has to make such mistakes and correct herself). But, unfortunately for a theory of learning from negative instances, these required errors do not occur in the speech of the child or the speech of the mother (for evidence, see, e.g., (18)). Thus the argument that learning is by and large from positive information only, adopted both by investigators such as Gleitman and by learnability theorists, derives primarily from this analysis of the course of learning: many important properties of language are acquired without error, and thus negative reinforcement (for mistakes) cannot be invoked as their explanation.

Currently, the best hope for a theory of language learning seems to come from a) such quantitative estimates of the child's exposure conditions as Gleitman and other psychologists have presented (see particularly (11, 13, 18, 21)), mated with b) analyses of "universal grammar," or the shared properties of natural languages, culled by linguists from examination of the languages of the world, and c) formal theories of learnability ((25, 26), and Osherson and Weinstein, this volume). We already know from Gold's (12) theorem and from the work of Wexler and Hamburger (14, 27) that a transformational grammar cannot be acquired from positive instances (grammatical sentences) alone. Yet, as just stated, the idea that children are materially aided by negative instances (corrections of their bad attempts, etc.) does not go through on the empirical facts. Then how are the children to learn? Wexler and Culicover's move is to revise the input situation in a plausible way: the child is assumed to receive not just sentences, but sentence/interpretation pairs (i.e., utterances accompanied by extralinguistic evidence about their meaning) as his input for learning; moreover, transformational grammars are assumed to embody formal and substantive constraints, detailed by these theorists, that have evolved on learning considerations (i.e., Wexler and Culicover postulate language-specific endowments that rescue the learner from the pitfalls which open-minded induction may place in his path). Osherson and Weinstein proceed on a rather different tack by examining certain (innate) constraints on memory and processing resources which, by their very nature, are to rein in the child's conjectures about the language to which he is being exposed, hence making the right inductions from the data base inevitable. Perhaps the most central differences between these two approaches to learnability, then, are that the Wexler group postulates specifically linguistic constraints in the child learner, while Osherson and Weinstein consider how general properties of learning conspire to yield language growth.

Speaking more generally, the essence of learnability theory is the specification of collections of languages that can be identified in the limit (cf. (12)) by devices that instantiate innate linguistic organizing principles (25) and psychologically realistic constraints on innate inductive strategies and computational resources (Osherson and Weinstein, this volume). Empirical constraints (boundary conditions) are required to approach the ethology of language learning; viz., Gold's model requires a perfect text as input, yet the child must converge to a hypothesis on contaminated data. How contaminated? And how will particular types of contamination throw the device off course? A child's conjectures about the language under analysis are constrained by memory limitations, i.e., (s)he does not have access to boundless past linguistic data. A child's current hypothesis must, therefore, be determined by the current grammar plus a fixed amount of past data. How large a set is available?

How can we place the growth of language qua functional system in correspondence with the growth and selection of the correlated neuronal substrate? The standard story, derived from language impairment after focal lesions in adult life and from the varied histology of cortical areas, suggests that a restricted number of regions in the left hemisphere (Broca's area, Wernicke's area, the arcuate fasciculus, the angular gyrus, and the supplementary motor area) are critically and differentially implicated in the expression and comprehension of language. Once the language system is committed to these particular anatomical substrates in adult life, what possibility is there of relearning when knowledge and skill are lost (either destroyed or made otherwise unavailable) consequent upon brain damage? Tight correlation between functionally and anatomically defined modules in the adult would seem to suggest that relearning will not be possible in a qualitatively and quantitatively "natural" form. The evidence of recovery from aphasia-inducing lesions supports this supposition. Typically, we see a period of "spontaneous" recovery (interpretable without recourse to notions of relearning) that may extend for four to six months, after which time spontaneous changes in performance level are slight. The rate and extent of this recovery is extremely variable from patient to patient. Reports of the effects of therapy (i.e., controlled practice) are mixed. At best, improvement seems slight compared with the effort expended, and it appears that even limited improvement may be restricted to specifically trained items and may fail to generalize to either new constructions or situations. Patients report fatigue consequent upon speaking having become a "conscious" activity (analogous to conversing in a second language?). The overall pattern is consistent with the relevant skills being of a qualitatively different type from the original "automatized" fluency. One case can serve to highlight the distinction. An aphasic woman manifested lexico-syntactic problems which included the inability to distinguish between sentences of the form, "The duck is easy/eager to eat." With either form she was uncertain as to whether the duck would eat or be eaten. After some weeks of ad hoc and quite unsuccessful training with these constructions, she began to respond with perfect accuracy. On being asked how she did the task, she replied that on hearing "easy" she visualized the word with an arrow pointing backwards, and on hearing "eager" she visualized the arrow pointing forwards (Morton and Hatfield, unpublished data). Similar constructions with, for example, "happy" and "simple" were dealt with by the same strategy. Clearly, we are dealing with an effective but ad hoc device that cannot be integrated into real-time parsing. (In other modules, one might think of the detailed, laborious, feature by feature analysis, whereby the prosopagnosic patient "relearns" to discriminate and recognize faces.)

Let us now turn to the effects of (relatively) focal lesions in maturing organisms, an issue that speaks to the differentiation of the nervous system and is relevant to selective theories of learning (4). Brown (this volume) notes that the effect of a presumptively "constant" lesion may change not merely up until maturity but throughout the life span. Between the ages of 4 and 60+ the predominant type of aphasia changes from mutism and agrammatism, through nonfluent aphasia with phonetic and articulatory errors, to fluent (conduction) aphasia and severe jargon aphasia. Such response to damage suggests that an initially "diffuse" organization of the neuronal substrate polarizes across the life span. What mechanisms, either of selective stabilization or maturational growth, could account for such a progression? Even more dramatically, what kind of "shifting" organ are we dealing with when language apparently transfers reasonably well to the right hemisphere after extensive damage to the left at an early age? "Decrease of plasticity" with age is a description of the phenomenon, not an explanation. Our evidence for a language module converges, but only grossly.

Study of the aphasias is usually organized around language as a knowledge-directed skill. Study of the amnesias has served to partition knowledge and skill. This partitioning is based on the convergence of at least three lines of theorizing: a) the (partial) equivalence of man and monkey; b) the relation between psychological models and the underlying neurobiology; and c) the (computational or philosophical?) distinction between declarative knowledge and procedural skill.

With respect to the first point, research on man and monkey is claimed to be mutually reinforcing (to reclaim the word from the professionals). Thus Mishkin (17) comments that "the strongest argument" in favor of localizing the storehouse for perceptual memories "within the highest levels of the cortical sensory processing areas" derives from the "clinical syndrome of global amnesia, following either medial temporal, or medial thalamic injury" ((11), p. 93). Old memories are spared, but the patient is "unable to lay down new memories of people, places, and events." Research on the monkey brain can (if monkeys have memories in addition to learned habits) provide a more complete and precise means of mapping from cognitive models to underlying anatomy. One interesting aspect of this latter anatomical work concerns the (inevitable) complexity of the representations of information flow in the monkey brain. In diagramming the areas involved in a relatively simple perceptual memory task, Mishkin draws eight anatomical modules (four of them actually in boxes!) and three feedback loops. One would expect the psychological (functional) model of this task to be at least as complex (cf. "another box and pair of arrows," Postman (19)), although one should not expect too close a relationship between the anatomic and psychologic modules, nor perhaps between man and monkey (if the linguistic system is utilized in learning nonlinguistic tasks).

Mishkin's work with the monkey leads to a distinction between "cognitive" mechanisms, mediated by the limbic system, that lead to the establishment of memories, and "behavioristic" mechanisms, mediated by the extra-pyramidal system, that underlie the formation of habits. Thus monkeys with limbic lesions are severely impaired in tasks involving one-trial recognition memory whilst retaining the ability to acquire motor and perceptual discrimination habits with the same objects and rewards.

Students of the human amnesic syndrome (Squire, this volume) see this distinction as analogous (identical?) to the pattern of preserved and impaired performance that follows damage to the medial temporal region (as in patient H.M.) or to the disencephalic region (as in patients with Korsakoff's disease). Mishkin's distinction between memories and habits is then equated with the distinction between declarative knowledge and procedural skill. Items within the former domain can be learned quickly by normal subjects (down to one trial), those within the latter more slowly with repetition. Declarative knowledge requires time over a period of years to consolidate through the continued good agency of the limbic system. Procedural knowledge neither requires the limbic system for establishment nor has the same need for consolidation (Squire, this volume).

Behaviorally, the declarative/procedural distinction may seem clear. For example, amnesics cannot remember what happened or what was told to them five minutes ago, and patients undergoing ECT suffer retrograde amnesia for events.

But it is also the case that amnesics can learn certain skills such as mirror reading, and this skill, when learned by ECT patients, survives the ECT treatment (22). The problem arises with the conceptualization of its taxonomy. Let us take one example, the Tower of Hanoi problem. This involves three pegs. On one of the pegs are five discs of different sizes, in order of size, with the largest at the bottom. The subject's task is to transfer all five discs to another peg while conforming to two rules: a) only one disc is to be moved at a time, and b) no disc can be put on top of a smaller disc. The amnesic patients learn this task ((8), also Cohen, unpublished dissertation) and can perform perfectly well if they are given starting positions which could not normally occur. They would do this despite claiming little or no knowledge of what was expected of them at the beginning of the session. Such performance requires the power of a recursive algorithm. It is difficult to describe as a habit, and it is not clear to what extent it can be described as procedural. Indeed, the only feature in common between this task and the others that can be learned by the amnesic is that the means to a solution can be achieved (by normal subjects as well as by amnesics) without any explicit knowledge as to how it is done. However, the Tower of Hanoi can be characterized as a conceptual task that involves some mental operations other than strictly perceptual or motor processes. As such, it may differ in some interesting ways from the other tasks that can be learned by the amnesic which are purely perceptual (such as mirror reading and jigsaw puzzle solving) or perceptuomotor (such as pursuit-motor tracking and mirror writing).

Exploration of the Tower of Hanoi problem and other similar tasks should serve to sharpen the distinction between what amnesic patients can and cannot do. Here lies our best chance to specify tightly the nature of the putative memory systems that depends on the integrity of the brain regions damaged in amnesia.

We close with a coda on the nature of cognitive theory, some (disputed) aspects of some aspects thereof. The success of interweaving results from research on the human amnesic syndrome and on monkey anatomy is clear: the human data is conceptually richer, and the lesion site can be better controlled in monkeys. Which (if any) other aspects of current theorizing in human cognitive psychology will be of value to scientists concerned with the behavior of other species was debated. Our hope was that the relatively precise models developed to account for linguistic competence and performance (1) might help to shed light on the application of the concepts of "innateness," "specialization," and "preparedness" in other species and in other areas of human cognition. In particular, the proposal that natural skills are organized into encapsulated modules offers the possibility of a new kind of comparative cognition - that within a species. In a specific case, the question is: Does a structurally defined capacity discovered in one kind of skill play a role in other behaviors? If it does, then we may conclude that the capacity is a component which can be a constituent of more than one skill. Such a finding would not disprove the claim that natural skills are encapsulated, that is, operationally isolated. However, it would show that there may be a stock of structures and principles available to the assembled in different ways to form distinct skills.

Consider, for example, the "A-over-A principle" (and its descendants in the linguistic literature). This principle states that a rule affecting a particular kind of grammatical unit applies to the highest potential unit of that kind in a hierarchy. Suppose the rule is one which optionally allows the deletion of any unit in a sequence of 3 units, e.g., A, B, and C. This would allow AB, AC, and BC as variants on ABC. Suppose further that such sequences can be embedded in each other, for example, if ABC and DEF are both sequences, then so is A(DEF)BC, etc. The A-over-A principle requires that when we apply the single-unit-deletion rule, it cannot apply only to the embedded sequence. That is, a possible version of A(DEF)BC would be A(DEF)C, but not A(DF)BC.

This example may seem abstract, but it can be mapped onto behavioral/perceptual units in various task settings. For example, one could present subjects with a pursuit tracking task in which the letters are instantiated as locations on a screen; one could ask subjects to discriminate sequences of tones or colors. In each case, the question would be whether sequence options that follow the A-over-A principle are easier to master or use than sequence options that do not follow this principle. Such studies in nonlinguistic domains could verify (or fail to verify) the extralinguistic validity of such principles in children and adults. The end result of a programmatic investigation might be a picture of which linguistic principles are available to other skills, whether this varies with age, with language dysfunctions, and so on. In the context of this workshop, we see such an enterprise as using the results of linguistics to enrich the tools for investigating structural laws in human skills outside of language.

Of course, this method can also be used in the manner more traditional to comparative behavioral research. Animals can be trained to produce and discriminate sequences of stimuli. There is no reason, in principle, why the force of structural principles governing sequences cannot be studied in such paradigms. This suggestion gently turns the traditional experimental comparative method on its head - rather than studying unnatural laboratory-inspired tasks across species (e.g., reversal learning), we propose to study structures inspired by the close examination of language, a complex but natural kind of behavior (2). Yet such utility as may devoutly be wished for will expire unless the nature and purpose of cognitive theory is understood. First, we should draw attention to the diversity of levels at which theory may be formulated. A glance at the papers in this volume that discuss language reveals theorizing at the logicomathematical level (Osherson and Weinstein), the linguistic level (Gleitman, Petersen), and the biological level (Brown). A further level, intermediate between the latter two - information-processing - is not represented. The nature of the relationships between such levels has been the subject of considerable recent discussion (15, 23).

With regard to the most centrally "psychological" of these levels - functional analysis in information-processing terms - a few more detailed remarks may be appropriate. First, one important aspect of information-processing models is heuristic (or pragmatic). Different notations can be used to address different kinds of questions. Thus, in relation to memory representations, one could equally well express theoretical claims within the framework of, say, associative nets or production rules. Both notations are sufficiently powerful to encompass any pattern of data, but not necessarily with equal ease. The advantage of any explicit information-processing formulation is just that it facilitates relating the wide variety of experimental and observational data. Within models formulated in these terms, tasks must be fully analyzed before the data can be properly interpreted. Thus various short-term memory paradigms, such as serial and free recall, are now treated as significantly different tasks. The same applies (though with, regrettably, more exceptions) to tasks involving single words, such as perceptual recognition and semantic judgment tasks. The resulting models occasionally look haphazard and unconstrained, but the evaluative criteria are clear. Models are accepted only if they give an economical account of data from distinct sources. That is, there has to be convergence.

What all such models have in common is that they represent what is going on in the head, not what is going on "outside." At the same time, the relationship between the elements of the models and the underlying biology is often construed as indifferent. A particular functional unit may or may not correspond to a particular (punctate) anatomical locus. Thus, it might be convenient to represent the store and control circuitry for visual memory as a single "box," even knowing that one is cortical and the other subcortical. Since Brown (this volume) suggests, on the basis of lesion data, that particular areas of the brain change the details of their functioning with age even through adulthood, the problem of ascertaining the anatomical locus and physiological mechanism of functional units becomes yet more complicated. The task of mapping functions into the brain still remains, but the nature of the mapping is an empirical issue. Two related questions then emerge concerning these mapping functions: Can neuroscience help in formulating interesting questions in functional psychology? Must psychological theory be constrained by any facts of neuroscience? Neuroscience is arriving at a stage of sophistication where patterns of neuronal connectivity, system properties of neuronal aggregates, and other phenomena can suggest processing capabilities in a rather direct fashion. For example, in animal work the functional relationship between sensitization and classical conditioning was sharpened when the deep connections between them were worked out at the neural level in invertebrates. An example of mutual support between functional and anatomical data comes from vocalization in birds. One group (chickens, doves) makes no use of environmental sounds in developing species-specific vocalization. This group has no discrete neural circuitry in the forebrain specialized for vocalization and no known integration of cells sensitive to sound and cells involved in vocalization. Another group of birds (songbirds) depend to some extent on hearing conspecific sounds in developing the species song. This group has discrete neural pathways in the forebrain (a song-control system). This system also receives auditory input, and auditory neurons are intermixed among motor neurons in at least two nuclei of the circuit (Konishi, personal communication). If anatomy can relate to function in such a simple and direct way as these facts suggest, then the facts of neuroanatomy could constrain psychological theory in a rather general way. The fact that brain regions involved in language comprehension and production are localized together in a part of the left hemisphere could suggest that the language faculty may not penetrate other cognitive functions to the same extent that it would if these brain regions were widely distributed.

Given the nature of cognitive models, we can now return to the topic of the workshop. We suggest that the question of how learning theory and natural behavior can be reconciled is that the relationship is to be determined primarily through functional representations of what the organism's nervous system does, not what "wetware" it is composed of. With simple organisms, such as Aplysia, this relationship can readily be established in terms of its neurophysiology and neurochemistry. With humans it can for the most part only be done abstractly; that is, in terms of information-processing predicates with which the behavior can be comprehendingly expressed. Thus, suppose it turned out that all human synapses were equivalent to the Aplysia's qua their mode of functioning, and suppose that all the behavior of such a synapse were expressible in terms of learning theory (i.e., as in various versions of "Hebb-type" synapses). Suppose further that we had all the human neurobiological information there was to have at the level of cell-circuit neuroscience. We might then have an account of natural human behavior, and that account might be couchable in learning theoretic terms, but we would not have an explanation in terms of the question we really wanted to ask: How does the system work as in information-processing device?

That is, we would still need systems-level information about how the nervous system performs its computational task. We would need a psychological theory to relate and map onto the neural information.

For the moment, it seems that fruitful research with steadily improving methods is going on, largely independently, at the levels of animal learning, behavior theory, and human cognition. There may turn out to be interesting parallels between some of the animal and human work (2, 17). Some members of the group would argue that these parallels can only be captured by formal, information-processing models; others doubted that behavioral evidence alone can judge whether the similarities are more than analogies. For this latter subgroup, increasing attention to neurobiological research on both animals and humans offers new hope of uncovering deeper correspondences.

REFERENCES

(1) Baker, C.L., and McCarthy, J.J., eds. 1981. The Logical Problem of Language Acquisition. Cambridge, MA: MIT Press.

(2) Bever, T.G. 1983. The road from behaviorism to rationalism. In Animal Cognition, eds. H.L. Roitblat, T.G. Bever, and H.S. Terrace. Hillsdale, NJ: Erlbaum.

(3) Brown, R., and Hanlon, C. 1970. Derivational complexity and order of acquisition in child speech. In Cognition and the Development of Language, ed. J.R. Hayes. New York: Wiley.

(4) Changeux, J.-P. 1974. Some biological observations relevant to a theory of learning. In Current Problems in Psycholinguistics, eds. F. Bresson and J. Mehler, pp. 281-287. Paris: CNRS.

(5) Chomsky, N. 1975. Reflections on Language. New York: Random House.

(6) Chomsky, N. 1980. Rules and Representations. New York: Columbia University Press.

(7) Chomsky, N. 1981. Lectures on Government and Binding. Dordrecht: Foris.

(8) Cohen, N.J., and Corkin, S. 1981. The amnesic patient H.M.: Learning and retention of a cognitive skill. Soc. Neurosci. Abstr. 7: 235.

(9) Collins, A.M., and Quillian, M.R. 1972. Experiments on semantic memory and language comprehension. In Cognition in Learning and Memory, ed. L.W. Gregg, pp. 117-137. New York: Wiley.

(10) Fodor, J.A. 1983. The Modularity of Mind. Cambridge, MA: MIT Press.

(11) Gleitman, L.; Newport, E.; and Gleitman, H. 1984. The current status of the Motherese Hypothesis. J. Child Lang. 11(1): 43-80.

(12) Gold, E.M. 1967. Language identification in the limit. Inf. Control 10: 447-474.

(13) Goldin-Meadow, S. 1982. The resilience of recursion: a study of a communication system developed without a conventional linguistic model. In Language Acquisition: State of the Art, eds. E. Wanner and L.R. Gleitman. Cambridge: Cambridge University Press.

(14) Hamburger, H., and Wexler, K. 1975. A mathematical theory of learning transformational grammars. J. Math. Psychol. 12: 137-177.

(15) Marr, D. 1982. Vision. San Francisco: W.H. Freeman.

(16) Miller, G.A.; Galanter, E.; and Pribram, K.H. 1960. Plans and the Structure of Behavior. New York: Henry Holt.

(17) Mishkin, M. 1982. A memory system in the monkey. Phil. Trans. Roy. Soc. Lond. B 298: 85-95.

(18) Newport, E.; Gleitman, H.; and Gleitman, L.R. 1977. Mother, I'd rather do it myself: Some effects and non-effects of maternal speech style. In Talking to Children: Language Input and Acquisition, eds. C.E. Snow and C.A. Ferguson. Cambridge: Cambridge University Press.

(19) Postman, L. 1975. Verbal learning and memory. Ann. Rev. Psychol. 26: 291-335.

(20) Shallice, T. 1979. Case study approach in neuropsychological research. J. Clin. Neuropsychol. 1: 183-211.

(21) Shatz, M. 1982. On mechanisms of language acquisition: Can features of the communicative environment account for development? In Language Acquisition: State of the Art, eds. E. Wanner and L.R. Gleitman. Cambridge: Cambridge University Press.

(22) Squire, L.; Cohen, N.J.; and Zouzounis, J. 1984. Preserved memory in retrograde amnesia: sparing of a recently acquired skill. Neuropsychologia 22: 145-152.

(23) Stabler, E.P. 1983. How are grammars represented? Behav. Brain Sci. 6: 391-402.

(24) Teuber, H.-L. 1955. Physiological psychology. Ann. Rev. Psychol. 6: 267-296.

(25) Wexler, K. 1982. A principle theory for language acquisition. In Language Acquisition: State of the Art, eds. E. Wanner and L.R. Gleitman. Cambridge: Cambridge University Press.

(26) Wexler, K., and Culicover, P. 1980. Formal Principles of Language Acquisition. Cambridge, MA: MIT Press.

(27) Wexler, K., and Hamburger, H. 1973. On the insufficiency of surface data for the learning of transformational languages. In Approaches to Natural Languages, eds. K. Hintikka, J. Moravcsik, and P. Suppes. Dordrecht: Reidel.

List of Participants with Fields of Research

BATESON, P.P.G.
Sub-Dept. of Animal Behavior
University of Cambridge
Madingley
Cambridge CB3 8AA, England

The development of behavior with special reference to the ontogeny of social and sexual preferences, and the origins of individual differences; the neural basis of imprinting

BEVER, T.G.
Dept. of Psychology
373 Schermerhorn Hall
Columbia University
New York, NY 10027, USA

Linguistics and psychology

BICKER, G.
Institut für Tierphysiologie
der Freien Universität Berlin
Königin-Luise-Strasse 28-30
1000 Berlin (West) 33

Invertebrate neurobiology

BISCHOF, H.-J.
Lehrstuhl Verhaltenspsychologie
Fakultät für Biologie
Universität Bielefeld
Morgenbreede 45
4800 Bielefeld 1, F.R. Germany

Behavioral and neurophysiological research on imprinting in birds

BOLLES, R.C.
Dept. of Psychology
University of Washington
Seattle, WA 98195, USA

Animal conditioning and motivation

BROWN, J.W.
Dept. of Neurology
New York University
Medical Center
530 First Avenue
New York, NY 10016, USA

Neuropsychology

CAREW, T.J.
Dept. of Psychology
Yale University
P.O. Box 11A Yale Station
New Haven, CT 06520, USA

Cellular analysis of learning and memory

CHANGEUX, J.-P.
Neurobiologie Moléculaire
Institut Pasteur
28, rue du Docteur Roux
75724 Paris Cedex 15, France

Molecular mechanisms of synapse formation, selective theory of learning

DELIUS, J.D.
Psychologisches Institut
der Ruhr Universität Bochum
Universitätsstrasse 150
4630 Bochum, F.R. Germany

Animal learning, animal perception, physiological psychology

ESTES, W.K.
Dept. of Psychology and
Social Relations
Harvard University
33 Kirkland Street
Cambridge, MA 02138, USA

Human learning and memory, learning theory, mathematical models in psychology

FISCHBACH, K.-F.
Institut für Genetik und
Mikrobiologie
der Universität Würzburg
Röntgenring 11
8700 Würzburg, F.R. Germany

Neurogenetics of the visual system of Drosophila melanogaster, including genetic analysis of habituation and sensitization of the landing response

GIBBON, J.
New York State Psychiatric
Institute
Columbia University
722 West 168th Street
New York, NY 10032, USA

Animal learning and timing

GOULD, J.L.
Dept. of Biology
Princeton University
Princeton, NJ 08544, USA

Ethology

GROSSMANN, K.E.
Institut für Psychologie
der Universität Regensburg
Universitätsstrasse 31
8400 Regensburg, F.R. Germany

Developmental psychology, the social basis of emotional and competence development, the meaning of intersubjectivity in infancy development

HEARST, E.
Dept. of Psychology
Indiana University
Bloomington, IN 47405, USA

Animal learning and conditioning, human discrimination learning and problem solving, history of psychology

HEINRICH, B.
Dept. of Zoology
University of Vermont
Burlington, VT 05405, USA

Behavioral and physiological ecology

HEISENBERG, M.A.
Institut für Genetik und Mikrobiologie
der Universität Würzburg
Röntgenring 11
8700 Würzburg, F.R. Germany

Neurogenetics of Drosophila behavior, neuroanatomy and development of structural brain mutants

HOLLAND, P.C.
Dept. of Psychology
University of Pittsburgh
Pittsburgh, PA 15260, USA

Behavioral properties of learning and memory in animals

HOLLIS, K.L.
Dept. of Psychology
Mt. Holyoke College
South Hadley, MA 01075, USA

The biological function of Pavlovian conditioning phenomena; more generally, interactions of species-specific and learned behaviors

HUBER, W.G.H.
Abt. Neurologie
RWTH Aachen
Goethestrasse 27-29
5100 Aachen, F.R. Germany

Neurolinguistics, aphasia

IMMELMANN, K.
Lehrstuhl Verhaltensphysiologie
Fakultät für Biologie
Universität Bielefeld
Morgenbreede 45
4800 Bielefeld 1, F.R. Germany

Behavior development in birds, mammals, and fish, particularly imprinting and other forms of early learning

JENKINS, H.M.
Dept. of Psychology
McMaster University
Hamilton, Ontario L8S 4K1, Canada

Animal learning, human judgement

KONISHI, M.
Div. of Biology 216-76
California Institute of Technology
Pasadena, CA 91125, USA

Neuroethology

KROODSMA, D.E.
Dept. of Zoology
Morrill Science Center
University of Massachusetts
Amherst, MA 01003, USA

Ecology and evolution of avian vocal learning

LEA, S.E.G.
Dept. of Psychology
University of Exeter
Washington Singer Laboratories
Exeter EX4 4QG, England

Concept learning in animals and humans, the relation of ecology and economics to psychology

LINDAUER, M.
Zoologisches Institut II
der Universität Würzburg
Röntgenring 10
8700 Würzburg, F.R. Germany

Orientation in the Earth's magnetic field together with problems of chronobiology, mechanisms of learning in honey bees, "tradition" in honey bees

MARKL, H.S.
Fakultät für Biologie
Universität Konstanz
Postfach 5560
7750 Konstanz 1, F.R. Germany

Animal behavior, social behavior in insects

MARLER, P.
The Rockefeller University
Field Research Center
Tyrrell Road
Millbrook, NY 12545, USA

Animal behavior

MARSHALL, J.C.
Neuropsychology Unit
The Radcliffe Infirmary
Woodstock Road
Oxford OX2 6HE, England

Neurobiology of higher cognitive functions, disorder of spoken and written language

MENZEL, E.W., JR.
Dept. of Psychology
State University of New York
Stony Brook, NY 11794, USA

Primate behavior

MENZEL, R.
Institut für Tierphysiologie
der Freien Universität Berlin
Grunewaldstrasse 34
1000 Berlin (West) 41

Learning of invertebrates, color vision

MISHKIN, M.
Laboratory of Neuropsychology
Building 9, Room 1N107
National Institutes of Mental Health
Bethesda, MD 20205, USA

Neurobiology of learning and memory in primates

MORTON, J.
Cognitive Development Unit
Medical Research Council
17 Gordon Street
London WC1H 0AH, England

Development of cognitive abilities

PETERSEN, M.R.
Dept. of Psychology
Indiana University
Bloomington, IN 47405, USA

Sensory and perceptual bases of primate vocal communication

QUINN, W.G.
Dept. of Biology
Princeton University
Princeton, NJ 08544, USA

Genetic analysis of learning and memory in Drosophila

RAUSCHECKER, J.P.
Max-Planck-Institut für
Biologische Kybernetik
Spemannstrasse 38
7400 Tübingen, F.R. Germany

Plasticity and functional organization of the mammalian visual system

REVUSKY, S.
Dept. of Psychology
Memorial University of Newfoundland
St. John's, Newfoundland A1B 3X9
Canada

Learned associations between drug states

SAHLEY, C.L.
Dept. of Psychology
Yale University
Box 11A Yale Station
New Haven, CT 06520, USA

Psychobiology, physiological and behavioral analysis of learning in invertebrates

SHETTLEWORTH, S.J.
Dept. of Psychology
University of Toronto
Toronto, Ontario M5S 1A1, Canada

Animal learning in a functional context (foraging, food storage, and memory)

SINGER, W.
Max-Planck-Institut
für Hirnforschung
Deutschordenstrasse 46
6000 Frankfurt/Main 71, F.R. Germany

Development and functional organization

SQUIRE, L.R.
Veteran's Administration
Medical Center
3350 La Jolla Village Drive
San Diego, CA 92161, USA

Neuropsychology of memory

STADDON, J.E.R.
Dept. of Psychology
Duke University
Durham, NC 27706, USA

Adaptive behavior, learning, artificial intelligence, evolution and behavior

TERRACE, H.S.
Dept. of Psychology
418 Schermerhorn Hall
Columbia University
New York, NY 10027, USA

Animal cognition, animal memory

WAGNER, A.R.
Dept. of Psychology
Kirtland Hall
Yale University
New Haven, CT 06520, USA

Learning theory

WEINSTEIN, S.
Dept. of Philosophy
University of Pennsylvania
305 Logan Hall CN
Philadelphia, PA 19104, USA

Mathematical models of inductive inference and language acquisition, philosophy of psychology

WERKA, T.F.
Nencki Institute of
Experimental Biology
Pasteura 3
02-093 Warsaw, Poland

Physiology of the limbic system, defensive behavior of animals

ZIELINSKI, K.
Nencki Institute of
Experimental Biology
Pasteura 3
02-093 Warsaw, Poland

Defensive conditioning, stimulus control, prefrontal cortex, amygdala

Subject Index

Author Index

Dahlem Workshop Reports

Life Sciences Research Reports (LS)

LS 1 The Molecular Basis of Circadian Rhythms.
Editors: J. W. Hastings, H.-G. Schweiger

LS 2 Appetite and Food Intake.
Editor: T. Silverstone

LS 3 Hormone and Antihormone Action at the Target Cell.
Editors: J. H. Clark et al.

LS 4 Organization and Expression of Chromosomes.
Editors: V. G. Allfrey et al.

LS 5 Recognition of Complex Acoustic Signals.
Editor: T. H. Bullock

LS 6 Function and Formation of Neural Systems.
Editor: G. S. Stent

LS 7 Neoplastic Transformation: Mechanisms and Consequences.
Editor: H. Koprowski

LS 8 The Bases of Addiction.
Editor: J. Fishman

LS 9 Morality as a Biological Phenomenon.
Editor: G. S. Stent

LS 10 Abnormal Fetal Growth: Biological Bases and Consequences.
Editor: F. Naftolin

LS 11 Transport of Macromolecules in Cellular Systems.
Editor: S. C. Silverstein

LS 12 Light-Induced Charge Separation in Biology and Chemistry.
Editors: H. Gerischer, J. J. Katz

LS 13 Strategies of Microbial Life in Extreme Environments.
Editor: M. Shilo

LS 14 The Role of Intercellular Signals: Navigation, Encounter, Outcome.
Editor: J. G. Nicholls

LS 15 Biomedical Pattern Recognition and Image Processing.
Editors: K. S. Fu, T. Pavlidis

LS 16 The Molecular Basis of Microbial Pathogenicity.
Editors: H. Smith et al.

LS 17 Pain and Society.
Editors: H. W. Kosterlitz, L. Y. Terenius

LS 18 Evolution of Social Behavior: Hypotheses and Empirical Tests.
Editor: H. Markl

LS 19 Signed and Spoken Language: Biological Constraints on Linguistic Form.
Editors: U. Bellugi, M. Studdert-Kennedy

Physical and Chemical Sciences Research Reports (PC)

PC 1 The Nature of Seawater (out of print)

PC 2 Global Chemical Cycles and Their Alteration by Man.
Editor: W. Stumm

Distributor for LS 1–19 and PC 1+2:
Verlag Chemie, Pappelallee 3, 6940 Weinheim,
Federal Republic of Germany

MIX
Papier aus verantwortungsvollen Quellen
Paper from responsible sources
FSC® C105338

If you have any concerns about our products,
you can contact us on
ProductSafety@springernature.com

In case Publisher is established outside the EU,
the EU authorized representative is:
Springer Nature Customer Service Center GmbH
Europaplatz 3, 69115 Heidelberg, Germany

Printed by Libri Plureos GmbH
in Hamburg, Germany